APPLICATIONS OF BIODEGRADABLE AND BIO-BASED POLYMERS FOR HUMAN HEALTH AND A CLEANER ENVIRONMENT

AAP Research Notes on Polymer Engineering Science and Technology

APPLICATIONS OF BIODEGRADABLE AND BIO-BASED POLYMERS FOR HUMAN HEALTH AND A CLEANER ENVIRONMENT

Edited by

Iuliana Stoica, PhD
Omari Mukbaniani, DSc
Neha Kanwar Rawat, PhD
A. K. Haghi, PhD

AAP | APPLE
ACADEMIC
PRESS

First edition published 2022

Apple Academic Press Inc.
1265 Goldenrod Circle, NE,
Palm Bay, FL 32905 USA
4164 Lakeshore Road, Burlington,
ON, L7L 1A4 Canada

CRC Press
6000 Broken Sound Parkway NW,
Suite 300, Boca Raton, FL 33487-2742 USA
2 Park Square, Milton Park,
Abingdon, Oxon, OX14 4RN UK

Library and Archives Canada Cataloguing in Publication

Title: Applications of biodegradable and bio-based polymers for human health and a cleaner environment / edited by Iuliana Stoica, PhD, Omari Mukbaniani, DSc, Neha Kanwar Rawat, PhD, A.K. Haghi, PhD.
Names: Stoica, Iuliana, editor. | Mukbaniani, O. V. (Omar V.), editor. | Rawat, Neha Kanwar, editor. | Haghi, A. K., editor.
Series: AAP research notes on polymer engineering science and technology.
Description: First edition. | Series statement: AAP research notes on polymer engineering science and technology | Includes bibliographical references and index.
Identifiers: Canadiana (print) 20210241136 | Canadiana (ebook) 20210241179 | ISBN 9781771889766 (hardcover) | ISBN 9781774639399 (softcover) | ISBN 9781003146360 (ebook)
Subjects: LCSH: Biopolymers. | LCSH: Polymers—Biodegradation.
Classification: LCC TP248.65.P62 A67 2022 | DDC 660.6/3—dc23

Library of Congress Cataloging-in-Publication Data

Names: Stoica, Iuliana, editor. | Mukbaniani, O. V. (Omar V.), editor. | Rawat, Neha Kanwar, editor. | Haghi, A. K., editor.
Title: Applications of biodegradable and bio-based polymers for human health and a cleaner environment / edited by Iuliana Stoica, Omari Mukbaniani, Neha Kanwar Rawat, A.K. Haghi.
Other titles: AAP research notes on polymer engineering science and technology.
Description: First edition. | Palm Bay, FL : Apple Academic Press, 2022 | Series: AAP research notes on polymer engineering science and technology | Includes bibliographical references and index. | Summary: "The world faces significant challenges as the population and human consumption continue to grow while nonrenewable fossil fuels and other raw materials are depleted at ever-increasing rates. This informative volume provides a technical approach to address these issues using green design and analysis. It takes an interdisciplinary look at concepts that can be applied across engineering disciplines in the development of products, processes, and systems to minimize environmental impacts across all life cycle phases. Topics include polymers for pollutant removal, wood-based biopolymers, bio-based polymers for drug formulations, biomaterial-based medical implants, biodegradabilty of biopolymer materials, bio-based polymers for food packaging applications, biodegradable polymers for tissue engineering applications, and more. Key features of the book: Presents updated information in the field of biodegradable biopolymers Provides an up-to-date summary of the varying market applications of biopolymers characterized by biodegradability and sustainability Includes case studies that illustrate the sustainable development process from a materials perspective Applications of Biodegradable and Bio-Based Polymers for Human Health and a Cleaner Environment is a valuable resource for academic and industrial researchers who are interested in new materials, renewable resources, sustainability, and polymerization technology. It will also prove useful for advanced students interested in the development of bio-based products and materials, green and sustainable chemistry, polymer chemistry, and materials science"-- Provided by publisher.
Identifiers: LCCN 2021027383 (print) | LCCN 2021027384 (ebook) | ISBN 9781771889766 (hardback) | ISBN 9781774639399 (paperback) | ISBN 9781003146360 (ebook)
Subjects: MESH: Biopolymers | Biomedical Technology | Green Chemistry Technology | Biodegradation, Environmental
Classification: LCC QP801.B69 (print) | LCC QP801.B69 (ebook) | NLM QT 37.5.P7 | DDC 572/.33--dc23
LC record available at https://lccn.loc.gov/2021027383
LC ebook record available at https://lccn.loc.gov/2021027384

ISBN: 978-1-77188-976-6 (hbk)
ISBN: 978-1-77463-939-9 (pbk)
ISBN: 978-1-00314-636-0 (ebk)

AAP RESEARCH NOTES ON POLYMER ENGINEERING SCIENCE AND TECHNOLOGY

The AAP Research Notes on Polymer Engineering Science and Technology reports on research development in different fields for academic institutes and industrial sectors interested in polymer engineering science and technology. The main objective of this series is to report research progress in this rapidly growing field.

Gennady E. Zaikov, DSc
Head, Polymer Division, N. M. Emanuel Institute of Biochemical Physics,
Russian Academy of Sciences; Professor, Moscow
State Academy of Fine Chemical Technology, Russia; Professor,
Kazan National Research Technological University, Kazan, Russia

Books in the AAP Research Notes on Polymer Engineering Science and Technology series

- **Functional Polymer Blends and Nanocomposites:**
  A Practical Engineering Approach
  Editors: Gennady E. Zaikov, DSc, Liliya I. Bazylak, PhD, and A. K. Haghi, PhD

- **Polymer Surfaces and Interfaces: Acid-Base Interactions and Adhesion in Polymer-Metal Systems**
  Irina A. Starostina, DSc, Oleg V. Stoyanov, DSc, and Rustam Ya. Deberdeev, DSc

- **Key Technologies in Polymer Chemistry**
  Editors: Nikolay D. Morozkin, DSc, Vadim P. Zakharov, DSc, and
  Gennady E. Zaikov, DSc

- **Polymers and Polymeric Composites: Properties,**
  Optimization, and Applications
  Editors: Liliya I. Bazylak, PhD, Gennady E. Zaikov, DSc, and A. K. Haghi, PhD

- **Applied Research on Polymer Composites**
  Editors: Pooria Pasbakhsh, PhD, A. K. Haghi, PhD, and Gennady E. Zaikov, DSc

- **High-Performance Polymers for Engineering-Based Composites**
  Editors: Omari V. Mukbaniani, DSc, Marc J. M. Abadie, DSc, and
  Tamara Tatrishvili, PhD

- **Compositional Analysis of Polymers**
  Editors: Aleksandr M. Kochnev, PhD, Oleg V. Stoyanov, DSc,
  Gennady E. Zaikov, DSc, and Renat M. Akhmetkhanov, DSc

- **Computational and Experimental Analysis of Functional Materials**
  Editors: Oleksandr Reshetnyak, DSc, and Gennady E. Zaikov, DSc

- **Applications of Biodegradable and Bio-Based Polymers for**
  Human Health and a Cleaner Environment
  Editors: Iuliana Stoica, PhD, Omari Mukbaniani, DSc,
  Neha Kanwar Rawat, PhD, and A. K. Haghi, PhD

ABOUT THE EDITORS

Iuliana Stoica, PhD
Scientific Researcher, Department of Polymer Materials Physics,
"Petru Poni" Institute of Macromolecular Chemistry, Romania

Iuliana Stoica, PhD, is currently a Scientific Researcher in physics at Romanian Academy, "Petru Poni" Institute of Macromolecular Chemistry, Department of Polymer Materials Physics. She received her PhD from the Department of Polymer Physics and Structure of the Romanian Academy at the same institute. She joined a postdoctoral fellowship program at Politehnica University of Bucharest, Faculty of Applied Chemistry and Materials Science, Department of Bioresources and Polymer Science. Her area of scientific activity is focused on characterization of a wide range of polymers, copolymers, polymeric composites, and polymeric mixtures. She was a main or co-author for over 95 papers in peer-reviewed ISI journals, and she has contributed as an author to several book chapters in the field of polymer and materials science. She was a member of the organizing and program committees of several scientific conferences. She was also reviewer for a number of prestigious journals in the field of polymer science.

Omari Mukbaniani, DSc
Full Professor, Ivane Javakhishvili Tbilisi State University,
Faculty of Exact and Natural Sciences, Department of Chemistry;
Chair of Macromolecular Chemistry; Director of the Institute of
Macromolecular Chemistry and Polymeric Materials at TSU,
Tbilisi, Georgia

Omari Mukbaniani, DSc, is Full Professor at the Ivane Javakhishvili Tbilisi State University (TSU), Faculty of Exact and Natural Sciences, Department of Chemistry; and Chair of Macromolecular Chemistry, Tbilisi, Georgia. He is also Director of the Institute of Macromolecular Chemistry and Polymeric Materials at TSU, and a member of the Academy of Natural Sciences of Georgia. For several years, he was a member of the advisory board of the journal *Proceedings of Iv. Javakhishvili Tbilisi State University* (Chemical Series) and contributing editor of the journal *Polymer News, Polymers Research Journal,* and *Chemistry and Chemical Technology.* His research interests include polymer chemistry, polymeric materials, and chemistry of organosilicon compounds. He is the author of more than 480 publications, 25 books, monographs, and 10 inventions. He created the *International Symposium on International Symposium on Polymers and Advanced Materials ICSP&AM,* which takes place every two years in Georgia. In 2018, he was a Chair of 26th World Annual Forum on Advanced Materials PolyChar 26.

Neha Kanwar Rawat, PhD
Researcher, Materials Science Division, CSIR-National Aerospace Laboratories, Bangalore, India

Neha Kanwar Rawat, PhD, is a recipient of a prestigious DST Young Scientist Postdoctoral Fellowship and is presently a researcher in the Materials Science Division, CSIR-National Aerospace Laboratories, Bangalore, India. She received her PhD in chemistry from Jamia Millia Islamia (a Central University), India. Her main interests include nanotechnology-nanostructured materials synthesis and characterization; her main focus is on green chemistry, novel sustainable chemical processing of nano-conducting polymers/nanocomposites, conducting films, ceramics, silicones, matrices: epoxies, alkyds, polyurethanes, etc. She also pursues her interest in fusing new technology in areas that include electrochemistry, organic–inorganic hybrid nanocomposites, and protective surface coatings for corrosion inhibition and MW shielding materials. She has published numerous peer-reviewed research articles in journals of high repute. Her contributions have led to many chapters in international books published with the Royal Society of Chemistry, Wiley, Elsevier, Apple Academic Press, Nova U.S., and many others in progress. She has been working on many prestigious research and academic fellowships in her career. She is a member of many groups, including the Royal Society of Chemistry and the American Chemical Society (USA) and a life member of the Asian Polymer Association.

A. K. Haghi, PhD

Professor Emeritus of Engineering Sciences, Former Editor-in-Chief, International Journal of Chemoinformatics and Chemical Engineering and Polymers Research Journal; Member, Canadian Research and Development Center of Sciences and Culture.

A. K. Haghi, PhD, is the author and editor of 200 books, as well as 1000 published papers in various journals and conference proceedings. Dr Haghi has received several grants, consulted for a number of major corporations, and is a frequent speaker to national and international audiences. Since 1983, he served as professor at several universities. He is former Editor-in-Chief of the *International Journal of Chemoinformatics and Chemical Engineering* and *Polymers Research Journal* and is on the editorial boards of many international journals. He is also a member of the Canadian Research and Development Center of Sciences and Cultures (CRDCSC), Montreal, Quebec, Canada. He holds a BSc in urban and environmental engineering from the University of North Carolina (USA), an MSc in mechanical engineering from North Carolina A &T State University (USA), a DEA in applied mechanics, acoustics and materials from the Université de Technologie de Compiègne (France), and a PhD in engineering sciences from Université de Franche-Comté (France).

CONTENTS

CONTRIBUTORS

Cristobal N. Aguilar
Bioprocesses and Bioproducts Research Group, Food Research Department, School of Chemistry, Autonomous University of Coahuila, Saltillo, 25280 Coahuila, Mexico

Miguel A. Aguilar-Gonzalez
Center for Research and Advanced Studies of the National Polytechnic Institute (CINVESTAV-IPN) Unit-Saltillo, C.P. 25900 Ramos Arispe, Mexico

Raluca Marinica Albu
"Petru Poni" Institute of Macromolecular Chemistry, Laboratory of Physical Chemistry of Polymers, 41A Grigore Ghica Voda Alley, 700487 Iasi, Romania

Mohd Danish Ansari
Laboratory of Green Synthesis, Department of Chemistry, University of Allahabad, Allahabad 211002, India

Alexandra Bargan
"Petru Poni" Institute of Macromolecular Chemistry, Aleea Gr. Ghica Voda 41A, Iaşi 700487, Romania

Andreea Irina Barzic
"Petru Poni" Institute of Macromolecular Chemistry, Laboratory of Physical Chemistry of Polymers, 41A Grigore Ghica Voda Alley, 700487 Iasi, Romania

S.I. Bhat
Material Research Laboratory, Department of Chemistry, Jamia Millia Islamia, New Delhi 110025, India

Luminita Ioana Buruiana
"Petru Poni" Institute of Macromolecular Chemistry, 41A Grigore Ghica Voda Alley, 700487 Iasi, Romania

Maria Cazacu
"Petru Poni" Institute of Macromolecular Chemistry, Aleea Gr. Ghica Voda 41A, Iaşi 700487, Romania

M. P. Chandresh
Department of Polymer Science and Technology, Sri Jayachamarajendra College of Engineering, JSS Science and Technology University, Mysuru 570 006, India

Mónica L. Chávez-Gonzalez
Bioprocesses and Bioproducts Research Group, Food Research Department, School of Chemistry, Autonomous University of Coahuila, Saltillo, 25280 Coahuila, Mexico
Laboratory of Chemistry and Biochemistry, School of Agronomy, Autonomous University of Nuevo Leon, General Escobedo, C.P. 66050 Nuevo León, Mexico

Bianca-Iulia Ciubotaru
Department of Inorganic Polymers, "Petru Poni" Institute of Macromolecular Chemistry, Aleea Grigore Ghica Voda, 41 A, 700487 Iasi, Romania

Tânia F. Cova
Department of Chemistry, University of Coimbra, CQC, 3004-535 Coimbra, Portugal

Sajad Ahmad Dar
Department of Physics, Government Motilal Vigyan Mahavidyalya College, Bhopal 462008, India

Abu Darda
Department of Applied Science and Technology, Faculty of Engineering and Technology,
Jamia Millia Islamia, New Delhi 110025, India

Daniela Filip
Department of Polyaddition and Photochemistry, "Petru Poni" Institute of Macromolecular Chemistry,
Aleea Grigore Ghica Voda, 41 A, 700487 Iasi, Romania

Roberto Garcia
Department of Chemistry, University of Coimbra, CQC, 3004-535 Coimbra, Portugal

Rinky Ghosh
Department of Materials Science and Engineering, Indian Institute of Technology Kanpur,
Kanpur 208016, India

Mohd Irfan
Material Research Laboratory, Department of Chemistry, Jamia Millia Islamia, New Delhi 110025, India

Purnima Jain
Department of Chemistry, Netaji Subhas Institute of Technology, Dwarka Sector 3, University of Delhi,
New Delhi 110021, India

Haroon Khan
Department of Pharmacy, Abdul Wali Khan University, Mardan 23200, Pakistan

Halima Khatoon
Material Research Laboratory, Department of Chemistry, Jamia Millia Islamia, New Delhi 110025, India

Rabia Kouser
Material Research Laboratory, Department of Chemistry, Jamia Millia Islamia, New Delhi 110025, India

Shrikaant Kulkarni
Adjunct Professor, Vishwakarma University, Pune, India

Cristian Logigan
Chemical Company SA, 14 Chimiei Bvd., 707252 Iasi, Romania

Nica Simona Luminita
Chemical Company SA, 14 Chimiei Bvd., 707252 Iasi, Romania

Doina Macocinschi
Department of Polyaddition and Photochemistry, "Petru Poni" Institute of Macromolecular Chemistry,
Aleea Grigore Ghica Voda, 41 A, 700487 Iasi, Romania

Guillermo Cristian G. Martínez-Avila
Bioprocesses and Bioproducts Research Group, Food Research Department, School of Chemistry,
Autonomous University of Coahuila, Saltillo, 25280 Coahuila, Mexico
Laboratory of Chemistry and Biochemistry, School of Agronomy, Autonomous University of Nuevo
Leon, General Escobedo, C.P. 66050 Nuevo León, Mexico

Dina Murtinho
Department of Chemistry, University of Coimbra, CQC, 3004-535 Coimbra, Portugal

Alberto A. C. C. Pais
Department of Chemistry, University of Coimbra, CQC, 3004-535 Coimbra, Portugal

Deepak Poddar
Department of Chemistry, Netaji Subhas Institute of Technology, Dwarka Sector 3, University of Delhi, New Delhi 110021, India

Delia Mihaela Rata
Faculty of Medical Dentistry, "Apollonia" University of Iasi, Iasi 700399, Romania

Neha Kanwar Rawat
Materials Research Laboratory, Department of Chemistry, Jamia Millia Islamia, New Delhi 110025, India
Department of Chemistry, Haryana Education Services, Haryana 127021, India

S. Roopa
Department of Polymer Science and Technology, Sri Jayachamarajendra College of Engineering, JSS Science and Technology University, Mysuru 570 006, India

Hozeyfa Sagir
Department of Chemistry, Paliwal P.G. College, Shikohabad 205135, India

G. Santhosh
Department of Mechanical Engineering, NMAM Institute of Technology, Nitte, Karnataka 574110, India

Anjana Sarkar
Department of Chemistry, Netaji Subhas Institute of Technology, Dwarka Sector 3, University of Delhi, New Delhi 110021, India

Marcela Savin
"Antibiotice" SA, Valea Lupului Alley, 707410 Iasi, Romania

Siddaramaiah
Department of Polymer Science and Technology, Sri Jayachamarajendra College of Engineering, JSS Science and Technology University, Mysuru 570006, India

I. R. Siddiqui
Laboratory of Green Synthesis, Department of Chemistry, University of Allahabad, Allahabad 211002, India

Amit Kumar Singh
Department of Chemistry, Dyal Singh College, University of Delhi, New Delhi 110021, India

B. Sowmya
Materials Science Division, CSIR—National Aerospace Laboratories, Old Airport Road, Kodihalli, Bengaluru 560017, India

George Ştiubianu
"Petru Poni" Institute of Macromolecular Chemistry, Aleea Gr. Ghica Voda 41A, Iaşi 700487, Romania

Elena Stoleru
Petru Poni" Institute of Macromolecular Chemistry, Physical Chemistry of Polymers Laboratory, 41A Grigore Ghica Voda Alley, 700487 Iasi, Romania
"Alexandru Ioan Cuza" University of Iasi, Faculty of Chemistry, 11 Carol I Blvd, 700506 Iasi, Romania

Darieo Thankachan
Department of Materials Science and Engineering, Indian Institute of Technology Kanpur, Kanpur 208016, India

Artur J. M. Valente
Department of Chemistry, University of Coimbra, CQC, 3004-535 Coimbra, Portugal

Deepak Kumar Verma
Agricultural and Food Engineering Department, Indian Institute of Technology Kharagpur,
Kharagpur 721302, India

Priti Yadav
Department of Chemistry, Netaji Subhas Institute of Technology, Dwarka Sector 3, University of Delhi,
New Delhi 110021, India

Mirela-Fernanda Zaltariov
Department of Inorganic Polymers, "Petru Poni" Institute of Macromolecular Chemistry,
Aleea Grigore Ghica Voda, 41 A, 700487 Iasi, Romania

ABBREVIATIONS

ABS	acrylonitrile-butadiene-styrene
ACCS	amidoxime-chelating cellulose
AGUs	anhydro-D-glucopyranose units
AOPs	advanced oxidation processes
BC	bacteria cellulose
BOPLA	bi-axially oriented PLA
BPA	bisphenol A
CA	cellulose acetate
CA	citric acid
CAB	CA butyrate
CAGR	compound annual growth rate
CBZ	carbamazepine
CCS	carbon capture and storage
CD	cyclodextrin
CDI	4,4'-dicyclohexylmethane diisocyanate
CEC	cyanoethylated cellulose
CMC	carboxymethyl cellulose
CMC	carboxymethylated cotton
CMF	craniomaxillofacial
CNC	cellulose nanocrystal
CNTs	carbon nanotubes
CPCs	cardiac progenitor cells
CR	Congo red
Cs	chitosan
CS	chondroitin sulfate
CTAB	cetyltrimethylammonium bromide
CTR	citric acid
CV	crystal violet
CVFF	consistent valence force-field
DBD	dielectric barrier discharge
DBTDL	dibutyltindilaurate
DMAEMA	dimethyl aminoethyl methacrylate
DMAP	dimethyl aminopyridine
DMF	dimethylformamide

EB	electron beam
ECE	epoxide chain extender
ECM	extracellular matrix
ECP	electronic chemical potential
EN	electronegativity
EoL	end-of-life
EP	electrostatic potential
EPI	epichlorohydrin
EVA	ethylene vinyl acetate
FCZ	fluconazole
FDCA	furandicarboxylic acid
FET	field effect transistor
FTIR	Fourier transformed infrared spectroscopy
GA	glutaraldehyde
GC	gas chromatographic
GF	gas foaming
GMA	glycidyl methacrylate
GO	graphene oxide
GOx	glucose oxidase
GPDMS	glycidoxypropylmethyldiethoxysilane
GPTMS	glycidoxypropyltrimethoxysilane
HA	hyaluronic acid
HA	hydroxyapatite
HDI	1,6-hexamethylene diisocyanate
HDPE	high-density polyethylene
HEC	hydroxyl ethyl cellulose
HEMA	hydroxy ethyl-methacrylate
HMF	(hydroxymethyl)-furfural
HNTs	halloysite nanotubes
HOMO	Highest occupied molecular orbital
HPC	hydrohypropyl cellulose
HPMC	hydroxypropyl methylcellulose
HP-β-CD	hydroxypropyl-β-cyclodextrin
HTC	hydrothermal carbonization
ionic liquids	ILs
IPN	interpenetrating network
LA	lactic acid
LA	lactobionic acid
LCAs	life cycle assessments

LDHs	layered double hydroxides
LEP	low-voltage electrospinning patterning
LF	landfilling
LUMO	lowest unoccupied molecular orbital
MA	maleic anhydride
MAA	methacrylic acid
MB	methylene blue
MBA	methylene-bis-acrylamide
MC	methyl cellulose
MC	Monte Carlo
MD	molecular dynamics
MDI	4,4'-diphenylmethane diisocyanate
MH	minocycline hydrochloride
MIPs	molecular imprinted polymers
MM	molecular mechanics
MMA	methylmethacrylate
MMT	montmorillonite
MNPs	magnetic nanoparticles
MO	methyl orange
MOFs	metal-organic frameworks
MR	methyl red
MS	mass spectrometry
MSNs	mesoporous silica nanoparticles
MCR	multicomponent reaction
MW	molecular weights
MWCNT	multiwalled carbon nanotube
NC	nanocellulose
NCH	nanocomposite hydrogel
NDI	1,5-naphtalene diisocyanate
nHA	nanohydroxyapatite
NMR	nuclear magnetic resonance
NO	nitric oxide
NPs	nanoparticles
OEGDMA	oligo (ethylene glycol) dimethacrylate
PAA	polyacrylic acid
PAHs	polyaromatic hydrocarbons
PALF	pineapple leaf fibers
PBA	poly(butylene adipate)diol
PBS	poly(butylene succinate)

PC	polycarbonates
PCL	poly(ε-caprolactone)
PCMX	chloroxylenol
PDI	1,4-phenylene diisocyanate
PDI	polydispersity index
PDLLA	poly(D,L-Lactic acid)
PDMS	polydimethylsiloxane
PE	polyethylene
PE	propolis extract
PEG	polyethylene glycol
PEGA	poly[(ethylene glycol)adipate]
PEGMC	poly(ethylene glycol)maleate citrate
PEI	polyethyleneimine
PEO	poly(ethylene oxide)
PET	polyethylene terephthalate
PET	polyethylene terephthalate
PGA	poly(glycolic acid)
PHA	polyhydroxyalkanssssoate
PHB	polyhydroxybutyrate
PL	particle leaching
PLA	poly(L-latic acid)
PLGA	poly(lactic-co-glycolic acid)
PMHS	polymethylhydrosiloxane
PMMA	polymethyl methacrylate
PP	plaster of paris
PP	polypropylene
PPC	poly(propylene carbonate)
PPEs	polyphosphoesters
PS	pine sawdust
PS	polystyrene
PU	polyurethane
PVA	poly(vinyl alcohol)
PVP	poly(vinyl pyrrolidone)
QM	quantum molecular
QSAR	quantitative structure–activity relationships
RHF	restricted Hartree Fok
ROP	ring-opening polymerization
RP	rapid prototyping
SA	sodium alginate

SAHs	super-absorbent hydrogels
SC	solvent casting
SD	sodium deoxycholate
SEM	scanning electron microscopy
SPIONs	superparamagnetic iron-oxide nanoparticles
SRHG	slide ring hydrogel
SWCNT	single-walled CNT
TBHP	*tert*-butyl hydroperoxide
TC	tetracycline
TDI	2,4-toluene diisocyanate
TE	tissue engineering
TEA	triethanolamine
TEOS	tetraethoxysilane
TFA	trifluoroacetic acid
TG	thermogravimetry
TMC	trimethylene carbonate
TSF	Tussah silk fibroin
UV	ultraviolet
VEGF	vascular endothelial growth factor
VPTT	volume phase transition temperature
WCE	wet chemical etching
XPS	extruded polystyrene
Z14G	zearalenone-14-glucoside

PREFACE

This book is addressed to all researchers and scientists as well as students of graduate and postgraduate levels at universities and technical institutes who wish to keep abreast of advances in the biodegradable polymers and biopolymers.

During the past 50 years, polymers have totally changed human life, and the polymers industry has developed durable materials increasingly adapted to specific uses. In response to demands for temporary applications, scientists are now working on polymeric systems that can serve as environmental-friendly materials and undergo degradation at controlled and predetermined rates after task completion. Biodegradable polymers are ever-expanding market materials finding focused applications in sectors where biodegradability, together with the performances they attain during use, offers systematic environmental benefits.

They represent a highly promising solution, since they have the potential to overcome environmental concerns such as the decreasing availability of landfill space and the depletion of petrochemical resources, and also offer a sustainable alternative option to mechanical and chemical recycling.

Biodegradable polymers have experienced strong growth recently and are set to make further inroads into markets traditionally dominated by conventional thermoplastics in future. Demand is being driven by a number of factors. The cost of biodegradable polymers has come down considerably while at the same time standard thermoplastic prices have increased considerably. Now, some classes of biodegradable polymers are price competitive with polymers such as PET.

Meanwhile, bio-based polymers and composites are systematically describing the green engineering, green chemistry, and manufacture of bio-based polymers and composites derived from plants.

Biopolymers have gained increasing importance in the development of numerous environmentally friendly materials for a wide variety of applications. The development of strategies such as blending and structural modifications are important factors to consider. The resulting advanced materials have the potential to compete with traditionally used nonbiodegradable polymers; thus, eventually replacing them.

This volume gives a detailed insight on the various applications and uses of bio-based polymers highlighting the significance of such materials for the future use, looking at and realizing the relevance of bio-materials in the future due to the adverse effects of the materials currently in use. It provides readers with the required information on the bio-based polymers and the related aspects on the importance of such environmental friendly materials in the present world.

In this volume, some critical issues and suggestions for future work are discussed, underscoring the roles of scientists and researchers for the future of the new green materials through value addition to enhance their use.

PART I
Biopolymers-Based Advanced Materials

CHAPTER 1

WOOD-BASED BIOPOLYMERS AS ACTIVE ELEMENTS IN NEW GREEN SILICONE COMPOSITES

GEORGE ŞTIUBIANU, ALEXANDRA BARGAN*, and MARIA CAZACU

"Petru Poni" Institute of Macromolecular Chemistry, Aleea Gr. Ghica Voda 41A, Iaşi 700487, Romania

Corresponding author. E-mail: anistor@icmpp.ro.

ABSTRACT

Composite materials made with silicones have tunable porosities and functionalities. Such materials have been studied as a new class of carbon dioxide sorbents since their properties make its highly efficient for capturing and storage of carbon dioxide, providing a pathway for capturing and thus reducing the anthropogenic emissions of carbon dioxide. Wood-based biopolymers (lignin and cellulose) are the go-to materials for the development of sustainable, environmentally friendly and commercially viable composites. Silicone composites with wood-based biopolymers are biodegradable and use a naturally abundant, carbon-rich material. These characteristics are perfectly fit for researchers interested in introducing easily degradable green bio-based structures in silicones as a method for the development of materials consistent with the concept of sustainable development. This chapter reviews and analyzes the results of the authors and those reported in the literature on the above-mentioned directions.

1.1 INTRODUCTION

Composite materials having silicone in their composition with adjustable porosities and functionalities represent an important class of CO_2 sorbents. The progress of porous carbons from various types of biomass is an economic, sustainable, and environmentally friendly strategy. Because wood is a naturally abundant, biodegradable, and carbon-rich raw material that can be used as a resource for obtaining functional porous carbons. An important application of porous carbon derived from cellulose, hemicelluloses, and lignin-wood-based biopolymers is their use in CO_2 capture.

The main cause of global warming is this increase in carbon dioxide emissions. The excess emission of CO_2 into the atmosphere is the principal reason for the current global climate changes and associated problems, like global warming, sea-level rise and ocean acidification, threatening human survival and development. These effects raise the necessity to reduce CO_2 emission and to control the atmospheric CO_2 level, which leads to efforts dedicated to the capture and storage of CO_2 from the atmosphere. Carbon capture and storage (CCS) is considered a necessary approach to reduce CO_2 emission. CCS covers a group of technologies including capture, compression, transportation and permanent storage of CO_2. CO_2 capture is the first and the most important step. Of the several technologies being developed for conventional CO_2 capture processes, postcombustion capture of CO_2 is a technique that can be easily fitted to existing power plants. Amine scrubbing, an absorption technique using aqueous alkanolamines solution to remove CO_2 from gas streams, has been applied industrially in natural gas purification and postcombustion capture of CO_2 for more than 50 years. The chemical reaction of CO_2 with amine forms stable carbamates, offering a high efficiency for CO_2 capture and separation. However, it suffers from several drawbacks such as high-energy penalties for the amine regeneration, risk of amine leakage, and corrosion to the employed equipment.

Large efforts have been made to the development of adsorption-driven separation techniques using porous materials as solid sorbents for CO_2 capture, which is considered an alternative approach to the amine scrubbing process. Porous materials are a type of solid containing pores, usually interconnected, or channels possessing a high surface area: zeolites, activated carbons, mesoporous oxides, and emerging metal-organic frameworks (MOFs), covalent-organic frameworks, porous organic polymers, etc.

The applicability of advanced composite materials with a hierarchical structure that conjugate MOFs with macroporous materials is limited by their inferior mechanical properties. Tu et al. (2020) studied a universal green synthesis method for the in situ growth of MOF nanocrystals within wood substrates are introduced. Nucleation sites in various MOFs are created by a sodium hydroxide treatment (applicable to different wood species). The resulting MOF/wood composite presents hierarchical porosity with 130 times larger specific surface area compared to native wood. Assessment of the CO_2 adsorption capacity demonstrates the efficient utilization of the MOF loading along with similar adsorption capability to that of pure MOF. The compression and tensile tests show superior mechanical properties. The functionalization strategy presents a sustainable platform for the obtaining of multifunctional MOF/wood-derived composites with potential applications in environmental- and energy-related fields.

Different from the chemical absorption approach, a typical adsorption process starts by attracting CO_2 because of the much weaker CO_2-adsorbent interactions taking place in the above-described process than in CO_2-amine interactions/chemical reactions, the adsorption-driven separation technique requires significantly lower amounts of energy in the regeneration procedure compared with the amine-scrubbing process. Area, pore size, and adsorption sites of the porous material, while the adsorption selectivity is influenced by molecular sieving, thermodynamic equilibrium, and kinetic effect. Other factors such as cost, stability, and processability of the material should also be considered when it comes to practical applications. CO_2 adsorption capacity mainly depends on intrinsic characteristics of the porous material, while the adsorption selectivity is influenced by molecular sieving, thermodynamic equilibrium, and kinetic effect.

Zeolites, crystalline microporous aluminosilicates, are used for industrial CO_2 capture, but this class of materials loses the adsorption capacity and selectivity for CO_2 under humid conditions. This is the reason why the operation cost is increased (the flue gas must be dried before passing through the zeolite sorbents). Using MOFs with the adapted crystal structure and pore size functional groups is a very known way, but these types of materials have low stability and high manufacturing cost.

Silicone materials have been studied as capturing CO_2 materials, among them, amino-silicones recently proved to be a highly efficient absorber of this. Amines containing siloxanes have several advantages over the classic organic amines such as high thermal stability of the siloxane bond, low volatility, and low viscosity which allows their use as such, without the need for dissolution/dilution with water or organic solvents. This behavior of amines helps to reduce the amount of heat energy needed for CO_2 release and for absorber regeneration. The solid adsorbents as compared to the aqueous solutions of amine are easy to handle, can save energy, and are not causing corrosion difficulties.

Since the 1990s, mesoporous silica materials were extensively investigated. The most used silica material is MCM-41, due to its combination of superior properties, as high surface area, thermal stability, and variable porous volume depending on the surfactant used for the synthesis reaction. The pore sizes can have dimensions between 2 and 50 nm. The mesoporous silica has a high adsorption capacity for CO_2, CH_4, N_2, H_2, and O_2 and this capacity can be improved for CO_2 adsorption by different methods: adjusting the synthesis parameters, functionalizing the mesoporous silica, or impregnate the material. Amine–silica hybrid/composite materials can exhibit high and fast CO_2 adsorption, low energy consumption, and a good selectivity. A very important feature is that these materials are tolerant to moisture. The amine–silica hybrid/composite materials can be obtained by grafting or impregnation. Impregnation is a physical process, while the grafting means to synthesize an amine–silica hybrid material through the formation of a covalent bond between organic-amines and silica. Amine-functionalized silica materials were investigated for their use as CO_2 sorbents by Li et al., Gunathilake et al., Sanz-Perez et al., Loganathan and Ghoshal, Le et al., Kishor and Ghoshal, Gholami et al., and other from a different point of view. For example, Sanz-Perez et al. studied the reuse and recycling of amine–silica materials as CO_2 adsorbents after their lifespan. The conclusion of their work was that the materials obtained using the impregnation of the calcined samples maintained their CO_2 adsorption properties more than six cycles without changes in their efficiency for CO_2 adsorption, and the materials obtained by grafting the calcined samples yield to smaller amine efficiency during CO_2 capture.

Porous carbons with high surface areas, adjustable pore sizes, and a very good tolerance to acidic, basic, and humid environments are presented to be ideal sorbents for CO_2 capture. Porous carbons can be made from sustainable biomass precursors providing an environmentally friendly and sustainable manner for the development of CO_2 sorbents. Wood, which is the largest biomass resource on earth, has been used for tools, fuels, and buildings throughout human history. Cellulose, hemicellulose, and lignin are biopolymers that represent the main components of wood and many other plants. From a chemistry point of view, cellulose is a linear polysaccharide consisting of a repeated D-glucose unit with the formula of $(C_6H_{10}O_5)n$. Hemicellulose is a branched polysaccharide containing different sugar monomers. Lignin is an aromatic polymer network of crosslinked phenylpropane-derived lignols. All these three biopolymers have been employed as bases for making porous carbons due to their rich carbon content, natural abundance, and low cost.

Cellulose can be transformed into porous carbon material using a one-step pyrolytic carbonization process under N_2 or Ar atmosphere and physical or chemical activation of the carbonized solid increasing its surface area and enhancing its CO_2 adsorption capacity. In the literature, some examples are presented as Heo et al. studied a series of porous carbons derived from commercial cellulose fibers after three stages: prepyrolysis in N_2 atmosphere at 200 °C, carbonization in N_2 atmosphere at 750–800 °C, and physical activation with vapors. When steam was activated, that led to an increase in CO_2 adsorption capacity and CO_2-over-N_2 selectivity.

Xu et al. (2019) treated Cladophora cellulose (nanofibrous cellulose extracted from algae), in a one-step carbonization/activation approach at 900 °C under N_2 or CO_2 atmosphere. The CO_2 adsorption capacity of the porous carbons was increased by the CO_2 activation approach. Another similar study made by Zhuo et al. was to prepared porous carbons by carbonization/ activation of cellulose aerogels under CO_2 and N_2 atmosphere. The CO_2-activated porous carbon had a higher surface area, higher volume of micropores, and higher CO_2 adsorption uptake than the N_2-carbonized porous carbon. The obtained results suggest that CO_2 or steam activation is an efficient approach to prepared cellulose-based porous carbons with high CO_2 adsorption capacities.

Chemically activated carbons have much higher surface areas and show higher CO_2 adsorption capacities. A study made by Sevilla et al. reported the chemical activation of hydrothermally carbonized cellulose by KOH. The carbon exhibited a high CO_2 adsorption rate and very good adsorption recyclability. Wang et al. used hemp stem hemicellulose as bases and prepared well-shaped porous carbon spheres by a hydrothermal carbonization (HTC) process and a subsequent KOH activation procedure. The obtained porous carbon spheres had high surface areas of up to 3062 m^2 g^{-1} and high CO_2 adsorption capacities of up to 5.63 mmol g^{-1} (1 bar, 273 K). Hao et al. reported on the treatment of lignin under HTC conditions at 360–385 °C. The activation of the hydrocarbon by KOH result in a highly porous carbons with surface areas of up to 2875 m^2 g^{-1} and high CO_2 adsorption capacities of up to 6.0 mmol g^{-1} (1 bar, 273 K).

The porosity and CO_2 adsorption behaviors of wood-biopolymer-derived porous carbons are influenced by multiple parameters, like a high KOH/carbon ratio, the contact between the lignin precursor and the activating agent, the way of biopolymers are activated, etc. Sangchoom et al. showed that activation at a high KOH/carbon ratio led to an increase in the surface area and total pore volume for lignin-derived porous carbon and a decrease in the CO_2 adsorption capacities and the ultramicropore volume. Balahmar et al. obtained lignin-based porous carbons by a new mechanochemical activation method starting from compaction of lignin precursors and KOH at a high pressure of 740 MPa before the thermal activation. Hu et al. made N-doped and N-free carbon aerogels using an activation/carbonization of cellulose aerogels at high temperatures under NH_3 or N_2 atmosphere with 40% higher CO_2 adsorption capacity than that of the N-free carbon aerogel (4.99 vs 3.56 mmol g^{-1} at 1 bar, 298 K).

Demir et al. and Saha et al. made N-doped porous carbons from lignin. All these materials presented relatively high CO_2 adsorption capacities up to 8.6 mmol g^{-1}. This high CO_2 adsorption capacity can be justified by different interactions between the N-containing species and the CO_2 molecules. In another study, Meng et al. presented a hypercrosslinking of organosolv lignin with formaldehyde dimethyl acetal as crosslinker in a Friedel–Crafts reaction. Pyrolysis of the hypercrosslinked lignin made microporous carbons with higher CO_2 capacity and high selectivity. All these studies confirm that by introducing crosslinking structures into biopolymers can promote the formation of micropores in the new derived materials. The formation of large amounts of ultramicropores and N-doping structure in the porous carbons are efficient manners to reach a high CO_2 adsorption capacity. Based on the reported data, it is expected that 1 ton of biopolymer-based porous carbon could capture up to 50 kg of CO_2 at real conditions for postcombustion capture of CO_2 from flue gas.

1.2 SILICONE COMPOSITES WITH WOOD-BASED BIOPOLYMERS

The combination of natural polymers from renewable resources and siloxane derivatives in the same system is considered to be a challenge and a promising manner for obtaining new materials with interesting and adjustable properties. Copolymers, networks, composites, ceramics, nanoparticles, or layered structures can be obtained based on the precursors from the two materials classes.

The latest developments in technology have given rise to a demand for new materials with improved functions and properties. Scientists have found that mixtures of materials may have improved properties as compared with those of the individual materials taken independently. They were inspired by natural materials consisting of organic and inorganic elements distributed at macro- or nano-scale. The inorganic component gives the mechanical strength and the overall structure of natural objects while in the same time provides the organic link between the inorganic and soft tissue. Mixing organic and inorganic components has been made since antiquity for making vivid and bright colors, by testing new mixtures of inorganic pigments and other organic or inorganic components to form paints. Hybrid materials are not a new invention but a continuous development since ancient times. Polymeric materials, like natural polysacharides and lignins, and synthetic silicones, are two well-known materials classes, each of these with specific fields of use. The difference between their chemical structures is what makes some of their properties totally different. The combination of natural polymers from renewable resources and siloxane derivatives in the same system by different types of interactions (Scheme 1.1) represents a challenge and a promising way for making new materials with interesting and adjustable.

Inorganic–organic hybrids can be used in many domains of chemistry and have the capability of being

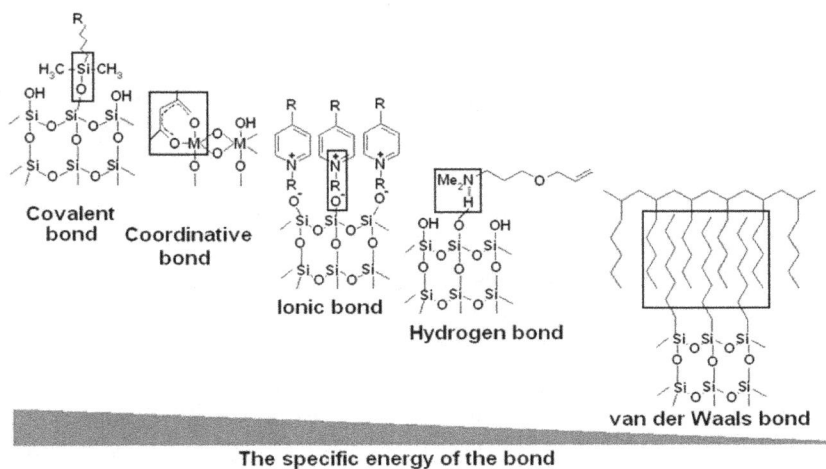

SCHEME 1.1 Types of interactions met in hybrid materials and their relative strength.

easily processed and can be projected at the molecular scale. There are many possibilities for composition and structure of hybrid materials:

Matrix: crystalline ↔ amorphous; organic ↔ inorganic

Basic elements: molecule ↔ macromolecule ↔ particle ↔ fiber

Interactions between components: strong↔weak.

The most interesting property of the hybrid materials is their processability. If pure solid inorganic materials require treatment at high temperature for processing, the hybrid materials can be manipulated in a similar manner to polymers due to high organic content and can be adapted in any shape, in bulk, or in films.

The explicit advantages of organic–inorganic hybrids are: (1) they can combine in a good manner the various properties of organic and inorganic components in a single material; (2) the facility of creating multifunctional materials (by incorporation of inorganic particles or clusters which have specific optical, electronic, or magnetic properties in organic polymer matrices—as an example). Hybrid composite materials are formed or composed of heterogeneous elements. The development of siloxane–lignocellulose hybrid materials is justified by the concentration of properties of two different materials in a single new material (Scheme 1.2).

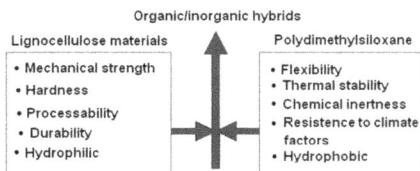

SCHEME 1.2 The characteristics of conferred to lignocellulose–PDMS hybrid materials.

Fibers based on natural polymers have many advantages over synthetic fibers, such as:

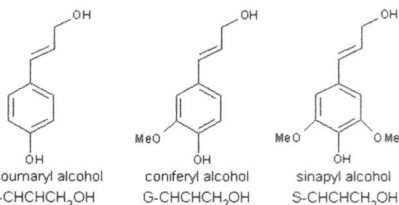

p-coumaryl alcohol
H-CHCHCH$_2$OH

coniferyl alcohol
G-CHCHCH$_2$OH

sinapyl alcohol
S-CHCHCH$_2$OH

1. Plant fibers are a regenerative raw material and therefore its availability is perpetual.
2. When reinforced natural materials were used to a fuel or disposed of in landfill site, the amount of CO_2 emitted is neutral as compared to the quantity of CO_2 accumulated during growth.

Abrasiveness nature of natural fibers is much lower than that of glass fibers or other hard synthetic fibers, presenting an advantage in terms of technical material recycling or processing of composite materials.

Plastics reinforced with natural fibers by using biodegradable polymers as base materials are friendly for the environment.

Polyorganosiloxanes, with the most important representative of their, polydimethylsiloxane (PDMS) (Scheme 1.3), present useful and interesting properties like very low surface energy, excellent gas and water vapor permeability, good heat stability, resistance to UV radiation, chemical resistance, good behavior under extreme conditions of temperature, minimal effect of temperature on properties, low-temperature flexibility and biocompatibility. Siloxane materials have instead poor mechanical properties due to weak intermolecular interactions between polymer chains that limit their use in specific applications. The most important reaction used for crosslinking siloxanes in commercial applications is based on the condensation of silanol groups with the formation of siloxane bonds. A regular system contains PDMS terminated with silanol groups, a crosslinking agent, and a catalyst. Radical crosslinking of siloxanes is performed with peroxides or high-energy radiations. Another way of crosslinking is based on hydrosilylation reaction where a liquid silicon rubber base, a catalyst, and curing agent are mixed, poured in a mold, and heated to elevated temperatures when the reaction takes place between vinyl ($SiCH=CH_2$) and hydrosilane (SiH) groups and the liquid mixture becomes a solid, crosslinked elastomer.

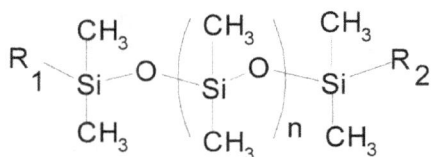

SCHEME 1.3   PDMS structure: R_1, R_2 are end chain groups such as: OH, H, $-CH=CH_2$.

After vulcanization, the obtained systems can lead to slightly reticulated silicone films, capable of releasing perpendicular, and shearing loads at high temperatures and also to flexible yet strong silicone elastomer which keeps their elastic properties on various temperatures. Some researchers demonstrated the use of block copolymer silicone-alkyl resin in composites with wood for UV and weather protective coating with efficacy in both natural and artificial weathering conditions, leading to minimized negative impact of UV radiation and weathering for panel products and bamboo composites (Sahoo et al., 2011).

Lignin has been used in a simple and effective process for the preparation of lignin-silicone composites via extrusion and compression molding. The reaction mix consists of hydrosilanes, metal-free Lewis acid tris(pentafluorophenyl)borane, and raw lignin particles, with no solvents and no pre-treatment required. After extrusion at room temperature and compression molding at high temperature, lead to ecofriendly foams, due to release of hydrogen in reactions of lignin-silicone and silicone cross-linking. An environmentally friendly one-pot method was used to prepare nanocomposites with cellulose matrix and up to 50% silica nanoparticles, using a percolation approach and a solvent exchange strategy. The nanocomposites demonstrated fungi decay resistance following test according to ASTM D-1413 standard test method (Rodriguez-Robledo et al., 2018). The efficiency of nanocomposites based on silicones and wood have led established silicone manufacturing companies, to launch silicones for wood-plastic composites with improved performance: GENIOPLAST WPC (Wacker) and (Amplify Si PE 1000) Dow. Such silicone materials enhance the mechanical properties and water-repellency of the composites and allow higher throughput for extrusion with lower melt temperature and less thermal degradation with lower production costs. Wang et al. prepared composites with silicone elastomer filled with vinyl-silane treated aluminum hydroxide as adhesive with flame-retardant capabilities for plywood (Wang, 2018). The material showed in tests higher burn-though resistance and thermal barrier efficiency and lower flame spread and heat release rate, and improved fire resistance as compared to the plywood treated with polyurethanes. The addition of cellulose fibers or glass fibers suppressed the delamination and cracking of the composite plywood and promoted the formation of an effective thermal barrier during combustion.

1.3   SILANES AS COUPLING AGENTS FOR VARYING THE DOMAIN OF LIGNOCELLULOSIC MATERIALS

The mechanical strength of a composite can be upgraded by creating an interphase between the two constituent materials when several components remain interpenetrated (in wood, bone, or cuticle of insects). A method used for creating an interphase region lies in creating links between the interface and component materials with a coupling agent.

There are three mechanisms of coupling for lignocellulosic material/thermoplastics materials: mechanical blockage, adsorption adhesion, and chemical binding. Mechanical blockage is a combination of swallow type tail and the two components do not react with one another but are held in position by geometric constraints. Mechanical blockage has the limitation that the two components have a different interface. Another way for coupling is by intermolecular forces and because they are short-range forces, there has to be an intimate association between components; the phenomenon of wood fibers wetting must exist. Depending on the components, chemistry different interactions are possible, such as acid-base, London dispersion, hydrogen bonds, etc. This is the mechanism of adhesion in wood-based composite systems (Scheme 1.1). The third mechanism which can be applied is covalent binding by direct reaction of components at the interface by free radical or ionic reactions. There are many coupling agents used for reinforcing thermoplastic composites (alkoxysilanes, isocyanates, and anhydrides).

The mechanism of coupling for three of the most used agents implies interaction and reaction with hydroxyl groups on the surface of cellulose (Scheme 1.4). Coupling agents contain portions consisting of benzyl or aliphatic groups (covalently attached to the reactive ends of the molecule) which should be able to link the interface between wood fiber and thermoplastic matrix. Highly polar cellulose fibers are inherently incompatible with hydrophobic polymers.

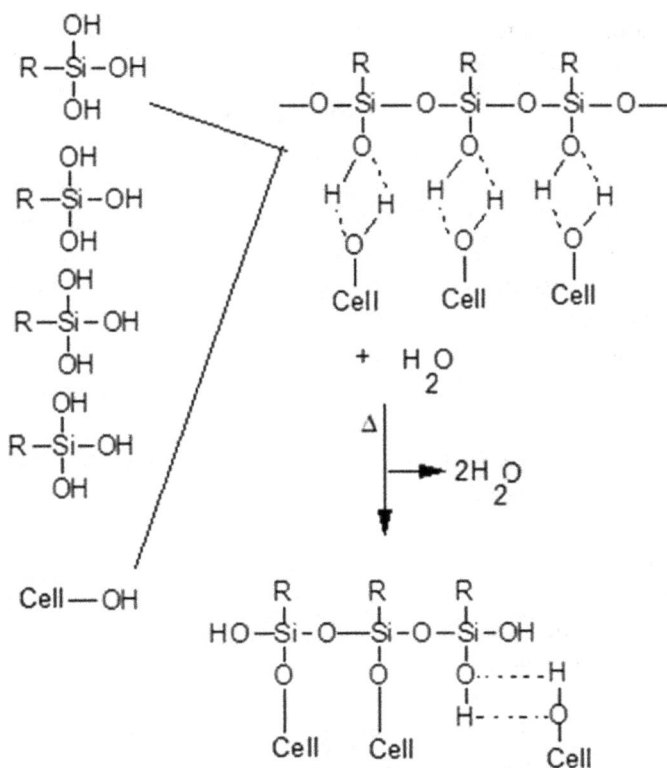

SCHEME 1.4 Cellulose reacted with alkoxysilane.

Silane coupling agents are usually used to make lignocellulosic materials compatible with other polymers. These coupling agents present a structure as one in Scheme 1.5.

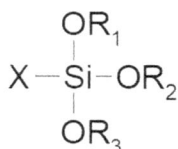

$$X - \underset{\underset{OR_3}{|}}{\overset{\overset{OR_1}{|}}{Si}} - OR_2$$

SCHEME 1.5 Structure of silanes, where R_1, R_2, and R_3 are different or identical alkyl groups (most frequent $-CH_3$, $-CH_2CH_3$, $-CH_2CH_2CH_3$) and X is a functional group.

Silane coupling agents can be applied by the following four different manners:

1. Aqueous solutions, from dilute solutions in pure water or mixtures of water and polar solvents such as alcohols, are used in industrial processes because of environmental reasons. Using this method, multilayer structures are obtained.
2. Organic solutions, from dilute solutions in dry organic solvents, tend to form monolayer structures, in this case, there is not any condensation in the solution.
3. Integral mixtures, filler materials in particles form in liquid resins, are modified using coupling agents by adding silane to the mixed substances. The filler material can be treated in situ (when they are grinded with elastomers).
4. Vapor phase, volatile silanes compounds can be applied to inorganic surfaces in the vapor phase under reduced pressure, experiments demonstrated that silanes applied using this method form a perfect monolayer structure.

Close hydroxyl groups lead to strongly adsorbed polysiloxane. The amorphous cellulose reacts with polysiloxane through hydrogen bonds. The cristallinity of the exposed surface influence the spacing of the $-OH$ groups. After adsorption of polysiloxanes on the fiber surface, alkyl groups connected to silicon in the main chain can interact to interact with the thermoplastic matrix and that is the reason why the fiber surface is changed so that the area becomes compatible with the thermoplastic matrix.

Inorganic whiskers used as reinforcement material with application in high-performance composites and nanocomposites attracted a big interest. The development of polymer whiskers—needle-like polymer single crystals where the chain axis is along the needle—for the reinforcement of plastics is a way that shows great potential.

The first stable suspensions of colloidal-sized cellulose crystals were obtained by Rånby and Ribi using a method involving sulfuric acid hydrolysis of wood and cotton cellulose. Other authors prepared native cellulose suspensions via acid hydrolysis from numerous sources (bacterial cellulose, tunicate cellulose, microcrystalline cellulose, softwood pulp, sugar beet primary cell wall cellulose, and cotton). The first researchers who used cellulose whiskers from the tunic of tunicate Microcosmus fulcatus as reinforcing nanofillers in a copolymer (styrene-butyl acrylate) were Favier et al. Researchers blended such nonflocculated suspensions with water-soluble polymers and latexes. These blends produced nanocomposite structures with enhanced mechanical properties after drying. Applications for whisker suspensions are found in the field of complex fluids. Other scientists have reported the use of these whiskers as particles for plastic reinforcement and gel-forming and thickening agent.

Cellulose whiskers can be used to enhance the properties of siloxane rubbers. First, the whiskers have to be functionalized with silane functions. Gousse et al. developed procedures for chemical modification of the OH groups at the surface of cellulose whiskers with different silanes:

isopropyldimethylchlorosilane (IPDMSiCl), n-butyldimethylchlorosilane, n-octyldimethylchlorosilane, n-dodecyldimethylchlorosilane (Scheme 1.6) (the cellulose microfibrils have an average size of 3 nm).

Chotirat et al. have studied the effect of silane coupling agents in polymer matrix composites and natural fiber filler, using the materials: acrylonitrile-butadiene-styrene (ABS); particles of sawdust as the reinforcement for ABS matrix: (1) 9.1% by mass (low content of sawdust composite) and (2) 33.3% wt (high content of sawdust composite); silane coupling agents (concentration range from 0.0% to 2.0% wt relative to the mass of sawdust): (i) 3-methacryloxypropyl trimethoxysilane. In this study, the mechanism for adhesion-strengthening of the composite resulted from the chemical reaction of coupling agent and particles of sawdust and from dipole-dipole interactions between C=O group of coupling agents and –C≡N group from ABS. Because of the nature of interactions with ABS matrix, 3-methacryloxypropyl trimethoxysilane

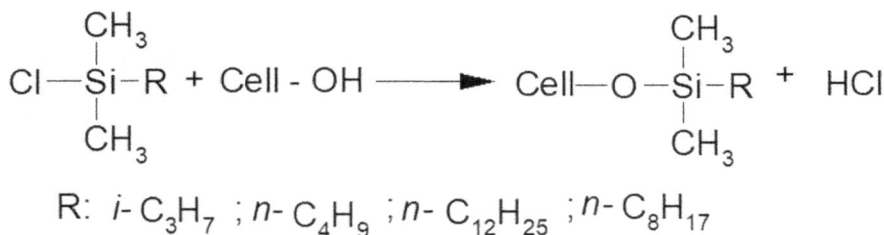

$$Cl-\underset{\underset{CH_3}{|}}{\overset{\overset{CH_3}{|}}{Si}}-R \ + \ Cell-OH \ \longrightarrow \ Cell-O-\underset{\underset{CH_3}{|}}{\overset{\overset{CH_3}{|}}{Si}}-R \ + \ HCl$$

$$R: \ i\text{-}C_3H_7 \ ; n\text{-}C_4H_9 \ ; n\text{-}C_{12}H_{25} \ ; n\text{-}C_8H_{17}$$

SCHEME 1.6 Silylation of cellulose whiskers.

is a "secondary agent for chemical coupling." (ii) N-2(aminoethyl)3-aminopropyl trimethoxysilane.

For this coupling agent, the adhesion mechanism consists of its chemical reaction with sawdust particles and the formation of covalent N–C bonds between –NH$_2$ groups from the coupling agent and –C≡N groups of ABS, so N–2 (aminoethyl) 3-aminopropyl trimethoxysilane leads to stronger interfacial links between ABS and sawdust in comparison with the former coupling agent: 3-methacryloxypropyl trimethoxysilane.

In addition to alkoxysilanes, other silane coupling agents were used. Thus, Zollfrank reported that trimethylchlorosilane was efficient for covering both interior and exterior of wood fibers from beech wood by substituting the proton of OH groups from the wood surface. Wood fiber material was dried before reaction, which took place in an environment of anhydrous pyridine and tetrahydrofuran and the silane coupling agent is trimethylsilyl chloride. Energy dispersive X-ray spectroscopy (EDX) and FTIR studies demonstrated the formation of cellulose derivative with trimethylsilyl ether groups. SEM study confirmed that silica was not restricted to the fiber surface but also penetrated inside the fibers.

Matuana et al. evaluated the effects of aminosilanes for surface treatment on the wood fibers in the PVC/wood–fiber composites studying the adhesion between PVC and laminated wood veneers. Aminopropyltriethoxysilane, dichlorodiethylsilane, phthalic anhydride, and maleated polypropylene were used as coupling agents for testing their efficiency. The degree of chemical modification of wood surfaces was characterized using complementary surface analytical techniques: X-ray photoelectron spectroscopy and surface tension measurements. The effect of aminosilanes as coupling agents was also tested by Balasuriya et al. for interface modification of composites based on raw wood flakes and high-density polyethylene (HDPE).

A matrix consisting of HDPE was modified by reaction with maleic anhydride in a twin-screw extruder and compounded with wood flakes to produce composites. In order to improve the adhesion between the matrix and reinforcing fibers, wood flakes were modified by reaction with aminosilane coupling agent in an aqueous medium before compounding with HDPE. The DSC and FTIR spectroscopy provided the evidence for the existence of a polyethylene-silane-grafted wood structure. This acts as a compatibilizer for wood flakes and PE. The mechanical strength tests showed important changes in tensile strength, ductility, and Izod impact strength.

Scanning electron micrographs provide evidence for strong interactions between the wood flakes and the matrix agent. Overall results indicate that 1–3 wt.% silane treatment on wood flakes provides wood polyethylene composites with the optimum properties. Sèbe et al. have modified maritime pine wood using a generic method that allows the introduction of a variety of silicones in wood. After Soxhlet extraction, the wood was esterified with maleic anhydride and allyl glycidyl ether and then subjected to hydrosilylation reaction in the presence of Karstedt catalyst. The authors chose the hydrosilylation reaction because it is a versatile method of forming Si-C bonds in organosilicon chemistry. Kokta et al. used a variety of silanes (such as vinyl tri(2-methoxyethoxy) silane, gamma-methacryloxypropyltrimethoxy silane, beta-(3,4-epoxy cyclohexyl) ethyltrimethoxy silane, gamma-glycidoxy propyltrimethoxysilane and gamma-aminopropyltrimethoxysilane) as coupling agents for composites of chemithermomechanical pulp and sawdust as reinforcing materials and polyvinyl chloride matrix (Kokta, 1990); thus obtaining materials with good dispersion of the cellulose fibers in the polymer matrix.

Sisal fibers were modified with silane coupling agents (namely 3-aminopropyltriethoxysilane (APS) and N-(2-aminoethyl)-3-aminopropyltrimethoxysilane) to determine the mechanism of bond formation. The research has shown a layer of film consisting of siloxane and polysiloxane was formed on the fiber surface by silane adsorption (Zhou, 2014).

The use of silane coupling agent gamma-methacryloxypropyl trimethoxy silane as crosslinking agent has significantly improved the bonding properties of starch-based wood adhesive. Poplar wood fibers were modified with silane coupling agent (3-glycidyloxypropyl)trimethoxysilane) for composites with high-density polyethylene matrix. A high-temperature hot air (HTHA) treatment promotes the hydration of the wood fibers and improves the mechanical properties of the composites.

The bonding mechanism and the interface structure in wood-plastic composites with wood fibers treated with silane coupling agents was studied (Rao, 2018) and it was confirmed the formation of covalent bonding with the functional hydroxyl groups of wood flour and crosslinking with polyethylene matrix molecules. The crosslinking under high temperature and pressure gives rise to improved fluidity of the polyethylene matrix, and thus improve its hydrodynamic flow in interface region for better bonding.

The silane coupling agents, gamma-aminopropyl triethoxy silane, and trimethoxy vinyl silane were used for the modification of the bonding interface between the WPC and the material of the wood veneer. This treatment achieved a significant increase of water resistance with the contact angle between distilled water and the veneer surface increased from 46 to greater than 120° after treatment (Liu, 2019).

Others researchers modified the surface of cellulose pine fibers with silane derivatives bearing specific functionalities. $-NH_2$, $-SH$, long aliphatic chain, and methacrylic group and used this as reinforcer for polypropylene matrix composites. The mechanical analysis showed an increase in Young's and flexural moduli, by 12% and 130%, respectively.

Cellulosic fibers from himalayan Eulaliopsis binate were subjected to alkaline treatment and then were treated with vinyltrimethoxysilane as coupling agent for the preparation of composites with synthetic polymers. This treatment affects the physicon-chemical properties of the cellulose fibers and improves the capability for forming of ecofriendly composites.

Microfibrillated cellulose fibers modified with silane coupling agent with NH_2 functional group were formed into sheets which were used for the preparation of composites with acrylic resin polypropylene glycol diacrylate (Ifuku, 2015).

The structure of the microfibrillated cellulose with interconnected fibrils and microfibrils with nano-order-scale has extremely large surface area and submicron pore size compared to conventional cellulose, thus improving the effects of silane coupling agent and the compatibility between the fibers and matrix and the dispersibility of fibers within the matrix. The mechanical tests showed Young's modulus was significantly increased to more than 70%.

Silane coupling agents were used for the modification of the surface of cellulose fibers and gold nanoparticles for preparation of nanocomposites with gold nanoparticles on the surface of cellulose fibers (Van Rie, 2017). The composite materials were tested for a variety of applications: catalysis, biosensor, Surface-Enhanced Raman Scattering, metal adsorption, enzyme immobilization, optical properties, conductivity and antioxidant activity.

The silane agents (3-aminopropyltriethoxysilane and 3-(trimethoxysilyl)propyl methacrylate) were elected for modification of spent kraft fibers and sisal fibers, and subsequent preparation of composites with polylactic acid matrix. The use of silane coupling agents led to composite samples with improved stiffness and brittleness with respect to PLA, while keeping similar value for tensile strengths (14.45 and 24.61 J/m, respectively) compared with composites using non-treated fibers.

Zhu et al. prepared composites with lignin from corn stalks and poly(L-lactic) acid. Lignin was treated with three coupling agents, namely 3-aminopropyltriethoxysilane, gamma-glycidoxypropyl trimethoxysilane and gamma-methylacryl-oxypropyltrimethoxysilane (Zhu, 2015). The treatment resulted in significant increases of tensile strength and Young modulus, as well as elongation of break compared with initial non-treated composites.

1.4 SILOXANE/LIGNOCELLU-LOSE HYBRID MATERIALS

A wide amount of lignin is obtained as a byproduct of the pulp and paper industry and conventionally has been used as an energy source. Lignin is a three-dimensional amorphous aromatic natural biopolymer, easily available and not expensive. Lignin produces a large amount of char by heating at high temperature in an inert atmosphere—a basic aspect of flame retardant additives, because char reduces the combustion rate of polymeric materials. The presence of the phenolic groups in lignin can give thermo-oxidative and light stability to a chemical system in which it is incorporated. Lignin is a low density, low abrasive, and low-cost material.

One of the possible applications for lignin is its use as a filler in the formulation of polymeric materials for increasing their content of renewable resources. With some polymers, partially or completely biodegradable composites can be achieved. Efforts have been made to integrate lignin into different polymer materials: poly(ethylene terephthalate), poly(vinyl alcohol) and polypropylene, poly(vinyl chloride), polyethylene and polystyrene, or natural rubber. Lignin powder was included as filler for the polysiloxane matrix, the motivation for this being the use of cheap renewable materials from biomass for obtaining new materials. Different techniques such as scanning electron microscopy, dynamic mechanical analysis, tensile strength tests, X-ray diffraction analysis, thermogravimetric analysis, and differential scanning calorimetry were used for studying the obtained composites and the results were compared with those obtained on a reference sample prepared with a standard diatomite and pure crosslinked PDMS. Telysheva et al. obtained siliceous lignins by using monomeric and oligomeric organosilicon compounds as modifiers (Telysheva, 2009). The Si compound guest molecule influences both the microsurface and the bulk structure of the lignin host matrix. Organic–inorganic hybrids with lignin and TEOS were prepared. The examination of the new materials with the SEM technique demonstrated that in the micrometer range, silica was stored on lignin surface in form of isolated or aggregated nanoparticles. Some of the particles were clustered into mixed aggregates.

Interest in organic–inorganic hybrid materials appeared because of the unique characteristics of these materials, closely related to the role of interfacial forces, and surface chemistry of molecular layers as the size of the dispersed phase increases.

An organic component containing a functional group that permits attachment to an inorganic network can behave as a network modifier agent, with the final structure having the inorganic network modified only by the organic group. Phenyltrialkoxysilanes are used for the following reasons:

1) if they have a reactive functional group, then they act as a network functionalizer (Scheme 1.7);
2) they can modify the silica network by the sol–gel reaction of trialkoxysilane groups without supplying the functional groups for further chemical reactions;
3) an organic segment with two or three anchor groups will provide materials in which the inorganic group becomes a constituent of the hybrid network.

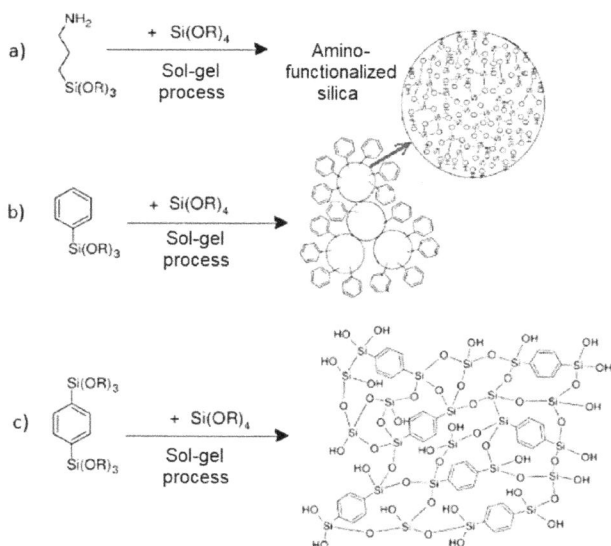

SCHEME 1.7   The role of organic groups functionalized trialkoxysilanes in sol–gel process. Lopattananon et al. prepared hybrid materials based on natural rubber and native cellulose fibers from soft wood, treated with the agent bis(triethoxysilylpropyl)tetrasulfide (Lopattananon, 2011). The tensile modulus improved, as well as the scorch safety. The Mooney viscosity decreased with increasing cellulose fiber content and the surface modification of fibers by coupling agent improved the processing characteristics of the composites.

Hybrid materials were prepared with poly(lactic-*co*-glycolic acid) (PLA/PLGA) blends matrix and cellulose nanocrystals (CNCs) with organophilic silica as nanoparticles (dos Santos, 2015). The mechanical performance of films was improved with the addition of CNCs, and was further enhanced by the incorporation of silica, indicating that silica decreases the surface tension between PLA-cellulose and PLA/PLGA cellulose (Ramesh, 2014). The hybrid materials prepared with cellulose and polyhedral oligomeric silsesquioxanes functionalized with amino groups demonstrated antimicrobian properties against Bacillus cereus (F4810/72) and *Escherichia coli* (ATCC 35150).

1.5 SILANE-CELLULOSE: CROSSLINKED STRUCTURES BASED ON CELLULOSE AND SILANES

Organosilanes are a favored family of reagents utilized at an industrial scale for modifying the surface of fibrous materials and for ensuring through different structures available useful purposes to best harmonize the chemical modification of the fiber surface with the nature of the macromolecular matrix. Many researchers have analyzed the reactions of –OH groups (from cellulose) with both $\equiv$Si-OR unchanged groups and $\equiv$Si–OH hydrolysate groups from silanes.

Studies made by Castellano et al. have demonstrated for different siloxanes that they are adsorbed on the surface of cellulose fibers and did not confirm the existence of chemical coupling at room temperature. Heating to 120 °C has determined few degrees of condensation $\equiv$Si–OH + $\equiv$C–OH, which carry to permanent chemical grafting of silane species on the surface of cellulose fibers. These reactions were used silanes which have a reactive group (–C=C–, –NH$_2$) capable to react as a comonomer unit for in situ polyaddition or polycondensation of monomer units with the matrix. In this manner, the authors achieved the covalent bonding between fiber and matrix. Herrera-Franco and collaborators showed that prehydrolized siloxanes react with cellulose fiber surface just at high temperature ($\equiv$Si–OH + $\equiv$C–OH). In another study, Mathias et al. proved the progress of the reaction through UV–VIS spectroscopy and diffuse reflectance FTIR .

The sol–gel technique involves the conversion of an inorganic network (from solution) in a colloidal suspension (sol) and its gelation for making a network in a continuous liquid phase. This process involves partial hydrolysis of metal alkoxydes (with intermediate species that contain hydroxyl groups), then condensation of hydroxyl groups and residual alkoxy groups, which is conducted to a three-dimensional network.

Organic/inorganic hybrid materials prepared by this technique can be made in two different ways:

1) The first one consists of mixing organic polymers with a metallic alkoxide among which the most used is tetraethoxysilane (TEOS). When the sol–gel process occurs the inorganic mineral is deposited in the organic polymer matrix forming hydrogen bonds between the organic phase and inorganic phase.

2) The second method is realized by the introduction of triethoxysilyl groups in the organic polymer before sol–gel reactions with TEOS. Like an example, it can be mentioned the use of isocyanato-propyltriethoxysilane for making an organic polymer with both ends ethoxysilanes and urethane or urea link; another example is the copolymerization of the organic monomer with vinyl-trietoxysilane. These types of hybrids derived from the sol–gel process having functional groups are studied and developed for different applications (nonlinear optical, coatings, and contact lenses).

Although ceramics have very good mechanical and optical properties, high values for surface rigidity, module, strength, transparency and refractive index, organic polymers possess flexibility, low density, strength, and formability. Functional organic–inorganic hybrid materials are obtained when these materials are combined in an effective way. In nature, plants utilize Ca and Si and amorphous silica shell is made in diatoms and in rice plants. Silica in amorphous form, in rice plants, is taken, transferred and precipitated in the matrix polysaccharides, making a mineral deposit of silica. This deposit reinforces leaves and stems in plants. In this way, the photosynthesis is improved due to the light diffusion by amorphous silica nanoparticles and reduced water evaporation in the stoma. Inorganic particles dispersed in a polymer matrix give hardness, brittleness, and transparency, meanwhile, the nature of the host organic polymer is important for density, free volume, and thermal stability of the hybrid.

Through the starting materials for the sol–gel process are natural or synthetic polymers, metal alkoxides-$M(OR)_n$ (where M = Si, Ti, Al, etc., R = CH_3, C_2H_5, C_3H_7, etc.), and a small amount of acid or alkaline catalyst. Hydrolysis—(1) and condensation—(2) reactions are responsible for the polymerization of inorganic precursors. Below is presented the polymerization of TEOS in acidic conditions (Scheme 1.8).

$$Et-O-\underset{\underset{OEt}{|}}{\overset{\overset{OEt}{|}}{Si}}-OEt + H_2O \xrightarrow[-EtOH]{} H-O-\underset{\underset{OEt}{|}}{\overset{\overset{OEt}{|}}{Si}}-OEt \quad (1)$$

$$y\ H-O-\underset{\underset{OEt}{|}}{\overset{\overset{OEt}{|}}{Si}}-OEt \xrightarrow[-nEtOH]{} Et-O-\underset{\underset{OEt}{|}}{\overset{\overset{OEt}{|}}{Si}}\left[-O-\underset{\underset{OEt}{|}}{\overset{\overset{OEt}{|}}{Si}}\right]_n-OH \quad (2)$$

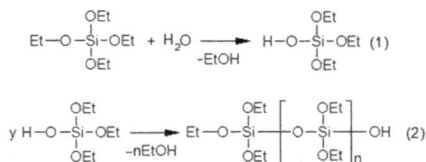

SCHEME 1.8 Polymerization of inorganic precursors for TEOS.

Sol–gel synthesis reaction also involves the following major steps: gelation, aging, drying, stabilization, and densification. The structure and properties of organic–inorganic hybrids obtained by the sol–gel method are dependent on the conditions of synthesis.

Sónia Sequeira et al. have used the sol–gel technique for obtaining and characterizing cellulose/silica hybrid material, starting from cellulose sulphate in Eucalyptus globulus and TEOS. Heteropolyacids $(H_3PMo_{12}O_{40}, H_4SiW_{12}O_{40}, H_4SiMo_{12}O_{40},$ and $H_3PW_{12}O_{40})$ or mineral acids (HCl, HNO_3, H_3PO_4, and H_2SO_4) were used as the catalyst. The syntheses were made at room temperature. Shoichiro Yano et al. have obtained organic–inorganic hybrids made with hydroxypropyl cellulose (HPC) and siloxane. HPC was dissolved in ethyl alcohol with TEOS, HCl, and H_2O, and the mixture was left to carry out sol–gel reaction and evaporation of the solvent, resulting in an optically transparent hybrid film that has been heated (60 °C) for a week to accelerate curing. Nanocrystalline cellulose (NCC) derived from empty fruit bunches (EFB) was cross-linked with 3-(2-aminoethylamino) propyl-dimethoxymethylsilane (AEAPDMS) (Mohd, 2016). The material has potential for carbon dioxide (CO_2) capture.

Crosslinked mats of electrospun fibers of poly(lactic acid) (PLA) and nanocrystalline cellulose (NCC) prepared by acid hydrolysis of microcrystalline cellulose were cross-linked with vinyltrimethoxy silane (VTMS) (Rahmat, 2016). The use of nanocrystaline cellulose reduces the diameter of the electrospun fibers, while silane crosslinking of PLA increases the mean fiber diameter. DSC thermograms also revealed that silane grafting caused a reduction in mobility of polymer segments, and consequently reduction of crystallinity. The chemical crosslinking improved the mechanical properties of nanofibers. The cytotoxicity tests showed the nanocomposite can be used for hard tissue engineering applications.

1.6 CELLULOSE DERIVATIVES MODIFIED WITH PREFORMED SILOXANES

Bourges Xavier et al. grafted hydroxyethylcellulose (HEC) with 3-glycidoxypropyltrimethoxysilane (GPTMS) or 3-glycidoxypropyl-methyldiethoxysilane (GPDMS) and the products were characterized for possible biomedical applications. The graft involved a Williamson reaction between the free hydroxyl function of HEC and the epoxy group of the silane. The product was in gel form at basic pH (>12.3), and in an aqueous solution when the grafted silanes are in ionic form. If the pH decreases, sodium silano-late is converted into silanol. The silanol groups interact, and the gel is converted into a crosslinked form.

Different researchers studied a self-hardening gel as an injectable bone substitute. Hydroxypropyl methylcellulose can form gels and was a good candidate for the synthesis of silylated hydroxyl-propyl methylcellulose by reaction with different volumes of 3-GPTMS. This study allowed the calculation of silane percentage and the reaction yield. The silylated-hydroxypropyl methylcellulose powders are soluble at basic pH. The silylated hydroxylpropyl methylcellulose

obtained possess the property of self-hardening as a function of the pH. The hardening of silylated-hydroxylpropyl methylcellulose was studied as a function of grafted silane percentage and temperature.

In order to obtain membranes for gas separation (CO_2/N_2) Fareha Zafar Khan et al. obtained silyl ethers of ethyl cellulose by reacting various chlorosilanes (chlorotri-methylsilane, chlorotriethylsilane, chloro-dimethylisopropylsilane, chlorodimethyl-*tert*-butylsilane, chlorodimethyl-*n*-octylsilane, chlo-rodimethylphenylsilane) with residual hydroxy groups of ethyl cellulose. These derivatives of ethyl cellulose displayed improved solu-bility in nonpolar solvents because of the substitution of hydroxy groups (Scheme 1.9) and were soluble in common organic solvents. The silylated derivatives were employed for membrane fabrication by casting the toluene solution onto a Petri dish

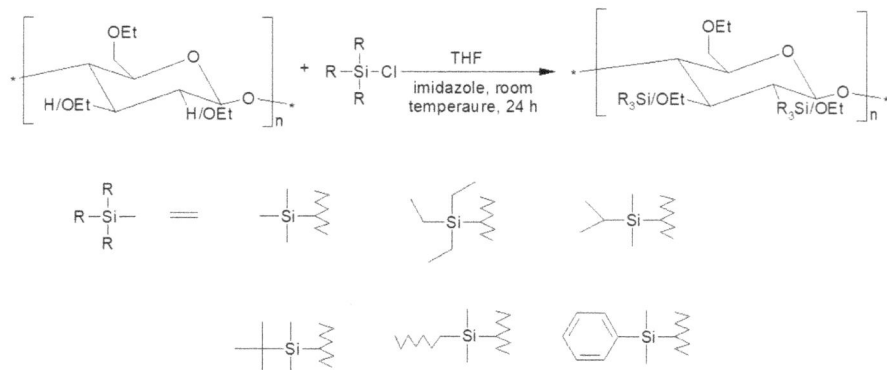

SCHEME 1.9   Silylation of ethyl cellulose with different chlorosilanes.

followed by slow evaporation. Copolymer films with CEC were homogeneous and transparent and had good adhesion to glass substrates.

Manjunath Kamath et al. made cellulose copolymers by crosslinking cyanoethylated cellulose with reactive monomers or prepolymers of polysiloxane (Scheme 1.10). Their study was caused by the properties of cyanoethylated cellulose (CEC)—a cellulose derivative, such as stability toward air-oxidation, solution processability, and water repellency. The reactive mixture was cast as films (cured at 125 °C). This polymer matrix, with reactive siloxane, has improved the thermal stability of copolymer product.

Liang et al. tested the biocidal efficacy against *Staphylococcus aureus and Escherichia coli* O157:

H7 of copolymers incorporating N-halamine siloxane and quaternary ammonium salt siloxane units coated on cotton swatches. The authors obtained different copolymers starting from a siloxane polymer (poly(3-chloropropylsiloxane) (PCPS) prepared from the monomer 3-chloropropyltriethoxysilane). They made a homopolymer (poly[3-(5,5-dimethylhydantoinylpropyl) siloxane] (PHS)) by reacting PCPS with the potassium salt of 5,5-dimethylhydantoin. The authors obtained a quaternary ammonium cation homopolymer, namely (poly[3-dimethyldodecylammoniumsiloxane chloride](PQS)), by reacting PCPS with dimethyldodecylamine using DMF as solvent at 100°C. The authors also synthesized two hydantoinyl/quaternary ammonium

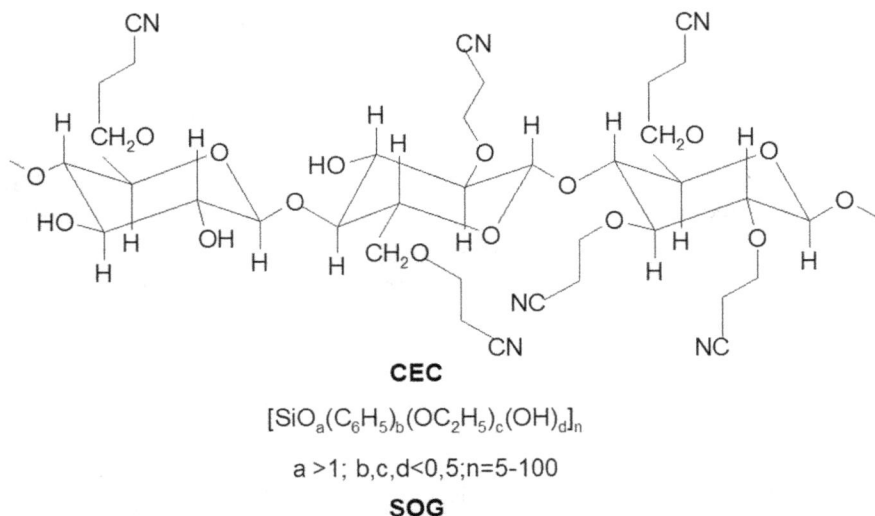

CEC

$[SiO_a(C_6H_5)_b(OC_2H_5)_c(OH)_d]_n$

a >1; b,c,d<0,5;n=5-100

SOG

SCHEME 1.10   Components of crosslinked CEC copolymer films.

cation siloxane copolymers: poly[3-(5,5-dimethylhydantoinylpropyl)siloxane-*co*-3-dimethyldodecylam-moniumpropylsiloxane chloride] (PHQS) using two different procedures: a one-stage and a two-stage method, with the salts 5,5-dimethy-lyhydantoin, and dimethyldodecyl-amine being added.

These types of copolymers are important for applications in which the aqueous media is preferred over organic solvents in the time of coating procedures. The polymers were coated on cotton swatches by soaking in baths with 0.15 mol L^{-1} of each compound dissolved in distilled water. For polymers with low solubility in water, a mixture ethanol/water was used. Then the coated swatches were mold and then chlorinated by soaking in the aqueous solution of NaOCl household bleach buffered to pH 7. These polymer-coated cotton swatches have proved to be stable at the action of water. They possessed biocidal efficacy.

Mohamed Hashem et al. used siloxane polymers with the purpose to obtain a cotton fabric with improved wrinkle-free and softness properties.

Carboxymethylated cotton (CMC) fabric was cationized with 3-chloro-2-hydroxypropyl trimethyl ammo-nium chloride (69%). After that, the ionically crosslinked cotton fabric was processed with an aqueous solution of amino-functional silicon microemul-sion softener (SiE—Scheme 1.11) (0%–20%) at pH 4 by pad-dry-cure technique (Scheme 1.12).

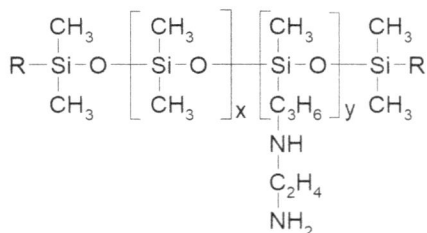

SCHEME 1.11   Structure of aminosilicone softener (SiE): R = OH, OCH$_3$, OCH$_2$CH$_3$.

CMC with negatively charged carboxymethyl groups and posi-tively charged protonated amino groups from SiE molecule crosslink forming a semi-interpenetrated network (semi-IPN). The formation of Si–O–Si–cellulose (Scheme 1.13) complex was confirmed by FTIR analysis. The cotton fabrics treated with SiE have higher surface smoothness and reduced protrusion

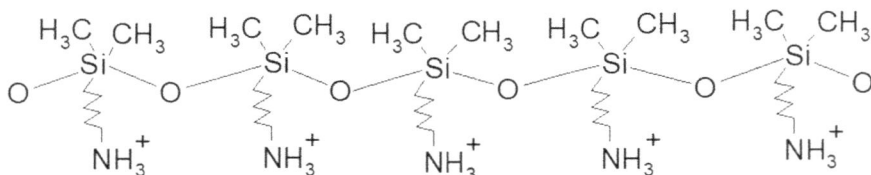

SCHEME 1.12   CMC fabric and protonated amino silicon softener molecule.

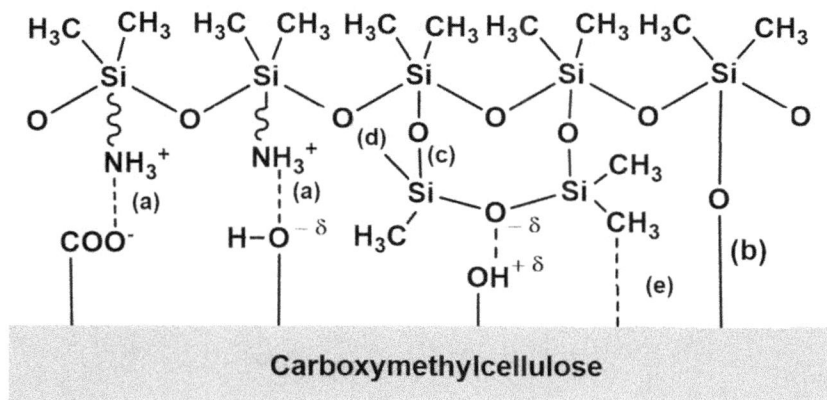

Carboxymethylcellulose

SCHEME 1.13 Schematic representation for semi-interpenetrated network (semi-IPN) made inside the cotton fibers: (a) ionic interaction; (b) covalent ether bond between cellulose and silicone polymer with amino groups; (c) covalent ether bonds between silicone polymer with amino groups molecules; (d) hydrogen bond; (e) van der Waals interaction.

of loose fibers, ditches, and grooves compared with the untreated ones. Such a structure determines durable wet and dry wrinkle resistance properties and softness. The measure of fabric wrinkle resistance and softness is dependent on factors such as crosslinking degree, intrachain/interchain crosslinks ratio, the concentration of amino silicon softener, pH, and state of cotton during crosslinking. The crosslinking of the cellulose has importance for possible applications. Also, crosslinks limit the swelling of the cellulose derivatives and protect the preexisting hydrogen bonds against the water destructive and dissolving action. Cellulose, as a polar polymer, has a different response to crosslinking toward the rubbery polymers. The most used crosslinking systems for cellulose were based on N-methylol chemistry. These systems have as disadvantages strength loss and the possibility to release formaldehyde, a known human carcinogen. This is why such a method is now no longer and formaldehyde-free crosslinking methods are searched.

Other studies involve Si–H end- or side-functionalized siloxanes as crosslinkers for cellulose derivatives. Cellulose acetate (CA) was reacted with different amounts of 1,1,3,3-tetramethyldisiloxane in presence of Karsedt's catalyst [3]. As solvent for the reaction was utilized dry acetone. The cellulose derivative was crosslinked as a result of dehydrocoupling reaction between Si–H and C–OH groups leading to the formation of Si–O–C bond (Scheme 1.14). The model reactions were controlled online by ^{1}H-NMR spectroscopy and hydrogen dosing.

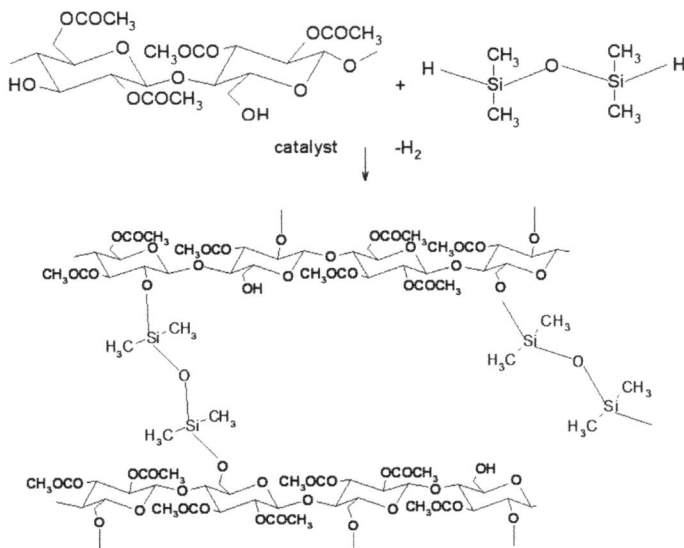

SCHEME 1.14   The crosslinking of cellulose acetate with siloxane (Stiubianu et al. 2010).

Different reactants ratios were used and correlated with the surface properties and mechanical properties of the film processed cellulose networks. CA was reacted in different ratios with poly[dimethyl(methyl-H) siloxane] (PMHS) containing 25% mole Si–H groups along the chain. In the presence of Karstedt catalyst, the dehydrocoupling reaction between Si–H and C–OH groups took place with the formation of the Si–O–C bond, proved by FTIR spectra (Scheme 1.15) resulting in a crosslinked cellulose derivative. The networks were studied using different techniques to find the morphology, thermal, dielectric, and surface properties acquired in dependence with the ratio between the two components. Cellulose with allyl functional groups was also modified with poly[dimethyl(methyl-H) siloxane] and 1,1,3,3-tetramethyldisiloxane (Stiubianu, 2012) by hydrosilylation reaction. Also cellulose acetate was modified with a, ω-bis(carboxy-propyl)oligosiloxane and 1,3-bis(carboxypropyl)tetramethyldisiloxane (Stiubianu, 2011) by solution polycondensation at room temperature. The reactions led to uniform films with hydrophobic character and reduced sorption of water vapors relative to initial unmodified cellulose films. Microcrystalline cellulose (MCC) was modified with poly(methylhydro) siloxane (PMHS) in a hexane as solvent (Xie, 2018). The resulted films possessed superhydrophobic character.

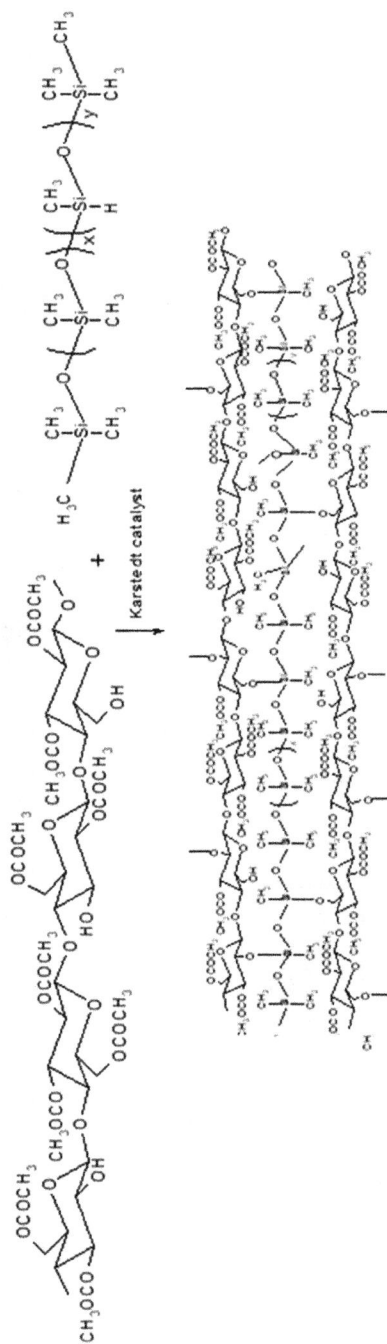

SCHEME 1.15 The crosslinking of CA by PMHS.

1.7 COMMON ROUTES FROM BIOMASS TO POROUS CARBONS

The transformation of biomass to porous carbons consists of two processes: carbonization and activation. An established carbonization process is realized in an inert atmosphere at high temperatures of 400–1000 °C. The hydrogen and oxygen atoms are released from the biomass to form H_2O, H_2, CH_4, CO gases, while the carbon atoms are condensed as solid residues with increased carbon content, these processes involving some complex reactions (e.g., dehydration, condensation, and isomerization). The yield and carbon content of the solid is dependent on some factors like carbonization temperature, heating rate, thermal stability of the biomass, resistance time, and chemical structure. An increased carbonization temperature and long resistance time form chars with a reduced yield but elevated carbon content. HTC is other manner that has been used in the conversion of biomass to valuable carbon materials. The treatment of biomass under HTC conditions, at moderate temperatures (<300 °C) and self-generated pressures, forms hydrochars with increased C/H and C/O ratios. The HTC process has several advantages: it operates under moderate temperatures, reducing the energy consumption; second, wet biomass precursors do not need drying and can be directly exposed to the HTC treatment (the hydrothermal process takes place in water). The changing of biomass by the HTC process yields to carbon-rich hydrochars. The carbonaceous materials can be complexed with new components like noble metal nanoparticles, magnetic nanoparticles, and electrochemically active species during the HTC process for obtaining functional nanocomposites with desired nanostructures and compositions with a variety of physiochemical properties and functionalities.

The HTC method not only offers an energy-saving and environmentally friendly approach for biomass carbonization but also contributes in a significant way to the development of carbon-based functional nanomaterials. The chars obtained are nonporous. Further activation (physically or chemically) of the chars at high temperatures is a required step for the preparation of highly porous carbon materials. The physical activation method uses oxidizing atmosphere like CO_2, air or steam, while chemical activation use activating agents like KOH, NaOH, $KHCO_3$, K_2CO_3, $ZnCl_2$ or H_3PO_4. Some moieties of the chars are inclined to be oxidized and dehydrated by the oxidizing gases or chemical reagents in the activation process, creating rich micro- and meso-pores.

The two activation methods make porous carbons with large differences in porosity. Usually, physical activation processes create porous carbons with moderate surface areas (<1000 m^2 g^{-1}) and narrow micropores (useful for, e.g., CO_2/N_2 and CO_2/CH_4 separation). The chemical activation methods can increase the surface area (up to >3000 m^2 g^{-1}) and pore volume of the porous carbons making it useful for applications in gas storage, water treatment, electrochemical supercapacitors, etc. Chemical activation techniques permit the conversion of biomass to highly porous carbons in a one-step carbonization/activation process, facilitating the production process without involving some procedures. The chemical activation methods suffer from important environmental disadvantages due to their dependence on large amounts of corrosive activating agents.

1.8 SYNTHESIS OF POROUS CARBONS FROM WOOD-BASED BIOPOLYMERS

Pyrolysis and hydrothermal treatment have been used in the carbonization of wood-based biopolymers. The carbonization process and the obtained carbonaceous solid have been well studied with the assistance of various analytical techniques. The pyrolysis behavior and analyze the gaseous products can be studied using the thermogravimetry (TG) and gas chromatography (GC)/mass spectrometry (MS). Yang et al. have conducted TG analyses for cellulose, hemicellulose, and lignin under N$_2$ atmosphere. These studies showed that hemicellulose having the lowest thermal stability among the three biopolymers due to its amorphous structure and low molecular weight, starts to decompose at 220 °C. Cellulose decomposes between 315 °C and 390 °C. The high thermal stability of cellulose can be assigned to its crystalline structures, strong hydrogen bonding between cellulose chains, and high molecular weight. Lignin has a wide and flat TG curve in the range of 250–900 °C. The slow pyrolysis rate and high thermal stability of lignin can be justified by its heavily crosslinked and aromatic-rich network. The solid yield of hemicellulose, cellulose, and lignin at a temperature of 900 °C was ~20%, 7%, and 40%, respectively.

Kwon et al. have studied the effects of gas atmosphere and heating rate on the pyrolytic carbonization of cellulose. They denoted that the use of different atmosphere (N$_2$ and CO$_2$) does not influence the solid yield at the same heating rate. The use of CO$_2$ for the carbonization improved the generation of gaseous products including H$_2$, CO, and CH$_4$ because CO$_2$ could promote the thermal cracking behaviors, indicating that the use of CO_2 as a reaction medium gave high thermal efficiency for biomass conversion. An increase of the heating rate from 10 °C min^{-1} to 500 °C min^{-1} decreased the solid yield from 17% to 8% at 900 °C, being coherent with the studies of other scientists (a high heating rate leads to a low solid yield for biomass conversion). Bommier et al. obtained a series of porous carbons with high surface areas by carbonization of filter paper under Ar atmosphere. Based on TG–MS studies that verified the carbonization process, the scientists proposed a self-activation mechanism (the obtained carbon material was in situ activated by the gaseous products from the decomposition of cellulose).

The detailed investigation of the thermal degradation process described in these studies permits us to understand the conversion mechanisms and also to optimize the carbonization parameters (temperature, heating rate, resistance time, and atmosphere for the synthesis of carbon materials with desired properties). The HTC process has the advantage of operating at moderate temperatures, forming hydrochars with a rich variety of organic functional groups, allowing for the study of their molecular details by various spectroscopic and electron microscopic tools with the purpose to investigate the details of the carbonization mechanism.

Falco et al. investigated the effects of processing temperature and time on the morphology and chemical structure of the HTC hydrochars generated from cellulose. X-ray diffraction, solid-state 13C nuclear magnetic resonance (NMR) spectroscopy, and scanning electron microscopy studies of the hydrochars showed that the fibrous and crystalline structure of the cellulose remained unaffected by hydrothermal treatment at low operating temperatures (<160 °C). With the increase of the temperature, the fibrous structure started to decompose with the formation of spherical particles. For the conversion of cellulose to hydrochars under HTC conditions, two possible ways were proposed:

1. One path involved (a) hydrolysis of cellulose into glucose, (b) dehydration of glucose into hydroxymethylfurfural (HMF), (c) polymerization and polycondensation of HMF into polyfuranic chains, (d) intramolecular condensation, dehydration, and decarboxylation of polyfuranic chains into aromatic-rich carbon network via reactions.

2. The other one was the direct aromatization of cellulose that forms aromatic carbon networks at higher HTC processing temperatures (200–280 °C), similar to a standard pyrolysis process. Other routes have been proposed for HTC conversion of

various biopolymers by means of infrared, Raman, and X-ray photoelectron spectroscopy. These studies support the understanding of chemical reactions occurring in the HTC process, making it possible to tailor the nanostructure, porosity, and functionality of carbon materials.

Deng et al. applied TG–MS tools to study the process of carbonization/activation of cellulose, hemicellulose, and lignin in the presence of an activating agent of $KHCO_3$. The TG analysis showed that $KHCO_3$ decomposed at 200 °C and $KHCO_3$ accelerated the pyrolysis of biopolymers. At temperatures higher than 400 °C, the byproducts of $KHCO_3$ catalyzed the activation process with the formation of H_2 and CH_4. This study proposed a possible structure evolution of the biopolymers in the carbonization/activation process. Multiple studies have proved that cellulose can be transformed into porous carbons in a single step of carbonization under inert gas without the use of any oxidizing gases or activating agents. Porous carbon was studied for application in batteries, supercapacitors and catalysis, due to its large specific area. Japanese cedar (Cryptomeria japonica) milled into a powder of 32–63 μm was flash heated (800 °C, heating speed 50 °C/s) and subsequently kept for low-temperature heat treatment (380 °C). The resulted material was a porous carbon with shape-controlled with high porosity of over 80% and high Brunauer–Emmett–Teller surface area of 670 m²/g (Kurosaki, 2008). Pd nanoparticles were deposited in the pores, for catalytic nichel plating. Nanoparticles of MnO were deposited in the pores of porous carbon in order to form a MnO/C nanocomposite material as the anode of Li-ion battery (Yang, 2014). The battery built with this anode had discharge capacity of 952 mAh $\times$ g^{-1} at current density of 0.1 A $\times$ g^{-1} with stable cycling performance over 100 charge/discharge cycles. Doping with nitrogen a composite of porous carbon and PANI nanoparticles leads to materials with capacitance of 347 F $\times$ g^{-1} at 2 A $\times$ g^{-1} and energy density of 44.4 Wh $\times$ kg^{-1} at 922 W $\times$ kg^{-1} (Yu, 2015). This work opened the way for production of supercapacitor materials from low-cost, environmentally friendly, and renewable wood wastes as raw materials.

Other researchers have prepared hierarchical porous carbon as electrode material for fuel cells, batteries, and supercapacitors. The porous carbon was prepared from yellow pine wood, with the first step at low temperature (240 °C), followed by high temperature treatment (900 °C) (Luo, 2017). The hierarchical porous carbon was tested as cathode for a Li-O_2 battery. 3D graphene-like porous carbon was prepared from precursor lignin separated and purified from alkaline pulp black liquor and pretreated using rapid freeze-dry technique (Du, 2018). This method provides an economical and environment friendly approach for graphene application in large scale.

Enzyme activity was explored for hydrolysis of the partial cellulose in bulk raw wood to form a vast network of nanopores, and then dope nitrogen atoms onto the carbon skeletons during the subsequent pyrolysis process. The cellulose-digested, carbonized wood plates are mechanically strong, have high conductivity, and contain natural ion-transport channels (Peng, 2019). Therefore, the material can be employed directly as metal-free electrode without carbon paper, polymer binders, or carbon black. The resulting carbons exhibit excellent catalytic activity with respect to the oxygen reduction and oxygen evolution reactions as metal-free cathodes in zinc-air batteries, they result in a specific capacity of 801 mAh $\times$ g^{-1} and an energy density of 955 Wh $\times$ kg^{-1} with the long-term stability of the batteries being as high as 110 h. This work paves the way for the ready conversion of abundant biomass into high-value engineering products for energy-related applications.

A new concept for preparation of porous carbon electrodes with a large specific surface area uses wood scraps activated with CO_2, then soaked into KOH solution to create new pores, and finally modified with HNO_3 (Zhang, 2019). The wood carbon slices have specific surface area of 703.5 m^2 $\times$ g^{-1}. The single electrode material has high specific capacitance of 285.6 F $\times$ g^{-1} and energy density of 38.0 mWh $\times$ cm^{-3}, more than twice the capacitance of the supercapacitor based on biomass carbon materials modified by diluted HNO_3.

1.9 CERAMIC MATERIALS DERIVED FROM WOOD BIOPOLYMERS AND SILICONE

Ceramics are inorganic materials, nonmetallic, processed, or consolidated at high temperatures. Classes of materials usually regarded as ceramics are oxides, nitrides, borides, carbides, silicides, and sulfides. Intermetallic compounds such as aluminates and beryllates are also considered ceramic materials, the same as phosphides, antimonites, and arsenates.

Wood is a naturally grown composite material of the complex hierarchical cellular structure. Both hardwoods (deciduous wood-botanically classified as dicotlyedonous angiosperms) and softwoods (coniferous wood or gymnosperms) are formed of elongated tubular cells (sclerenchyma cells) aligned with the axis of the tree trunk. Different scientists have studied the design of new ceramic structures by imitating cell tissue anatomy of native ligno-cellulose structures such as wood, fiber, or leaf area. The predominant chemical constituents of wood are C (50% wt), O (44% wt), H (6% wt), and other trace elements (1% wt). Wood cell structure anisotropy can be used as a template to make hier-archical cellular ceramic showing micro-, meso-, and nanostructures pseudomorph with the initial porous tissue skeleton. These structures can have dimensions varying from nanometer (cell wall fibrils) to milli-meters (growth rings patterns). For example, in order to generate hydro-phobic, self-cleaning plant surfaces (Lotus effect), which possess antiadhesive behavior caused by different cuticule microstructures were studied. Other example is the formation of fibers of Al_2O_3, TiO_2, SiC, and Si_3N_4 from natural fibers, such as sisal, jute, hemp, and the formation of thin yarn of SiC of rice husks and coconuts. Earlier studies on converting wood into ceramic were focused on the liquid infiltra-tion of pyrolysed carbon matrix with sols of TEOS at low temperature or silicon melts at high temperatures. The infiltrate was converted to SiC by pyrolysis in an inert atmosphere (for TEOS) or by reaction with carbon (silicon). Different species of trees such as oak (*Quercus robur*), maple (*Acer pseudoplatanus*), beech

(*Fagus sylvatica*), ebony (*Diospyros celeb*), balsa (*Ochroma pyramid*), and pin (*Pinus sylvestris*) were modi-fied to isomorphic cellular silicon carbide ceramic. Some science people developed the processing scheme for the conversion of wood tissue into a microcellular C/SiOC ceramic composite material in one single high-temperature step. The porous structure of the native wood tissue was used for infiltration with a low viscous preceramic polymer and subsequently pyrolysed by annealing in an inert atmosphere. Polymeric ceramic precursors such as silicone resins (e.g., poly-silsesquioxanes—$RSiO_{1.5}$) can be converted into an amorphous silicon oxycarbide (SiOC-phase) by thermal treatment in an inert atmosphere at temperatures above 600 °C. The preceramic polymers are materials for near-net shape manufacturing of complex parts.

Other authors used polymethylhy-drosiloxane (PMHS) of low viscosity and high ceramic yield as preceramic polymer, beech, and pinewood as biomorphous ceramic template. During infiltration and curing of PMHS, the SiH-functional groups of the polymeric precursor reacted with the OH-groups of the lignin and cellulose from the native wood in a dehydrogenative condensation reaction to form SiO–ether bonds. Properties for loading in an axial direction of wood pseudomorph SiC ceramics reach to higher values for all mechanical properties (strength, elastic modulus, and failure strain, fracture patterns) when compared with values for loading on radial and tangential directions. This is caused by the uniaxial pore channel orientation in the starting wood material. The researchers are interested in high temperature and corrosion-resistant biomorph ceramics, which have a structure similar to that of native lignocellulose and consist of carbides, nitrides, or oxides, because native tissue have unique structural characteristics combined in cell anatomy: hierarchy, selectivity, and anisotropy; available in an infinite variety of structures; natural plant-growth can be adjusted through chemical and physical methods to develop the optimal function of body structures such as pore size distribution; it is assumed that the native tissue is in a state of optimal mechanical balance; preprocessing of lignocellulose materials by delignification and surface treatments of cellulose fibers permits a variety of macrostructure of cellulosic fibers to be exploited for the construction of lightweight ceramic structures.

Conversion of lignocellulose materials in ceramic products is starting from native plant tissue or from preprocessed technical structures. Direct reproduction of the tissue structure must solve the problem of macro-, meso-, and micro-scale heterogeneity of natural biostructures and that is the reason why products such as paper, boards, and wooden matches are used, as they can offer precursors which are homogeneous at the macroscale, wood fibers are oriented perpendicular to the direction of pressing or filtering applied during processing.

For the transformation of native biopolymer materials to oxide and nonoxide ceramics, the basic techniques used include:

1. Pyrolytic decomposition resulting in a porous carbon replica which is further reacted to obtain carbide phases or may be infiltrated with sols or reactive salts which can then be oxidized to obtain oxide products.

2. Infiltration of preprocessed technical or natural products with gaseous or liquid organometallic or metaloorganic chemical precursors and consecutive oxidation for removing the free carbon phase. The carbon template can be fabricated in the desired form before conversion to composite ceramic.

1.9.1 FORMATION OF CARBON MATRIX OF CELLULOSE

The mechanisms involved in the transformation of cellulose to carbon (graphite) are: (1) desorption of adsorbed water until 150 °C, (2) removing bound water from pulp between 150 °C and 240 °C, (3) breaking chains and breaking the C–O and C–C bonds from pyranose ring, obtaining: H_2O, CO, and CO_2 between 240 °C and 400 °C, (4) aromatization with the formation of layers of graphite at temperatures above 400 °C. Wood heating in an inert atmosphere will make the same products that are obtained at separate pyrolysis of the three major components. Pyrolysis takes place in several steps: first hemicelluloses break at 200–260 °C, followed by cellulose at 240–350 °C, and lignin at 280–500 °C.

Converting carbon biopolymers to carbon gives rise to a special anisotropic structure of the carbon phase with a low degree of crystalline order—turbostatic carbon, a very fine-grained carbon powder with the carbon atoms aligned in layers.

1.9.2 TRANSFORMING CARBON CERAMIC MATRIX

Infiltration and reaction with liquid or gas Si and SiO can form wood tissue pseudomorph β-SiC:

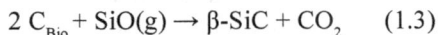

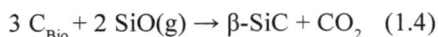

$$C_{Bio} + Si(l) \rightarrow \beta\text{-SiC} \qquad (1.1)$$
$$C_{Bio} + Si(g) \rightarrow \beta\text{-SiC} \qquad (1.2)$$
$$2\,C_{Bio} + SiO(g) \rightarrow \beta\text{-SiC} + CO_2 \qquad (1.3)$$
$$3\,C_{Bio} + 2\,SiO(g) \rightarrow \beta\text{-SiC} + CO_2 \qquad (1.4)$$

- reaction (1.1) leads to ceramic composites SiC/Si with a content of Si in cellular pores less than 30 μm diameter;
- SiC formation with gaseous reactants according to Equations (1.2)–(1.4) gives a single-phase SiC material.

1.9.3 DIRECT INFILTRATION AND CONVERSION REACTION

Wood cell wall matrix has an important contribution to the sorption capacity of cellulose wall. Low-temperature processing avoids the formation of a carbon matrix seen at pyrolysis treatment. That involves native tissue infiltration with monomer, oligomer, or polymer precursor compounds based on $Si(OR)_4$, $R'_x Si(OR)_{4-x}$, or $[R_2SiX]_n$, where X = O (siloxane), NH (silazane), CH_2 (carbosilane). A prerogative of this method is the fact that because of the chemical coupling with hydroxyl groups of glucose units, weight loss, and contraction that happen during heat treatment can be reduced. After pyrolysis, the final oxidation in the air goes to a material containing silicon as Si–O(–X) with high porosity.

The reaction of a carbon network to form a carbon phase can be useful for the formation of composite materials with interpenetrated networks (IPNS). Dry wood surface impregnation with solutions of low viscosity preceramic polymer and then the conversion to a ceramic residue by local heating of the surface leads to new types of wood-ceramic materials: the bulk volume of timber stays as native wood and ceramic surface provide significantly improved wear resistance and strength. Cellular ceramic material having homogeneous or heterogeneous anisotropic structures of pores is good in the field of filter for high-temperature exhaust gas and supports for catalysts in environmental technologies and energy, bioinert and corrosion-resistant immobilization carriers for living cells, microbes, and enzymes in medicine and biotechnology. Cellular ceramics from heterogeneous tissue are used to design multimedia cell biocatalytic support structures. Multimodal size structures of the pore can be useful in the food industry, fermentation, or wastewater treatment. It is possible to obtain an advanced reactor substrate with large tube pores for medium transport and small pores where the biocatalytically active organisms are fixed. Immobilization of microbes or enzymes requires accesible pore sizes from 10 to 500 μm. Media for cells with an open structure of pores can become interesting for ex vivo molecular reactor systems with biological cells and artificial organs in medicine.

Overall, silicon carbide ceramic with wood-like microstructure can be prepared from various wood types (oak, pine, poplar, cereal straw) can be prepared either by carbothermal reduction process in argon atmosphere where biomass is impregnated with silica and treated at high temperature (1400–1600 °C) or by siliconization of carbonized wood in silicon vapor. The SiC materials are porous with large surface area (>10 m^2/g) and SiC whiskers of 50–400 nm in diameter and 5–20 mm in length within the tubular pores (Qian, 2004; Shin, 2005; Yushin, 2006; Karyasa 2016).

Ceramics with the homogeneous porous structure are good candidates for filters; catalysts support layers and aerator structures. Filter structures require well-defined single-mode pore size to assure a certain size required for good filter selectivity. Porous devices with permeable beams can be used as irrigators in industrial agriculture. Multifunctional ceramic membrane reactors are other potential uses. Perm- and non-perm-selective catalytically active membranes surrounded by a bed of catalyst particles are good for separation of the gas, O_2–N_2 separation for production of pure O_2. Parenchymal tissue-derived ceramics can be used as supports for catalysts and devices reservoir fluid/gas. Due to irregular star-like anatomy, turbulent flow is favored in these structures, leading to a longer reaction time for catalysts deposited on the surface. This is of interest for devices like miniature reactors for using them in chemical reaction technology.

1.10   CARBON FIBERS

The literature provides methods for producing carbon fibers starting from cellulose with improved mechanical strength, which involves the use of rayon yarn coated with a polysiloxane and a stage of rapid pyrolysis. The effects of this additive prevent the collapse of mechanical strength of rayon yarn and improve the tensile strength after the formation of carbon residue. Toward the behavior of polysiloxanes at high temperature, many studies treat their thermal degradation and it is well-known that the mechanism of redistribution takes place around the silicon atoms.

Birot et al. studied the interactions between cellulose and polysiloxane at high temperatures during the formation of a carbon network. Because of the low amount of polysiloxane used in the process and low yield of carbon a similar chemical system was used: cellulose was substituted with its structural unit cellobiose: $C_{12}H_{22}O_{11}$-(4-O-(β-D-glucopyranosyl)-β-D-glucopyranose, Scheme 1.16), presuming that chemical interaction has to take place in an early stage of thermal conversion as cellulose leads to chain breaks

and the formation of glucopyranose oligomers units when heating at temperatures above 240 °C. In an analogous way, an MQ resin (M, D, T, and Q = mono-, di-, tri-, and quaternary coordination of oxygen around silicon in silicones: M = Me_3SiO, D = Me_2SiO_2, T = $MeSiO_3$, Q = SiO_4 units) was employed as organosilicone compound because researchers found before that this resin has an important role in improving the mechanical strength of carbon fibers with the added benefit of assuring a high efficiency in residue from pyrolysis (65% at 1000 °C), allowing structural characterization. A resin T was obtained by hydrolysis of methyltrichlorosilane. Another resin DQ was made by cohydrolysis of dimethyldichlorosilane and tetrachlorosilane. The similar TGA curves of cellobiose and cellulose fibers in an inert gas atmosphere led the researchers to the conclusion that cellobiose is an accessible model of cellulose. The experimental residual masses evidence of cellobiose/MQ resin between 485 °C and 655 °C were higher than calculated values, implying that the chemical binding between the residue of glucopyranose degradation and

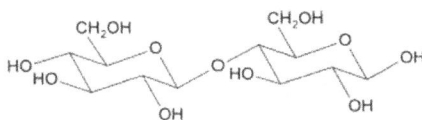

SCHEME 1.16   Structure of cellobiose.

siloxane fragments took place in this temperature range. The pyrolysis residue of the resin-treated at 1200 °C seems to be formed of T and Q units (Schemes 1.17 and 1.18). For cellobiose/MQ resin mixture, the resin redistribution reactions were delayed due to the presence of cellobiose, as T and D sites showed only at temperatures above 555 °C. M units are involved in the redistribution reactions.

Since $R_3Si–OH$ functions cannot survive at elevated temperatures in the bulk of residue, this leads to the conclusion that new more thermally stable places were produced progressively, with close chemical shifts. They may happen in ways that imply the condensation of initial T^{OH} groups with remaining OH groups in the cellobiose pyrolysis tar and also from breaking the Si–O bonds by the remaining alcohol groups (Scheme 1.19). Siloxane bonds are undergoing breaking reactions with alcohols and phenols at temperatures above 200 °C. In the

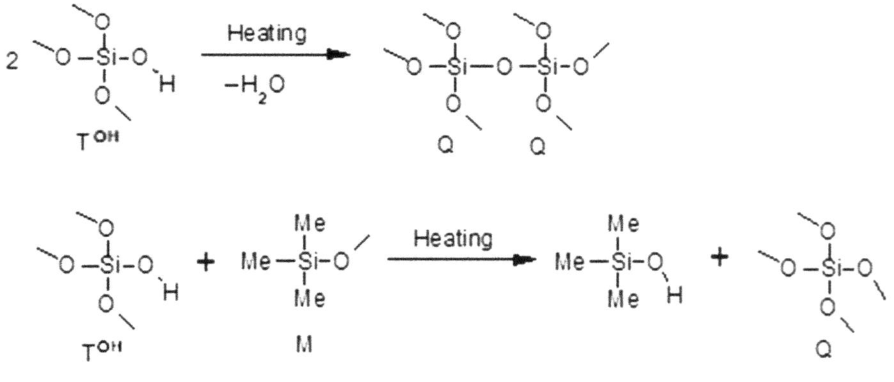

SCHEME 1.17   Formation of Q fragments.

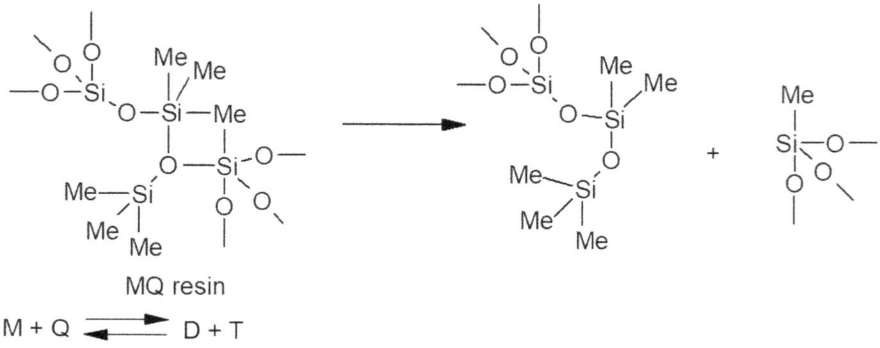

MQ resin

$$M + Q \rightleftharpoons D + T$$

SCHEME 1.18   The formation of D and T fragments.

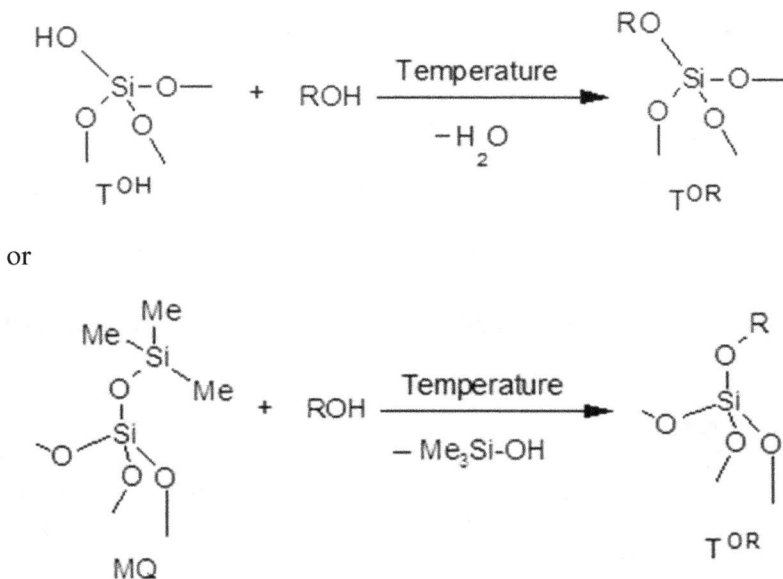

SCHEME 1.19   The formation of T^{OR} fragments.

pyrolized (temperatures above 550 °C) samples, M^{OR} was formed from the reaction of type D sites and the remaining OH groups of carbon tar pyrolysis (Scheme 1.20).

1.11   MEMBRANES AND IPNS

A permselective membrane is considered an ion-exchange material that permits ions of a specific electrical

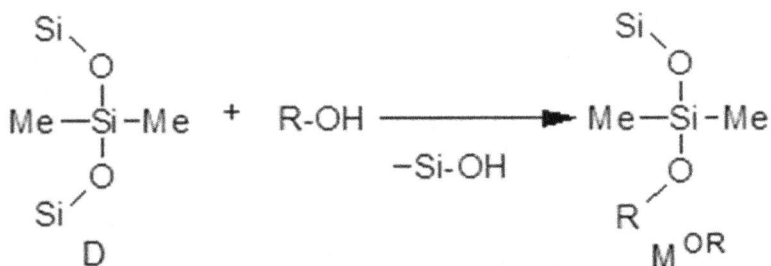

SCHEME 1.20   The formation of M^{OR} fragments.

sign to enter and pass through it. The performance of separation membranes is evaluated with oxygen permeability coefficient, PO_2 ($\times 10^{-10}$ cm^3 (STP) cm^3 (STP) cm^{-2} s^{-1}cm Hg^{-1} (= barrier), oxygen solubility coefficient, $(SO_2$: $\times10^{-3}$ cm^3 (STP) cm^3 (STP) cm^{-3}cm Hg^{-1}), diffusion coefficient of oxygen $(DO_2$: $\times10^{-6}$ cm^{2}s^{-1}), oxygen separation factor ($\alpha = PO_2/PN_2$). Target values for a good separation membrane are PO_2 >100 and α >3.0.

Polydimethylsiloxane (PDMS) has the highest permeability coefficient for oxygen PO_2 but the lowest permselectivity α for oxygen. PDMS is a promising material for the separation of organic compounds from water. Li et al. have made composite membranes with polydimethylsiloxane (PDMS) on the support of CA with the prewetting method for separation of methanol, ethanol, n-propanol, and acetone from water. The study of operating parameters on pervaporation performance led researchers to the conclusion that PDMS–CA composite membranes have a permeation flow higher than that of membranes with or without the PDMS support reported in the literature and have good selectivity for ethanol. Membranes were prepared by grafting of vinyltrimethoxysilane (VTMS) to −OH groups of cellulose acetate (CA) and the subsequent condensation of hydrolyzed silane methoxy groups formed a polymer network (Achoundong, 2013). The permeabilities for pure carbon dioxide and hydrogen sulfide were more than an order of magnitude higher than the starting cellulose acetate polymer. The membrane

modified with silicone had lower glass transition temperature, lower crystallinity, and higher flexibility than neat CA due to the vinyl substituent provided by VTMS. This helps reducing brittleness of the material, which would provide an excellent material for membranes with asymmetric structure.

A new method for preparation of membranes for organic solvent nanofiltration and ethanol pervaporation is based on the introduction of trimethylsilyl groups in the structure of cellulose, followed by preparation of blend membranes with polydimethylsiloxane. Then the hydrolysis of trimethylsilyl groups leads to emergence of nanopores for separation and filtration (Puspasar, 2018). The blending of PDMS with cellulose results in a 100% increase of the separation factor relative to pure PDMS.

An interpenetrated network is a material made of a combination of two or more polymer networks synthesized together. Interpenetration of the two crosslinked polymers goes to forced miscibility compared with normal blends and the obtained materials have good dimensional stability. The general objective of such associations is to find polymer materials with better mechanical properties, increased resistance to degradation, and possibly a favorable combination of the properties of their constituents. Semi-IPNS are composed of a linear polymer trapped in a different or similar polymer network. Fichet et al. tried to attach silicones with a network of thermoplastic CA butyrate (CAB) in interpenetrated network architecture for improving the mechanical properties of the silicone network. The IPNS made of CAB and telechelelic α,ω-divinyl-polydimethylsiloxane through *one pot-one shot* process was used in which all components were mixed together and networks were formed at the same time by independent mechanisms of reaction. The PDMS network is synthesized from the di-vinyl-PDMS and trimethylolpropane tris(3-mercaptopropionate) as crosslinker by the radical thiol-ene process. CAB network is crosslinked through its OH groups with a pluri-isocyanate Desmodur N3300. The alcohol-isocyanate reaction is catalyzed by dibutyltindilaurate (DBTDL) leading to urethane crosslinks. The reagents used are capable to react in different ways—some reactions conducting to the formation of desired IPNS:

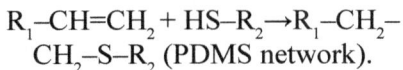

$$R_3-OH + R_4-NCO \rightarrow R_3-O-CO-NH-R_4 \text{ (CAB network)}$$

$$R_1-CH=CH_2 + HS-R_2 \rightarrow R_1-CH_2-CH_2-S-R_2 \text{ (PDMS network)}.$$

Also, different reactions can take place to different degrees and will interfere with the above reactions, disturbing the formation of the networks:

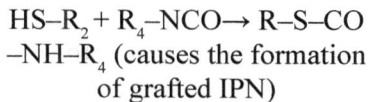

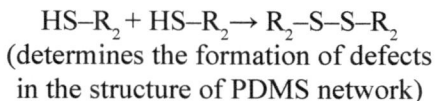

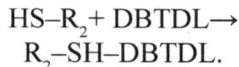

$$HS-R_2 + R_4-NCO \rightarrow R-S-CO-NH-R_4 \text{ (causes the formation of grafted IPN)}$$

$$HS-R_2 + HS-R_2 \rightarrow R_2-S-S-R_2$$ (determines the formation of defects in the structure of PDMS network)

$$HS-R_2 + DBTDL \rightarrow R_2-SH-DBTDL.$$

The final morphology of the networks resulted from an overlay of the relative curing speeds of the two networks and the speed of phase separation. The authors investigated the kinetics of the PDMS network and the IPN formation by middle and near IR spectroscopy (by the disappearance of the final link C=C (6136 cm^{-1}) and of crosslinker H–S (2256 cm^{-1}) groups). Two types of PDMS-CAB semi-IPNs were made: one where CAB was not crosslinked and one where PDMS was not crosslinked. In the case of semi-IPNs, translucent to opaque materials were prepared. For all semi-IPNs, the analysis demonstrated the noncrosslinked component presented phase separation.

1.12 CONCLUSIONS AND PERSPECTIVES

This chapter relieves the utilization of the wood-based biopolymers cellulose, hemicellulose, and lignin as start materials together with silicone-based compounds for preparing sustainable and efficient porous carbons as CO_2 sorbents. Such porous carbons can be made at a large scale at a low cost for practical industrial applications thanks to the abundance and low price of these biopolymers. The advantages of using such durable porous carbons for industrial CO_2 capture include high CO_2 adsorption capacity, high physiochemical stability, easy regeneration, fast adsorption/desorption rate, low operation cost, and low manufacturing cost.

The combination of lignocelluloses with polysiloxanes leads to materials with interesting combinations of properties. The silanes are the best coupling agents in lignocellulose/thermoplastic systems. Crosslinked structures based on silanes and cellulose are waterproof and fireproof materials that can be used in household and commercial applications. Cellulose derivatives modified with preformed siloxanes have found applications in the medical field. There have been important steps in the preparation of ceramic materials based on wood and siloxane, in preparation of carbon fibers, and in preparation of IPNS based on cellulose derivatives and polydimethylsiloxane. Lignin, an important component of biomass, was used as a filler for siloxanes. Studies have shown that by choosing the appropriate preparation conditions, it is possible to obtain materials with clearly defined characteristics. All these achievements open the way

for new research in the field of hybrid siloxane-lignocellulosic materials with applications in the capture and storage of CO_2. The wood-based biopolymers can be used as active elements in obtaining new green silicone composites.

KEYWORDS

- siloxane
- lignin
- cellulose
- CO_2 capture

REFERENCES

1. Bioenergy with Carbon Capture and Sequestration, (Chapter 4). In: Negative Emissions Technologies and Reliable Sequestration: A Research Agenda, 2019, https://www.nap.edu/read/25259/chapter /6

2. Xu, C.; Stromme, M.; Sustainable porous carbon materials derived from wood-based biopolymers for CO_2 capture. *Nanomaterials* 2019, 9, 103, doi:10.3390/nano9010103.

3. Stiubianu, G.; Nicolescu, A.; Nistor, A.; Cazacu, M.; Varganici, C.; Simionescu, B.; Chemical modification of cellulose acetate by allylation and crosslinking with siloxane derivatives. *Polymer International* 2012, DOI: 10.1002/pi.4189.

4. Stiubianu, G.; New materials developed with lignocellulose and siloxane derivatives, (Chapter 5). In: Recent Developments in Silicone-Based Materials, Editor: Cazacu, M., 2010, Nova Science

Publishers, Incorporated: Hauppauge, NY, ISBN 978-1-61668-624-6.

5. Gunnarsson, M.; Theliander, H.; Hasani, M.; Chemisorption of air CO_2 on cellulose: an overlooked feature of the cellulose/NaOH(aq) dissolution system. *Cellulose* 2017, 24, 2427–2436, DOI: 10.1007/s10570-017-1288-8.

6. Bernard, F.L.; Rodrigues, D.M.; Polesso, B.B.; Donato, A.J.; Seferin, M.; Chaban, V.V.; Vecchia, F.D.; Einloft, S.; New cellulose based ionic compounds as low-cost sorbents for CO_2 capture. *Fuel Processing Technology* 2016, 149, 131–138, http://dx.doi.org/10.1016/j.fuproc.2016.04.014.

7. Garcia-Gutierrez, P.; Cuellar-Franca, R.M.; Reed, D.; Dowson, G.; Styring, P.; Azapagic, A.; Environmental sustainability of cellulose-supported solid ionic liquids for CO_2 capture. *Green Chemistry* 2019, DOI: 10.1039/c9gc00732f.

8. Reed, D.G.; Dowson, G.R.M.; Styring, P.; Cellulose-supported ionic liquids for low-cost pressure swing CO_2 capture. *Frontiers in Energy Research* 2017, 5, 13, doi: 10.3389/fenrg.2017.00013.

9. Florin, N.H.; Harris, A.T.; Hydrogen production from biomass coupled with carbon dioxide capture: the implications of thermodynamic equilibrium. *International Journal of Hydrogen Energy* 2007, 32, 4119–4134, DOI: 10.1016/j.ijhydene.2007.06.016.

10. Zhuo, H.; Hu, Y.; Tong, X.; Zhong, L.; Peng, X.; Sun, R.; Sustainable hierarchical porous carbon aerogel from cellulose for high-performance supercapacitor and CO_2 capture. *Industrial Crops and Products* 2016, 87, 229–235, http://dx.doi.org/10.1016/j.indcrop.2016.04.041.

11. Gebald, C.; Wurzbacher, J.A.; Tingaut, P.; Zimmermann, T.; Steinfeld, A.; Amine-based nanofibrillated cellulose as adsorbent for CO_2 capture. *Environmental Science & Technology* 2011,

45, 9101–9108, dx.doi.org/10.1021/es202223p.

12. Bernard, F.L.; Rodrigues, D.M.; Polesso, B.B.; Chaban, V.V.; Seferin, M.; Vecchia, F.D., Einloft, S.; Development of inexpensive cellulose-based sorbents for carbon dioxide. *Brazilian Journal of Chemical Engineering* 2019, 36(1), 511–521, dx.doi.org/10.1590/0104-6632.20190361s20170182.

13. Venturi, D.; Chrysanthou, A.; Dhuiege, B.; Missoum, K.; Baschetti, M.G.; Arginine/nanocellulose membranes for carbon capture applications. *Nanomaterials* 2019, 9, 877, DOI:10.3390/nano9060877.

14. Bargan, A.; Cazacu, M.; Aminosilicones as active compounds in the detection and capture of CO2 from the Environment. In: Smart Materials, Integrated Design, Engineering Approaches, and Potential Applications, A. Filimon, Ed., 2018, Apple Academic Press: Palm Bay, FL, USA.

15. Perry, R.J.; Aminosilicone systems for post-combustion CO_2 capture. In: Absorption-based Post-Combustion Capture of Carbon Dioxide, 2016, Elsevier: Amsterdam, 121–144.

16. Feng, X.; Simpson, A.J.; Wilson, K.P.; Williams, D.D.; Simpson, M.; Increased cuticular carbon sequestration and lignin oxidation in response to soil warming. Letters 2008, 1, 836, Doi: 10.1038/ngeo361.

17. Meng, Q.B.; Weber, J.; Lignin-based microporous materials as selective adsorbents for carbon dioxide separation. *ChemSusChem* 2014, 1–8, DOI: 10.1002/cssc.201402879.

18. Demir, M.; Tessema, T.D.; Farghaly, A.A.; Nyankson, E.; Saraswat, S.K.; Aksoy, B.; Islamoglu, T.; Collinson, M.M.; El-Kaderi, H.M.; Gupta, R.B.; Lignin-derived heteroatom-doped porous carbons for supercapacitor and CO_2 capture applications. *International*

Journal of Energy Research 2018, 42, 2686–2700, DOI: 10.1002/er.4058.

19. Saha, D.; Van Bramer, S.E.; Orkoulas, G.; Ho, H.C.; CO_2 capture in lignin-derived and nitrogen-doped hierarchical porous carbons. *Carbon* 2017, 121, 257–266, http://dx.doi.org/10.1016/j.carbon.2017.05.088.

20. Fierro, C.M.; Gorka, J.; Zazo, J.A.; Rodriguez, J.J.; Ludwinowicz, J.; Jaroniec, M.; Colloidal templating synthesis and adsorption characteristics of microporous-mesoporous carbons from Kraft lignin. *Carbon* 2013, 62, 233–239, http://dx.doi.org/10.1016/j.carbon.2013.06.012.

21. Saha, D.; Taylor, B.; Alexander, N.; Joyce, D.F.; Faux, G.I.; Lin, Y.; Shteyn, V.; Orkoulas, G.; One-step conversion of agro-wastes to nanoporous carbons: role in separation of greenhouse gases. *Bioresource Technology* 2018, 256, 232–240, https://doi.org/10.1016/j.biortech.2018.02.026.

22. Atta-Obeng, E.; Dawson-Andoh, B.; Felton, E.; Dahle, G.; Carbon dioxide capture using amine functionalized hydrothermal carbons from technical lignin. *Waste and Biomass Valorization* 2019, 10, 2725–2731, https://doi.org/10.1007/s12649-018-0281-2.

23. Hao, W.; Bjornerback, F.; Trushkina, Y.; Bengoechea, M.O.; Salazar-Alvarez, G.; Barth, T.; Hedin, N.; High-performance magnetic activated carbon from solid waste from lignin conversion processes. Part 1: their use as adsorbents for CO_2. *ACS Sustainable Chemistry & Engineering* 2017, 5, 3087–3095, DOI: 10.1021/acssuschemeng.6b02795.

24. Zhu, B.; Huang, J.; Lu, J.; Zhao, D.; Lu, L.; Jin, S.; Zhou, Q.; Worm-like hierarchical porous carbon derived from bio-renewable lignin with high CO_2 capture capacity. *International Journal of Electrochemical Science* 2017, 12, 11102–11107, DOI: 10.20964/2017.12.49.

25. Zhang, J.; Fleury, E.; Brook, M.A. Foamed lignin-silicone bio-composites by extrusion then compression molding. *Green Chemistry* 2015, 17, 4647–4656, https://doi.org/10.1039/C5GC01418B.

26. Rodriguez-Mirasol, J.; Cordero, T.; Rodriguez, J.J.; Activated carbons from CO_2 partial gasification of eucalyptus Kraft lignin. Energy & Fuels 1993, 7, 133–138.

27. Austin, A.T.; Ballare, C.L.; Dual role of lignin in plant litter decomposition in terrestrial ecosystems. PNAS, 2010, 107, 10, 4618–4622, www.pnas.org/cgi/doi/10.1073/pnas.0909396107.

28. Collins, M.N.; Nechifor, M.; Tanasa, F.; Zanoaga, M.; McLoughlin, A.; Strozyk, M.A.; Culebras, M.; Teaca, C.A.; Valorization of lignin in polymer and composite systems for advanced engineering applications—a review. *International Journal of Biological Macromolecules* 2019, 131, 828–849, https://doi.org/10.1016/j.ijbiomac.2019.03.069.

29. Petric, M.; Influence of silicon-containing compounds on adhesives for an adhesion to wood and lignocellulosic materials: a critical review. *Reviews of Adhesion and Adhesives* 2018, 6, 1, DOI: 10.7569/RAA.2018.097305.

30. Onofrei, M.D.; Filimon, A.; Cellulose-based hydrogels: designing concepts, properties, and perspectives for biomedical and environmental applications. In: Polymer Science: Research Advances, Practical Applications and Educational Aspects, Mendez-Vilas, A.; Solano, A., Eds., 2016, Formatex Research Center: Badajoz.

31. Yu, L.; Kanezashi, M.; Nagasawa, H.; Tsuru, T., Role of amine type in CO_2 separation performance. *Applied Science* 2018, 8, 1032, https://doi.org/10.3390/app8071032.

32. Yu, L.; Kanezashi, M.; Nagasawa, H.; Tsuru, T., Fabrication and CO_2

permeation properties of amine-silica membranes using a variety of amine types. *Journal of Membrane Science* 2017, 541, 447–456, http://dx.doi.org/10.1016/j.memsci.2017.07.024.

33. Stiubianu, G.; Cazacu, M.; Cristea, M.; Vlad, A. Polysiloxane-lignin composites. *Journal of Applied Polymer Science* 2009, 113, 4, 2313–2321.

34. Dobos, A.M.; Filimon, A., Bargan, A., Zaltariov, M.F., New approaches for the development of cellulose acetate/tetraethyl orthosilicate composite membranes: rheological and microstructural analysis. *Journal of Molecular Liquids* 2020, 113129, DOI: 10.1016/j.molliq.2020.113129.

35. https://www.co2.earth/daily-co2.

36. Kickelbick, G. Hybrid Materials. Synthesis, Characterization, and Applications and references given in the paper, Wiley-VCH Verlag GmbH & Co. KgaA: Weinheim, 2007.

37. Tu, K.; Puertolas, B.; Adobes-Vidal, M.; Wang, T.; Sun, J.; Traber, J.; Burgert, I.; Perez-Ramirez, J.; Keplinger, T., Green synthesis of hierarchical metal-organic framework/wood functional composites with superior mechanical properties. *Advance Science* 2020, 1902897, DOI: 10.1002/advs.201902897.

38. Li, K.M.; Jiang, J.G.; Tian, S.C.; Chen, X.J.; Yan, F.; Influence of silica-types on synthesis and performance of amine-silica hybrid materials used for CO_2 Capture. *The Journal of Physical Chemistry C* 2014, 118, 2454–2462, https://doi.org/10.1021/jp408354r.

39. Gunathilake, C.; Manchanda, A.S.; Ghimire, P.; Kruk, M.; Jaroniec, M.; Amine-modified silica nanotubes and nanospheres :synthesis and CO_2 sorption properties. *Environmental Science: Nano* 2016, 3, 806–817.

40. Hao, N.; Jayawardana, K.W.; Chen, X.; Yan, M.; One-step synthesis of amine-functionalized hollow mesoporous silica nanoparticles as efficient antibacterial and anticancer materials. *ACS Applied Materials & Interfaces* 2015, 7, 1040–1045, https://doi.org/10.1021/am508219g.

41. Soto-Cantu, E.; Cueto, R.; Koch, J.; Russo, P.S.; Synthesis and rapid characterization of amine-functionalized silica. *Langmuir* 2012, 28, 5562–5569, https://doi.org/10.1021/la204981b.

42. Sanz-Perez, E.S.; Dantas, T.C.M.; Arencibia, A.; Calleja, G.; Guedes, A.P.M.A.; Araujo, A.S.; Sanz, R.; Reuse and recycling of amine-functionalized silica materials for CO_2 adsorption. *Chemical Engineering Journal* 2017, 308, 1021–1033, https://doi.org/10.1016/j.cej.2016.09.109.

43. Nigar, H.; Garcia-Banos, B.; Penaranda-Foix, F.L.; Catala-Civera, J.M.; Mallada, R.; Santamaria, J.; Amine-functionalized mesoporous silica: a material capable of CO_2 adsorption and fast regeneration by microwave heating. *AIChE Journal* 2016, 62, 547–555, https://doi.org/10.1002/aic.15118.

44. Loganathan, S.; Ghoshal, A.K.; Amine tethered pore-expanded MCM-41: a promising adsorbent for CO_2 capture. *Chemical Engineering Journal* 2017, 308, 827–839, https://doi.org/10.1016/j.cej.2016.09.103.

45. Thi, Le M.U.; Lee, S.Y.; Park, S.J.; Preparation and characterization of PEI-loaded MCM-41 for CO_2 capture. *International Journal of Hydrogen Energy* 2014, 39, 12340–12346, https://doi.org/10.1016/j.ijhydene.2014.04.112.

46. Kishor, R.; Ghoshal, A. K., APTES grafted ordered mesoporous silica KIT-6 for CO_2 adsorption. *Chemical Engineering Journal* 2015, 262, 882–890, https://doi.org/10.1016/j.cej.2014.10.039.

47. Li, K.; Jiang, J.; Yan, F.; Tian, S.; Chen, X., The influence of polyethyleneimine type and molecular weight

on the CO_2 capture performance of PEI-nano silica adsorbents. *Applied Energy* 2014 136, 750–755, https://doi.org/10.1016/j.apenergy.2014.09.057.

48. Gholami, M.; Talaie, M.R.; Aghamiri, S.F.; CO_2 adsorption on amine functionalized MCM-41: effect of bi-modal porous structure. *Journal of Taiwan Institute of Chemical Engineering* 2016, 59, 205–209, https://doi.org/10.1016/j.jtice.2015.07.021.

49. Melendez-Ortiz, H.I.; Mercado-Silva, A.; Garcia-Cerda, L.A.; Castruita, G.; Perera-Mercado, Y.A.; Hydrothermal synthesis of mesoporous silica MCM-41 using commercial sodium silicate. *Journal of the Mexican Chemical Society* 2013, 57, 2, 73–79, https://doi.org/10.29356/jmcs.v57i2.215.

50. Jafari, T.; Jiang, T.; Zhong, W.; Khakpash, N.; Deljoo, B.; Aindow, M.; Singh, P.; Suib, S.; Modified mesoporous silica for efficient siloxane capture. *Langmuir* 2016, 32, 2369–2377, https://doi.org/10.1021/acs.langmuir. 5b04357.

51. Niu, M.; Yang, H.; Zhang, X.; Wang, Y.; Tang, A.; Amine-impregnated mesoporous silica nanotube as an emerging nanocomposite for CO_2 capture. *ACS Applied Materials & Interfaces* 2016, 8, 17312–17320, https://doi.org/10.1021/acsami.6b05044.

52. Costa, J.A.S.; Garcia, A.C.F.S.; Santos, D.O.; Sarmento, V.H.V.; Porto, A.L.; Mesquita, M.E.; Romao, L.P.C., *Journal of the Brazilian Chemical Society* 2014, 25, 2, 197–207, http://dx.doi.org/10.5935/0103-5053.20130284

53. Stone, E.J.; Lowe, J.A.; Shine, K.P.; The impact of the carbon capture and storage on climate. *Energy & Environmental Science* 2009, 2, 81–91, DOI: 10.1039/B807747A.

54. European Environment Agency, Technical report No. 14/2011.

55. Noll, W. Chemistry and Technology of Silicones, Academic Press: New York, NY, 1968.

56. Pearson, R. Introduction to the Toughening of Polymers. In: Toughening of Plastics. Advances in Modeling Experiments 2000, R. Pearson et al., Eds., Oxford University Press: Washington, DC, pp.1–11.

57. Lu, J.; Wu, Q.; McNabb, H.; Chemical coupling in wood fiber and polymer composites: a review of coupling agents and treatments. *Wood and Fiber Science* 2000, 32, 1, 88–104.

58. Gauthier, R.; Joly, C.; Coupas, A.; Gauthier, H.; Escoubes, M., Interfaces in polyolefin/cellulosic fiber composites : chemical coupling, morphology, correlation with adhesion and aging in moisture. *Polymer Composites* 1998, 19, 287–300, https://doi.org/10.1002/pc.10102.

59. Rånby, B.G.; Ribi, E.; Ultrastructure of cellulose. *Experientia* 1950, 6, 12–4, DOI: 10.1007/bf02154044.

60. Goussé, C.; Chanzy, H.; Excoffier, G.; Soubeyrand, L.; Fleury, E., Stable suspensions of partially silylated cellulose whisskers dispersed in organic solvents. *Polymer* 2002, 43, 2645–2651, DOI: 10.1016/S0032-3861/02/00051-4.

61. Siqueira, G.; Bras, J.; Dufresne, A., Cellulose whiskers versus microfibrils: influence of the nature of the nanoparticle and its surface functionalization on the thermal and mechanical properties of the nanocomposites. *Biomacromolecules* 2009, 10, 2, 425–432, DOI: 10.1021/bm801193d.

62. Li, Y.; Mai, Y. W.; Ye, L., Effects of fibre surface treatment on fracture-mechanical properties of sisal-fibre composites. *Composites Science and Technology* 2000, 60, 2037–2055, https://doi.org/10.1163/1568554053542151.

63. Herrick, F. W.; Casebier, R. L.; Hamilton, J. K.; Sandberg, K. R., Microfibrillated cellulose: morphology and accessibility. *Journal of Applied*

Polymer Science. Applied Polymer Symposium 1983, 37, 797–813.

64. Goussé, C.; Chanzy, H.; Cerrada, M.L.; Fleury, E.; Surface silylation of cellulose microfibrils: preparation and rheological properties. *Polymer* 2004, 45, 1569–1575, https://doi.org/10.1016/j.polymer.2003.12.028.

65. Chanzy, H. Aspects of cellulose structure. In: Cellulose Source and Exploitation. Industrial Utilization, Biotechnology and Physico-chemical Propertie 1990, Kennedy, J. F.; Phillips, G. O.; Williams, P. A. Eds, Ellis Horwood: New York, NY, pp. 3–12.

66. Chotirat, L.; Chaochanchaikul, K.; Sombatsompop, N. On adhesion mechanisms and interfacial strength in acrylonitrile-butadiene-styrene/wood sawdust composites. *International Journal of Adhesion and Adhesives* 2007, 27, 669–678, https://doi.org/10.1016/j.ijadhadh. 2007.02.001.

67. Zollfrank, C.; Silylation of solid beech wood. *Wood Science and Technology* 2001, 35, 183–189, https://doi.org/10.1007/s002260000071.

68. Matuana, L.; Balatinecz, J., Effect of surface properties on the adhesion between PVC and wood veneer laminates. *Polymer Engineering & Science* 1998, 38, 5, 765–773, https://doi.org/10.1002/pen.10242.

69. Balasuriya, P.; Ye, L.; Mai, Y.; Wu, J.J., Mechanical properties of wood flake-polyethylene composites. Part II : interfacial modification. *Journal of Applied Polymer Science* 2002, 83, 2505–2521, https://doi.org/10.1002/app.10189.

70. Plueddemann, E.P.; Silane Coupling Agents 1991. Plenum: New York, NY, Cap. 2.

71. Matias, M.C.; De La Orden, M.U.; Gonzales Sanchez, C.; Martinez Urreaga, J., Comparative spectroscopic study of the modification of cellulosic materials with different coupling agents. *Journal of Applied Polymer Science* 2000, 75, 256–266, https://doi.org/10.1002/(SICI)1097-4628(20000110)75:2<256::AID-APP8>3.0.CO;2-Z.

72. Castellano, M.; Gandini, A.; Fabbri, P.; Belgacem, M. N.; Modification of cellulose fibers with organosilanes: under what conditions does coupling occur? *Journal of Colloid and Interface Science* 2004, 273, 505–511, https://doi.org/10.1016/j.jcis.2003.09.044.

73. Yano, S.; Furukawa, T.; Kodomari, M. High Technology Composites in Modern Applications 1995, Paipetis, S.A.; Youtsos, A.G., Eds. University of Patras: Patras, Greece, p. 46.

74. Brinker, C.J.; Scherer, G.W.; Sol–Gel Science: The Physics and Chemistry of Sol–Gel Processing. Academic Press: San Diego, CA, USA, 1990.

75. Morton, J.; Cantwell, W.J.; Encyclopedia of Chemical Technology 1997, Kirk-Othmer ed., John Wiley & Sons: Boston, MA, Vol. 7, p. 1.

76. Sequeira, S.; Evtuguin, D.V.; Portugal, I.; Esculcas, A.P.; Synthesis and characterization of cellulose/silica hybrids obtained by heteropoly acid catalysed sol-gel process. *Materials Science and Engineering* 2007, C 27, 172–179, https://doi.org/10.1016/j.msec.2006.04.007.

77. Bourges, X.; Weiss, P.; Daculsi, G.; Legeay, G.; Synthesis and general properties of silated-hydroxypropyl methylcellulose in prospect of biomedical use. *Advances in Colloid and Interface Science* 2002, 99, 215–228, DOI: 10.1016/s0001-8686(02)00035-0.

78. Khan, Fareha Zafar; Sakaguchi, Toshikazu; Shiotsuki, Masashi; Nishio, Yoshiyuki; Masuda, Toshio; Synthesis, characterization and gas permeation properties of silylated derivatives of ethyl cellulose. *Macromolecules* 2006,

39, 6025–6030, https://doi.org/10.1021/ma060601w.

79. Kamath, M.; Mandal, B. K.; Cross-linked copolymers of cyanoethylated cellulose. *European Polymer Journal* 1996, 32, 3, 285–288, https://doi.org/10.1016/0014-3057(95)00143-3.

80. Liang, J.; Chen, Y.; Barnes, K.; Wu, R.; Worley, S.D.; Huang, T.-S.; N-hala-mine/quat siloxane copolymers for use in biocidal coatings. *Biomaterials* 2006, 27, 2495–2501, DOI:10.1016/j.biomaterials.2005.11.020.

81. Hashem, M.; Ibrahim, N.A.; El-Shafei, A.; Refaie, R.; Hauser, P.; An eco-friendly-novel approach for attaining wrinkle-free/soft-hand cotton fabric. *Carbohydrate Polymers* 2009, 78, 4, 690–703, DOI: 10.1016/j.carbpol.2009.06.004.

82. Stiubianu, G.; Cazacu, M.; Hamciuc, V.; Vlad, S., Silicone-modified cellulose. Crosslinking of the cellulose acetate with 1,1,3,3-tetramethyldisiloxane by Pt-catalyzed dehydrogenative coupling. *Journal of Polymer Research* 2010, 17, 837–845, DOI: 10.1007/s10965-009-9375-7.

83. Voronkov, M.G.; Mileshevitch, V.P.; Yuzhelevsich, Yu. A. The Siloxane Bond, Studies in Soviet Science 1978. Consultants Bureau: New York, NY.

84. Kendrick, T.C.; Parbhoo, B.; White, J.W.; The Chemistry of Organic Silicon Compounds 1989, Patai, S.; Rappoport, Z.; Eds. Wiley: Weinheim, Vol. 2, p. 1289.

85. Birot, M.; Pillot, Jean-Paul; Daudé, G.; Pailler, R.; Guette, A.; Plaisantin, H.; Loison, S.; Olry, P.; Simon, P.; Labrugère, C.; Pétraud, M.; Investigation of the pyrolysis mechanisms of cellobiose in the presence of a polysiloxane. *Journal of Analytical and Applied Pyrolysis* 2008, 81, 2, 263–271, https://doi.org/10.1016/j.jaap.2007.12.004.

86. Tomanek, A.; Silicones et Industry. Polytechnica: Paris, 1990.

87. Li, L.; Xiao, Z.; Tan, S.; Pu, L.; Zhang, Z., Composite PDMS membrane with high flux for the separation of organics from water by perevaporation. *Journal of Membrane Science* 2004, 243, 177–187, https://doi.org/10.1016/j.memsci.2004.06.015.

88. Bikson, B.; Nelson, J.K.; Composite membranes and their manufacture and use. Patent AP 4826599, 1989.

89. Williams, S.E.; Bikson, B.; Nelson, J.K.; Burchesky, R.D.; Composite membranes for enhanced fluid separation. Patent EP 0,286,091 B1, 1994.

90. Fichet, O.; Vidal, F.; Laskar, J.; Teyssié, D., Polydimethylsiloxane-cellulose acetate butyrate interpenetrating polymer networks: synthesis and kinetic study. Part I. *Polymer* 2005, 46, 37–47, DOI: 10.1016/j.polymer.2004.10.053.

91. Telysheva, G.; Lignocellulosics—Science Technology, Development and Use 1992, Kennedy, J.; Phillips, G.; Williams, P. Eds. Ellis Harwood: London, pp. 643–655.

92. Sahoo, S. C.; Sil, A.; Thanigai, K.; Pandey, C. N. Use of silicone based coating for protection of wood materials and bamboo composites from weathering and UV degradation, *J Indian Acad Wood Sci* 2011, 8(2), 143–147.

93. Zhang, J.; Fleury, E.; Brook, Michael A. Foamed lignin–silicone bio-composites by extrusion and then compression molding, *Green Chem.*, 2015, 17, 4647-4656.

94. Rodríguez-Robledo, M. Concepción, et al. Cellulose-silica nanocomposites of high reinforcing content with fungi decay resistance by one-pot synthesis, *Materials* 2018, 11, 575, 13.

95. Wang, W.; Zammarano, M.; Shields, John R.; Knowlton, Elizabeth D.; Kim, I.; Gales, John A.; Hoehler, M. S.; Li,

J. A novel application of silicone-based flame-retardant adhesive in plywood, *Mater Des.* 2018, 189.

96. Kokta, B. V.; Maldas, D.; Daneault, C.; Beland, P. Composites of polyvinyl chloride-wood fibers. III: Effect of wane as coupling agent, *J. Vinyl Technol.* 1990. 12, 146–153.

97. Zhou, F.; Cheng, G.; Jiang, Bo, Effect of silane treatment on microstructure of sisal fibers, *Appl. Surface Sci.* 2014, 292, 806–812.

98. Chen, L. et al. Enhancing the performance of starch-based wood adhesive by silane coupling agent (KH570), *Int. J. Biol. Macromol.* 2017, 104, A, 137–144.

99. Chen, F.; Han, G.; Li, Q.; Gao, X.; Cheng, W. High-temperature hot air/ silane coupling modification of wood fiber and its effect on properties of wood fiber/HDPE composites, *Materials* 2017, 10, 286, 17.

100. Rao, J.; Zhou, Y.; Fan, M. Revealing the interface structure and bonding mechanism of coupling agent treated WPC, *Polymers* 2018, 10, 266, 13.

101. Liu, Y.; Guo, L.; Wang, W.; Sun, Y.; Wang, H. Modifying wood veneer with silane coupling agent for decorating wood fiber/high-density polyethylene composite, *Construction and Building Materials* 2019, 224, 691–699.

102. Gironès, J.; Méndez, J. A.; Boufi, S.; Vilaseca, F.; Mutje, P. Effect of silane coupling agents on the properties of pine fibers/polypropylene composites, *J. Appl. Polym. Sci.*, 2007, 103, 3706–3717.

103. Thakur, M. K.; Gupta, R. K.; Thakur, V. K. Surface modification of cellulose using silane coupling agent, *Carbohydrate Polymers*, 2014, 111, 849-855.

104. Ifuku, S.; Yano, H. Effect of a silane coupling agent on the mechanical properties of amicrofibrillated cellulose composite, *Int. J. Biol. Macromol.* 2015, 74, 428-432.

105. Van Rie, J.; Thielemans, W. Cellulose–gold nanoparticle hybrid materials, *Nanoscale*, 2017, 9, 8525–8554.

106. González, D.; Santos, V.; Parajo, J. C. Silane-treated lignocellulosic fibers as reinforcement material in polylactic acid biocomposites, *Journal of Thermoplastic Composite Materials* 2011, 25(8) 1005–1022.

107. Zhu J.; Xue, L.; Wei, W.; Mu, C.; Jiang, M.; Zhou, Z., Modification of lignin with silane coupling agent to improve the interface of poly(L-lactic)acid / lignin composites, *BioResources* 2015, 10, 4315–4325.

108. Xu, G.; Yan, G.; Zhang, J. Lignin as coupling agent in EPDM rubber: thermal and mechanical properties, *Polym. Bull.* 2015, 72, 2389–2398.

109. 109. Song, Y.; Zong, X.; Wang, N.; Yan, N.; Shan, X.; Li, J. Preparation of divinyl-3-aminopropyltriethoxysilane modified lignin and its application in flame retardant poly(lactic acid), *Materials* 2018, 11, 1505, 13.

110. Lopattananon, N.; Jitkalong, D.; Seadan, M. Hybridized reinforcement of natural rubber with silane-modified short cellulose fibers and silica, *J. Appl. Polym. Sci.* 2011, 120, 3242–3254.

111. Telysheva, G.; Dizhbite, T.; Evtuguin, D.; Mironova-Ulmane, N.; Lebedeva, G.; Andersone, A.; Bikovens, O.; Chirkovaa, J.; Belkova, L.; Design of siliceous lignins-novel organic/inorganic hybrid sorbent materials. *Scripta Materialia* 2009, 60, 687–690, https://doi.org/10.1016/j.scriptamat.2008.12.051.

112. dos Santos, F. A.; Tavares, M. I. B. Development of biopolymer/cellulose/ silica nanostructured hybrid materials and their characterization by NMR relaxometry, *Polymer Testing* 2015, 47, 92–100.

113. Ramesh, S.; Kim, J.; Kim, J.-H. Characteristic of hybrid cellulose-amino functionalized poss-silica nanocomposite and

antimicrobial activity, *Hindawi Journal of Nanomaterials*, 2015, 936590, 9 pp.

114. Mohd, N. H. et al. Effect of aminosilane modification on nanocrystalline cellulose properties, *Hindawi Journal of Nanomaterials* 2016, 4804271, 8.

115. Rahmat, M.; Karrabi, M.; Ghasemi, I.; Zandi, M.; Azizi, H. Silane crosslinking of electrospun poly(lactic acid)/ nanocrystalline cellulose bionanocomposite, *Mater. Sci. Eng. C Mater. Biol. Appl.* 2016, 68, 397–405.

116. Stiubianu, G.; Racles, C.; Nistor, A.; Cazacu M.; Simionescu, B. C. Cellulose modification by crosslinking with siloxane diacids, *Cellulose Chem. Technol.*, 2011, 45, 157–162.

117. Xie, Y.; Cai, S.; Hou, Z.; Li, W.; Wang, Y.; Zhang, X.; Yang, W. Surface hydrophobic modification of microcrystalline cellulose by poly(methylhydro) siloxane using response surface methodology, *Polymers* 2018, 10, 1335, 10.

118. Kurosaki, F.; Koyanaka, H.; Tsujimoto, M.; Imamura, Y. Shape-controlled multi-porous carbon with hierarchical micro–meso-macro pores synthesized by flash heating of wood biomass, *Carbon* 2008, 46, 850-857.

119. Yang,C.; Gao, Q.; Tian, W.; Tan, Y.; Zhang, T.; Yang, K.; Zhu, L.; Superlow load of nanosized MnO on the porous carbon matrix from wood fibre with superior lithium ion storage performance, *J. Mater. Chem. A*, 2014, **2**, 19975-19982.

120. Yu, Shuai, et al. Synthesis of wood derived nitrogen-doped porous carbon-polyaniline composites for supercapacitor electrode materials, *RSC Adv.*, 2015, 5, 30943–30949.

121. Luo, Jingru et al. Free-standing porous carbon electrodes derived from wood for high-performance Li-O2 battery applications, *Nano Research* 2017, 10(12), 4318–4326.

122. Du, Qi-Shi, et al. Graphene like porous carbon with wood-ear architecture prepared from specially pretreated lignin precursor, *Diamond & Related Materials* 2018, 90, 109–115.

123. Peng, Xinwen, et al. Hierarchically porous carbon plates derived from wood as bifunctional ORR/OER electrodes, *Adv. Mater.* 2019, *31*, 1900341, 7.

124. Zhang, Sen, et al. High performance flexible supercapacitors based on porous wood carbon slices derived from Chinese fir wood scraps, *Journal of Power Sources* 2019, 424, 1–7.

125. Qian, J.; Wang, J.; Jin, Z. Preparation of biomorphic SiC ceramic by carbothermal reduction f oak wood charcoal, *Mater. Sci. Eng. A* 2004, 371, 229–235.

126. Shin, Y.; Wang, C.; Exarhos, G. J. Synthesis of SiC ceramics by the carbothermal reduction of mineralized wood with silica, *Adv. Mater*. 2005, 17, 73–77.

127. Vyshnyakova, K.; Yushin, G.; Pereselentseva, L.; Gogotsi, Y. Formation of Porous SiC Ceramics by Pyrolysis of Wood Impregnated with silica, *Int. J. Appl. Ceram. Technol.*, 2006, 3, 485–490.

128. Wayan Karyasa1, I Wayan Muderawan1, and I Made Gunamantha, Renewable silica-carbon nanocomposite and its use for reinforcing synthetic wood made of rice straw powders, *KnE Engineering*, 2016, 6.

129. Achoundong, C. S. K.; Bhuwania, N.; Burgess, Steven K.; Karvan, O.; Johnson, Justin R.; Koros William J. Silane modification of cellulose acetate dense films as materials for acid gas removal, *Macromolecules* 2013, 46, 14, 5584–5594.

130. Puspasari, T.; Chakrabarty, T.; Genduso, G.; Peinemann, K.-V. Unique cellulose/polydimethylsiloxane blends as advanced hybrid material for organic solvent nanofiltration and pervaporation membranes, *J. Mater. Chem. A*, 2018, **6**, 13685–13695.

CHAPTER 2

RHEOLOGICAL INSIGHTS IN DEVELOPMENT OF BIOPOLYMER SCAFFOLDS

ANDREEA IRINA BARZIC* and RALUCA MARINICA ALBU

"Petru Poni" Institute of Macromolecular Chemistry, Laboratory of Physical Chemistry of Polymers, 41A Grigore Ghica Voda Alley, 700487 Iasi, Romania

Corresponding author. E-mail: irina_cosutchi@yahoo.com

ABSTRACT

Among the materials employed for the fabrication of the supports for cell growth, a great interest was attributed to bio-based polymers. The chapter presents some essential aspects regarding the rheological behavior involvement of polymers in designing scaffolds with certain architectures, such as porous, fibrous, and hydrogel materials. Also, molecular modeling of the interactions of biodegradable and biocompatible polymers with adhesion proteins is performed for a proper understanding of the forces acting in such bio-systems. In some cases, the polymer alone does not meet the desired demands and it is further doped, thus the microstructure features of the material are modified as shown in the shear flow and viscoelastic characteristics. Along with rheological performance, a brief presentation of the practical importance of the described bio-based plastics in tissue engineering is done.

2.1 INTRODUCTION

In material science, during several decades, information extracted from the rheological investigation of simple or complex systems has provided a large amount of knowledge with respect to conformational variations or rearrangements of material elements at micro- and nano-level when they are subjected to mechanical forces [1]. These findings emphasized the significance of rheological character-istics of materials in their processing and functionality. Therefore, over the past years, there have been many studies dealing with the role of the flow behavior of numerous categories of materials, from low molecular weight substances [2, 3] to macromo-lecular systems in fluid or melt state [4–7]. In this context, rheology was proved to be a useful tool for exami-nation of the quality of polymer-based products designed for various indus-tries, like paints [8], food packaging [9], and electronics [10], but also for biomedical fields, including pharma-ceutics [11], hematology [12, 13], or tissue engineering [14, 15].

Polymer materials have gained increased importance in regenerative medicine, including tissue engi-neering, where they can be success-fully used as supports for cell growth [16–18]. They have some advantages over ceramics and metals, such as biodegradability, flexibility, and processability into specific shapes [18]. The ideal polymer substrate for cell growth must display certain essential features:

1. adequate interaction with cells to intensify cell adherence, proliferation, and secretion of extracellular matrix (ECM),
2. appropriate architecture that enables the effective mass transfer of gas, nutrients, and waste to attain acceptable cell viability,

3. tunable biodegradation rate to allow gradual vanishing of the supporting role of the scaffold material used in tissue regeneration,
4. comparable mechanical charac-teristics to those of the contact tissue,
5. good biocompatibility, and
6. suitable chemical surface modi-fication and typical morpho-logical features to support the viability of cells and other related processes.

In order to achieve a particular architecture of the scaffold, the polymer must be first solved in a carefully selected solvent and then processed accordingly to the desired morphological features, namely porous, fibrous, or microspherical. Given the rheological properties of cells and tissues [19], there have been many efforts to obtain perfect polymer structures adequate for biomedical purposes on both in vitro and in vivo uses, such as bone fillers [20–22], skin regeneration [23–25], muscle repair [26, 27], cartilage reconstruction [28], and constructed heart valves [29].

Polymer-based scaffolds can be imparted in many categories as a function of several factors, such as the origin of the fabrication material, the created architecture, material composition, and biocompatibility properties, including the scaffold action in the presence of the tissue.

A proposed scheme for polymer scaffold classification is depicted in Figure 2.1.

From the analysis of the sorts of polymer scaffolds, one may notice that bio-based polymer materials having 2D and 3D architecture are becoming more and more important in tissue regeneration. Bio-based plastics can be fabricated from materials entirely or partially collected from renewable resources (i.e., biological materials or biomass). The performance of this kind of plastics may fluctuate from one material to another. Among the properties that might differ are the rheological ones and through them the processing conditions and morphology of the final polymeric product. Also, in the processing stage, it interferes with the initial phase (melt, suspension, or viscous solution) from which one starts to produce the cell growth substrate. In addition, the best conditions in terms of concentration, temperature, solvent, and deformation to properly prepare the product can be established by performing rheological tests.

The chapter is presenting, based on the aforementioned background, some aspects concerning the relation between rheological behavior and processing of bio-based-polymers designed for cell culture goals. In such applications, it is very important to attain good interfacial adhesion

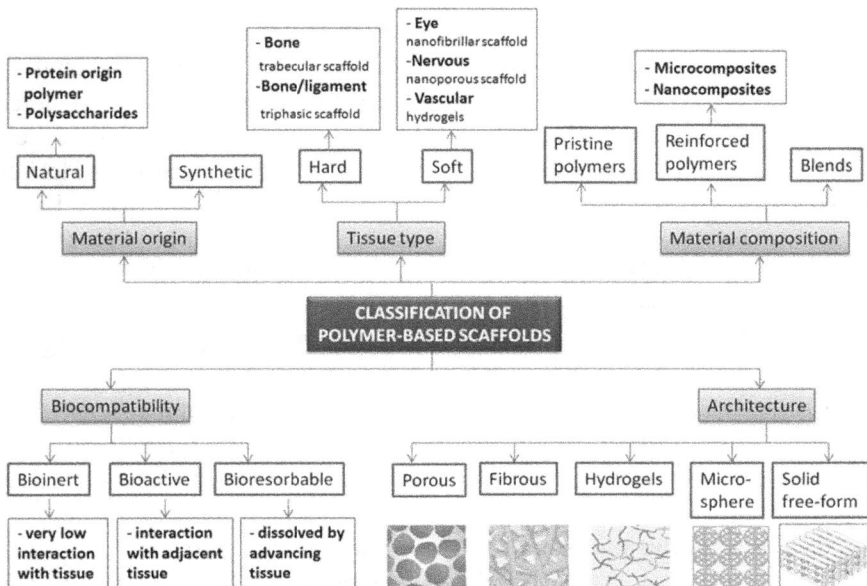

FIGURE 2.1 A proposed classification of polymer-based scaffolds as a function of several factors.

of cells to the polymeric coating. Since the biological medium tends to stick to the polymer surface via interactions with adhesion proteins employing specific cell receptors, which can be encountered on the cell membrane, it is very important to understand the interaction of such proteins with polymer support. For the stated purpose, molecular modeling showing the interaction of certain biodegradable polymers with two relevant adhesion proteins (collagen and fibronectin). Moreover, recent reports on the importance of rheological parameters for designing the scaffold architecture are reviewed. A short description of the practical use of the presented biodegradable polymer materials for cell growth supports is made.

2.2 MOLECULAR MODELING OF BIOPOLYMER INTERACTIONS WITH CELL ADHESION PROTEINS

For molecular modeling purposes, two kinds of polymer supports were selected owing to their widely studied biodegradability and compatibility with cells, namely hydroxypropyl methylcellulose (HPMC) [30, 31] and poly(vinyl alcohol) (PVA) [32, 33]. Knowing that cellular proliferation is ensured by the interaction with adhesion proteins (like fibronectin, collagen, and not only) with polymer surface [34], computational methods

were applied to simulate such interactions. Molecular modeling was employed in similar studies on other polymers with proteins [35–37].

In the first stage, the polymers and protein structures were subjected to an optimization procedure that involves the Polak Ribiere algorithm, which leads to the reduction of energy gradient until its value is smaller than 0.01 kcal mol^{-1} Å. Figure 2.2 presents optimized structures of the investigated compounds as obtained from molecular modeling with HyperChem software (demo version). In order to determine essential physical parameters, which are linked to the biological activities, it is mandatory to examine the levels of frontier orbitals energies. Highest occupied molecular orbital (HOMO) and lowest unoccupied molecular orbital (LUMO) were estimated by using Parametric Method 3 (denoted PM3) chosen from a semiempirical approach within Restricted Hartree Fok (RHF) formalism and selected single point setting. Figure 2.3 displays the image of the frontier orbitals and electrostatic potential surfaces (ESP) corresponding to each evaluated compound.

The magnitude of the HOMO and LUMO energies is influencing many parameters, such as ionization energy (E_I), electronegativity (EN), electron affinity (E_{aff}), electronic chemical potential (ECP), electrophilicity (E_{PH}), energy gap (E_g), chemical hardness (C_H), and chemical softness

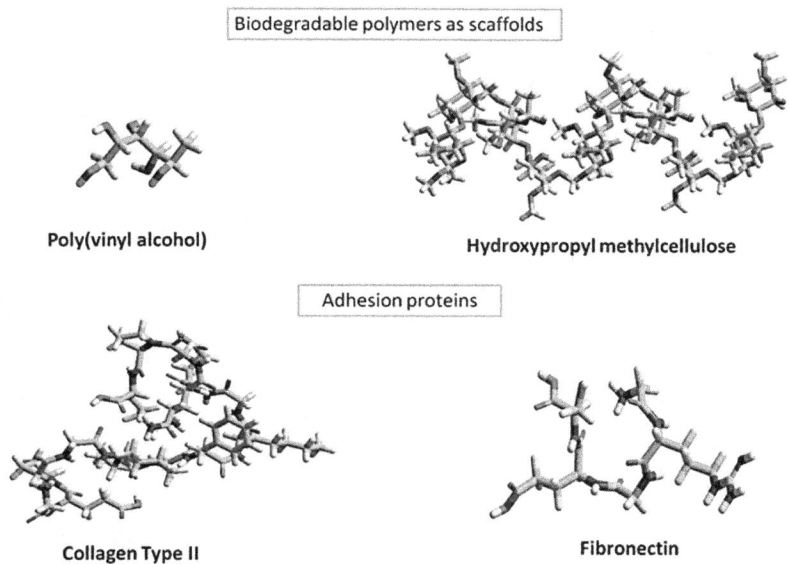

FIGURE 2.2 The tridimensional conformation of chosen polymers (structural units) and adhesion proteins (one structural unit) in free space.

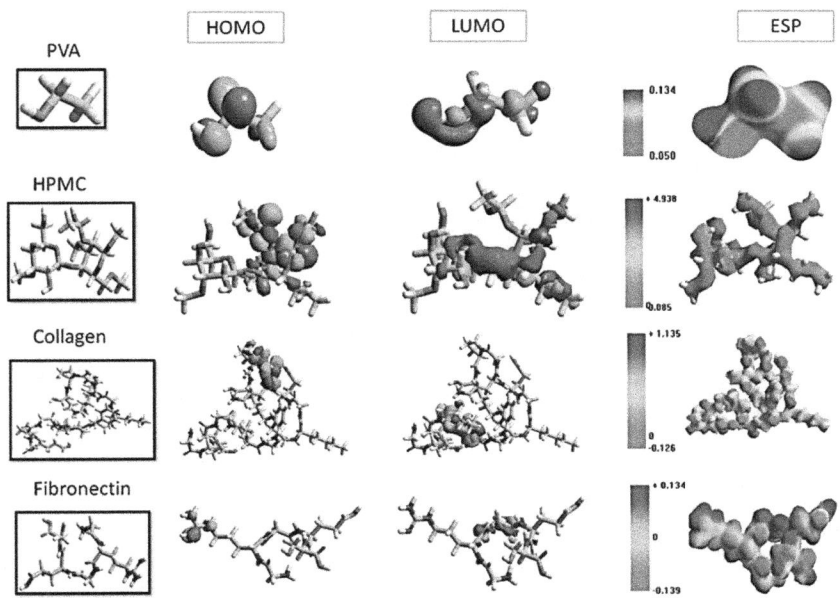

FIGURE 2.3 The images of the frontier orbitals and electrostatic potential surfaces (ESP) for PVA, HPMC, collagen, and fibronectin.

(C_S) [38]. All these data resulted from the computations performed on the selected compounds in free space are listed in Table 2.1.

It can be remarked that the higher value of E_g for HPMC in regard to PVA indicates that the first one is a harder and more stable molecule, but with a smaller reactivity. On the other hand, a low energy gap is favorable for an easier polarization of the macromolecule. If analyzing the adhesion proteins, it can be stated that fibronectin is more reactive than collagen type II. The electrophylicity index, which reveals the compound's stability upon accepting of other electron charges, is smaller for PVA and HPMC in comparison to the studied adhesion proteins.

The molecular descriptors of the investigated structures were theoretically estimated from the QSAR setting of the used software. The achieved information is displayed in Table 2.2.

TABLE 2.1 The Computed Values of HOMO, LUMO, E_g, E_I, E_{aff}, EN, ECP, C_H, C_S, E_{PH}

Physicochemical Parameters	Polymer Scaffold		Adhesion Proteins	
	PVA	HPMC	Collagen Type II	Fibronectin
HOMO	−10.9268	−9.8526	−9.3707	−8.5779
LUMO	3.2737	2.1891	−0.7131	0.2456
Eg	7.6531	7.6635	8.6576	8.3323
E_I	10.9268	9.8526	9.3707	8.5779
E_{aff}	−3.2737	−2.1891	0.7131	−0.2456
EN	3.8267	3.8318	5.0419	4.1662
ECP	−3.8267	−3.8318	−5.0419	−4.1662
C_H	7.1003	6.0209	4.3288	4.4118
C_S	0.1408	0.1661	0.2310	0.2266
E_{PH}	1.0311	1.2193	2.9362	1.9671

TABLE 2.2 The Molecular Descriptors of the Polymer Scaffolds and Adhesion Proteins

Molecular Descriptor	Collagen II	Fibronectin	PVA	HPMC
Surface area (grid), Å^2	1486.46	756.76	192.90	633.66
Volume, Å^3	3366.66	1359.99	242.16	1147.58
Hydration energy, kcal/mol	−34.85	−38.22	−5.83	−10.37
Log P	−7.90	−5.32	0.08	−1.65
Refractivity, Å^3	363.61	116.04	13.01	98.04
Polarizability, Å^3	145.41	46.62	5.08	39.90
Mass, amu	1470.64	504.50	46.07	442.46

The surface area and volume are ranging according to the structural peculiarities of each compound. The hydration energy is smaller for the adhesion proteins in comparison to the considered biodegradable polymer scaffolds. Collagen type II presents the highest polarizability and refractivity among the studied materials. The partition coefficient (log P) is denoting the hydrophilic or hydrophobic character of the studied materials. The negative values of log P obtained for all samples, except for PVA, reveal the hydrophilicity of these structures.

The interaction energy between the polymer and selected proteins was evaluated from the PM3 method within (RHF) formalism, by using equation (2.1) [39]:

$$\Delta E = E_{(S-P)} - (E_S + E_P) \quad (2.1)$$

where ΔE is the energy of interaction, $E_{(S-P)}$ is the binding energy of the scaffold (denoted S)—protein (denoted by P) complex, E_S is the binding energy of the scaffold and E_P is the binding energy of the adhesion protein.

Binding energy can be viewed as the quantity of energy necessary for disrupting the molecule into its atoms. A large ΔE value is corresponding to a bigger affinity toward the surrounding molecule and a more stable complex resulted. Figure 2.4 depicts the computed fluctuations in

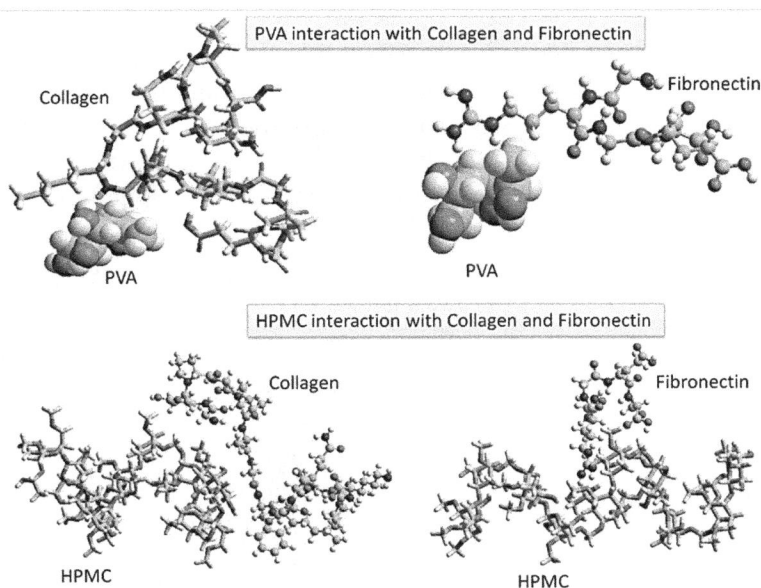

FIGURE 2.4 The schematic illustration of the conformational changes of HPMC and PVA (having five repeating units) upon interacting with adhesion proteins (collagen type II and fibronectin).

the spatial arrangement of a chain of five repeating units of PVA or HPMC during interaction with the adhesion proteins. One may notice that HPMC chain suffers a more visible modification in the presence of collagen or fibronectin in regard to PVA structure. In other words, in HPMC/ protein systems, the macromolecular cellulosic chain is more coiled when interacting with collagen and more stretched when adding fibronectin. The same observation, but to a less extent, can be made for PV/protein systems. This could be possibly explained by accounting for the distinct amount of hydrogen bonding occurring in each analyzed system.

Table 2.3 contains calculations concerning the thermodynamic parameters extracted using semiempirical methods from molecular modeling of the interactions among all biodegradable polymers and protein. It is widely known that macromolecules with the lowest

binding energy will display the largest binding affinity [40]. The energy of interaction was calculated according to equation (2.1) for all simulated systems. For PVA complexes, it was remarked that this parameter is smaller indicating that these systems present lower stability. Conversely, for HPMC complexes, the interaction energy between the polymer and protein is higher, resulting in a better stability for modeled molecules. Regardless of the polymer structure, in the presence of fibronectin, the magnitude of the interaction energy is below that obtained for the polymer systems with collagen. Thus, for cell culture purposes, it is desirable to initially perform molecular modeling in order to check the compatibility of the biopolymer with the adhesion proteins and implicitly with the cells of the tissue that is to be regenerated. Furthermore, the processing of the polymer support is related to

TABLE 2.3 The Computed Thermodynamic Parameters, in kcal/mol: the Total Energy, Binding Energy, and Interaction Energy of the Simulated Polymer–Protein Systems

	Total Energy	Binding Energy	Interaction Energy
PVA	−55237.93	−2264.23	–
HPMC	−709613.41	−29566.52	–
Collagen II	−422860.23	−20599.85	–
Fibronectin	−151443.62	−6494.49	–
HPMC/Collagen II	−1132594.98	−50287.71	121.34
HPMC/Fibronectin	−861170.19	−36174.16	113.15
PVA/Collagen II	−478130.74	−22896.67	32.59
PVA/Fibronectin	−206709.05	−8786.22	27.50

its conformational properties, which in turn will dictate the rheological behavior. So, both methods are related and such analyses must be performed to warrant the development of great quality scaffolds.

2.3 RHEOLOGICAL IMPLICATIONS IN PREPARING POLYMER SCAFFOLDS

As a function of their structural peculiarities, biodegradable polymers can be shaped from their fluid state into certain shapes and particular architectures. The involvement of rheological characteristics in designing several kinds of scaffolds is further presented in the next sections of the chapter.

2.3.1 POROUS POLYMER SUBSTRATES

Scaffolds with porous (foams, mesh, and sponge) architecture are largely studied since they enable through the interconnected structure (with or lacking a prevailing orientation) the nutrients transport to the cells. In addition, such substrate morphology stimulates ECM to grant cells to interact with the surrounding medium. It was reported that foams and sponges-like scaffolds exhibit higher mechanical stability than meshes, but the latter one has the advantage of larger space throughout the cell growth substrate [41, 42].

The dimensions of the created pores are vital for processes like cellular penetration or ECM deposition. Some scientists state the best performance of the scaffold is attained when the porosity is around 90%, which is affecting the mechanical performance [43]. However, many practical situations reveal that the ideal dimensions of the generated pores in the polymer materials should be adapted to the cultured cells and implicitly to the features of the regenerated tissues [44]. For instance, very small pores ranging in diameter between 10 and 44 μm are adequate for fibrous tissues [45]. Scaffolds with pores in the interval of 50–200 μm are more suited for smooth muscle tissue [46], whereas those with large pores 200–400 μm are favoring accommodation of the hard bone tissues [47]. Mouriño and Boccaccini [48] formulated a general rule for porous cell growth substrates, namely those that have pores that exceed 100 μm grant tissue development and vascularization, while the materials with finer pores around 2–50 nm are fitted for cell adhesion and resorbability. The pores which are redundantly large in diameter (above 400 μm) diminish the cell-to-cell contact ratio because the biological material experiences 2D growth pattern on the support rather than a 3D arrangement [42]. The features of the porous scaffolds (diameter, pore distribution, surface area/volume ratio, etc.) can be tuned via the selected solvent, phase separation conditions, and so on [41].

Mulder et al. [49] discussed the problem that the majority of porous scaffolds are mainly isotropic and this in discrepancy with native healthy tissues. Because of that, the reconstructed tissue displays no structural and functional connection with the initial tissue. To attain a proper regeneration, it is mandatory to have an additional differentiation stage upon resorption of the isotropic polymer support. This is less credible if the necessary plasticity for this is left present in an already final differentiated tissue. There would be a solution if the novel resulted tissues in the porous substrate would be able to differentiate directly into the anisotropic structuring of the initial native tissues.

Pawelec et al. [50] prepared porous scaffolds based on the idea that anisotropy of the cell growth support is impacted by nucleation, whereas the pore dimensions are related to the crystal development and annealing. The procedure of ice-templating is influenced by several factors, including the quantity and type of solutes. They used an ionic solute like sodium chloride and a nonionic one, such as sucrose which was inserted in collagen slurries. Depending on the polymer-solute interactions and also on the manner in which the growth kinetics is controlled, it is possible to set the scaffold architecture. Rheology tests revealed a pseudoplastic behavior for all specimens. Upon addition of nonionic solute, no significant changes in the solution behavior of collagen were noted. As a result, the shear flow curves indicate close viscosity values at all shear rates. In the presence of sucrose, the ice growth was diminished, producing smaller pores. On the other hand, ionic additives in the polymer slurry determine conformational changes and this is also reflected in the viscosity values, which are smaller in regard to samples with no additive or containing sucrose. This could be explained by taking into account that the entanglements of the collagen chains are decreasing [51–54]. The fewer entanglements in the collagen offset reduce the growth rate provoked by NaCl insertion in the slurry and inflicted ice crystals to become greater. The presence of ionic solute also determined an important lowering of both rheological moduli. For all samples, it was observed that the elastic modulus (G'') is higher than the viscous one (G'), regardless of the applied shear frequency. For the systems containing sucrose, the scaffold had finer pores (~70 μm) and upregulated cell attachment, while for systems with NaCl fibrous, scaffolds with larger pores (~120 μm) were accomplished. The latter display reduced attachment of chondrocytes.

Pietrucha [55] examined the modifications in denaturation and rheological behavior of materials with sponge architecture designed for tissue engineering. The denaturation temperature of pristine collagen is 69 °C and reaches 86 °C upon the addition of hyaluronic acid. These aspects are reflected in the rheological moduli up to 226 °C for investigated scaffolds owing to the removal of loosely and strongly bound water. The viscoelastic behavior of porous samples is affected by hydroxyapatite (HA) in the sense that there is noted a shift of the tan δ peak attributed to the process of decomposition occurring at elevated temperatures.

Sab and coworkers [56] employed foam replication technique for obtaining porous polyurethane scaffold, in which was embedded a slurry of betatricalcium phosphate (β-TCP) and carboxymethyl cellulose (CMC). In the rest stage, in the slurries, agglomerates tend to comprise a share of dispersant resulting superstructures that are disrupted under shearing leading to smaller structures with a higher ability to flow. All suspensions present very small low yield points, ranging in the interval of 0.43–3.77 Pa. The sample containing the highest CMC binder amount of 0.8% flows more easily and presents improved movement over pores revealing that the suspensions containing smaller percents of CMC present more solid structures as reflected in the level of yield point. The absence of fluctuations in shear stress over the impose hearing indicates an enhancement of the suspension stability in the presence of CMC. The analyzed suspensions displayed non-Newtonian pseudoplastic properties below 30 s^{-1} followed by a nearly constant viscosity domain. The slurry having 0.8% binder displays the largest viscosity, almost over the entire shearing range (10–100 s^{-1}), which can be understood as a fact that the suspension is more facile retained within the template pores. Furthermore, shear recovery experiments showed that all samples are thixotropic in nature. The degree of recovery is smaller as the CMC amount in the system is larger because it enables the circulation of the slurry in the bulk of the template. The rheology data show that β-TCP sample containing 0.8% of CMC which is most adequate for coating on the polyurethane sponge.

Song and collaborators [57] made porous scaffolds from polycaprolactone (PCL) using a green procedure relying upon supercritical CO_2 (scCO_2) melt-state foaming. In order to understand the relation between interconnectivity and pore dimensions for scaffold purposes, polymers of variable molecular weights (MW) were chosen. Frequency sweep analyses confirmed the classical relation between viscosity and molecular mass. As the latter is lower, the samples present Newtonian behavior (regardless of the frequency) since there are fewer chain entanglements. PCLs with high molecular mass have larger shear moduli than those with low molecular weight. In the case of PCLs with shorter chains, the magnitude of the viscous modulus is above the elastic one in the entire sweeping domain, so the G''/G' value is higher than 1. Literature data [58] show that for such cases the samples have reduced pore stability and also low foamability. So, polymers of low molecular weight are not ideal for obtaining foamed scaffolds. Given the bigger complex viscosity of PCL samples with high molecular mass, they exhibit better melt strength which in turn renders larger resistance to the deformation of porous architecture, as well for pore growth and disruption. Corroboration of rheological tests with thermal analyses proves that the specimens present a delayed crystallization as the molecular mass increases. Upon foaming, polymer supports having 27 kDa have the desired morphological characteristics in terms of pore diameter and interconnectivity, where the smallest melt strength is suitable for the production of interconnected macropore, and faster crystallization leads to adequate foamability.

Choi et al [59] prepared poly(lactic acid) (PLA)-based scaffolds with superior porosity and foamability by the polyester-epoxide reaction. The epoxide units present in the chain extender are able to react with the functional groups (OH, COOH) found at the chain end of PLA. Such procedure would enhance molecular weight and is expected to improve rheological features for extrusion foaming. The flow behavior of pristine PLA is affected by the amount of inserted epoxide chain extender (ECE). It was shown that molecular weight increases with ECE amount up to 1.5 wt% and above this value MW decreases since there are not enough reaction sites at end groups of PLA to react with remaining ECE. Moreover, the polydispersity index is diminished with the change in molecular mass implies that the cleavage of long macromolecules during thermal decomposition, while shorter chains appear particularly in the chain extension reaction [60]. Rheology tests were done at 200 °C, which corresponds to the extrusion foaming process. The complex viscosity (η^*) of initial PLA is Newtonian up to 10 rad s^{-1} and then decreases. The η^* of the modified PLA samples is decreasing as the angular frequency is greater. The system containing 1.5 wt% ECE has the biggest complex viscosity and hence the highest melt stability. From viscoelastic experiments, it was concluded that the elastic modulus is enhanced proportionally with the ECE percent up to 1.5 wt%. For this composition, the rheological parameters (η^* and G') are optimized, facilitating the pursued processing steps, namely high-temperature melting and thermal foaming. The cell culture assays on mouse fibroblasts of this foamed scaffold led to promising results for tissue engineering.

2.3.2 FIBROUS POLYMER SUBSTRATES

Fibrous polymer scaffolds are mainly composed of micro- and nano-fibers achieved by electrospinning, phase separation, or self-assembly procedures. If the polymer fibers have dimensions of nanometric level, they have the advantage that can reproduce the architecture of human tissues at the nanoscale. It was proved that nanofibrous scaffolds are suitable for directing the orientation of neurite growth by setting topographical clues influencing important processes, like cell differentiation and fate [61]. These nanofibers imitate the structure and features of the EMC while owing to their high aspect ratio and porosity, such materials determine higher cellular attachment in regard to microfibers [62, 63]. The large surface-area-to-volume ratio of the nanofibers together with their microporous morphology is ideal for promoting cellular adhesion, proliferation, migration, and differentiation, as desired in regenerative medicine [42]. Moreover, nanofibrous cell growth supports could enable quicker diffusion of incorporated substances

and cell infiltration [64]. These types of scaffolds have found utility for repairing musculoskeletal tissues, such as cartilage, ligament, bone, muscles, skin, and neural ones.

Among the fabrication methods of polymer fibers, the most employed one is electrospinning, which is conducted to the most reliable and promising results. In order to attain polymer fibers without defects, a detailed rheological analysis is required. The variations of viscosity with polymer amount in solution must be monitored to understand how chain interactions in liquid medium affect the fiber morphology. It was proved that almost double the value of entanglement concentration is the suitable concentration for the preparation of defect-free polymer fibers from solutions subjected to electrospinning [65, 66].

Literature reports [15, 67] the fabrication of scaffolds based on PLA doped with nano-hydroxyapatite/collagen (nHAC) and variable amounts of chitin fibers. Dioxane was employed as a pore-forming agent. They analyzed the rheological properties of these complex systems. Both complex modulus and complex viscosity were magnified by the presence of more chitin fibers. At the highest reinforcement level with chitin fibers, the complex modulus hardly ranges with angular frequency as a result of improved elasticity of sample. At the same level of doping, the complex viscosity displays the highest drop with frequency increasing in comparison with the less doped samples which are mainly Newtonian. This can be ascribed to the fact that the chitin component modifies the fluid character of specimens. On the other hand, the storage and loss moduli are more sensitive to shear frequency at the lower fiber content. Upon doping, the rheological moduli are increasing particularly G' augmenting the elastic character of the polymer fluid. At the highest chitin fiber content of 45% the G' overcomes G'' in the whole frequency domain. The invariability of these rheological parameters in the low-frequency zone is not only indicative of fiber intertwining but also shows the interaction of chitin with the matrix. Furthermore, the flatbase of shear moduli could also signify a yield point. So, in the situation where applied force is below the yield point stress, the sample will not present fluidity, but elasticity. If the deformation force exceeds the yield stress, the sample begins to flow. The augmentation of fibers content determines the shifting of flatbase of rheological moduli curves toward bigger frequencies. Analysis of loss tangent reveals that above 35% fiber content, this parameter is less affected by angular frequency and its values are subunitary, thus the samples have viscoelastic properties. This is not so advantageous materials' processing.

Li et al. [68] prepared scaffolds using doping and chemical crosslinking concepts combined with low-voltage electrospinning patterning (LEP). The solution dispersion rheology is an adequate indicator for the elucidation of fiber processability conditions in relation to the fabrication strategy for patterning. Thus, it was investigated how rheological characteristics of gelatin and its processing into fibers are changed upon introduction of homogenized decellularized matrix particles (dCMps). Shear oscillatory experiments were performed at variable gelatin amounts in the systems. The reported data indicated that upon increasing the gelatin quantity in the solution the slope of rheological moduli dependence on frequency decreases, while complex viscosity is more and more sensitive to applied frequency. Sampling complex viscosity results for all concentrations at 10 rad s^{-1} (extrusion corresponding rate) and 100 rad s^{-1} (patterning analogous rate), the variation of complex viscosity with concentration can provide essential information for processing. The trend-line ascribed to 10 rad s^{-1} plot displays an abrupt gradient in regard to the 100 rad s^{-1} one. Based on this, it is possible to state that as the gelatin amount increased, it determined a greater resistance to flow to a higher level during extrusion compared with patterning.

The described effect is the main factor explaining why solutions with distinct rheological behavior impose different LEP configurations, "extrusion-patterning" and "drag-patterning."

Mirtic and coworkers [69] performed a complex investigation involving scanning electron microscopy and rheological experiments on systems prepared from alginate mixed with variable MW poly(ethylene oxide) (PEO). Their study aimed to enhance electrospinability polymer-blend solution for achieving optimum scaffold formulation. At fixed 3.5% total polymer concentration and 8% PEO proportion in the system, it was noted that MW of PEO did not influence much the fiber diameter as seen in almost insignificant differences in rheological data depicting shear-thinning flow properties at low shear rates followed by an almost Newtonian zone. When varying total polymer concentration while keeping constant MW of PEO and PEO proportion, the viscosity of the system increased and the pseudoplastic features became more obvious. This is reflected in the morphological characteristics, namely higher diameter, the occurrence of microfibers, or transformation from beaded nanofibers to defect-free ones. However, the rheology data alone cannot distinguish the formation of microfibers or nanofibers. The nanofiber features are not the result of a sole composition parameter, and cumulative effects of the formulation compositions should be considered. At variable PEO proportions, it was seen that beaded nanofibers are formed at 6%–8% PEO, while smooth nanofibers are achieved at 12%–15% PEO. In any case, the solution parameters corresponding to beadless nanofibers are typical for the PEO molecular weight. The nanofiber diameter was not impacted by the PEO proportion. The prepared materials can be successfully used in tissue engineering.

2.3.3 HYDROGEL-BASED POLYMER SCAFFOLDS

Rheology is the most adequate tool to characterize gel properties, like gelation time, degree of crosslinking, or homogeneity [70]. Hydrogels have the benefit of injectability and biocompatibility, combined with controllable degradability. Given the absence of an inorganic counterpart in hydrogel composition, most of them have issues concerning the mechanical performance and also the stimulation of bone regeneration. As a consequence, a good alternative is to incorporate in the hydrogel, specific nanoparticles to enhance biomimetic features. Therefore, injectable polymer composites could have a behavior intermediary between a strong gel and a viscous fluid. Polymer-filler interactions are clear evidence by means of rheological analyses [71, 72]. The majority of such biocomposites are gels or viscoelastic solids with high elastic modulus.

Jiao et al. [73] performed rheological analyses of hydrogels based on poly(ethylene glycol)maleate citrate (PEGMC). The biodegradable scaffolds were reinforced with HA. Shear oscillatory experiments were done during the gelation at 37 °C and also several factors were monitored: maleic anhydride/citric acid ratios, crosslinker percent, and filler amount. As the elastic character of composite samples is becoming prevalent, it was noted that the loss tangent is monotonically decreasing and this emphasizes a liquid–solid transition. Crossover of the shear moduli is indicative of the gel point particularly when the relaxation exponent (denoted by n) has a value of 0.5 and tan δ is equal to unity. For constant 50 wt% of HA and 0.4 molar ratio of PEGMC, the exponent n equals 0.5, but for PEGMC (0.8), n value increases to 0.58. It was obtained a multiwave time sweep data for the biodegradable composites of diverse maleic acid amounts in which tan δ is invariant to shear frequency (close to 1) till gelation is finished. Tan δ curves for several frequencies concur at the gel-point. Consequently, the Winter–Chambon

criteria are adequate for the chemically crosslinked gels doped with HA. The complex viscosity diverged at the sol–gel transition. For HA reinforced samples PEGMC (0.6) and PEGMC (0.8), the viscosity is easily augmented upon doping, except for PEGMC (0.4)/HA which after gelation the viscosity is almost unchanged. The low viscosity of this hydrogel is facilitating its injection. Sweep oscillatory analyses were made during the crosslinking of samples showing at first a preponderant viscous character (liquid state prevailed) and as the reaction continued the material gained elastic features, resulting in a hydrogel composite. Both shear moduli increased but with different slopes determining their overlapping at a point that denotes the gelation time. This parameter is lowest for the highest PEGMC amount indicating that the vinyl group from maleic anhydride reacts faster with the alkene crosslinker. The rheological moduli of composites with 0.6 and 0.8 PEGMC at 1000 s are not meaningfully distinct. The closeness of the shear moduli reveals that the vinyl units placed in the middle of macromolecular chains of samples could not have reacted. Frequency sweep oscillatory tests were done on dried composites displayed like solids having an elastic modulus around 8 kPa for HA doped sample with PEGMC (0.4) and even larger.

Multiwave-analyses performed at 10, 50, and 100 rad s^{-1} on hydrogels exhibit a tan δ around unity, whereas the intersection point at which tan δ is no longer affected by frequency (linked to gelation time) increases from 200 to 500 s, when the crosslinker percent is reduced from 15 to 3% (w/v). The complex viscosity presents a sudden change as time increases delimiting the occurrence of sol–gel transition. Moreover, the viscosity of reinforced samples is significantly enhanced at 15% (w/v) crosslinker, while the gelation time increased up to 668 s as the crosslinker agent decreased to 3% (w/v). In order to prepare an injectable material, it better to have a lower gelation time. The reinforced hydrogels with larger crosslinker amounts display improved mechanical stability within the smaller period of time. Based on this, reduced crosslinker percent is adequate cell delivery to the targeted site, while concentrated crosslinker polymer systems are more suited for acellular cements placed in the bone cavity. When discussing the effect of HA amount, it was ensured a uniform distribution of particles in the PEGMC network at several reinforcement levels ranging between 30% and 70% HA. The complex viscosity was enlarged by the presence of filler material. Surprisingly samples containing 70% loading exhibit higher G' than G'' as a possible suppression of crosslinking induced by fillers. Rheological moduli of reinforced materials were higher up to a factor of 10 for 50%–70% loading in regard to unfilled sample. This could be explained if considering that HA particles are properly dispersed in the matrix and favor drawing chains together via ionic interactions forming crossover-points in the sample structure and as a result better mechanical strength. Close investigation of the viscoelastic characteristics of injectable hydrogel biocomposites is paramount for the preparation of scaffolds for bone tissue regeneration.

Polymer hydrogels for articular cartilage repairing were reported by Nanda et al. [74]. They used poly (vinyl alcohol)-chondroitin sulfate as scaffolds, where the PVA has the role to enhance bioadhesivity and chondroitin sulfate augments the glycosaminoglycan amount of ECM. Glutaraldehyde (GA) was used to ensure crosslinking of chondroitin sulfate and the selected polymer. The scaffold formulations were studied from a rheological point of view. The viscosity of samples subjected to different shearing levels (0–5 s^{-1}) tends to decrease under higher deformation rates. All aqueous samples present a pseudoplastic flow. The most pronounced steep viscosity curve was achieved for the system having 15 w/v (%) of PVA, 1.5 v/v (%) GA, and 100 mg chondroitin sulfate. This could be ascribed to the existence of free hydroxyl groups that are able to react with the cross-linking agent. The flow behavior of hydrogels is influenced by the PVA concentration and percent of cross-linker that has the role to optimize the flow properties. Data collected from in vivo tests indicate that the biodegradable hydrogels enabled the repairing of defects without inflammation. Thus, the doped PVA is a promising material with applicability in cartilage tissue engineering.

Zuidema et al. [75] developed a protocol concerning the rheological investigation of hydrogel scaffolds. The proposed approach enables evaluation of the hydrogel equilibrium modulus together with gelation time. The reported protocols are applied to certain biopolymers used in regenerative medicine. The procedure involves the following steps:

1) time sweep using random strain and frequency to estimate the gelation time polymer network in solution,
2) strain sweep to evaluate the linear viscoelastic zone of the material in regard to strain,
3) frequency sweep analyses to check the linear equilibrium modulus plateau of the polymer gel, and
4) time sweep with values attained from sweeps of strain and frequency in order to precisely

determine the equilibrium moduli and gelation time.

In this way, it is easy to achieve data on gel's strength and structure during network formation. Further nonlinear rheological analyses could be done to study samples' nonlinear behavior and fracture strength. Based on the outlined approach, one may obtain accurate information on equilibrium gel modulus and gelation time of a polymeric gel. The mentioned parameters are essential for tissue regeneration since the introduced hydrogel should not just interface mechanically with the adjacent tissue, but it must also proceed so in a clinically meaningful manner (here interferes gelation time).

2.4   CONCLUSION

Regenerative medicine, particularly tissue engineering, requires materials with specific architecture to render adequate results. Cellular growth supports made from biopolymers were proven to be suitable for such a purpose, but careful attention must be given to the processing stage. Molecular modeling is a good procedure to visualize the conformational changes of macromolecules in the presence of the adhesion proteins found on the cell membrane. Such computations are useful to select a polymer material cell system with optimum interfacial interactions. Furthermore, rheological parameters can be essential descriptors of the material behavior in precise processing conditions (temperature, solvent, deformation). Depending on the desired scaffold architectural features, the biopolymer fluid must exhibit typical flow behavior and viscoelastic properties.

Future prospects in this biomedical field should include biopolymer systems with partial ordering in aqueous environments. The anisotropy of such materials could be useful for mimicking the living cell features, providing better clinical results. The rheological properties of such systems can be tuned as a function of molecular weight, system composition, shearing force, and in some cases as a function of temperature.

ACKNOWLEDGMENT

This work is dedicated to the 154th Anniversary of the Romanian Academy of Sciences.

KEYWORDS

- bio-based polymer
- rheology
- molecular modeling
- interface
- scaffolds

REFERENCES

1. Massiera, G.; Van Citters, K.M.; Biancaniello, P.L.; Crocker, J.C.; Mechanics of single cells: rheology, time dependence, and fluctuations. *Biophys. J.* **2007**, *93*, 3703–3713.
2. Hiddessen, A.L.; Weitz, D.A.; Hammer, D.A.; Rheology of binary colloidal structures assembled via specific biological cross-linking. *Langmuir* **2004**, *20*, 6788–6795.
3. Al-Malah, K.I.; Azzam, M.O.J.; Abu-Jdayil, B.; Effect of glucose concentration on the rheological properties of wheat-starch dispersions. *Food Hydrocoll.* **2000**, *14*, 491–496.
4. Cosutchi, A.I.; Hulubei, C.; Ioan, S.; Rheological study of some epiclon-based polyimides. *J. Macromol. Sci. Part B.* **2007**, *46*, 1003–1012.
5. Cosutchi, A.I.; Hulubei, C.; Stoica, I.; Ioan, S.; Morphological and structural-rheological relationship in epiclon-based polyimide/hydroxypropylcellulose blend systems. *J. Polym. Res.* **2010**, *17*, 541–550.
6. Aho, J.; Edinger, M.; Botker, J.; Baldursdottir, S.; Rantanen, J. Oscillatory shear rheology in examining the drug-polymer interactions relevant in hot melt extrusion. *J. Pharm. Sci.* **2016**, *105*, 160–167.
7. Vlachopoulos, J.; Polychronopoulos, N.; Basic concepts in polymer melt rheology and their importance in processing. In: *Applied Polymer Rheology: Polymeric Fluids With Industrial Applications.* Kontopoulou, M.; Ed.; Wiley: Hoboken, NJ, **2012**, 1–28.
8. Sirqueira, A.; Júnior, D.; Coutinho, M.; Neto, A.; Silva, A.; Soares, B.; Rheological behavior of acrylic paint blends based on polyaniline. *Polímeros* **2016**, *26*, 215–220.
9. Modi, S.; Koelling, K.; Vodovotz, Y.; Thermal and rheological

properties of poly-(3-hydroxybutyrate-co-3-hydroxyvalerate) and poly(lactic acid) blends for food packaging applications. *Annual Technical Conference.* **2010**, *3*, 2179–2183.

10. Morgan, M.L.; Curtis, D.J.; Deganello, D.; Control of morphological and electrical properties of flexographic printed electronics through tailored ink rheology. *Org. Electron.* **2019**, *73*, 212–218.

11. Negrini, R.; Aleandri, S.; Kuentz, M.; Study of rheology and polymer adsorption onto drug nanoparticles in pharmaceutical suspensions produced by nanomilling. *J. Pharm. Sci.* **2017**, *106*, 3395–3401.

12. Chien, S.; Usami, S.; Taylor, H.M.; Lundberg, J.L.; Gregersen, M.I.; Effects of hematocrit and plasma proteins on human blood rheology at low shear rates. *J. Appl. Physiol.* **1966**, *21*, 81–87.

13. Quinto-Su, P.A.; Kuss, C.; Preiser, P.R.; Ohl, C.-D.; Red blood cell rheology using single controlled laser-induced cavitation bubbles. *Lab Chip.* **2011**, *11*, 672–678.

14. Yokpradit, A.; Tongloy, T.; Kaewpirom, S.; Boonsang, S.; A real-time rheological measurement for biopolymer 3D printing process. *Sensor. Mater.* **2018**, *30*, 2199–2209.

15. Li, X.; Feng, Q.; Dynamic rheological behaviors of the bone scaffold reinforced by chitin fibres. *Mater. Sci. Forum.* **2005**, 475-479, 2387–2390.

16. Stratton, S.; Shelke, N.B.; Hoshino, K.; Rudraiah, S.; Kumbar, S.G.; Bioactive polymeric scaffolds for tissue engineering. *Bioact. Mater.* **2016**, *1*, 93–108.

17. Yang, E.; Miao, S.; Zhong, J.; Zhang, Z.; Mills, D.K.; Zhang, L.G.; Bio-based polymers for 3D printing of bioscaffolds. *Polym. Rev.* **2018**, *58*, 668–687.

18. Hsu, S.; Hung, K.; Chen, C.; Biodegradable polymer scaffolds. *J. Mater. Chem. B* **2016**, *4*, 7493–7505.

19. Rostami, S.; Garipcan, B.; Rheological properties of biological structures, scaffolds and their biomedical applications. In: *Biological, Physical and Technical Basics of Cell Engineering.* Artmann, G.; Artmann, A.; Zhubanova, A.; Digel, I.; Eds.; Springer: Singapore, **2018**, 119–140.

20. Ge, Z.; Jin, Z.; Cao, T.; Manufacture of degradable polymeric scaffolds for bone regeneration. *Biomed. Mater.* **2008**, *3*, 022001 (1–11).

21. Ghassemi, T.; Shahroodi, A.; Ebrahimzadeh, M.H.; Mousavian, A.; Movaffagh, J.; Moradi, A.; Current concepts in scaffolding for bone tissue engineering. *Arch. Bone. Jt. Surg.* **2018**, *6*, 90–99.

22. Polo-Corrales, L.; Latorre-Esteves, M.; Ramirez-Vick, J.E.; Scaffold design for bone regeneration. *J. Nanosci. Nanotechnol.* **2014**, *14*, 15–56.

23. Huss, F.R.; Nyman, E.; Gustafson, C.J.; Gisselfält, K.; Liljensten, E.; Kratz, G.; Characterization of a new degradable polymer scaffold for regeneration of the dermis: in vitro and in vivo human studies. *Organogenesis* **2008**, *4*, 195–200.

24. Jeong, K.; Park, D.; Lee, Y.; Polymer-based hydrogel scaffolds for skin tissue engineering applications: a mini-review. *J. Polym. Res.* **2017**, *24*, 112 (1–10).

25. Kennedy, K.M.; Bhaw-Luximon, A.; Jhurry, D.; Skin tissue engineering: biological performance of electrospun polymer scaffolds and translational challenges. *Regen. Eng. Transl. Med.* **2017**, *3*, 201–214.

26. Fuoco, C.; Petrilli, L.L.; Cannata, S.; Gargioli, C.; Matrix scaffolding for stem cell guidance toward skeletal muscle tissue engineering. *J. Orthop. Surg. Res.* **2016**, *11*, 86.

27. Nakayama, K.H.; Shayan, M.; Huang, N.F.; Engineering biomimetic materials

for skeletal muscle repair and regeneration. *Adv. Healthc. Mater.* **2019**, *8*, e1801168.

28. Duarte Campos, D.F.; Drescher, W.; Rath, B.; Tingart, M.; Fischer, H.; Supporting biomaterials for articular cartilage repair. *Cartilage* **2012**, *3*, 205–221.

29. Morsi, Y.S.; Bioengineering strategies for polymeric scaffold for tissue engineering an aortic heart valve: An update. *Int. J. Artif. Organs.* **2014**, *37*, 651–67.

30. Sannino, A.; Demitri, C.; Madaghiele, M.; Biodegradable cellulose-based hydrogels: design and applications. *Materials* **2009**, *2*, 353–373.

31. Long, Y.; Zhao, X.; Liu, S.; Chen, M.; Liu, B.; Ge, J.; Jia, Y.G.; Ren, L.; Collagen-hydroxypropyl methylcellulose membranes for corneal regeneration. *ACS Omega.* **2018**, *3*, 1269–1275.

32. Halima, N.B.; Poly(vinyl alcohol): review of its promising applications and insights into biodegradation, *RSC Adv.* **2016**, *6*, 39823–39832.

33. Qi, B.; Yu, A.; Zhu, S.; Chen, B.; Li, Y.; The preparation and cytocompatibility of injectable thermosensitive chitosan/poly(vinyl alcohol) hydrogel. *J. Huazhong. Univ. Sci. Technolog. Med. Sci.* **2010**, *30*, 89–93.

34. Stoica, I.; Barzic, A.I.; Butnaru, M.; Doroftei, F.; Hulubei, C.; Surface topography effect on fibroblasts population on epiclon-based polyimide films. *J. Adhes. Sci. Technol.* **2015**, *29*, 2190–2207.

35. Panos, M.; Sen, T.Z.; Göktuğ Ahunbay, M.; Molecular simulation of fibronectin adsorption onto polyurethane surfaces. *Langmuir* **2012**, *28*, 12619–12628.

36. Shamloo, A.; Sarmadi, M.; Investigation on the adhesive characteristic of polymer-protein systems through molecular dynamics simulation and its correlation with cell adhesion and proliferation. *Integr. Biol.* **2016**, *8*, 1276–1295.

37. Gunamalai, L.; Molecular dynamics simulation studies on the interaction of type I collagen telopeptides with cyclodextrins. *Int. J. Curr. Res.* **2014**, *6*, 7826–7830.

38. Benchea, A.C.; Găină, M.; Dorohoi, D.O.; The computed thermodynamic parameters of salicylic acid. *Bul. Polytechnique Inst. Iasi*, **2016**, *62*, 41–51.

39. Suárez, J.C.; Miguel, S.; Pinilla, P.; López, F.; Molecular dynamics simulation of polymer–metal bonds. *J. Adhes. Sci. Technol.* **2008**, *22*, 1387–1400.

40. Tong, Z.; Xie, Y.; Zhang, Y. Molecular dynamics simulation on the interaction between polymer inhibitors and β-dicalcium silicate surface. *J. Mol. Liq.* **2018**, *259*, 65–75.

41. Dhandayuthapani, B.; Yoshida, Y.; Maekawa, T.; Kumar, D.S.; Polymeric scaffolds in tissue engineering application: a review. *Int. J. Polym. Sci.* **2011**, *2011*, 290602, 1–19.

42. Nikolova, M.P.; Chavali, M.S.; Recent advances in biomaterials for 3D scaffolds: a review. *Bioact. Mater.* **2019**, *4*, 271–292.

43. Shruti, S.; Salinas, A.J.; Lusvardi, G.; Malavasi, G.; Menabue, L.; Vallet-Regi, M.; Mesoporous bioactive scaffolds prepared with cerium-, gallium- and zinc-containing glasses. *Acta Biomater.* **2013**, *9*, 4836–4844.

44. Wei, G.; Ma, P.X.; Structure and properties of nano-hydroxyapatite/polymer composite scaffolds for bone tissue engineering. *Biomaterials* **2004**, *25*, 4749–4757.

45. Hulbert, S.F.; Young, F.A.; Mathews, R.S.; Klawitter, J.J.; Talbert, C.D.; Stelling, F.H.; Potential of ceramic materials as permanently implantable skeletal prostheses. *J. Biomed. Mater. Res.* **1970**, *4*, 433–456.

46. Lee, M.; Wu, B.M.; Dunn, J.C.; Effect of scaffold architecture and pore size on

smooth muscle cell growth. *J. Biomed. Mater. Res. A.* **2008**, *87*, 1010–1016.

47. Boyan, B.D.; Hummert, T.W.; Dean, D.D.; Schwartz, Z.; Role of material surfaces in regulating bone and cartilage cell response. *Biomaterials* **1996**, *17*, 137–146.

48. Mouriño, V.; Boccaccini, A.R.; Bone tissue engineering therapeutics: controlled drug delivery in three-dimensional scaffolds. *J. R. Soc. Interface.* **2012**, *7*, 209–227.

49. de Mulder, E.L.W.; Buma, P.; Hannink, G.; Anisotropic porous biodegradable scaffolds for musculoskeletal tissue engineering. *Materials (Basel)* **2009**, *2*, 1674–1696.

50. Pawelec, K.M.; Husmann, A.; Wardale, R.J.; Best, S.M.; Cameron, R.E.; Ionic solutes impact collagen scaffold bioactivity. *J. Mater. Sci. Mater. Med.* **2015**, *26*, 91–95.

51. Barzic, A.I.; Rusu, R.D.; Stoica, I.; Damaceanu, M.D.; Chain flexibility versus molecular entanglement response to rubbing deformation in designing poly (oxadiazolenaphthylimide)s as liquid crystal orientation layers. *J. Mater. Sci.* **2014**, *49*, 3080–3098.

52. Cosutchi, A.I.; Nica, S.L.; Hulubei, C.; Homocianu, M.; Ioan, S.; Effects of the aliphatic/aromatic structure on the miscibility, thermal, optical, and rheological properties of some polyimide blends. *Polym. Eng. Sci.* **2007**, *52*, 1429–1439.

53. Albu, R.M.; Hulubei, C.; Stoica, I.; Barzic, A.I.; Semi-alicyclic polyimides as potential membrane oxygenators: rheological implications on film processing, morphology and blood compatibility. *eXPRESS Polym. Lett.* **2019**, *13*, 349–364.

54. Barzic, A.I.; Soroceanu, M.; Albu, R.M.; Ioanid, E.G.; Sacarescu, L.; Harabagiu, V.; Correlation between shear-flow rheology and solution

spreading during spin coating of polysilane solutions. *Macromol. Res.* **2019**, *27*, 1210–1220.

55. Pietrucha, K.; Changes in denaturation and rheological properties of collagen–hyaluronic acid scaffolds as a result of temperature dependencies. *Int. J. Biol. Macromol.* **2005**, *36*, 299–304.

56. Sab, G.; Hesarak, S.; Hajisafari, M.; Utlization of rheological parameters for the prediction of β-TCP suspension suitability to fabricate bone tissue engineering scaffold through foam replication method. *J. Aust. Ceram. Soc.* **2018**, *54*, 587–599.

57. Song; C.; Luo; Y.; Liu; Y.; Li, S.; Xi, Z.; Zhao, L.; Cen, L.; Lu, E.; Fabrication of PCL scaffolds by supercritical CO_2 foaming based on the combined effects of rheological and crystallization properties. *Polymers* **2020**, *12*, 780 (1–14).

58. Xu, M.; Yan, H.; He, Q.; Chen, W.; Tao, L.; Ling, Z.; Park, C.B.; Chain extension of polyamide 6 using multifunctional chain extenders and reactive extrusion for melt foaming. *Eur. Polym. J.* **2017**, *96*, 210–220.

59. Choi, W. J.; Hwang, K.S.; Kwon, H.J.; Lee, C.; Kim, C.H.; Kim, T.H.; Heo, S.W.; Kim, J.-H.; Lee, J.-Y.; Rapid development of dual porous poly(lactic acid) foam using fused deposition modeling (FDM) 3D printing for medical scaffold application. *Mater. Sci. Eng.* C **2020**, *110*, 110693.

60. Corre, Y.-M.; Duchet J.; Reignier J.; Maazouz A.; Melt strengthening of poly (lactic acid) through reactive extrusion with epoxy-functionalized chains. *Rheol. Acta.* **2011**, *50*, 613–629.

61. Xia, H.; Chen, Q.; Fang, Y.; Liu, D.; Zhong, D.; Wu, H.; Xia, Y.; Yan, Y.; Tang, W.; Sun, X.; Directed neurite growth of rat dorsal root ganglion neurons and increased colocalization with Schwann cells on aligned

poly(methyl methacrylate) electrospun nanofibers. *Brain Res.* **2014**, *1565*, 18–27.

62. Hejazian, L.B.; Esmaeilzade, B.; Ghoroghi, F.M.; Moradi, F.; Hejazian, M.B.; Aslani, A.; The role of biodegradable engineered nanofiber scaffolds seeded with hair follicle stem cells for tissue engineering. *Iran. Biomed. J.* **2012**, *16*, 193–201.

63. Kazemnejad, S.; Khanmohammadi, M.; Baheiraei, N.; Arasteh. S.; Current state of cartilage tissue engineering using nanofibrous scaffolds and stem cells. *Avicenna J. Med. Biotechnol.* **2017**, *9*, 50–65.

64. WadeabJason, R.J.; Burdick, A.; Advances in nanofibrous scaffolds for biomedical applications: from electrospinning to self-assembly. *Nano Today.* **2014**, *9*, 722–742.

65. McKee, M.G.; Wilkes, G.L.; Colby, R.H.; Long, T.E.; Correlations of solution rheology with electrospun fiber formation of linear and branched polyesters. *Macromolecules* **2004**, *37*, 1760–1767.

66. Chisca, S.; Barzic, A.I.; Sava, I.; Olaru, N.; Bruma, M.; Morphological and rheological insights on polyimide chain entanglements for electrospinning produced fibers. *J. Phys. Chem. B.* **2012**, *116*, 9082–9088.

67. Tanodekaew, S.; Channasanon, S.; Kaewkong, P.; Uppanan, P.; PLA-HA scaffolds: preparation and bioactivity. *Procedia Eng.* **2013**, *59*, 144–149.

68. Li, Z.; Lei, I.M.; Davoodi, P.; Huleihel, L.; Huang, Y.Y.S.; Solution formulation and rheology for fabricating extracellular matrix-derived fibers using low-voltage electrospinning patterning. *ACS Biomater. Sci. Eng.* **2019**, *5*, 3676–3684.

69. Mirtic, J.; Balažic, H.; Zupancic, Š.; Kristl, J.; Effect of solution composition variables on electrospun alginate nanofibers: response surface analysis. *Polymers* **2019**, *11*, 692 (1–20).

70. Maas, M.; Hess, U.; Rezwan, K.; The contribution of rheology for designing hydroxyapatite biomaterials. *Curr. Opin. Colloid Interface Sci.* **2014**, *19*, 585–593.

71. Barzic, R.F.; Barzic, A.I.; Dumitrascu, Gh.; Percolation network formation in poly(4-vinylpyridine)/aluminum nitride nanocomposites: rheological, dielectric, and thermal investigations. *Polym. Compos.* **2014**, *35*, 1543–1552.

72. Barzic, A.I.; Temperature implications on the rheological percolation threshold in poly(4-vinylpyridine)/barium titanate nanocomposites. *Rev. Roum. Chim.* **2014**, *59*, 515–519.

73. Jiao, Y.; Gyawali, D.; Stark, J.M.; Akcora, P.; Nair, P.; Tran, R.T.; Yang, J.; A rheological study of biodegradable injectable PEGMC/HA composite scaffolds. *Soft Matter.* **2012**, *8*, 1499–1507.

74. Nanda, S.; Sood, N.; Reddy, B.V.K.; Markandeywar, T.S.; Preparation and characterization of poly(vinyl alcohol)-chondroitin sulfate hydrogel as scaffolds for articular cartilage regeneration. *Indian J. Mater. Sci.* **2013**, *2013*, 516021, 1–8.

75. Zuidema, J.M.; Rivet, C.J.; Gilbert, R.J.; Morrison, F.A.; A protocol for rheological characterization of hydrogels for tissue engineering strategies. *J. Biomed. Mater. Res. B Appl. Biomater.* **2013**, *102*, 1063–1073.

CHAPTER 3

REVEALING NOVEL FUNCTIONALITIES AND INHERENT PROPERTIES OF BIO-BASED POLYMERS FOR BIOMEDICAL AND SUSTAINABLE APPLICATIONS

DARIEO THANKACHAN[1], RINKY GHOSH[1], and NEHA KANWAR RAWAT,[2,3,*]

[1]Department of Materials Science and Engineering, Indian Institute of Technology Kanpur, Kanpur 208016, India

[2]Materials Research Laboratory, Department of Chemistry, Jamia Millia Islamia, New Delhi 110025, India

[3]Department of Chemistry, Haryana Education Services, Haryana 127021, India

*Corresponding author. E-mail: neharawatjmi@gmail.com

ABSTRACT

Biopolymers are aptly known as fascinating biodegradable polymers obtained from both renewable and nonrenewable sources. These polymers can either be produced from biological systems like microorganisms, plants, and animals or biological starting materials like corn, sugar, and starch through the chemical synthesis method. This category includes starch, polyhydroxyalkanoates which are produced by microbial degradation, and polylactic acid which is also categorized as synthetic polymers obtained from renewable resources. Their splendid properties owe their versatility for developing scaffolds for tissue engineering applications by promisingly tailored the degradation kinetics and mechanical integrity and provides a platform for drug encapsulations. Because of the effectiveness of synthetic polymers, they can be easily processed into various shapes with desired pore size and magnificent morphologic features helps in tissue regrowth. Drug delivery systems, orthopedic fixation devices such as rods, pins, and screws, and resorbable sutures are the major applications. Among all biopolymers, polyesters are extremely attractive due to their ease of degradation kinetics causes hydrolysis of ester linkages. The researchers across the globe are more concerned about the orthopedics diseases/injuries to find tissue-engineered solutions and subsequently the importance of the development of new polymeric materials that bunch up a number of demanding requirements. Bioplastics are environmentally friendly as it helps in eradicating the main cause of global warming, for example, emission of less carbon dioxide. Plastics are made from petroleum byproducts when they are burned, they release a huge amount of carbon dioxide into the atmosphere that leads to global warming. Moreover, bioplastics are economical as compared to petroleum-based plastics in terms of lower production costs due to the availability of cheaper raw materials, shorter process chains, and tailored raw materials. It can reuse more efficiently as it breakdown fastly and hence its recycling consumes less energy.

3.1 INTRODUCTION

Biodegradable polymers are vastly divided based on their origin, that is, natural and synthetic (Table 3.1). Polymers with synthetic origin have a wide range of applications and according to the need, they have the capability to tailor mechanical properties and altering the rate of degradation. At the same time, natural polymers due to their excellent biocompatibility seem to be attractive but due to their undesirable properties like batch-to-batch variation and antigenicity, they have not been fully investigated. Natural polymers with a functional group are considered a suitable candidate for tissue engineering applications and are less prone to produce toxic effects. The presence of such functional groups may sometimes be caused by undesirable immunological effects which promptly lead to contamination. Biodegradable polymers are versatile biopolymers that are rapidly replacing other classes of biomaterials such as ceramics, metals, and alloys [1–3]. Biodegradability is a term that can directly be correlated with the biopolymer functionality, and it specifically degrades under the action of microorganisms within a given time period and in a suitable environment. According to the definition of the Japan Bioplastics Association, biodegradability is the characteristic property of a material that undergoes microbiological degradation and break down into corresponding CO_2 and H_2O by-products, which can easily be recycled in nature. Biodegradation in other terms means disintegration under microbial actions. In the year 2003, survey accounts for sales of more than \$7 billion polymeric biomaterials due to their outstanding properties and faster degradation kinetics and worldwide accepted as a better alternative as compared to bioresistant plastics [4–6]. For polymers, the most generalized criteria to choose particular materials are on the basis of their toxicological profiles. On these accounts, plastics with less toxicological and immunological response are particularly taken into consideration which does not possess any adverse effects on the environment. By the end of 2007, the data recorded for the annual consumption of biodegradable polymers show an increment from 409 to an estimated 541 million pounds. By 2012, it was showing an annual growth rate of 17.3%, more specifically, it is also expressed in terms of compound annual growth rate that indicates a notable contribution toward energy efficiency and the most desired progress toward ecological balance [4–7]. In considering the biomedical applications, the requirement becomes more diverse in terms of degradation time, mechanical strength required, surface properties, degree of crosslinking, physicochemical parameters, tagging, and necessity of functional group subjected to surface modification. The purely mechanical function requires in addition to biocompatibility for some of the applications like bone repair and bone grafting. So that, polymers that have load withstanding property and faster degradation kinetics are promisingly used for clinical applications [8, 9]. Applications such as sutures, surgical dressings, and the like require varying degradation time, and strength depends on the type of injury and the type of tissue. For generating more biofunctional tissue and expeditious regrowth of the tissue, growth factors are playing a vital role in tissue engineering. However, various underlined features of natural polymers suggested are important in various technological fields and are affected by some limitations such as difficulty in processing, immunogenicity, potential risk of transferring origin related pathogens, and batch to batch variability. For drug delivery applications, the time of the release of drug at the targeted sites is the prime important factor. Commonly, the slower release rate of bioactive agents is showed by polymers having higher hydrophobicity. In this case, the requirement of the instant release of drugs that are triggered by external stimuli, polymers with functional groups in them is suitable.

Therefore, targeting moieties can be attached to it. Conversely, for the release of bioactive, the polymer is only the governing factor, then predictable release is of sizeable importance.

3.2   BIOPOLYMER DEFINITION

Biopolymers are biomaterials that are either classified as biodegradable or nonbiodegradable in terms of their degradation kinetics, which are conventionally derived from renewable and nonrenewable resources.

3.3   TYPES OF BIOPOLYMERS

1. Biopolymers, which are made from renewable bio-based raw materials, are biodegradable
2. Biopolymers which are obtained from fossil fuels are considered as biodegradable
3. Biopolymers which are obtained from bio-based raw materials are categorized as nonbiodegradable.

3.3.1   BIOPOLYMERS: BIO-BASED RAW MATERIALS (BIODEGRADABLE)

Here, the biopolymer is produced by biological systems like microorganisms, plants, and animals. Also, it is chemically synthesized

from biological starting materials like corn, sugar, and starch. This category includes starch or proteins, polyhydroxyalkanoates (PHAs)-based biopolymers are obtained from microbial actions, and polylactic acid (PLA)-based synthetic polymers are obtained from renewable resources. Among all biopolymers, starch and PHAs are widely used.

3.3.2   BIOPOLYMERS: BIO-BASED RAW MATERIALS (NONBIODEGRADABLE)

These are nonbiodegradable biopolymers that are obtained directly from biomass or renewable resources. Some of the examples under this category are polyamide 11 (specific polyamides from castor oil), natural rubber, polyesters-based biopropanediol, biopolyethylene (bio-LDPE, bio-HDPE), bio-polypropylene (bio-PP), and bio polyvinylchloride (bio-PVC).

3.3.3   BIOPOLYMERS: FOSSIL FUELS BASED RAW MATERIALS (NONBIODEGRADABLE)

This category includes synthetic aliphatic polyesters made from natural gas or crude oil and is ratified compostable and biodegradable: poly(ε-caprolactone) (PCL), poly(butylene succinate) (PBS), and certain aliphatic–aromatic copolyesters.

3.4 WHY BIOPOLYMERS?

Biopolymers will reduce the carbon footprint. Carbon dioxide equivalent is calculated by summing up the greenhouse gas emission, and accountable results are prevailing obtained from every stage of a product or service lifetime (production of material, manufacturing, use phase, end of life, disposal). The practice of biopolymers is necessary due to the limited availability of fossil fuel resources, and the main cause behind the boycott of petroleum-based plastics is that it has adverse effects on climate change.

3.5 WHY BIOPLASTICS?

Bioplastics are environmentally friendly and their production results in less emission of carbon dioxide, which in turn is the major consequence of global warming. When plastics are burned, they release a huge amount of carbon dioxide into the atmosphere, which leads to global warming. Moreover, bioplastics are economic since petro-based plastics, especially with rising oil prices and, have a lower production cost due to cheaper raw materials, shorter process chains, and tailored raw materials.

3.6 CARBON FOOTPRINT

The carbon footprints are mainly classified into two categories: primary and secondary footprints. The primary footprint is described as the total of direct CO_2 emissions produced by the burning of fossil fuels, and the secondary footprint is the indirect measure of CO_2 emissions associated with the manufacturing industries providing goods, services which are owned by individual or business consumes [10, 18].

3.7 CARBON OFFSETTING

Carbon offsetting is the reduction in the emissions of greenhouse gases or carbon dioxide made in pursuance of compensating for or to offset an emission made elsewhere. By developing alternative products, utilizing solar energy environmental friendly renewable practices can be followed [19].

3.8 COMPOSTABLE POLYMERS

According to ASTM D6002, compostable polymers are defined as those polymers which undergo biological decompositions in the presence of aerobic bacteria and fungi and eventually breaking down the hydrocarbon part into carbon dioxide, inorganic compounds, water, ammonia, and biomass, therefore, at the end leaves no toxic residue.

For plastics to be called compostable, it should meet three important criteria as described below:

1. Biodegrade—break down under aerobic conditions at the same rate as described for cellulose, into carbon dioxide, water, and biomass.
2. Disintegrate—material should be indistinguishable, and it is essential to be screened out and is not visible.
3. Eco-toxicity—compost can support plant growth and the degradation process leads to no toxic effects [11].

3.9   CRITERIA FOR BIODEGRADATION

The amount of CO_2 released over a certain period of time and can be determined from the rate of biodegradation of plastic. The ASTMD standard described that 60% hydrocarbon conversion takes place within 180 days of the experiment, which is observed in the case of single-use polymer and 90% conversion of carbon into carbon dioxide in the case of copolymers.

The comprehensive list of abbreviations used in Table 3.1 are as follows: PBAT = poly(butylene adipate-*co*-terephthalate); PBS=poly (butylene succinate); PBST = poly (butylene succinate-*co*-terephthalate); PBSL=poly(butylenesuccinate-*co*-lactide); PTMAT = poly(methylene adipate-*co*-terephthalate); PBSA = poly(butylene succinate adipate); PA11 = amino-undecanoic acid-derived polyamide; PA12 = laurolactam-derived polyamide; CAP = cellulose acetate propionate; CAB = cellulose acetate butyrate.

3.10   CLASSIFICATION OF BIOPOLYMERS (TABLE 3.2)

TABLE 3.1   Representing Significant Dissimilarities Between Biodegradable and Nonbiodegradable Polymers According to its Origin

Origin	Biodegradable	Nonbiodegradable
Bio-based	Starch, Chitosan, PLA, PHBV, CA, CAB, CAP, CN, PHB	PET, PTT, PA11, PA12, PE(LDPE)
Partially bio-based	Starch blends, PLA blends PBS, PBAT	PET, PBT, PVC, SBR, ABS, PU, PTT, epoxy resin
Fossil fuel-based	PVOH, PBS, PBSA, PBSL, PCL, PGA	PP, PS, PC, PE(HDPE, LDPE) ABS, PBT, PET, PS, PA 6, PA 6, 6 epoxy resin, synthetic rubber

TABLE 3.2 Representative Classification of Biopolymers

	Biodegradable				Nonbiodegradable
	Bio-based			Fossil-based	Bio-based
Plant		**Animal**	**Microorganisms**		
Cellulose and its derivatives (polysaccharide)	Chitin (polysaccharide)		PHAs (e.g., P4HB), PHB, PHBH, PHBx, PHBV	Poly(alkylene dicarboxylate)s (e.g., PBS, PBA, PBSA, PEA, PES, PBSE, PESE, PPS, PPF, PTMS, PTSE, PTT	PE(LDPE, HDPE) PP, PVC
Starch and its derivatives (monosaccharide)	Hyaluronan (polysaccharide)		Bacterial cellulose	PCL	PU
Lignin	Chitosan (polysaccharide)		PHF	PGA	PET, PPT
Alginate (polysaccharide)	Casein (protein)		Hyaluronan (polysaccharide)	PVOH	PC
Lipids (triglycerides)	Whey(protein)		Xanthan (polysaccharide)	POE	Poly(ether-ester)s
Corn, potato, wheat, soy, Pea(protein)	Collagen (protein)		Curdian (polysaccharide)	Polyanhydrides	Polyamides (PA410, PA610, PA1010, PA1012, PA11)
Gums (e.g., cis-1,4-polyisoprene)	Albumin (protein)		Pullulan (polysaccharide)	PPHOS	Polyester amides
PLA (from sugarcane or starch)	Leather (protein)				Epoxy
Carrageenan	Keratin, PFF (protein)		Silk (protein)		Unsaturated polyesters

3.11 VARIOUS BIOMEDICAL APPLICATIONS OF BIODE- GRADABLE POLYMERS (TABLE 3.3)

3.11.1 PROSTHETIC DEVICES AND IMPLANTS

During Roman's age and Egyptian's period, the use of biomaterials for medical implants is undivulged. They have used linen for sutures, metals, and woods for dental and toe applications. Later they realized the consequences of immunological rejection [20–22] by the body of the patient's that leads to implant-related infections. Under certain favorable biological conditions, for example, during implants, the success of materials in the body is determined by the strong interactions between host tissues and surgical procedures, and sterilization procedure. Adopting various therapeutic approaches to deal with the infections were not proven to be effective on the laboratory scale causes a dismal prognosis to the patients. By the end of the 20th century, monomers-containing glycolic acid and other hydroxyl acid groups were prohibited to use for polymers synthesis as the requisite polymers were too unpredictable under environmental conditions leading to degradation at a faster rate, therefore, failed as a product in terms of industrial uses. Biodegradable sutures made from degradable polymers was first reported in the 1960s [23] as it offers multitude advantages and was successful in fulfilling the physician's desire of having an implantable device, to evade the long-term effect of cytotoxic substances inside the body and eliminate the necessity for the second surgical event.

The concept of using degradable polymers for implant surgery has been a topic of interest for many years. The research works around the globe from the last few decades for the development of biomaterials, which would have a significant contribution in the field of medical science to employ as implants in the human body. The word biopolymers was coined to explain the biological performance and biocompatibility of the materials. Biopolymers have been recognized as a versatile material with promising properties, such as noncytotoxicity, half-life time, three-dimensional structure, and biocompatibility with a comprehensive acceptance in implantable devices. The development of therapeutic devices based on biopolymers is highly suitable as its degradation products are nonimmunogenic and do not reduce inflammatory response, therefore proven to be quite promising in its effectiveness. The implants are accompanied by regeneration/replacing/repairing of a damaged organ and traumatized tissues, its main objective to mimic a body part and assist in the healing process. The new and advanced escalation studies revealed the effectiveness and safety of employing biopolymeric materials in implants for curing recurrent disease and thus improving the quality and span life of the patients. Numerous clinical

TABLE 3.3 Representing Various Biodegradable Polymers and Their Promising Features and Characteristics in Biomedical Arenas

Polymer	Chemical Structure	Key Features/ Characteristics	Applications	Marketed Product
Poly (glycolic acid) [13]		First biodegradable synthetic polymer explored for biomedical applications. Excellent mechanical integrity and high crystallinity	Drug delivery, tissue engineering, sutures	Dexon biofix
Poly (lactic acid) [14, 15]		Chiral and exist in two optically active forms; L- and D-lactide. L-lactide is naturally occurring monomer	Tissue engineering, sutures, Drug delivery, bone fixtures	Thread soft, phantom soft, atridox
Poly (ε-capro-lactone) [16]		Bulk degrading polymer, forming an compatible blends with other polymers	Tissue engineering, sutures, drug delivery	Monacryl, capronor, synBiosys
Poly (phospha-zenes) [17]	OAr/PEG $\mathord{-}\!\!\left(\!\!\begin{array}{c} \mathrm{P}\!=\!\mathrm{N}\!-\! \end{array}\!\!\right)_{\!\!n}$ OAr/PEG	Ability to control rate and mode of degradation, biocompatibility, unique functionality	Nonload-bearing tissue engineering, drug delivery	
Gelatin	Consists of distribution of polypeptide fragments of different sizes	Often exhibit lot-to-lot variability	Device coatings, hemostats, implants	Surgifoam, gelfoam, cultiSpher-G
Fibrin	It is a protein matrix	Providing a substrate for cell adhesion, differentiation, and proliferation	Tissue engineering, biological adhesives	Tisseel
Ceramic, hydroxya-petite	$[Ca_{10}(PO_4)_6-(OH)_2]$	The main inorganic compound of teeth and bone, bone tissue engineering applications	Used as granules, paste, or porous blocks in dental and orthopedic fields	Hapex

TABLE 3.3   *(Continued)*

Polymer	Chemical Structure	Key Features/ Characteristics	Applications	Marketed Product
Chitin and chitosan	Chitin chitosan	Native chitin is insoluble in many organic solvents but a partially deacetylated form of chitosan shows solubility in water. It is biocompatible, biodegradable, nontoxic, and biofunctional	Absorbable sutures, artificial skin, water-soluble pro-drug. As a moisturizer in cosmetics	
Cellulose		Highly crystalline and high molecular weight polymer	Film-forming agent	
Starch		Acetylated starch have better structural fiber and more hydrophobic compared to native starch	Film formation. In solid usage form used as a disintegrant, binder, etc.	

studies are currently in progress to explore the more possible benefits of biodegradable polymers in various fields of medicinal science.

Designing prosthetic devices from pure biodegradable polymeric materials must ensure certain practicability issues, for example, low-elastic modulus and structural compatibility but mechanical strength is the important factor that ensures viability and durability. Furthermore, due to the low density of polymeric materials in prosthetics, improvement in agility and gait was noticed which also permit to seamlessly compensate the stiffness compared to stainless steel equivalents [25, 26]. Literature depicts the mechanical impact on the skeleton in terms of "stress experienced by bones and tendons/ligaments" during stretching, jumping, or any physical activities were calculated in the range of 4–80 MPa. Moreover, loads are variable parameters and are applied repeatedly and frequently over and over on hip joints, and the impact on it is measured in terms of stress cycles that are estimated to be of the order of 1×10^6 cycles per year [24].

A large group of polymers have been used for various applications due to their availability and can be molded into any complex shape and structure but sometimes it is way too flexible to meet the demands where mechanical strength is the predetermined factor, as well as the chances of leaching out of cytoreactive substances, are more in polymers. Therefore, the sterilization process is essential during the usage of polymers in medical applications. Biopolymers offering a number of advantages in biomedical applications mainly in terms of biocompatibility and entrapping capability of therapeutic agents on the surface and ability to stimulate proliferation, differentiation, cellular adhesion, and tissue regeneration. On the other hand, metals have certain drawbacks and some of them are listed below: highly corrosive, very high surface density and high stiffness, low biocompatibility, fabrication cost is high, difficult to molded, lack of bioadhesion, and release of metal ions over a period during implants may cause an allergic reaction to tissues and leads to severe infections ultimately causes the death of the patients.

3.11.1.1   IMPORTANCE OF BIODEGRADABLE POLYMERS IN PROSTHESIS AND IMPLANTS

3.11.1.1.1   Poly(lactide)

PLA consisting of a cyclic dimer of lactic acid prepared by ring-opening polymerization. PLA can be obtained from natural sources like corn and starch. It exists in two different enantiomeric forms: D- and L-lactide. The L-PLA is a semicrystalline polymer exhibiting high-load bearing capacity and have a high modulus of elasticity make them a suitable candidate for orthopedic fixations and bone implants. On the other hand, D-PLA is an amorphous polymer with the random distribution of monomers along the backbone of the polymeric chain know for exhibiting lower mechanical strength with high elongation and more rapid degradation rate make them attractive prospective for drug delivery applications. PLA is biocompatible and bioresorable materials (Table 3.4). Its hydrophobic characteristics and slower degradation rate makes it difficult during implants in the human body and was showed by Ishaug-Riley et al. have also successfully demonstrated the active encapsulation of chondrocytes on the surface porosity of PLA mesh preferably supported the cytocompatibilty and supported in cell proliferations [28].

TABLE 3.4   Lists of Biomedical Use of PLA and Its Different Optical Isomeric Forms [29]

Sr. No	Abbreviation	Polymer	Future Use
1.	PLA	Poly(lactic acid)	Tissue engineering and cell transplantations
2.	L-PLA	Poly(L-lactide)	Orthopedics surgery, bone implants, meniscus implant, sutures
3.	D-PLA	Poly(D-lactide)	Wound dressing, drug delivery, tissue engineering, fracture fixations, sutures

Kulkarni et al. reported the pioneering use of PLA in biodegradable surgical implant and it can be applicable for numerous versatile applications such as the bone implant, cardiovascular devices, and drug delivery—the controlled and release behavior of PLA make them a suitable candidate for drug release [27].

3.11.1.1.2 Poly(organo) phosphazene

Poly (organo) phosphazene having tremendous outstanding properties and better prospects for nerve implants. They are an emerging class of biopolymers and their unprecedented structure containing alternative nitrogen and phosphorous atoms along the backbone of polymeric chain enhances polymer properties, such as electrical conductivity, bioactivity, and stretchability provides a platform for better encapsulation of bioactive agents.

Recent advances in poly(organo) phosphazene offer a functionalization to the known hydrophilic sensitive group, such as alcohol-containing groups, imidazoles, or amino esters group possible through the controlled mode of synthesis, to improve the degradation behavior in water leads to the noncytotoxic formation of the byproducts, therefore, emerged as foremost significant in nerve regrowth and gene delivery [30]. Due to its novel molecular architectures, controlled degradability, and biocompatibility, they provide a requisite condition for the preparation of guided channels for nerve repair and regrowth and helps in elucidation of the cell proliferations. They find applications in vaccine delivery, biosensors, biodetection, tubular prosthesis, nerve regeneration, and drug release devices.

3.11.1.1.3 Silk Protein

Silk is obtained from the cocoons of the matured silkworm. Fibroin and sericin are major proteins extracted from the silk albumen with an average length of 1500 m (approx.) and are promisingly used as antimicrobial and antibacterial agents owes to their versatile properties like biocompatibility, biodegradability, noninflammatory, hemostatic prospects, and nonimmunogenicity [31]. The applications of silk protein started at an early stage in the arenas of prosthetic implants and regenerative medicine. China is the largest producer of silk followed by India, which discovered the sericulture practicing and usefulness of silk filaments nearly 3000 years ago. The important characteristics of silk fibers lie in their long-lasting strength and low shrinkage properties depend upon the chemical composition.

The main tissue engineering application for a decade is the seed in curing physiological disorders with long-lasting and devastating impact. These are mainly related to various nervous, spinal and bone defects, injuries, and deformations. More significantly, silk-based proteins are proven to be effective for bone tissue engineering.

3.11.1.1.4 Alginate

Alginate, a natural polymer, belongs to the family of unbranched polysaccharides, its main constituents are 1-4 linked β-D-mannuronic acid and L-guluronic acid. It is extracted from brown seaweeds in the form of *Laminaria hyperborea* and has a number of utilities in soft tissue implants and largely investigated in breast

implants possessing desirable anti-inflammatories and antimicrobial properties and having a high range of osmolality and radiolucency, which helps in maintaining the appropriate pH in the human body. Therefore, alginate gels listed as bioresorable and nonirritants to the human body parts according to the BIBRA Toxicological report and, therefore, proven to be an effective alternative for the replacement of silicone gel filling materials in breast implants as it suffers from marginal limitations. Alginate has been studied via in vivo methods to check the behavior of this polymer via animal modeling studies. The results of animal modeling were further applied to human bodies. The literature reports depicting once alginate-based scaffolds implemented into the dental pulp of the rats/mice, resulting in seeded cells differentiate into odontoblast and helps in stimulating calcification in the tooth [32, 33].

3.11.2 WOUND HEALING

Wounds caused permanent damage to the skin and significantly affect the anatomical structure of the skin caused by the results of the trauma in abrasions, avulsions, incisions, contusions, and lacerations or physiological disorders [39]. The effective initiators involved in the healing process are cytokines, macrophages, and fibroblasts. Macrophages are inflammatory cells that assist in the wound healing process and their important roles including inflammation promotion by removal of apoptotic cells which leads to cell escalation. Thus, stimulates angiogenesis and fibroplasia, which results in regrowth of cell strains compromised during injury/damage. New prospects of emerging technologies based on wound healing management systems attributed to the use of novel biodegradable polymers that offer numerous advantages such as regulating moist environment for re-epithelialization, providing antimicrobial activity, analgesic effect, appropriate gaseous exchange, and absorption of extra fluids or exudates are pivotal key factors. Biocompatibility of polymeric materials ensures the regeneration of neo-tissues without catalyzing any cytotoxic by-products formation during hydrolysis, therefore proven to be an ideal candidate for repairing tissue defects. A few years back, materials used for wound dressing are fats, honey, or plant fibers reinforced scaffolds to cover up the wounded area that prevents microbial infection and allow active passage of oxygen transfer but lack of could lead to infections and serious complications to the patients and many have died. The functional parameters of tissue repair are generally divided into four essential processes, as discussed in Figure 3.1.

During normal acute tissue injury, at the first stage of the healing process also known as platelet-derived growth factor: fibrin started plugging which helps in aggregating platelets and caused coagulation leads to hemostasis, which ultimately leads to tissue regeneration [12, 40]. The second stage ascribed the starting of a re-epithelialization within a few hours of an injury, induce the activation of keratinocytes and fibroblasts, and started migrating to the wounded sites, strongly influences the proliferation which enables the construction of an extracellular matrix (ECM) formation, and finally enhances the rate of wound closure. The gradual replacement of ECM with the collagenous intercellular substances stimulating the formation of new blood vessels also called angiogenesis. The next stage involves the generation of new granulation tissues and the transition between phases causes degradation of granulation tissues and regrowth of the dermis layer. The normal chronic wounds (diabetes) are generally characterized by a peripheral loss of nerve fibers and to some extent, permanent damage of a nerve tissue has been perceived with progressively decreased in the blood flow rate and increment in glycaemic levels [41]. However, in certain instances, diabetic patients may not develop neuropathy but permanent nerve damage in most of the cases

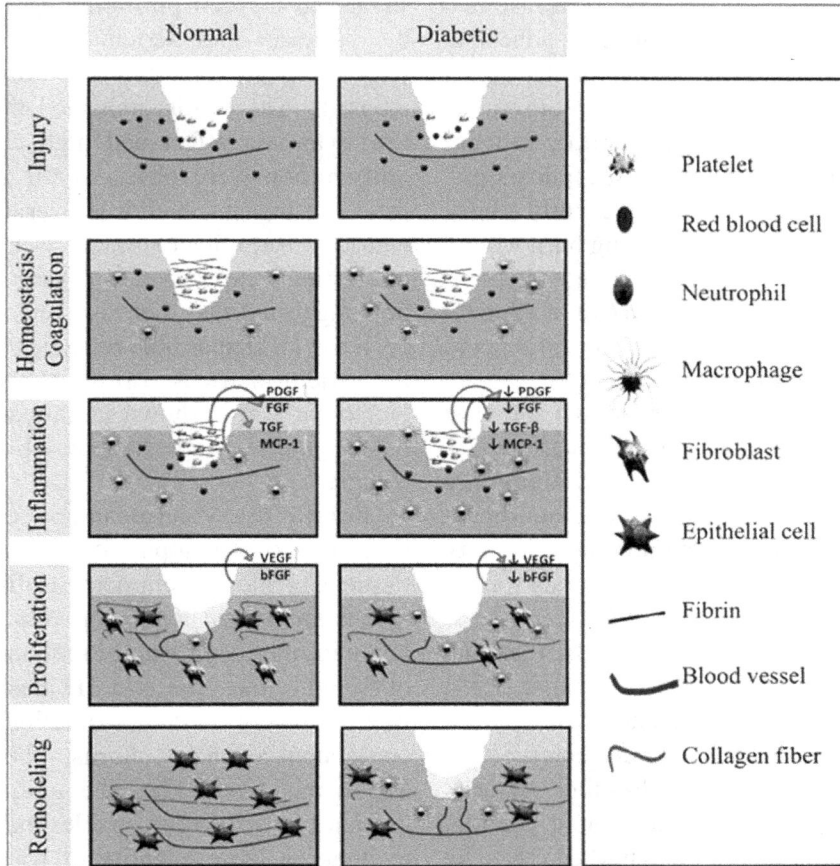

FIGURE 3.1 Schematic representation of the wound healing process and emphasize some major differences in the healing process of diabetic patients and a normal human being.

Source: Reprinted with permission from Ref. [43]. Copyright © 2013 Acta Materialia Inc. Published by Elsevier Ltd.

was reported and the main consequence of developing this serious medical condition is stated below, which is started appearing after 10–20 years of a diabetes diagnosis.

PLGA is a random copolymer of poly(lactic acid) and poly(glycolic acid). PLGA is the most investigated biopolymer and has very vast applications in the medical field as reports depict, owing to its facile synthesis and tunable mechanical and degradation rate make them an excellent candidate in the preparation of skin substitutes [34]. Scaffolds made from PLGA have some limitations for wound dressing application but a recent report refused this false argument. Literature demonstrated the enhancement in cell adhesion, proliferations, and modulation in degradation rate and epithelisation's can be well utilized for drug delivery and tissue engineering applications.

Fucoidan has recently been found to use for treating burn wound injury and has gained tremendous acceptance in the medicinal industry due to its antithrombic and antitumoral impression [35]. Recent reports have been shown the best results when tissue engineering, that is, clinical applications are concerned, due to its significant eliciting anticoagulating property and form novel biomaterials for the bioengineering.

Chitosan is polysaccharides that are obtained from the exoskeleton of crabs, shells, and other marine species. It has been emerged as potential wound dressing material for hemostasis and wound healing. Chitosan is considered to be cationic in nature that is ionically attracted toward negatively charged species and therefore enhances the rate of the healing process by regulating macrophage activity. Therefore, it reduces scar formation by restricting the formation of fibrin in wounds.

Chen et al. have shown the benefits of using chitosan than sericin for wound dressing applications and the evaluation was done by using Alamar blue Assay, which suggested the cell proliferations on the chitosan and sericin matrices with respect to different time intervals. The results indicated that low cytotoxic effect and high cell proliferation were observed till 28 days of constant studies on sericin matrices. But after two more weeks of testing, experimental observation suggested that there is a steady decrease in Alamar blue reduction in the sericin matrices with a relative increment that is shown in chitosan with high proliferative response [42] (Figure 3.2).

Keratin has widespread application in the development of skin tissue and most prominent usefulness in the wound healing process. It is a naturally occurring biopolymer obtained in the form of a 3D mesh-like structure and a functional component of hairs, horns, nails, and wools, etc. [37]. The unique morphology of keratin enables them to intake a large quantity of water which is an essential functioning of wound dressing materials and further helps in the reduction of biological fluids and blocking interactions with infectious agents [38]. Keratin is the most exploited wound dressing material extracted from ovine wool and effectively used in the treatment for severe wound disorder named as epidermolysis bullosa by reducing wound exudate and provide effective barriers against microbial attack.

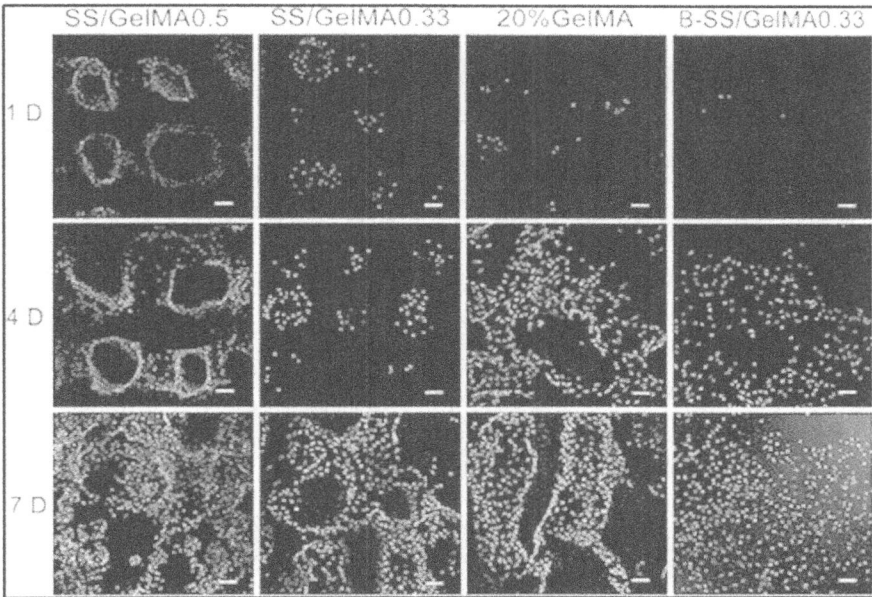

FIGURE 3.2 Schematic representation of cell-cultured (L929 cells) details on sericin matrices for days 1, 4, and 7 are shown using confocal microscopy [36].

3.12 FUTURE PROSPECTIVES AND CONCLUSION

Recently, more research works have been focussed on the bioresorable polymer and degradation mechanisms of the biodegradable polymers. Recent advances in biodegradable polymers (synthetic or natural) hold potential features in biomedical arenas and clinical applications. Future modifications of the chemical and physical properties of biodegradable polymers have led to the development of novel and creative biomaterials with a better understanding of the interaction between the host biomolecules and the implant interface. BIBRA Toxicological survey on biodegradable polymers has shown commendable output in the improvement in patients' health and tremendously recognized active biomaterials during the last few decades due to its excellent cytocompatibility, unique morphology, degradation kinetics, therefore an outshine alternative for existing polymers to be used as scaffolds for tissue regrowth and drug delivery applications. This ever-growing field needs a lot of innovations and collaborative efforts of expertise from multidisciplinary fields that are required to work closely along with these new scientific and bioengineering fields.

KEYWORDS

- bio-based polymers
- biocompatible
- bioplastics
- controlled release
- tunable properties
- biomedical applications
- global warming
- degradation kinetics

REFERENCES

1. Domb, A. J. N. K., Clinical Use and Clinical Development Biodegradable Polymers in Clinical Use and Clinical, Doi:10.1002/9781118015810.

2. Taylor, M., Tomlins, P., Sahota, T. (2017). ThermoresponsiveGels, Gels. 3, 4. https://doi.org/10.3390/gels3010004.

3. Sautter, C. (2007), Advances in Biochemical: Preface, Advances in *Biochemical Engineering/Biotechnology*, 107. https://doi.org/10.1007/978-3-540-71323-4.

4. Hopkins, R., Solutions for Tissue Engineering and Methods of Use, (n.d.), https://patentimages.storage.googleapis.com/43/d2/f2/f10ae3b0b9e6fd/CA2530416C.pdf.

5. Foox, M., Zilberman, M., Foox, M., & Zilberman, M. (2016). Expert Opinion on Drug Delivery: Drug Delivery from Gelatin-based Systems, 5247, Doi:10.1517/17425247.2015.1037272.

6. Journal, A. I., Kafshdooz, T., Kafshdooz, L., Akbarzadeh, A., Kafshdooz, T., Kafshdooz, L., Akbarzadeh, A., Hanifehpour, Y., & Joo, S. W. (2016). Applications of Nanoparticle Systems in Gene Delivery and Gene Therapy. Artificial Cells, Nanomedicine, and Biotechnology, 44, 581–587, Doi:10.3109/21691401.2014.971805.

7. Zhang, P., Han, N., Kou, Y., Zhu, Q., Liu, X., Quan, D., Chen, J., & Jiang, B. (2019). Tissue Engineering for the Repair of Peripheral Nerve Injury. Neural Regeneration Research, Doi:10.4103/1673-5374.243701.

8. Iu, X. I. L., Eter, P. X. M. A. (2004). Polymeric Scaffolds for Bone Tissue Engineering. Annals of Biomedical Engineering, 32, 477–486.

9. Navarro, M., Michiardi, A., Casta, O.ño, Planell, J. A., Interface, J. R. S., Navarro, M., Michiardi, A., & Castan, O. (2008). Biomaterials in Orthopaedics. Journal of the Royal Society Interface, 5, 1137–1158, Doi:10.1098/rsif.2008.0151.

10. Hottle, T. A., Bilec, M. M., & Landis, A. E. (2013). Sustainability Assessments of Bio-based Polymers. Polymer Degradation and Stability, 98, 1898–1907, Doi:10.1016/j.polymdegradstab. 2013.06.016.

11. Greene, J. (2007). Biodegradation of Compostable Plastics in Green Yard-waste Compost Environment. Journal of Polymers and the Environment, 15, 269–273, Doi:10.1007/s10924-007-0068-1.

12. Schmidt, C. E., & Leach, J. B. (2003). Neural Tissue Engineering: Strategies for Repair and Regeneration. Annual Review of Biomedical Engineering, 5, 293–347, Doi:10.1146/annurev.bioeng.5.011303.120731.

13. Neuenschwander, S., & Hoerstrup, S. P. (2004). Heart Valve Tissue Engineering. Transplant Immunology, 12, 359–365, Doi:10.1016/j.trim.2003.12.010.

14. Shah Mohammadi, M., Bureau, M. N., & Nazhat, S. N. (2014). Polylactic Acid (PLA) Biomedical Foams for Tissue Engineering. Woodhead Publishing Limited: Cambridge.

15. Agrawal, C. M., Huang, D., Schmitz, J. P., & Athanasiou, K. A. (1997). Elevated Temperature Degradation of a 50:50 Copolymer of PLA-PGA. Tissue Engineering, 3, 345–352, Doi:10.1089/ten.1997.3.345.

16. Yang, J., Wan, Y., Yang, J., Bei, J., & Wang, S. (2003). Plasma-treated, Collagen-anchored Polylactone: Its Cell Affinity Evaluation Under Shear or Shear-free Conditions. Journal of Biomedical Materials Research—Part A, 67, 1139–1147, Doi:10.1002/jbm.a.10034.

17. Andrianov, A. K. (2006). Water-soluble Polyphosphazenes for Biomedical Applications. Journal of Inorganic and Organometallic Polymers and Materials, 16, 397–406, Doi:10.1007/s10904-006-9065-4.

18. Peters, G. P. (2010). Carbon Footprints and Embodied Carbon at Multiple Scales. Current Opinion in Environmental Sustainability, 2, 245–250, Doi:10.1016/j.cosust.2010.05.004.

19. Lehmann, J. (2007). A Handful of Carbon, Nature, 447, 143–144. https://doi.org/10.1038/447143a..

20. Lu, L., Peter, S. J., Lyman, M. D., Lai, H., Leite, S. M., Tamada, J. A., et al. (2000). In Vitro and in Vivo Degradation of Porous Poly(DL-lactic-*co*-glycolic acid) Foams. Biomaterials, 21, 1837–1845.

21. Oh, S. H., Kang, S. G., Kim, E. S., Cho, S. H., & Lee, J. H. (2003). Fabrication and Characterization of Hydrophilic Poly(lactic-*co*-glycolic acid)/poly(vinyl alcohol) Blend Cell Scaffolds by Melt-molding Particulate-leaching Method. Biomaterials, 24(22), 4011–4021, https://doi.org/10.1016/S0142-9612(03)00284-9.

22. Grigoreva, M. V. (2013). Polyurethane Composites As Drug Carriers: Release Patterns. Biotechnologia Acta, 6(5), 41–48, https://doi.org/10.15407/biotech6.05.041.

23. Ashammakhi, N., & Rokkanen, P. (1997). Absorbable Polyglycolide Devices in Trauma and Bone Surgery. Biomaterials, 18(1), 3–9, https://doi.org/10.1016/S0142-9612(96)00107-X.

24. Chen, J., Shi, W., Norman, A. J., & Ilavarasan, P. (2002). Electrical Impact of High-speed Bus Crossing Plane Split. IEEE International Symposium on Electromagnetic Compatibility, 2, 861–865, https://doi.org/10.1109/isemc.2002.1032709

25. Hachisuka, K., Makino, K., Wada, F., Saeki, S., & Yoshimoto, N. (2007). Oxygen Consumption, Oxygen Cost and Physiological Cost Index in Polio Survivors: A Comparison of Walking Without Orthosis, with an Ordinary or a Carbon-fibre Reinforced Plastic Knee-ankle-foot Orthosis. Journal of Rehabilitation Medicine, 39(8), 646–650, https://doi.org/10.2340/16501977-0105.

26. Bartonek, Å., Eriksson, M., & Gutierrez-Farewik, E. M. Effects of Carbon Fibre Spring Orthoses on Gait in Ambulatory Children with Motor Disorders and Plantarflexor Weakness. Developmental Medicine & Child Neurology, 49, 615–620. https://doi.org/10.1111/j.1469-8749.2007.00615.x

27. Kulkarni, R. K., Pani, K. C., Neuman, C., & Leonard, F. (1966). Polylactic Acid for Surgical Implants. Archives of Surgery, 93, 839–843.

28. Ishaug-Riley, S. L., Okun, L. E., Prado, G., Applegate, M. A., & Ratcliffe, A. (1999). Human Articular Chondrocyte Adhesion and Proliferation on Synthetic Biodegradable Polymeric Films. Biomaterials, 20, 2245–2256.

29. Cheung, H. Y., Lau, K. T., Lu, T. P., & Hui, D. (2007). A Critical Review on Polymer-based Bio-engineered Materials for Scaffold Development. Composites Part B: Engineering, 38(3), 291–300, https://doi.org/10.1016/j.compositesb.2006.06.014.

30. Langone, F., Lora, S., Veronese, F. M., Caliceti, P., Parnigotto, P. P., Valenti, F., & Palma, G. (1995). Peripheral Nerve Repair Using a Poly(organo) phosphazene Tubular Prosthesis. Biomaterials, 16(5), 347–353, https://doi.org/10.1016/0142-9612(95)93851-4.

31. Li, M., Ogiso, M., & Minoura, N. (2003). Enzymatic Degradation Behaviour of Porous Silk Fibroin Sheets. Biomaterials, 24, 357–365.

32. Fujiwara, S., Kumabe, S., & Lwai, Y. (2006). Isolated Rat Dental Pulp Cell Culture and Transplantation with an Alginate Scaffold. Okajimas Folia Anatomica Japonica, 83, 15–24.

33. Kumabe, S., Nakatsuka, M., Kim, G. S., June, S., Aikawa, F., Shin, J. W., et al. (2006). Human Dental Pulp Cell Culture and Cell Transplantation with an Alginate Scaffold. Okajimas Folia Anatomica Japonica, 82, 147–156.

34. Baoyong, L., Jian, Z., Denglong, C., & Min, L. (2010). Evaluation of a New Type of Wound Dressing Made from Recombinant Spider Silk Protein Using Rat Models. Burn, 36, 891–896.

35. Murakami, K., Aoki, H., Nakamura, S., Nakamura, S., Takikawa, M., Hanzawa, M., Kishimoto, S., Hattori, H., Tanaka, Y., Kiyosawa, T., Sato, Y., & Ishihara, ML. (2010). Hydrogel Blends of Chitin/chitosan, Fucoidan and Alginate as Healing-impaired Wound Dressings. Biomaterials, 31(1), 83–90.

36. Mir, M., Ali, M. N., Barakullah, A., Gulzar, A., Arshad, M., Fatima, S., & Asad, M. (2018). Synthetic Polymeric Biomaterials for Wound Healing: A Review. Progress in Biomaterials, 7(1), 1–21, https://doi.org/10.1007/s40204-018-0083-4.

37. Nayak, S., Dey, S., & Kundu, S. C. (2013). Skin Equivalent Tissue-engineered Construct: Co-cultured Fibroblasts/keratinocytes on 3D Matrices of Sericin Hope Cocoons. PLoS One, 8(9), e74779.

38. Basu, P., Uttamchand, N. K., & Inderchand, M. (2017). Wound Healing Materials—A Perspective for Skin Tissue Engineering. Current Science, 112, 2392–2404.

39. Boateng, J. S., Matthews, K. H., Stevens, N. E., & Eccleston, G. M. (2008). Wound Healing Dressings and Drug Delivery Systems: A Review. Journal of Pharmaceutical Sciences, 97, 2892–2923.

40. Wilgus, T. A. (2008). Immune Cells in the Healing Skin Wound: Influential Players at Each Stage of Repair. Pharmacological Research, 58, 112–116.

41. Liu, H., Wang, C., Li, C., Qin, Y., Wang, Z., Yang, F., et al. (2018). A Functional Chitosan-based Hydrogel as a Wound Dressing and Drug Delivery System in the Treatment of Wound Healing. RSC Advances, 8(14), 7533–7549, https://doi.org/10.1039/c7ra13510f.

42. Chen, C. S., Zeng, F., Xiao, X., Wang, Z., Li, X. L., Tan, R. W., et al. (2018). Three-dimensionally Printed Silk-Sericin-Based Hydrogel Scaffold: A Promising Visualized Dressing Material for Real-Time Monitoring of Wounds. ACS Applied Materials and Interfaces, 10(40), 33879–33890, https://doi.org/10.1021/acsami.8b10072.

43. Moura, L.I.F., Dias, A.M.A., Carvalho, E., and de Sousa, H.C. Recent advances on the development of wound dressings for diabetic foot ulcer treatment—A review. Acta Biomaterialia, 9(7). 7093-7114

CHAPTER 4

BIO-BASED POLYMERS FOR LIPOSOMAL DRUG FORMULATIONS

MIRELA-FERNANDA ZALTARIOV[1*], BIANCA-IULIA CIUBOTARU[1], MARCELA SAVIN[2], DANIELA FILIP[3], and DOINA MACOCINSCHI[3]

[1]Department of Inorganic Polymers, "Petru Poni" Institute of Macromolecular Chemistry, Aleea Grigore Ghica Voda, 41 A, 700487 Iasi, Romania

[2]"Antibiotice" SA, Valea Lupului Alley, 707410 Iasi, Romania

[3]Department of Polyaddition and Photochemistry, "Petru Poni" Institute of Macromolecular Chemistry, Aleea Grigore Ghica Voda, 41 A, 700487 Iasi, Romania

*Corresponding author. E-mail: zaltariov.mirela@icmpp.ro.

ABSTRACT

This chapter covers the impact of different classes of bio-based polymers in the medical field focused on their utilization in different types of liposomes and other nanomaterials. Currently, a number of drugs based on conventional and PEGylated liposomal formulations are clinically used for the treatment of various fungal and viral infections, cancer, pain, and photodynamic therapy, while many others have reached superior phases of medical trials. New strategies employed to produce liposome formulations based on biocompatible, mucoadhesive, and biodegradable polymers, their use as carriers or drug delivery systems with improved drug solubility and controlled distribution, prolonged- and sustained-release ability, as well as their properties responsible for their performance are presented, highlighting the future developments of such assemblies as the next generation of therapeutic agents.

4.1   INTRODUCTION

Liposomes were first reported by Bangham in 1960 who observed the natural self-assembly of the phospholipids into multilayer vesicles (unilamellar or multilamellar) in aqueous solutions and demonstrated that these vesicles are capable of mimicking a cell membrane (structure, adhesion, or permeation) [1, 2]. Firstly, liposomes functioned as models for biological membranes, regarding the barrier properties and the way to interact with biomembranes. During the self-assembly process, the phospholipids as well as other amphiphiles, separate the hydrophilic inner core, in which drugs can be encapsulated, from the aqueous phase, forming closed spherical lipid vesicles with a diameter of 400 nm–2.5 mm [3]. Particular attention was accorded to the unilamellar vesicles very useful in medical applications as nanocarriers or nanoscale drug delivery systems of both, hydrophilic compounds from the water core or hydrophobic molecules through inclusion into the lipid layers.

The first compounds integrated with the lipid bilayer were enzymes proving the ability of the liposomes to protect them from protease in the serum and the body's immune response, thus preventing the severe allergic reactions which took place during the administration of "free" drugs [4].

More than that, liposomal antibiotics, antiviral and antifungal drugs, vaccines, or cosmetics as topical formulations have been approved as pharmaceuticals, which attests to the development of these technologies as innovative approaches in the prevention and treatment of certain diseases [4].

By liposomes, drug absorption is lower depending on the composition of lipidic layers and their integrity over time and the release is targeted at the site of action. These characteristics control the drug dosage that can be administrated often intravenous [5].

Today, liposomes are considered the most efficient drug delivery systems, being part of many anticancer drugs reducing their toxic effects. In some cases, the encapsulation in liposomes of anticancer drugs, such as cisplatin, resulted in a decrease of the activity, and development of the drug resistance leading to recurrence. In other cases, by intravenous administration, the liposomes were able to maintain the antitumor activity and to reduce the toxicity. PEGylated liposomes are now accepted as the most stable and long-term circulating drug carriers (20–30 h in the human body), the molecular mechanism involved in their protective action being determined by the conformation flexibility (free rotation of each molecular entity) in a solution of the polymer molecule, which prevents the interaction with plasma protein, opsonization, and capturation in liver and spleen. The steric protection can be insured by biocompatible, soluble, flexible, and hydrophilic polymers, such as poly(vinyl alcohol) (PVA), poly(vinyl pyrrolidone) (PVP), polyacrylamide (PAA), Pluronic F127 that can be incorporated into the liposome bilayer to improve their mucoadhesive [6] and to prolong their circulation [7]. It has been proven that the conjugation of these polymers with nonphospholipid molecules had a positive impact on the protective efficacy, the circulation time in the body, and opsonization. Thus, liposomes coated with palmitate-linked PVP or PVA had better stability than PEGylated ones and a higher circulation persistence. Also, poly(amino acids)-based liposomes (polyglutamic acid, poly (hydroxyethyl-L-asparagine or poly (hydroxyethyl-L-glutamine)-succinyldioctadecylamine (PHEG-DODASuc))) have been studied for their complete biodegradable nature, reducing the accumulation in different organs, having an extended blood circulation half-life [8]. In addition, super-hydrophilic zwitterionic polymers have shown promising results by ensuring stronger hydration, weak interactions with lipid bilayers, and improved liposome stabilization. Various polymers including poloxamers, dextran, polyacrylic, and sialic acid derivative have been used to exploit the passive targeting of liposomes [9].

4.2 LIPOSOMAL PHARMACEUTICAL FORMULATION

The first reported compounds encapsulated in liposomes were lysozyme, β-fructofuranosidase, and chlorophyll a as excellent models for membranes or photosynthetic membrane, having the purpose to protect enzymes from serum proteases and to modulate the immune response. Through these actions, liposomes have shown that they can protect the body from allergic reactions. Another purpose of their development was to reduce the toxic effects of some drugs, especially anticancer drugs [4]. The first encapsulated anticancer drug was cisplatin used in the clinic since 1978. Cisplatin had some limitations associated with severe side effects (neurotoxicity, hepatotoxicity, and nephrotoxicity) and drug resistance is an insufficient alternative [10]. The liposomal formulation of cisplatin proved to be less toxic for kidneys but caused a decrease in antitumor activity. In some cases, in addition to intravenous administration, the liposomal formulations of drugs could be administrated orally proving an increased efficiency in the treatment of oral ulcers (radioactive triamcinolone acetonide palmitate), skin diseases (lidocaine), intravitreal diseases (antibiotics and antiviral drugs), etc.

Since 20 years ago, a number of chemotherapeutic liposomes have passed the preclinical and clinical studies with positive impacts and have been approved by Food and Drug Administration as treatments in simple or combined anticancer therapy, while an impressive number of liposomes are tested in clinical trials and are about to receive the approval. Simultaneously with the diversification of liposome types, a large number of technologies have been developed and designed to ensure the selectivity of liposome formulations on cancer cells or accumulation of liposomes at a targeted site by including signal molecules that are recognized by specific receptors: transferrin or glucose-vitamin C complex [11]. Beyond these, pH-, temperature-sensitive, magnetic, or multifunctional liposomes have been studied. The next generation of liposomes consists of the utilization of lipid-based nanomaterials, such as solid lipid nanoparticles, self-assembled lipid nanocarriers, and lipid–polymer hybrid nanoparticles [12]. Currently, most liposomal formulations have received approval for anticancer treatment, infectious diseases, viral infectious, while about 28 products are under clinical evaluation (Table 4.1) [13].

4.3 CLASSIFICATION OF LIPOSOMES

Liposomes can be classified according to several criteria, as shown below.

First, the composition consists of amphiphilic phospholipids (classical liposomes), neutral lipids and surfactants (niosome), and nonphospholipids (vesicle). The lipid bilayer of the liposomes can be natural or synthetic phospholipids that self-associate into bilayers through interactions from water molecules and hydrophobic phosphate groups. Better known are the natural phosphatidylcholine, known as lecithins, especially those from eggs and soy being chemically inert, without charge, and easily accessible. The most used phospholipids are presented in Table 4.2.

Phosphatidylcholine and phosphatidylethanolamine, as major constituents of the biological membranes, are found in both, plants and animals. Cholesterol is also an essential element in the preparation of liposomes, being arranged with the hydroxyl groups toward the aqueous phase. Its major role is to diminish the flexibility of the liposomal membrane, to confer a semipermeable character for soluble molecules, and to improve the liposomal membrane in a biological medium, by preventing the loss of the phospholipids. Its absence leads to interactions with blood proteins (albumin, transferring, and

TABLE 4.1   Examples of Approved Liposomal Formulations

Liposomal Formulation (Market)	Drug Class	Indication	Observations and References
Doxil® (in the US) or Caelyx (in Europe)	Doxorubicin	Kaposi sarcoma	The first approved PEGylated liposomes [14, 15]
DaunoXome®	Doxorubicin	Kaposi sarcoma	[16]
Depocyt®	Cytarabine	Lymphomatous meningitis	[17]
Myocet®	Doxorubicin	Metastatic breast cancer	Non-PEGylated liposomes [18]
Abelcet®	Amphotericin B	Infectious diseases	Complex with two phospholipids in a 1:1 drug-to-lipid molar ratio [19]
AmBisome®	Amphotericin B	Infectious diseases	*Lower incidence of nephro-toxicity than Abelcet* [20]
Amphotec®	Amphotericin B cholesteryl sulphate	Infectious diseases	[19]
Epaxal® and Inflexal®	Virosomal vaccines	Viral infections	The first virosome-adjuvanted vaccine for hepatitis A [21, 22]
Vyxeos®	Daunorubicin and cytarabine	Acute myeloid leukaemia	The first low-cholesterol product [23]
Visudyne®	Verteporfin	Photodynamic therapy	[24]
DepoDur®	Morphine sulfate	Pain management	[25]

lipoproteins) destabilizing the liposomes, and reducing their ability to transport and release drugs [12].

The aqueous phase of the liposomes contains water or an aqueous solution of the drugs, which can be encapsulated by in situ liposome formation in a saturate aqueous solution of a soluble drug or in organic solvents followed by solvent extraction operations in case of lipophilic drugs, or pH gradient techniques. Often, anticancer drugs are encapsulated into liposomes, the clinical trials indicating their ability to improve chemotherapy effects and to reduce toxicity. In addition to these, a wide range of active principles can be loaded: enzymes, immune vectors, anti-inflamators, polynuceotides, insulin, antiviral, antibiotics or chelating agents, hormones, radiopharmceuticals, vitamins, etc. [26].

1. Conventional liposomes are the first generation of liposomes

TABLE 4.2   Some Phospholipids Used in Liposomal Formulations

Product (lipid) Name	Lipid Structure	Specific Role
Phosphatidylcholine, PC (egg)		Frequently used in liposome formulations
Phosphatidylethanolamine		Forms "open" vesicles, can bind different molecules to liposomes
Cholesterol		Reduces the permeability of the phosphatidylcholine vesicles
Distearoyl phosphatidylcholine		Less permeable for the aqueous phase
Lysophosphatidylcholine		Increases the liposomal permeability and has a positive impact on the interactions with cells
Phosphatidylserine		Confers negative charge to the liposomes

used in pharmaceutical applications. The main constituent is natural phospholipids or lipids, such as distearoyl phosphatidylcholine and egg phosphatidylcholine. These liposomes are not stable in plasma, having a short-blood circulation half-life, and are quick removed by the mononuclear phagocyte system. They cannot carry out their role efficiently.

2. Long circulating or "stealth" liposomes (PEGylated liposomes) obtained by surface—modified with PEG of 1000–5000 Da is

one of the most used liposomal formulations for drug delivery applications. The conjugation with PEG conferred superior properties: high biocompatibility, low toxicity and immunogenicity, high hydrophilicity, and ability to release active molecules at target sites [27].

3. Targeted liposomes or immunoliposomes contain ligands or signal molecules (antibodies, growth factors, folic acid, glycoproteins, etc.) on the surface, which are recognized by specific membrane receptors

(overexpressed receptors such as folate receptors). The presence of the ligand ensures a higher accumulation in a specific location, but often in healthy cells with increased sensitivity which indicates a lack of selectivity. The conventional, stealth, and targeted liposome technologies are already clinically approved. The new generations of liposomes are stimuli-responsive and gene-based liposomes [28, 29].

4. Cationic liposomes also called lipoplex used as nonviral delivery vectors by self-assembly through electrostatic interactions with negatively charged DNA or cell membrane molecules (glycoproteins and proteoglycans) and thus being able for cellular uptake and intracellular delivery [30, 31].

5. pH-sensitive liposomes able to facilitate the drug delivery in targeted sites under the influence of different environmental factors, the bilayers consist of dioleoylphosphatidylethanolamine stabilized with oleic acid/cholesteryl hemisuccinate [32].

6. Liposome-based theranostic agents combine the therapeutic effect of the encapsulated drug with an imaging (dye) organic component inside the lipid bilayers that are specific modified with PEG for stealth characteristics and a predefined targeting ligand (folate or transferrine) and aptamer for directed release to a specific site. Such liposomes are used not only in cancer treatment but also in the detection and monitoring of the release of drug [33].

7. Liposome-based vaccines are a way to design vaccines for different administration routes (oral, topical, or mucosal) [34].

Second, morphology is unilamellar with a size range of 50–250 nm and multilamellar (two or more lipid layers) vesicles with a size range of 500–5000 nm;

Third, dimensions are of small (20–40 nm in diameter, with a loading capacity of 0.1%–1%) and large (50–5000 nm in diameter) vesicles [28]. The size of the liposomes is one of the most significant parameters which controls their biodistribution and the elimination rate. Small vesicles are subjected to a short-term circulation being eliminated with predilection by urinary excretion, while the large vesicles are seized in the liver (200–400 nm) and spleen (68–340 nm) and eliminated by opsonization by activating the mononuclear phagocytic system. Another disadvantage of small liposomes is insufficient drug loading capacity. The optimal dimensions of liposomes to achieve a suitable circulation time and tumor accumulation must be between 100 and 300 nm. Also, if the composition of the lipid bilayer is not biodegradable then the liposomes develop toxic effects by accumulation in the liver or spleen so that PEGylated liposomes are able to reduce the nonspecific binding of plasma proteins at the surface.

Fourth, the surface charge is also an important characteristic of liposomes which can be evaluated by zeta potential. A neutral surface charge denotes a colloidal instability with a tendency to agglomeration, while an increase of the surface charge, either positive or negative (zeta potential value), determines a faster rate of elimination and often no effect on tumor uptake. The studies revealed that only a slightly (positively or negatively) charge could lead to optimal results, by establishing an equilibrium between the stability and the clearance of liposomes via plasma proteins interactions which in the first phase can lead to a high release of the encapsulate drug [12].

4.4 ADVANTAGES AND LIMITATIONS IN UTILIZATION OF LIPOSOMES

The major advantages of using liposomes are as follows:

1. Improve the solubility of the drugs.
2. Reduce the toxicity of the drug by preventing direct interactions with proteins from the biological medium.
3. A long-term circulation process that will ensure a higher efficiency.
4. Direct the drugs to a specific site (cells/organs).
5. Increase the bioavailability of the drugs and the therapeutic efficiency by decreasing the elimination rate.
6. The potential to combat the multidrug resistance frequently occurring in cancer, which is a major factor in decreasing chemotherapeutical efficiency [35].

Limitations of liposome for clinical applicability are determined by certain factors:

1. The impossibility for oral administration route.
2. Low-loading capacity and encapsulation efficiency.
3. A fast release of the drug by interaction with blood proteins.
4. Great instability leads to aggregation.
5. Issues related to reproducibility.
6. The scaling for clinical studies.
7. Toxicity resulted from liposome composition, charge or size, or as a result of the interaction with blood proteins [35].

Once these factors are overcome, other issues related to their instability, short half-time, and rapid clearance need to be managed. Often, during the fabrication process, a degradation of the encapsulated drugs can occur by exposure to adverse conditions:

heat, sonication, and organic solvents. Another aspect is about the possibility to control the drug release from liposomes with the preservation of the bioactivities of the drugs.

Several strategies have been studied to address the following limitations:

1. Size adjustment to 200–250 nm which proved a higher loading efficiency of drugs.
2. Surface modification to improve the circulation of liposomes in the body.
3. Selective response of the liposomes in the biological medium (stimuli-responsive at the acidity, temperature, light, etc.).
4. Surface functionalization with "recognizing vectors" to direct the drugs to a target site [36].

An advantageous way is to use polymer-based systems instead of the liposomes to ensure a prolonged release, an increased efficiency, and a biocompatible and biodegradable system with low toxicity [28].

To prevent the instability of the liposomes, the fast release and rapid oxidation of the lipid bilayer, covering with a bioadhesive and polymeric membrane around liposomes is a good alternative that can facilitate the targetability (active based on ligand-receptor recognition or passive by accumulation in tumors by diffusion process) of the liposomes [26, 37].

The key factors of such coating materials are biocompatibility and biodegradability so that their metabolites are nontoxic and can be easily cleared from circulation. First, the small particles (0–30 nm) are cleared by excretion, while the particles with a size >30 nm are cleared by opsonization. Therefore, it is important to design liposomes that provide an optimal drug concentration at a target site and to maintain a constant therapeutic dose by reducing clearance, increasing the drug stability, and reducing its toxic effects in healthy organs [26]. The clinical application of these materials is limited by the unexpected factors that could determine an immune response and formation of thrombus when expose to blood, so that the chemical inertness is the first limitation in the development of blood-contacting devices.

Natural product-based liposomal drug delivery systems remain a safe alternative that combines the liposomes and the technology to produce matrices or scaffolds for a prolonged drug release rate. In these attempts, collagen-based liposomal complexes proved good biocompatibility and degradability upon implantation, improved storage stability, a prolonged drug release rate, and an increased therapeutic efficacy mainly due to the collagen scaffold obtained by chemical crosslinking or physical treatments.

Gelatin-based liposomal systems showed a controlled release of the bioactive agents: drugs, growth factors, or proteins by combination with PEG to obtain porous scaffolds. Chitosan-based hydrogels have been used for embedding liposomes by in situ gelling formulations to ensure a sustained drug release over a prolong period of time. Fibrin-, alginate-, and dextrane-based liposomal complexes have been also designed mainly due to the biodegradability, low toxicity, and nonimmunogenicity of the matrices, which will ensure the controlled and slow delivery of the liposomes and the stabilization in the aqueous media. Calcein, insulin, or inter-leukin-2 are few examples of large multilamellar liposomes that have been evaluated in vivo by using calcium-crosslinking alginate and dextrane hydrogels.

Beside these, liposomal drug delivery systems based on synthetic polymers are also reported. Carbopol is widely used as a pharmaceutical carrier due to its bioadhesivity, biocompatibility, low toxicity, ability to inhibit the activity of some enzymes in the gastrointestinal tract, so that it is a promising candidate for incorporating liposomes, being able to promote a pH-dependent release. In addition, PVA gained a high interest in such formulations being biodegradable, biocompatible, with low toxicity, bioadhesive,

proving excellent film-forming ability or gel formation by chemical crosslinking being used to incorporate chemotherapeutic antibiotics (ciprofloxacin) [28].

Prefabricated polymeric scaffolds have become an advantageous alternative for drug delivery systems and regenerative tissue, being predefined at highly reproducible shapes and sizes and may constitute an ideal temporary depot of drugs in order to enhance the therapeutic efficacy. These are not induced by any adverse response because of their appropriate surface chemistry, biodegradable and bioresorbable properties designed according to therapeutic needs. More than that, liposomes-based polymeric scaffolds can ensure a sustained drug release over prolonged periods of time.

This chapter presents some new strategies to develop bio-based polymers as scaffolds or matrices for liposome encapsulation referring to mucoadhesive polymeric gels based on hydroxypropyl methylcellulose (HPMC), poly(acrylic acid) (PAA), or sodium alginate (SA), mucoadhesive polymeric blends based on polyurethanes (PU) and hydrohypropyl cellulose (HPC) and bionanocomposite membranes based on PU, extracellular matrix (EM), and silver nanoparticles (AgNPs). Structural particularities, as well as their morphology, bio- and muco-adhesiviness, moisture, and ability to incorporate biological molecules are highlighted.

Biodegradable polymers, such as chitosan, HPMC, SA, gellan gum, have been applied in different pharmaceutical formulations as delivery systems or carriers for different drugs. HPMC is a common polymer for the preparation of hydrophilic matrices, having good stability, versatility, and suitability for loading and release of biological molecules. Preparation of hydrogel blend systems based on interpolymer interactions, such as HPMC and PAA or SA, is an excellent alternative to improve the adhesion with the mucosal for a target release with low toxicity and side-effects, as it happens in intravenous administration. Interpolymeric scaffolds based on alginate and chitosan, carboxymethyl cellulose acetate, carboxymethyl cellulose acetate butyrate, HPMC and fumaric acid or cellulose acetate, and ethyl cellulose have proven to be very effective for the solubilization and delivery of some insoluble drugs: nystatin, sulfadiazine, dipyridamole, etc. [38]. In particular, blends of HPMC and PAA proved to be efficient in the loading and local release of fluconazole (FCZ) or FCZ/β-cyclodextrin inclusion complex [39]. HPC is also used in pharmaceutics due to its nontoxicity, high performance in the residence time with the mucosa tissue maintaining a therapeutic concentration of the drugs. Over these features, its low cost, the natural availability, and the possibilities to be modified by attaching different functional groups, increase the interest to be applied as polymeric blends for scaffolds or matrices for liposomes, and other drug formulations. Polymeric blends of HPC and PU have proven large applicability as microfibers and nanofibers with applications in transdermal drug delivery systems, ocular mucoadhesive release systems or membranes, having haemocompatibility properties [40, 41]. The biocompatibility and tolerance of such polymeric systems can be improved by the addition of elements of EM, such as collagen, elastin, chondroitin sulfate (CS), hialuronic acid (HA), which have a positive impact on the mucoadhesiviness and local release of drugs [42].

4.5 EXPERIMENTAL

4.5.1 MATERIALS

HPMC (Aldrich), hydroxypropyl-cellulose (Aqualon; Hercules Inc., Wilmington, USA), Carbopol 934 or PAA (Serva FeinBiochemica Heidelberg, Germany), SA (Aldrich), triethanolamine (TEA) (Aldrich), glycerol (CREMER OLEO, Germany), Poloxamer 407, $PEO_{101}PPO_{56}PEO_{101}$ (BASF, Germany),

Anise oil [43], FCZ (Aldrich), b-cyclodextrin (Fluka, Germany), Poly[(ethylene glycol)adipate] (PEGA) (Aldrich), poly(ethylene glycol)-block-poly(propylene glycol)-block-poly(ethylene glycol) (Pluronic L-61) (Aldrich), poly(butylene adipate)diol (PBA) (Aldrich), sodium deoxycholate (SD) (Aldrich), methylene dicyclohexyl diisocyanate ($H_{12}MDI$) (Aldrich), diphenylmethane diisocyanate (MDI) (Aldrich), polyethylene glycole (PE) (Aldrich), hydrolyzed collagen (Aldrich), CS (Aldrich), HA (Aldrich), and $AgNO_3$ (Fluka) were used as received.

4.5.2 METHODS

4.5.2.1   ATTENUATED TOTAL REFLECTANCE FOURIER TRANSFORM INFRARED SPECTROSCOPY

Attenuated total reflectance Fourier transform infrared (ATR-FTIR) spectra were measured on a Bruker Vertex 70 spectrometer (Bruker Optics, Ettlingen, Germany). The spectra were recorded in the range of 4000–6000 cm^{-1} with a resolution of 4 cm^{-1} at room temperature.

Bioadhesion and mucoadhesion tests as the maximum force of detachment and the work of adhesion (the area under the force/distance curve) have been determined by using a

TA.XT Plus texture analyzer (Stable Micro Systems, UK) in triplicates. The sample was analyzed in the form of films of 10 mm diameter. Small portions (about 4 cm^2) of cellulose membrane (for bioadhesion test) and pig stomach (for mucoadhesion test) were placed on the sample holder and the samples were attached to a mobile cylindrical component by a double-face tape. Then, the cylindrical holder was lowered slowly with a speed of 1 mm s^{-1} until it reached the cellulose membrane or stomach tissue with a determinate contact force of 1 gf and a contact time of 30 s.

4.5.2.2   SCANNING ELECTRON MICROSCOPY

The surface morphology of the hydrogel and polymeric blends samples was investigated with an ESEM Quanta 200 Scanning Electron Microscope (The Netherlands) operating at 20 kV in low vacuum mode, with integrated EDS system, GENESIS XM 2i EDAX with SUTW detector.

4.5.2.3   DYNAMIC WATER VAPOR SORPTION MEASUREMENTS

Water vapors sorption capacity of the samples at 25 °C in the 0%–90%

relative humidity range was investigated by using the fully automated gravimetric analyzer IGAsorp (Hiden Analytical, Warrington, UK). Before sorption measurements, the samples were dried at 25 °C in flowing nitrogen (250 mL min^{-1}) until the weight of the sample remains constant.

4.6 RESULTS AND DISCUSSION

4.6.1 OBTAINING OF POLYMERIC GELS BASED ON HPMC, PAA, OR SA

Two different polymeric gels based on HPMC and PAA and HPMC and SA containing anise oil have been prepared according to a pre-established recipe, as shown in Table 4.3.

A schematic representation of the polymeric networks HPMC/PAA and HPMC/SA can be seen in Figure 4.1.

4.6.2 PREPARATION OF POLYMERIC BLENDS BASED ON PU AND HPC

PU was prepared from Pluronic L 61, PBA, methylene dicyclohexyl diisocyanate (H$_{12}$MDI), and SD using dimethylformamide (DMF) as a solvent. The blends have been obtained by mixing a solution of PU and HPC in different ratios (20/80, 50/50, and 80/20) in DMF at room temperature over a period of time. After the evaporation of the solvent, the blends were dried in a vacuum oven for 48 h resulting in blended films. The chemical composition of the blends is shown in Figure 4.2.

TABLE 4.3 The Composition of the Polymeric Gels F1A-F8A Based on HPMC, PAA or SA, and Anise Oil

Precursors, g	F1A	F1A_ 1:2	F1A_ 1:3	F2A	F2A_ 1:2	F2A_ 1:3	F3A	F4A	F5A	F6A	F7A	F8A
PAA	0.1	0.1	0.1	0.1	0.1	0.1	0.1	0.1	–	–	–	–
HPMC	0.1	0.1	0.1	0.2	0.2	0.2	0.3	0.4	0.1	0.2	0.3	0.4
Sodium alginate (SA)	–	–	–	–	–	–	–	–	3	3	3	3
TEA	1	1	1	1	1	1	1	1	–	–	–	–
Glycerol	10	10	10	10	10	10	10	10	–	–	–	–
Distilled water	100	80	80	100	80	80	100	100	100	100	100	100
q.s. Anise oil	0.16	19.4	19.4	0.16	19.4	19.4	0.16	0.16	0.15	0.15	0.15	0.15
Poloxamer	–	1	1	–	1	1	–	–	–	–	–	–
FCZ	–	1	1	–	1	1	–	–	–	–	–	–
β-Cyclodextrin	–	2	3	2	3	–	–	–	–	–	–	–

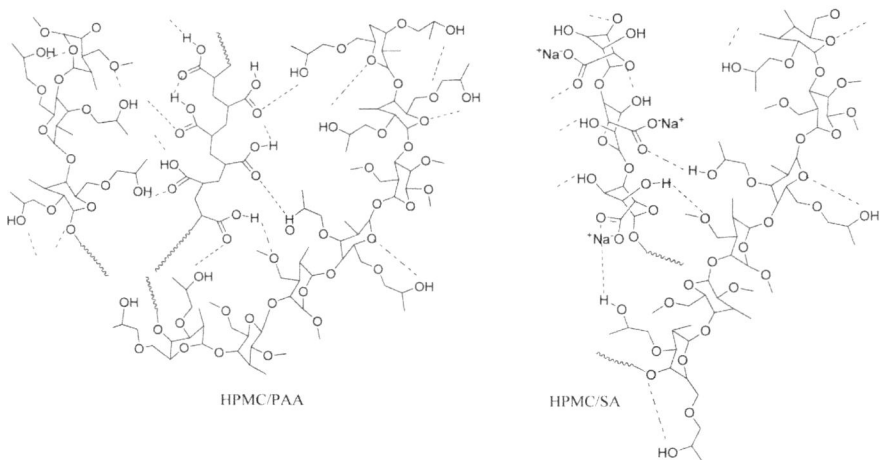

FIGURE 4.1   Schematic representation of the polymeric networks based on HPMC and PAA or SA.

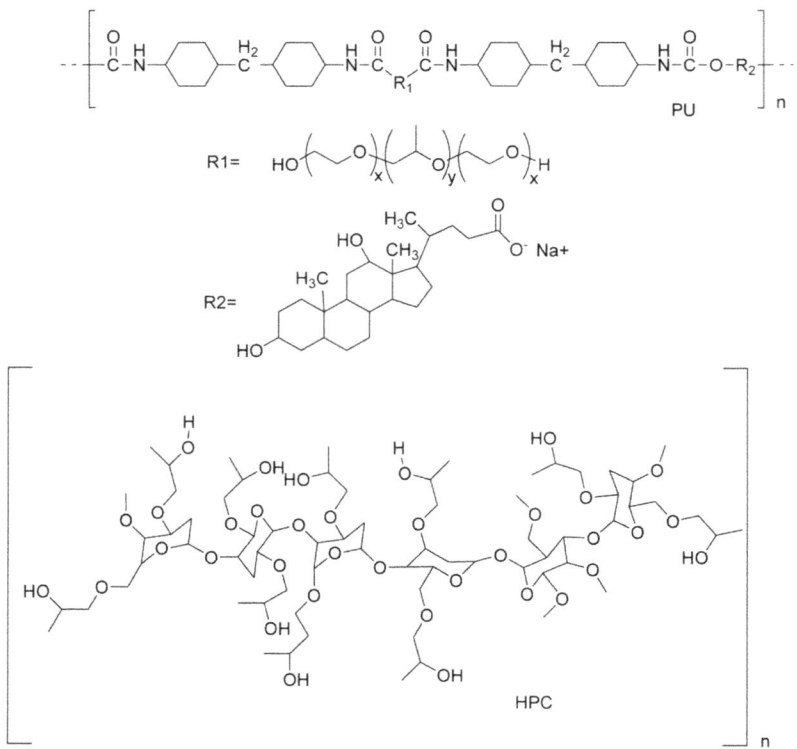

FIGURE 4.2   The chemical structure of the polymers used for blend preparation.

4.6.3 PREPARATION OF BIONANOCOMPOSITE MEMBRANES BASED ON POLYURETHANE, EM, AND SILVER NANOPARTICLES (AgNPs)

Polyurethane was obtained by the reaction between PEGA, polyethylene glycole (PE), and diphenyl MDI in molar ratio 1:4:5. The solution of PU was then treated in two steps with a mixture of hydrolyzed collagen and elastin 10:1 and with components of the EM, CS, or HA 0.1% followed by the addition of 0.5% or 1.5% AgNPs (resulted from the reduction of $AgNO_3$ in the presence of DMF). The polymeric films resulted from the evaporation of DMF and drying at vacuum for 55 h. The chemical structure of the polyurethane can be seen in Figure 4.3.

4.6.4 FTIR SPECTROSCOPY

FTIR spectroscopy was used to evidence the inter- and intra-molecular interactions between HPMC/PAA, HPMC/SA, and HPC/PU.

The most representative bands which indicate the formation of interpolymer complexes can be seen in the 3800–3000 cm^{-1} spectral range and are assigned mainly to –OH intra- and inter-molecular stretching vibrations between HPMC and PAA/SA networks (Figure 4.4).

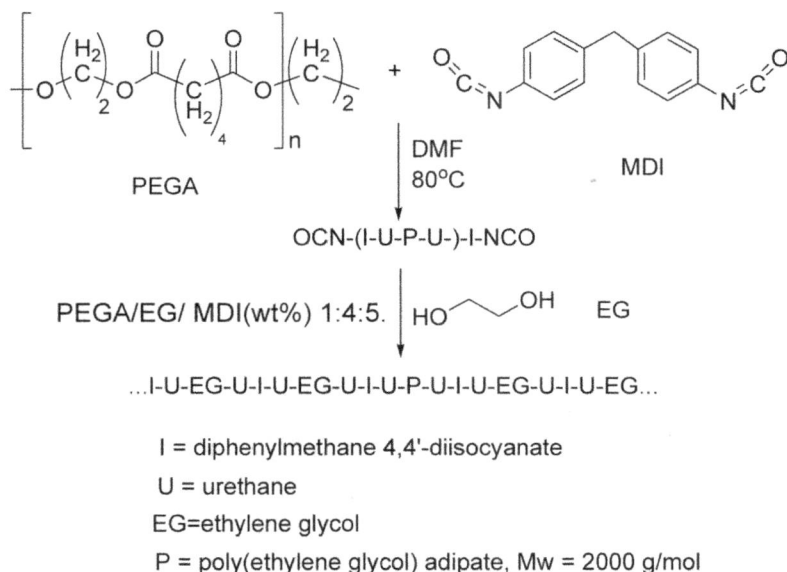

PEGA/EG/ MDI(wt%) 1:4:5.

OCN-(I-U-P-U-)-I-NCO

...I-U-EG-U-I-U-EG-U-I-U-P-U-I-U-EG-U-I-U-EG...

I = diphenylmethane 4,4'-diisocyanate

U = urethane

EG=ethylene glycol

P = poly(ethylene glycol) adipate, Mw = 2000 g/mol

FIGURE 4.3 The chemical structure of the polyurethane used for the preparation of bionanocomposite membranes.

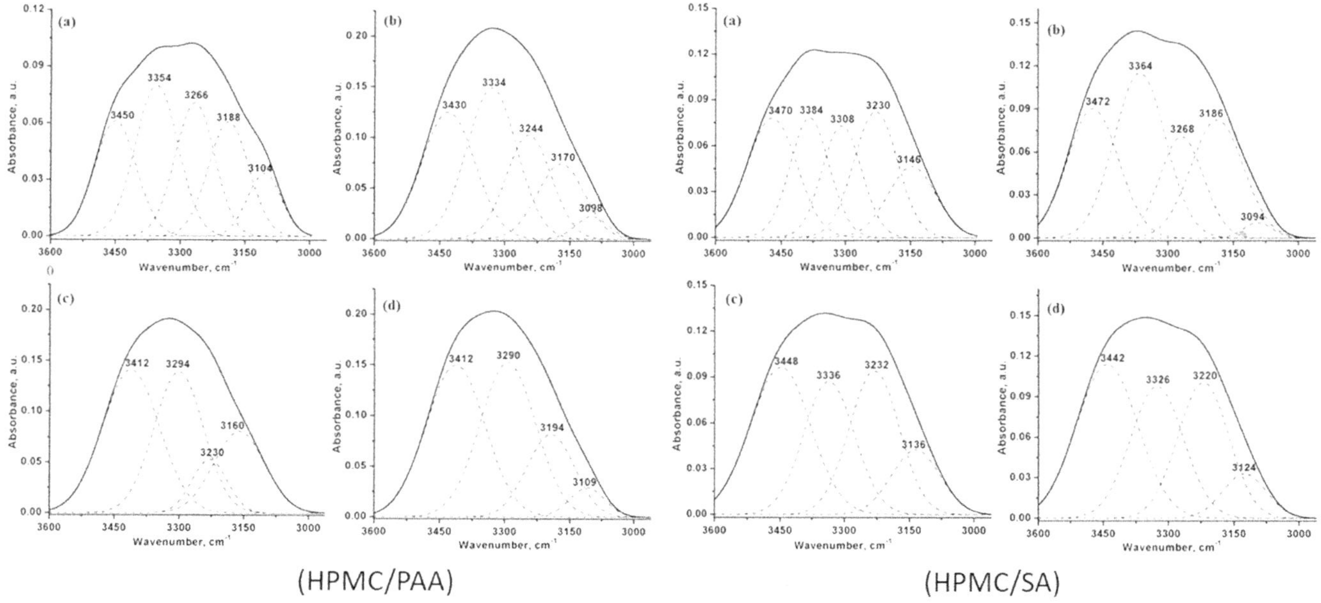

FIGURE 4.4   ATR-FTIR deconvoluted spectra in the 3800–300 cm^{-1} spetral range of the HPMC/PAA (a—F1A, b—F2A, c—F3A, d—F4A) blends and HPMC/SA blends (a—F5A, b—F6A, c—F7A, d—F8A).

By using Sederholm equations [44], the H-bond distances are calculated according to relation (4.1):

$$\Delta v \ (cm^{-1}) = 4.43 \times 10^3 \times (2.84 - R) \qquad (4.1)$$

where $\Delta v = v - v_0$, v_0 is the standard wavenumber corresponding to free O–H groups located at 3650 cm^{-1}, v is the stretching wavenumber of the H-bonded O–H groups in the IR spectra of the analyzed samples (Table 4.4).

TABLE 4.4 The Energy and the Distances of the Hydrogen Bonds of Studied Hydrogels Based on HPMC/PAA and HPMC/SA

Sample	Hydrogen-bonded O–H v, cm^{-1}	E_H (kJ)	R (Å)
F1A	3354	21.29	2.773
F1A_1:2	3384	19.13	2.779
F1A_1:3	3380	19.41	2.779
F2A	3334	22.73	2.768
F2A_1:2	3372	19.99	2.777
F2A_1:3	3392	18.55	2.781
F3A	3294	25.60	2.759
F4A	3290	25.89	2.758
F5A	3384	19.13	2.779
F6A	3364	20.56	2.775
F7A	3336	22.58	2.769
F8A	3326	23.30	2.766

The energy of the H bonds (E_H) was determined according to the following equation [45]:

$$E_H = \frac{1}{k}\frac{v_0 - v}{v_0} \qquad (4.2)$$

where $1/k$ is a constant equal with 2.625×10^2 kJ.

The structural flexibility of PAA is the main factor that could determine the formation of stronger interactions within the polymer networks as compared with those based on HPMC and SA. Because of the dynamic character of the H bonds, due to their weaker character as compared with the covalent bonds, the supramolecular polymeric complexes (HPMC-PAA- and HPMC-SA-based hydrogels) are kinetically more labile, dynamically more flexible, and thermodynamically less stable than classical covalent polymers, which make them suitable as scaffolds for liposomal formulations.

In HPC/PU blends, the same spectral region (3800–3000 cm^{-1}) evidenced the characteristic bands assigned to the "free" and hydrogen bonding N–H groups and those characteristic for intra- and intermolecular stretches of hydroxyl groups (Figure 4.5). Based on the position of these bands, the H-bond distances specifically for O–H and N–H groups have been calculated and the results are presented in Table 4.5.

PU component in the blend leads to the redshift of the O–H intramolecular and intermolecular interactions suggesting a larger number of interacting segments between HPC and PU. The increased

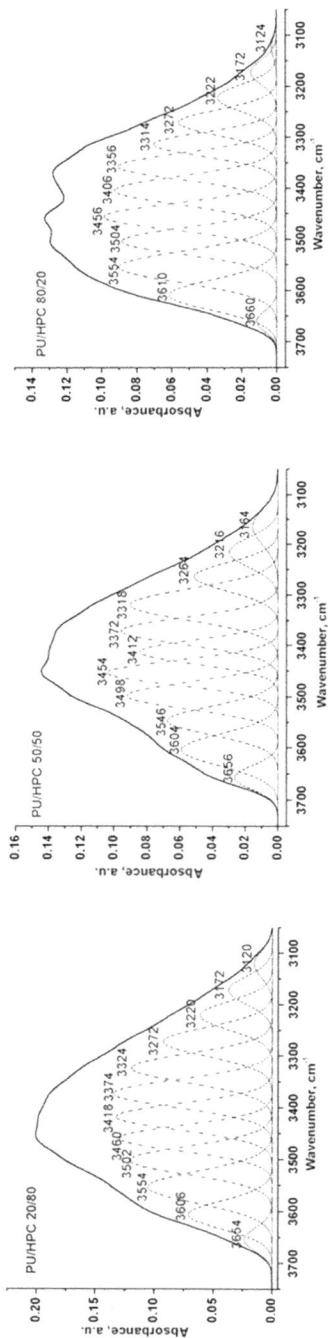

FIGURE 4.5   ATR-FTIR deconvoluted spectra of the HPC/PU blends in the 3800–3000 cm^{-1} spectral range.

TABLE 4.5   The Calculated H-Bond Distances and H-Bond Energy for HPC/PU Blends

Sample	Hydrogen-bonded O–H		
	ν (cm⁻¹)	E_H (kJ)	R (Å)
PU	–	–	–
HPC	3406	17.54	2.785
PU/HPC 20/80	3374	19.85	2.777
PU/HPC 50/50	3380	19.42	2.779
PU/HPC 80/20	3400	17.98	2.783
	Hydrogen-bonded N–H		
PU	3352	9.22	2.987
HPC	–	–	–
PU/HPC 20/80	3324	11.33	2.936
PU/HPC 50/50	3318	11.78	2.925
PU/HPC 80/20	3314	12.09	2.918

number of vibrations in this region is also explained by the presence of both ester and ether groups involved in H-bonds within the PU backbone. The N–H stretches have two main contributions: "free" (nonhydrogen bonded) and hydrogen-bonded N–H groups. In HPC/PU blends, the H-bonded N–H vibrations shift systematically in frequency from 3324 to 3314 cm⁻¹ suggesting an increase in the H-bonds content [46]

4.6.5 BIOADHESION AND MUCOADHESION PROPERTIES OF THE POLYMERIC SCAFFOLDS

Adhesion properties of all polymeric blends and bionanocomposite PU have been evaluated on cellulose membrane at pH 7.4 (bioadhesion) and on mucosa tissues (stomach at pH 2.6, bladder at pH 5). The bio- and muco-adhesiveness have been estimated by the detachment force, and the work of adhesion and the results are shown in Table 4.6.

The mucoadhesiviness of the polymeric blends based on HPMC/PAA has been evaluated by determination of the interfacial tension between blood and the blend surface according to the method described by Gafitanu et al. [47]. F1A and F2A proved to be long-term compatible with blood so that these hydrogel blends have been tested as drug delivery systems for FCZ [39, 48].

The HPMC/SA blends, as well as the nanocomposite PU, have been studied at pH 5 on mucosa bladder and pH 7.4 on cellulose membrane,

TABLE 4.6   The Results of the Bioadhesion and Mucoadhesion Tests of the Polymeric Blends

Sample	Bioadhesion		Mucoadhesion	
	Detachment Force (gF)	Work of Adhesion (N × s)	Detachment Force (gF)	Work of Adhesion (N × s)
F5A	10	0.0097	7.4	0.0054
F6A	12.7	0.0154	6.1	0.0062
F7A	12.6	0.234	10.6	0.0064
F8A	12.2	0.0226	9	0.0107
PU/HPC 20/80	0.556	0.0179	0.0853	0.0076
PU/HPC 50/50	0.3804	0.0145	0.0676	0.0032
PU/HPC 80/20	0.0735	0.0054	0.0676	0.0037
HPC	0.7649	0.0082	–	–
PU	9.3	0.0285	7.2	0.0091
PU-CS	5.4	0.0086	7.7	0.0104
PU-HA	9.6	0.0124	8.3	0.0063
PU 0.5%Ag	4.7	0.0070	5.7	0.0111
PU 1.5%Ag	10.2	0.0251	4	0.0053
PU-CS 0.5%Ag	7	0.0122	11	0.0146
PU-CS 1.5%Ag	3.7	0.0056	23	0.0311
PU-HA 0.5%Ag	4.5	0.0066	6.8	0.0055
PU-HA 1.5%Ag	5.7	0.0096	12.1	0.0175

while the HPC/PU blends were evaluated at pH 2.6 corresponding to stomach mucosa.

The results indicated in Table 4.6 highlighted that HPMC/SA are more muco- and bio-adhesive than polyurethane nanocomposites and blends, mainly due to the intermolecular interactions, which can be established with the mucosa or cellulose membrane [38].

Other factors that can influence the mucoadhesiviness of the polymeric blends can be explain by understanding the mechanisms involved in this phenomenon [49]. One of them is the wettability of the blends occurring at the first contact with the mucosa, which is influenced by the hydrophilic component of the belnd (HPC or HPMC). Also, the rigidity of the polyurethane structure due to the intermolecular H bonds reduces the contact surface with mucosa.

The bionanocomposite PU revealed higher values of the detachment force and work of adhesion by the introduction of the extracellular components (CS and HA). Intermolecular interactions with the mucosa bladder through these components can explain this behavior. The most mucoadhesive nanocomposite is PU–CS 1.5% AgNPs, followed by PU–HA 1.5% AgNPs and PU–CS 0.5% AgNPs suggesting that other structural characteristics of the CS and HA are involved in the mucoadhesion mechanisms [42].

4.6.6 SCANNING ELECTRON MICROSCOPY ANALYSIS

The HPMC/PAA [39] and HPMC/SA morphology was evaluated by scanning electron microscopy (SEM) micrographs (Figure 4.6). Freeze-dried scaffolds showed a highly interconnected 3D networks, with the largest pores when the HPMC content is higher. The presence of numerous H-bonds interactions between the HPMC and PAA or SA could explain the formation of a stronger connected architecture

through which is possible to estimate the swelling and the release of the encapsulated drugs.

4.6.7 DYNAMIC VAPOR SORPTION ANALYSIS

The moisture properties of the polymeric blends are essential for their storage and manipulation as bioactive formulations. The sorption process depends firstly on the structural characteristic of the samples, their hydrophilic/hydrophobic balance, the humidity, and temperature or exposure time. The polymeric gels HPMC/PAA and HPMC/SA revealed a porous morphology with 3D interconnected pores larger with the increased content of HPMC. In these cases, the higher sorption capacity is explained by the presence of the polar groups (OH and COOH) that are involved in the H-bonds formation (Table 4.7). The same behavior

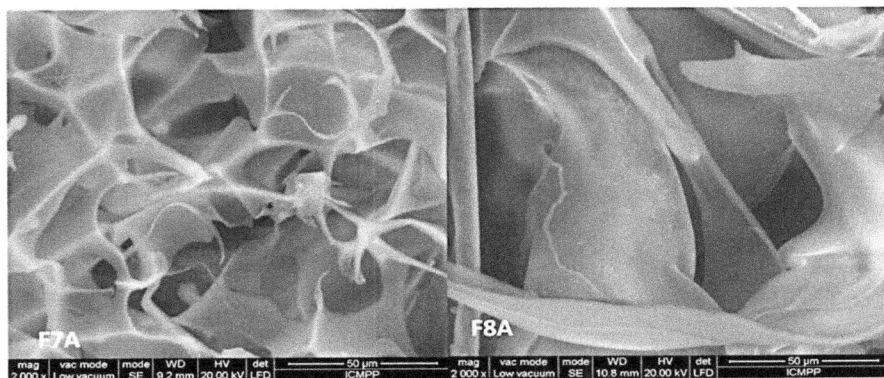

FIGURE 4.6 SEM micrographs of the HPMC/SA blends.

was observed in the case of HPC/PU blends, where the higher sorption capacity value was observed at 80% content of HPC in blend. The water sorption capacity of the polymeric blends has a positive impact on the mucoadhesive properties. Water is involved in the dynamic process of H-bonds formation and ensures the equilibrium between breaking and reorganization proceses which control the bioadhesion phenomenon [50]. Also, a higher wettability of the sample reduces the mucodhesiviness mainly due to the strong affinity for water which disintegrates the blend, by breaking the interactions with the mucosa constituents.

TABLE 4.7   The Values of the Water Sorption Capacity of the Polymeric Blends

Sample	Water Sorption Capacity, wt%
F1A	7.45
F2A	13.30
F3A	26.90
F4A	28.71
F5A	30.85
F6A	27.63
F7A	36.22
F8A	33.07
HPC	23.9
PU/HPC 20/80	22.46
PU/HPC 50/50	16.55
PU/HPC 80/20	7.6
PU	6.8

4.6.8 DRUGS LOADING AND RELEASE STUDIES

The polymeric gels HPMC/PAA and HPMC/SA have been evaluated for drug encapsulation and release in vitro. The HPMC/PAA samples FIA and F2A have been tested for encapsulation of FCZ/β-cyclodextrin inclusion complex. Cyclodextrin was used as an alternative for FCZ solubilization, stabilization, transportation, and release, being an efficacy delivery system haemobiocompatibility, nontoxicity, and good adhesiveness [39].

The encapsulation of FCZ (1% in β-cyclodextrin inclusion complexes at 1:2 and 1:3 ratios) (Table 4.3) in the polymeric gels was evaluated by FTIR-ATR and thermogravimetrical analysis [51] and the release was monitored by UV–vis spectroscopy.

The release results evidenced that our gels ensured a controlled release of FCZ over a few days. The highest released concentration (92%) was observed after 40 h, while in more reported cases this concentration was release after 2 h [39].

In another approach, we encapsulated ruthenium complexes in the polymeric blends HMPC/SA [38] showing the ability of these polymeric gels to function as drug delivery systems and opening up many opportunities toward new bio-based polymeric drug formulations. Metal complexes, particularly those with antiproliferative activity, are often poorly water-soluble, which makes difficult their antitumor screening and administration in vivo. Palladium(II) complexes with antitumor activity on HCT116 cancer cells were successfully administrated after encapsulation in cellulose nanoparticle composite [36], proving that the conjugation in these systems induces an increase of the aqueous stability and reduces the toxicity.

4.7   CONCLUSION

Liposomes have become one of the most efficient drug carriers and delivery systems, contributing to the solubilization of many drugs, reducing the toxicity, ensuring a long-term circulation of the drug, increasing the bioavailability of the drugs, and the therapeutic efficiency. Clinical application of the liposomes is often limited by the immune response so that many technologies have been reported to improve their haemocompatibility, the most known being PEGylation to produce "stealth" liposomes with long-term circulation and targeted liposomes technology, which are clinically approved.

To enhance the therapeutic efficiency and overcome the limitations, functionalized formulations have been proposed, including cationic liposomes, pH-sensitive liposomes, liposome-based theranostic agents, or liposome-based vaccines. Recent advancements in the development of liposomes have been reported by conjugation with biodegradable and biocompatible products (scaffolds) to produce liposomal complexes with a prolonged drug release rate, which is of great interest.

In summary, we proposed the development of a combined system, as a model for a sustained release of the drug, replacing the short-term circulation of the liposomes. Since such materials (polymeric blends or biocomposites) are implantable, it may be useful in the future for the management of many diseases.

ACKNOWLEDGMENTS

This work was supported by a grant of Ministry of Research and Innovation, CNCS-UEFISCDI, project number PN-III-P1-1.1-PD-2016-1027 (Contract 5/2018). The financial support of European Social Fund for Regional Development, Competitiveness Operational Programme Axis 1—Project "Petru Poni Institute of Macromolecular Chemistry—Interdisciplinary Pol for Smart Specialization through Research and Innovation and Technology Transfer in Bio(nano)polymeric Materials and (Eco)Technology," InoMatPol, ID P_36_570, Contract 142/10.10.2016, cod MySMIS:107464, is also gratefully acknowledged.

KEYWORDS

- liposomes
- bio-based polymers
- drug delivery system
- biodegradation

REFERENCES

1. Siepmann, J.; Faham, A.; Clas, S.-D.; Boyd, B. J.; Jannine V.; Bernkop-Schnürch, A.; Zhao, H.; Lecommandoux, S.; Evans, J. C.; Allen, C.; Merkel, O. M.; Costabile, G.; Alexander, M. R.; Wildman, R. D.; Roberts, C. J.; Leroux, J.-C.; Lipids and Polymers in Pharmaceutical Technology: Lifelong Companions. *Int. J. Pharm.* **2019**;558:128–142.

2. Sheikhpour, M.; Barani, L.; Kasaeian, A.; Biomimetics in Drug Delivery Systems: A Critical Review. *J. Control. Release.* **2017**;253:97–109.

3. Svenson, S.; Carrier-based Drug Delivery. ACS Symposium Series, American Chemical Society, Washington, DC, **2004**, pp. 2–23.

4. Leung, A. W. Y.; Amador, C.; Wang, L. C.; Mody, U. V.; Bally, M. B.; What Drives Innovation: The Canadian Touch on Liposomal Therapeutics. *Pharmaceutics* **2019**;11:124.

5. Zhang, Y.; Heidari, Z.; Su, Y.; Yu, T.; Xuan, S.; Omarova, M.; Aydin, Y.; Dash, S.; Zhang, D.; John, V.; Amphiphilic Polypeptoids Rupture Vesicle Bilayers To Form Peptoid–Lipid Fragments Effective in Enhancing Hydrophobic Drug Delivery. *Langmuir* **2019**;35:15335–15343.

6. Schattling, P.; Taipaleenmäki, E.; Zhang, Y.; Städler, B.; A Polymer Chemistry Point of View on Mucoadhesion and Mucopenetration. *Macromol. Biosci.* **2017**;17:1700060.

7. Torchilin, V. P.; Shtilman, M. I.; Trubetskoy, V. S.; Whiteman, K.; Milstein, A. M.; Amphiphilic Vinyl Polymers Effectively Prolong Liposome Circulation Time in Vivo. *Biochim. Biophys. Acta.* **1994**;1195:181–184.

8. Nag, O. K.; Awasthi, V.; Surface Engineering of Liposomes for Stealth Behavior. *Pharmaceutics* **2013**; 5: 542–569.

9. Gao, W.; Hu, C. M.-J.; Fang, R. H.; Zhang, L.; Liposome-like Nanostructures for Drug Delivery. *J. Mater. Chem. B.* **2013**;1:6569–6585.

10. Hannon, M. J.; Metal-based Anticancer Drugs: From a Past Anchored in Platinum Chemistry to a Post-Genomic Future of Diverse Chemistry and Biology. *Pure Appl. Chem.* **2007**;79:2243–2261.

11. Teleanu, D. M.; Chircov, C.; Grumezescu, A. M.; Volceanov, A.; Teleanu, R. I.; Blood-Brain Delivery Methods Using Nanotechnology. *Pharmaceutics* 2018;10:269; doi:10.3390/pharmaceutics 10040269

12. Yingchoncharoen, P.; Kalinowski, D. S.; Richardson, D. R.; Lipid-based Drug Delivery Systems in Cancer Therapy: What Is Available and What Is Yet to Come. *Pharmacol. Rev.* **2016**; 68(3):701–787.

13. Bulbake, U.; Doppalapudi, S.; Kommineni, N.; Khan, W.; Liposomal Formulations in Clinical Use: An Updated Review. *Pharmaceutics* 2017; 9:12; doi:10.3390/pharmaceutics 9020012

14. Barenholz, Y.; Doxil®—The First FDA-Approved Nano-Drug: Lessons Learned. *J. Control. Release.* **2012**;160: 117–134.

15. Tefas, L. R.; Sylvester, B.; Tomuta, I.; Sesarman, A.; Licarete, E.; Banciu, M.; Porfire, A.; Development of Anti-proliferative Long-Circulating Liposomes Co-encapsulating Doxorubicin and Curcumin, Through the Use of a Quality-By-Design Approach. *Drug Des. Devel. Ther.* **2017**;11:1605–1621.

16. Forssen, E. A.; The Design and Development of DaunoXome® for Solid Tumor Targeting in Vivo. *Adv. Drug Deliv. Rev.* **1997**;24:133–150.

17. Glantz, M. J.; Jaeckle, K. A.; Chamberlain, M. C.; Phuphanich, S.; Recht, L.; Swinnen, L. J.; Maria, B.; LaFollette, S.; Schumann, G. B.; Cole, B. F.; Howell, S. B.; A Randomized Controlled Trial Comparing Intrathecal Sustained-Release Cytarabine (DepoCyt) to Intrathecal Methotrexate in Patients with Neoplastic Meningitis from Solid Tumors. *Clin. Cancer Res.* **1999**;5(11):3394–3402.

18. Swenson, C. E.; Perkins, W. R.; Roberts, P.; Janoff, A. S.; Liposome Technology and the Development of Myocet™ (Liposomal Doxorubicin Citrate), *The Breast* **2001**;10:1–7.

19. Clemons, K. V.; Stevens, D. A.; Comparison of Fungizone, Amphotec, AmBisome, and Abelcet for Treatment of Systemic Murine Cryptococcosis. *Antimicrob. Agents Chemother.* **1998**; 42:899–902.

20. Stone, N. R. H.; Bicanic, T.; Salim, R.; Hope, W.; Liposomal Amphotericin B (AmBisome®): A Review of the Pharmacokinetics, Pharmacodynamics, Clinical Experience and Future Directions. *Drugs* **2016**;76(4):485–500.

21. Bovier, P. A.; Epaxal: A Virosomal Vaccine to Prevent Hepatitis A Infection. *Expert Rev. Vaccines.* **2008**; 7(8):1141–1150; doi: 10.1586/ 14760584.7.8.1141.

22. Mischler, R.; Metcalfe, I. C.; Inflexal V a Trivalent Virosome Subunit Influenza Vaccine: Production. *Vaccine* **2002**; 20: B17–23.

23. Alfayez, M.; Kantarjian, H.; Kadia, T.; Ravandi-Kashani, F.; Daver, N.; CPX-351 (vyxeos) in AML. *Leuk. Lymphoma.* **2020**;61(2):288–297.

24. Bressler, N. M.; Bressler, S. B.; Photodynamic Therapy with Verteporfin (Visudyne): Impact on Ophthalmology and Visual Sciences. *Invest. Ophthalmol. Vis. Sci.* **2000**; 41:624–628.

25. Gambling, D.; Hughes, T.; Martin, G.; Horton, W.; Manvelian, G.; A Comparison of Depodur, a Novel, Single-Dose Extended-Release Epidural Morphine, with Standard Epidural Morphine for Pain Relief After Lower Abdominal Surgery. *Anesth. Analg.* **2005**; 100(4): 1065–1074.

26. Malam, Y.; Loizidou, M.; Seifalian, A. M.; Liposomes and Nanoparticles: Nanosized Vehicles for Drug Delivery in Cancer. *Trends Pharmacol. Sci.* **2009**;30:592–599; doi:10.1016/j.tips. 2009.08.004.

27. Zalipsky, S.; Synthesis of an End-Group Functionalized Polyethylene Glycol-Lipid Conjugate for Preparation of Polymer-Grafted Liposomes. *Bioconjugate Chem.* **1993**;4:296–299.

28. Mufamadi, M. S.; Pillay, V.; Choonara, Y. E.; Du Toit, L. C.; Modi, G.; Naidoo, D.; Ndesendo, V. M. K.; A Review on Composite Liposomal Technologies for Specialized Drug Delivery. *J. Drug Deliv.* **2011**;2011: 939851; doi:10.1155/2011/939851.

29. Yuba, E.; Liposome-based Immunity-Inducing Systems for Cancer Immunotherapy. *Mol. Immunol.* **2017**;98:8–12.

30. Woodle, M. C.; Scaria, P.; Cationic Liposomes and Nucleic Acids. *Curr. Opin. Colloid Interface Sci.* **2001**;6: 78–84.

31. Tyagi, P.; Santos, J. L.; Macromolecule Nanotherapeutics: Approaches and

challenges. *Drug Discov. Today.* **2018**; 23(5):1053–1061.

32. Roux, E.; Lafleur, M.; Lataste, E.; Moreau, P.; Leroux, J.-C.; On the Characterization of pH-Sensitive Liposome/ Polymer, Complexes. *Biomacromolecules* **2003**;4:240–248.

33. Vijayan, V. M.; Muthu, J.; Polymeric Nanocarriers for Cancer Theranostics. *Polym. Adv. Technol.* **2017**; 28: 1572–1582.

34. Roopngam, P. E.; Liposome and Polymer-based Nanomaterials for Vaccine Applications. *Nanomed. J.* **2019**;6(1):1–10.

35. Palazzolo, S.; Bayda, S.; Hadla, M.; Caligiuri, I.; Corona, G.; Toffoli, G.; Rizzolio, F.; The Clinical Translation of Organic Nanomaterials for Cancer Therapy: A Focus on Polymeric Nanoparticles, Micelles, Liposomes and Exosomes. *Curr. Med. Chem.* **2018**;25: 4224–4268.

36. Yaroslavov, A. A.; Sybachin, A. V.; Sandzhieva, A. V.; Zaborova, O. V.; Multifunctional Containers From Anionic Liposomes and Cationic Polymers/Colloids. *Polym. Sci. C.* **2018**;60: S179–S191.

37. Zaltariov, M. F.; Ciubotaru, B.-I.; Verestiuc, L.; Peptanariu, D.; Macocinschi, D.; Filip, D.; Ruthenium(II) Complexes With Cytotoxic Activity Embedded in Hydroxypropyl Methylcellulose/Sodium Alginate Mucoadhesive Hydrogels. *Cellulose Chem. Technol.* **2019**; 53(9–10):869–878.

38. Gafitanu, C. A.; Filip, D.; Cernatescu, C.; Rusu, D.; Tuchilus, C. G.; Macocinschi, D.; Zaltariov, M.-F.; Design, Preparation and Evaluation of HPMC-based PAA or SA Freeze-Dried Scaffolds for Vaginal Delivery of Fluconazole. *Pharm. Res.* **2017**;34:2185–2196.

39. Gencturk, A.; Kahraman, E.; Gungor, S.; Ozhan, G.; Ozsoy, Y.; Sarac, A. S.; Polyurethane/Hydroxypropyl Cellulose Electrospun Nanofiber Mats as Potential Transdermal Drug Delivery System: Characterization Studies and in Vitro Assays. *Artif. Cell. Nanomed. B.* **2017**; 45 (3):655–664.

40. Raschip, I. E.; Vasile, C.; Macocinschi, D.; Compatibility and Biocompatibility Study of New HPC/PU Blends. *Polym. Int.* **2009**;58:4–15.

41. Macocinschi, D.; Filip, D.; Paslaru, E.; Munteanu, B. S.; Dumitriu, R. P.; Pricope, G. M.; Aflori, M.; Dobromir, M.; Nica, V.; Vasile, C.; Polyurethane–Extracellular Matrix/Silver Bionanocomposites for Urinary Catheters. *J. Bioact. Compat. Polym.* **2015**;30(1):99–113.

42. Lawless, J.; The Encyclopedia of Essential Oils, J. Lawless, ed. Conari Press Publisher, San Francisco, **2013**, pp. 38–39.

43. Pimentel, G. C.; Sederholm, C. H.; Correlation of Infrared Stretching Frequencies and Hydrogen Bond Distances in Crystals. *J. Chem. Phys.* **1956**;24:639.

44. Struszczyk, H.; Modification of Lignins. III. Reaction of Lignosulfonates with Chlorophosphazenes. *J. Macromol. Sci.* **1986**;23:973–992.

45. Coleman, M. M.; Lee, K. H.; Skrovanek, D. J.; Painter, P. C.; Hydrogen Bonding in Polymers. 4. Infrared Temperature Studies of a Simple Polyurethane. *Macromolecules* **1986**;19:2149–2157.

46. Gafitanu, C. A.; Filip, D.; Cernatescu, C.; Ibanescu, C.; Danu, M.; Paslaru, E.; Rusu, D.; Tuchilus, C. G.; Macocinschi, D.; Formulation and Evaluation of Anise-based Bioadhesive Vaginal Gels. *Biomed. Pharmacother.* **2016**;83: 485–495.

47. Ruckenstein, E.; Gourisankar, S. V.; Surface Restructuring of Polymeric Solids and its Effect on the Stability of the Polymer–Water Interface. *J. Colloid Interface Sci.* **1986**;109:557–566.

48. Mortazavi, S. A.; Smart, J. D.; An Investigation into the Role of Water

Movement and Mucus Gel Dehydration in Mucoadhesion. *J. Control. Release* **1993**;25:197–203.

49. Zaman, M. A.; Martin, G. P.; Rees, G. D.; Mucoadhesion, Hydration and Rheological Properties of Non-Aqueous Delivery Systems (NADS) for the Oral Cavity. *J. Dentistry.* **2008**; 3 6:351–359.

50. Zaltariov, M.-F.; Filip, D.; Varganici, C.-D.; Macocinschi D.; ATR-FTIR and Thermal Behavior Studies of New Hydrogel Formulations Based on Hydroxypropyl Methylcellulose/ Poly(Acrylic Acid) Polymeric Blends. *Cellulose Chem. Technol.* **2018**; 52 (7–8):619–631.

51. Mirmashhouri, B.; Tehrani, A. D.; Eslami-Moghadam M.; Divsalar, A.; Preparation and Characterization of Palladium(II) Complex/Cellulose Nano-whisker as a New Potent Anticancer Prodrug. *Cellulose Chem. Technol.* **2018**; 52(1–2):27–33.

CHAPTER 5

BIOMATERIAL-BASED IMPLANTS: A MEDICAL DEVICE TO RECONSTRUCT THE BIOLOGICAL STRUCTURE

G. SANTHOSH[1*] and B. SOWMYA[2]

[1]*Department of Mechanical Engineering, NMAM Institute of Technology, Nitte, Karnataka 574110, India*

[2]*Materials Science Division, CSIR—National Aerospace Laboratories, Old Airport Road, Kodihalli, Bengaluru 560017, India*

Corresponding author. E-mail: drsanthug@gmail.com.

ABSTRACT

Biodegradable polymers have great feasibility in encapsulating the drugs for controlled drug delivery, like implants and a wide range of biomedical applications. The application and use of these polymers have seen steady growth. Natural polymers are the first biodegradable biomaterials that are well established in many of the biomedical uses. These natural polymers undergo enzymatic degradation. In the late 1980s, the use of synthetic polymers has started. In the field of medicine, biodegradable synthetic polymers are broadly used as surgical aids like nerve guided tubes, reabsorbable sutures, hemostats, and orthopedic implants. However, the application of synthetic polymers is still in the beginning stage which is expected to be expanded in other clinical applications. Nonmedical applications of biodegradable polymers include food packaging. Owing to their unique properties and ease of modification to achieve the desired properties, biodegradable polymers are capable of reducing the use of metals, alloys, and ceramics as biomaterials. The research on biodegradable polymers has created novel polymeric materials and formulations that lead to the development of biomaterials for various medical applications. For instance, cardiovascular stents are now made of metals, although a few months of lumen support may be necessary to retain vessel potency. A biodegradable stent is designed to function for a specific duration by providing adequate support that solves restenosis complications. Alternatively, in crucial circumstances, a smart stent releases antirestenosis agents contained in the polymeric matrix to the neighboring tissues to restore its stability. Biodegradable polymers are favorable for biomedical applications such as scaffolds, gene carriers, controlled drug delivery, implants, in a few medical devices, etc.

5.1   INTRODUCTION

Biodegradation is the ability of the material to decompose by the action of microorganisms or any living organism. Biodegradable polymers degrade either by enzymatic or hydrolytic reaction forming natural endogenous by-products and these products are taken away from the body. In addition, the by-products obtained from degradation reactions must be nontoxic, metabolic, and easy to clean [1–4].

Biodegradable polymers are "materials of today" and one of the pivotal sources to synthesize biomaterials. A biomaterial is "any natural or synthetic substance engineered to interact with biological systems

to direct medical treatment" [5]. They do not need any removal or surgical interventions. Biodegradable polymers are eco-friendly in nature that makes them applicable for household products [6]. The rate at which biodegradable polymers degrades is similar to the process of healing and regeneration. Biodegradable polymers must possess adequate mechanical properties to suit the targeted applications. These polymeric materials should be biocompatible, exhibit ease of processing, stable enough to sustain the fabrication process, and should not elicit immune reactions [7–9].

There are several applications of biodegradable polymers such as large and small implantable devices as shown in Figure 5.1. These devices include bone screws, bone plates, contraceptive reservoirs, and stents; small implants such as sutures, staples, drug delivery vehicles.

Metals, ceramics, polymers, and composites are some of the biomaterials used to fabricate implants that are in close contact with living tissues [10, 11]. The metallic alloys possess good biocompatibility and mechanical properties by virtue of which they are used as fracture fixations over years, and the most commonly used implant materials include titanium, cobalt, stainless steel, and magnesium alloys [12–14]. After the removal of internal metallic fixations from the bones,

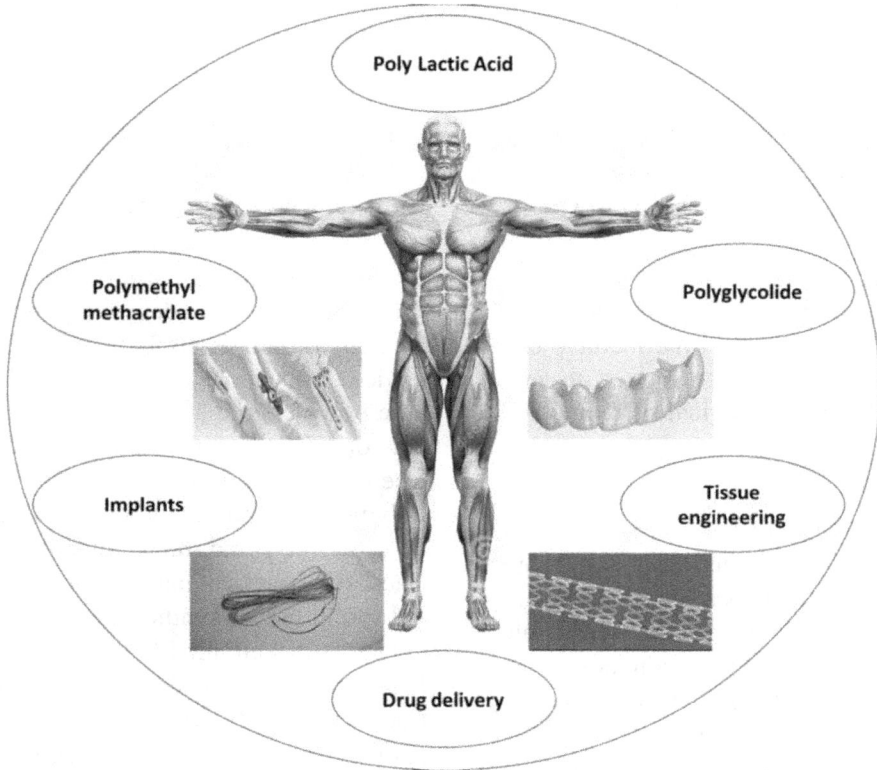

FIGURE 5.1   Resorbable materials and their biomedical applications.

it was found that the treated bones were weaker than the normal bones [15, 16]. This was attributed due to the residual holes of the screw and also the destructive effect on bones caused by rigid plates due to atrophy in the implant site. This anomaly occurred due to the dissimilarity of modulus of elasticity between the implant and the bone [10]. Metallic implants have compatibility issues when used as implants, they cause an inflammatory reaction, infections due to toxic corrosion materials [17]. An appeal to compensate for the inconveniences caused by metallic implants ended due to the use of biodegradable polymers as medical devices. The low-elastic modulus and their light-weight nature are said to be the main reason for this paradigm shift [18, 19]. The biodegradable devices are designed to degrade and transfer the load progressively to the bone. Biodegradable polymers ease the problems that existed in metallic implants [20, 21].

5.2   TECHNICAL REQUIRE-MENTS OF POLYMERS AS IMPLANTABLE MATERIALS [22]

The biomaterials can be defined as any natural or synthetic materials that are used in the medical device, modified, and designed to interact with the biological system.

1. Biostability: the material should not enable the soluble compounds to enter the living system unless it is intentional, for example, drug delivery systems.
2. Implants must be sterilizable.
3. It should remain nontoxic and nonallergic in response to the surrounding tissues.
4. Materials should be easily processable into the final product shape.
5. Implantable materials must have adequate mechanical and physical strength to serve the intended purpose and until surrounding tissues heal.
6. Materials should be biocompatible, namely, implanted material should not adversely affect the surrounding tissues and should not interact with the biological fluids.
7. Implants should not be degraded by a living system.

8. It should have an appropriate shelf life.

5.3   TYPICAL BIODEGRADABLE POLYMERIC MATERIALS IN BIOMEDICAL DEVICES

In recent decades, a variety of polymers have been widely used and studied as biomaterials for biomedical device applications. The biodegradable polymers are studied from both the origin—synthetic and natural. The collagen and gelatin are natural polymers that possess poor stability, compatibility, and biodegradability. These drawbacks of natural polymers led to the use of synthetic polymers because of their modifiable properties favoring biomedical applications. Poly(lactic-*co*-glycolic acid) (PLGA), polylactic acid (PLA), and polyurethanes (PURs) exhibit exceptional compatibility.

5.3.1   SYNTHETIC POLYMERS

Synthetic polymers have gained more attention as compared to other polymers. Unique properties such as biocompatibility, cell adherence, controlled drug release, and stimuli-response are sometimes necessary for biodegradable polymers.

Biosynthetic polymers are of three classes such as aliphatic, semi-aromatic, and aromatic. The most widely used biosynthetic polymers include polyphosphazenes, polyesters, PURs, polyanhydrides, and polyphosphoesters.

Polyesters synthesized by polycondensation are the most prominent thermoplastic biomaterial having ester link in their structural unit. While all polyesters are degradable due to reversible esterification, diversity, and synthetic versatility, hence these are presented as very unique polymers [23]. Among many polyesters, PLA and poly(glycolic acid) (PGA) are significantly studied biopolymers.

In the 1960s, the first-ever synthetic glycolide-based suture was developed. Since then many aliphatic biopolyesters were designed with good biocompatibility and degradation time [24]. Few commercially used polyester based devices are shown in Figure 5.2.

PURs and poly(ether urethanes) are considered biostable biopolymers. These polymers are expansively studied as long-term medical devices due to their exceptional mechanical properties, hence used as cardiac pacemakers and vascular grafts. These biostable polymers

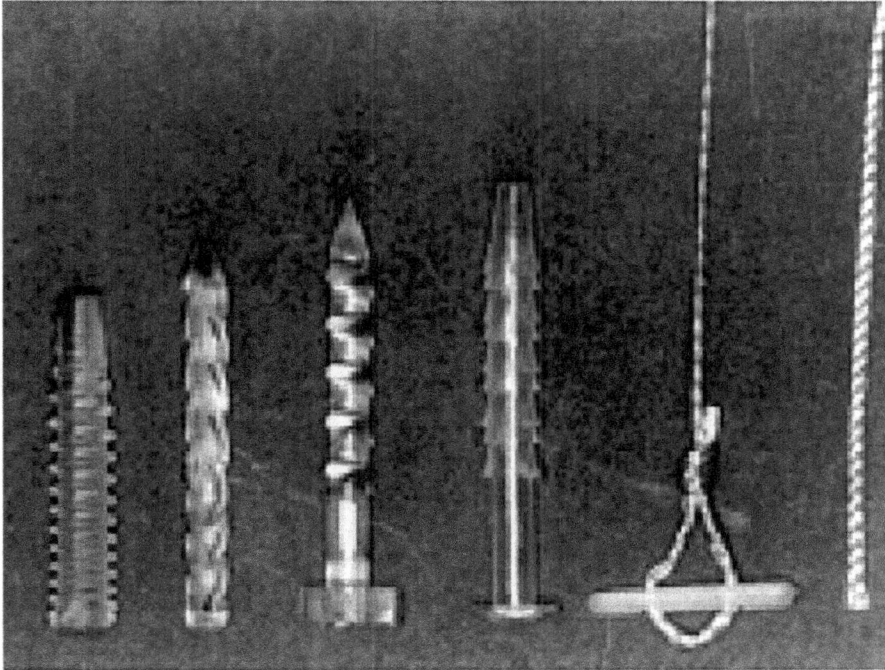

FIGURE 5.2 Commercially developed polyester devices [24]: (a) clear fix screw (b) Arthrex dart, (c) bionx meniscus arrow, (d) linvatec biostinger, (e) Smith & Nephew T-fix, (f) 2-0 ethibond suture.

are synthesized by polycondensation reaction between alcohols and amines [25].

Polyanhydrides are prepared by melt condensation of diacid esters. The polyanhydride polymers are designed especially for drug delivery owing to their surface eroding behavior. However, the presence of sensitive anhydride bonds in the chain makes polyanhydride as hydrolytically labile polymer. The hydrolytic nature of the polyanhydrides makes them a suitable material for drug delivery devices [26].

Polyphosphazenes are hybrid polymers developed by cationic polycondensation [27]. The presence of phosphorus–nitrogen side groups in the backbone of polyphosphazenes makes them potential candidates in biomedical application. Polyphosphazenes are used as drug delivery vehicles [28], matrices for bone tissue engineering [29], by the virtue of their biocompatibility and controlled rate of degradation [30, 31–35].

Polyphosphoesters are one of the promising synthetic polymers having phosphorus in their backbone [36]. However, polyphosphoesters are developed by polycondensation and polyaddition reactions [37]. Polyphosphoesters degrade due to the hydrolytic and enzymatic action.

PLLA, PLGA, polyether ketone, and polymethylmethacrylate are widely used biomaterials, amongst PLA is one of the outstanding choices for biomedical devices. PLA exhibits excellent biocompatibility with host tissues, hydrophobicity, and biodegradability [38–45]. Another biomaterial that gained more attention is PLGA. The potential features of this synthetic polymer include its safety, mechanical property, cell adhesion, and, last but not least, the rate of degradation [46]. This well-known synthetic biopolymer is prepared by ring-opening copolymerization of PLA and PGA; hence the degradation rate is decided by the variation of the PLA and PGA. Accordingly, PLGA is a favorite choice among researchers as bone implants [47].

PLA is one of the widely used biomaterials derived from corn starch and sugar cane, the diversity of PLA made them suitable for fabricating screws, pins, and plates. The favorable qualities of PLA due to its high mechanical strength with good glass transition temperature make it an ideal candidate for biomedical device fabrication [48, 49].

The aliphatic polyesters include polylactide along with its copolymers are with various properties, such as strong biocompatibility, biodegradability. These attractive characteristics have made polyesters as one of the ground-breaking materials for biomedical and pharmaceutical applications. They are commonly used as bioabsorbable surgical instruments in orthopedics, soft tissue repair, and synthetic grafts [50]. PLA as well as its copolymers with glycolic and hydroxy acids are of particular interest and are promoted in orthopedic applications. After 1965, the use of absorbable materials has been common in place of conventional metallic products. In 1973, the PLA sutures were used for the first time to repair the crack. Recently, in orthopedic and cranio-maxillofacial (CMF) surgery applications, polylactic acid-polyglycolic acid (PLA-PGA) and bioabsorbable materials have been widely used. The PLA and/or PGA based devices with ultra-high-strength implants are also used for fracture stabilization, bone grafting, and tendons/ligament reattachment.

PLA reduces the risk of preimplant osteoporosis and the risks of infection. The intrinsic existence of PLA/PGA copolymers allows promising application in the field of orthopedics for a slow release of bioactive agents. Such materials remove osteopenia that occurs due to additional surgery, but detrimental to cell proliferation [51].

The composites based on PLA are effective scaffold materials that promote cell proliferation [52]. The PLA- and PGA-based bioresorbable scaffolds are used for cell seeding which is then cultivated for the proliferation of tissues which is dependent on the type of polymeric material in the scaffold as well as the dynamics of the bioreactor fluid [53]. Polymeric biocides are antimicrobial polymers that exhibit antimicrobial activity in order to prevent the growth of microorganisms such as bacteria, fungi, or protozoans. Polymers are currently being developed to imitate antimicrobial peptides that are used for destroying bacteria by the immune systems of living organisms. In general, these polymers are synthesized by using an alkyl or acetyl linker to bind or inject an antibacterial drug onto a polymer backbone. Antimicrobial polymers make antimicrobial agents more effective and selective with a less environmental hazard. The appropriate polymeric antimicrobials used to prevent the growth of microorganisms in potable water in water sanitation are PLGA- and PLA-related polymers. Research is actively done to explore an efficient biodegradable material for the ureteral stent. The PLA-based composites are broadly used as ureteral stents to treat ureteral injuries. After ureteral repair, a biodegradable SR-PLA 96 stent with an appropriate expansion potential can be used for stenting [54]. Studies have centered on screening the use of poly-L,D-lactide copolymer (SRPLA 96; L/D ratio 96/4) tool for kidney-functioning ureteral stenting [55].

5.3.2 NATURAL POLYMERS

Natural polymers are produced in nature during the growth cycles of all species. These are called as building blocks of life. Natural polymers are considered chemicals, but they are not just products of the convenience of mother nature. These are more than likely a normal manifestation of the increased complexity of species. One must remember that not all-natural polymers can be completely synthesized because of their complex structure. The interest in natural polymers dates back to the early part of civilization to provide treatments for pain and recreational use. Further, natural polymer-based products are used in pharmaceuticals.

The natural polymers basically include peptides, proteins, and enzymes. The certain natural polymer includes polypeptides, and polysaccharides are considered as ideal biomaterials for implants because they are cheap and result in good biological activity. The biopolymers have highly ordered structures, called a ligand, which contain an extra-cellular matrix necessary to bind with cell receptors. Natural polymers form highly organized structures and at different stages direct the tissues, triggering an immune response.

5.3.2.1 PROTEINS/ POLYPEPTIDES

Proteins have been used in biomedical applications for a long period. Proteins based on amino acids are known to have improved mechanical and physical properties. Proteins

consisting of more than 20 different amino acid copolymer complexes are referred to as polypeptides. Proteins can be used as drug delivery devices, biomedical implants, and tissue engineering scaffolds [56].

Gelatin, collagen, and albumin are structural proteins of various tissues. Gelatin is made up of 19 amino acids that are connected by peptides and can be hydrolyzed by a number of proteolytic enzymes to generate its constituent amino acids or peptide components. Gelatin is a water-soluble, biodegradable polymer with broad commercial, medicinal, and biomedical applications, used for the coating and microencapsulation of various drugs [57, 58] and for the preparation of biodegradable hydrogels [58, 59]. Collagen is the richest collagenous material of all proteins in animals. The medical applications of fibrillar collagen include drug delivery systems in ophthalmology and tissue culture. Collagen also acts as a support structure of various soft and hard tissue. Albumin is a blood protein and is the carrier for molecules that have low water solubility. Albumin is a stable and safe biomaterial suitable for drug delivery and surgical tissue adhesive [60].

5.3.2.2 POLYSACCHARIDES

Polysaccharides are abundantly available and commonly used natural biopolymer materials. They are more suitable for conjugation. Polysaccharides are either positively charged or negatively charged. Cellulose and starch are key polysaccharides of interest as biomaterials, due to their simple and economical resources, such as cellulose, starch, chitin and chitosan, alginate, pectin, heparin, and hyaluronic acid [61].

Cellulose is an omnipresent polysaccharide with high crystallinity and molecular weight. Chemically modified cellulose is soluble in both water and organic solvents [62, 63]. Chitin is the second most abundant biopolymer after cellulose. Chitin is chemically related to cellulose which is a nitrogen-containing polysaccharide. Chitin is insoluble in most of the solvents [64]. The water-soluble chitin is named chitosan that is derived by deacetylation of chitin.

Starch is a linear polymer with D-glucopyranoside repeating units. Acetylated starch is more popular as compared to native starch due to its biodegradability, and its renewable nature makes it a promising biomaterial in tissue engineering [65]. The details of various synthetic and natural polymers as medical devices and their advantages and disadvantages are listed in Table 5.1.

This chapter focuses on the use of polylactide/lactic acid (PLA) as permanent or temporary implants. Lactic acid (LA) is a naturally occurring and synthetically produced and widely used compound [70–74]. The low molecular weight PLA

was first synthesized by the French chemist using the monomers of LA by polycondensation method [76]. Later low-cost, commercially viable, improved PLA was produced with the high molecular weight with high mechanical properties making them suitable for commercial applications [78].

5.4 BIODEGRADABLE IMPLANTS-BASED ON PLA

The biodegradable implants can be categorized into permanent and temporary devices [70–74]. If permanent implants are used repeatedly then one would not undergo replacement surgery [75]. In contrast to this, some implants should degrade after the tissue (soft or hard) heals and regains enough physical strength [76–78]. PLA is one such polymer that supports and degrades over the time giving space to the indigenous tissues and bones to regrow [79]. Table 5.2 summarizes different PLA-based devices and their degradation profiles.

5.4.1 MAJOR FIELDS OF APPLICATION

5.4.1.1 ORTHOPAEDICS

Bones of the human body or any other living body have distinctive structures and actions, in case of

TABLE 5.1 Conventional Biodegradable Polymers and Their Use in Medical Field [66–68]

Polymers	Devices	Advantages	Disadvantages
Natural biodegradable polymers			
Polysaccharides e.g. Alginates	• Wound healing • Tissue engineering • Delivery of bioactive agents	• Noninflammatory • Nontoxic • Biocompatible	• Poor mechanical strength • Poor biodegradability • Lack of cell adhesion
Starch	• Tissue engineering scaffolds • Drug delivery and also bioactive compounds	• Nontoxic • Bioavailable	• Poor thermal stability • Not easily processible • Moisture sensitive
Cellulose [69]	• Skin burns, drug delivery, wound healing, implants, and tissue engineering	• Complete biodegradation • Excellent biocompatibility	
Chitin and Chitosan	• Tissue engineering • Drug carriers • Absorbable sutures	• Excellent biocompatibility and mechanical properties • Ease of functionalization • Biodegradable	• Less mechanical strength • Low elongation
Biopolymers			
Collagen	• Drug delivery, cartilage, skin, cornea, bone, and blood vessels	• Biodegradability, availability, and low cost	• Mild immunogenicity
Gelatin	• Drug delivery, wound dressing, and tissue engineering	• Biodegradability, availability, and low cost	
Fibrin	• Numerous tissue engineering scaffold and carrier vehicle for bioactive molecules	• Excellent biodegradability, biocompatibility and injectability	

TABLE 5.1 (Continued)

Polymers	Devices	Advantages	Disadvantages
Silk fiber	• Sutures	• Good flexibility and elasticity	
		• Excellent tensile strength	
		• Controlled biodegradability	
Hyaluronic acid	• Scaffolds, drug carriers, and wound dressing	• Good structural and biological properties	
	• Temporary implants		
		Synthetic biodegradable polymers	
Aliphatic polyester, e.g., PLA and PGA	• Sutures	• Ease of processing	
	• Tissue engineering scaffolds	• Good mechanical, chemical and physical properties	
	• Wound closure staples	• Tuneable biodegradability	
	• Orthopaedic fixation devices	• Biocompatible	
Polycaprolactone	• Drug delivery system	• Nontoxic	• Hydrophobicity hinders cell adhesion
	• Long-term implants	• Higher permeability to drugs	• Slower rate of biodegradation
	• Tissue engineering scaffolds	• Extremely good biocompatibility	
Poly(β-hydroxybutyrate)	• Long-term tissue engineering scaffolds	• Good processability	• Hydrophobic
		• Biodegradability and biocompatibility	
Poly(phosphoesters)	• Drug delivery	• Excellent biocompatibility	
	• Gene carriers	• Good cytocompatibility	
	• Tissue engineering scaffolds		

TABLE 5.1   *(Continued)*

Polymers	Devices	Advantages	Disadvantages
Polyphosphazenes	• Drug delivery	• Flexibility • Good mechanical and physical properties	
Poly(orthoesters)	• Drug delivery vehicles	• Hydrophobic • Specific design flexibility for drug delivery applications • pH-sensitive	• Poor mechanical properties
Polyanhydrides	• Drug delivery	• Biodegrades by surface erosion mechanism • Biocompatible • Biodegradable	• Uneven drug release
Polyether urethanes (PEU)	• Catheters, artificial heart, ligaments, pacemaker leads, etc.	• Biocompatible • Biostable • Good flexibility and durability • Good biological performance	

TABLE 5.2 PLA-based Formulations Degradation Profiles

Degradation Time	Degradation Condition	Usage	PLA	Reference
Complete degradation from 42–70 days	In vivo	Sutures	Fibers	[80]
Leftovers/fragments are found postimplantation (after 38 weeks)	In vivo	PLA bone implants	Sheet	[81]
After 12 months about 9% of material is lost	In vitro	PLA screws	Screws	[82]
About 70% molecular wt., of the PLA implants is lost	In vivo	PLA bone implants	Plates	[83]

bone fracture or bone loss, it is very important to replace the fractured bone. Bone replacement is a complicated process that involves special polymeric materials for bone substitution. Synthetic polymers are the most favorable biomaterials as artificial bone structures. The artificial structures are designed to mimic the structural property of the natural bones. Synthetic biopolymers/biomaterials play a crucial role in orthopedic treatment. Synthetic polymers are widely used in hip and joint replacements, although various other biomaterials are used in devices such as bone and ligament fixation plates.

PLA and its copolymers used in tissue engineering revealed better results [84, 85]. Three-dimensional printed PLA membranes of 100 μm of thickness and pore diameter varying from 200 to 375 μm exhibited better proliferation and differentiation. The stacking of membranes provides better results for nonbearing bone

tissues [86, 87]. Titanium was one of the major metallic implant materials used to treat bone fractures. Titanium-based implants are not degradable but a second surgery was necessary to remove the implant from the body. The drawback of titanium-based implant directed to the discovery of PLA-based devices such as screws, pins, plates, and suture anchors. These bioabsorbable PLA-based implants are popular in clinical use especially in the orthopedic areas [88–95]. PLA and its copolymers are used as carriers for bone morphogenetic proteins. PLA and bone morphogenetic proteins were mixed and used as a bone graft substitute to boost bone repair [96, 97].

5.4.1.2 CRANIOMAXILLO-FACIAL

The design, development, and evaluation of the use of resorbable materials in CMF surgery can be viewed as part of fracture fixation to heal facial fractures. This includes healthy therapeutic applications and studies of bioabsorbable polymeric materials with biochemical properties. Recently researchers are attracted to developing appropriate bioabsorbable materials for fracture repair in CMF surgery. These materials are identical to metal tools but are safe and easy to handle in clinical applications. Such products are proved safe for clinical applications. Numerous materials such as PLA, PGA, and polydioxanone are commonly used in orthognathic surgery, followed by upper facial skeleton fracture therapy. In CMF surgery, lactide and glycolide polymers and copolymers are used as fixation materials, due to their biocompatibility and gradually loses strength. Hence these materials do not need secondary surgery for material removal [98]. These are competent as drug delivery devices for different therapeutic small molecular weight agents, chemotherapeutic drugs of high molecular weight, antibiotics, and peptide hormones. Several biodegradable nanoparticles are essential for therapeutic agent delivery in cytosolic form. PLA screws and fixation plates with improved mechanical properties were used to fix jaw fractures without any additional support. The PLA implants used were prepared by controlled copolymers ratio. Figure 5.3 shows the PLA implants used in jaw fracture fixation.

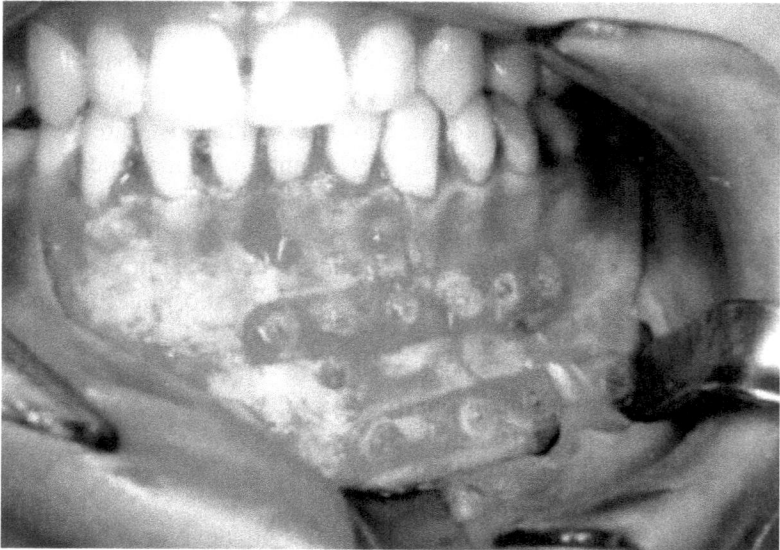

FIGURE 5.3   PLA-based biodegradable plates and screws.

Source: Reprinted with permission from Ref. [99]. © 2002 Elsevier.

5.4.1.3   CARDIOVASCULAR

The number of implants used as cardiovascular stents sometimes does not perform or meet the patients and doctor's requirements, to overcome this problem synthetic biomaterials are investigated by conducting more biocompatibility tests. The synthetic stents after serving the purpose (intravascular dilator) should degrade by the biofluids present in the body. PLA and polyethylene terephthalate PET biostents are patented by a research group for re-establishing the blood flow in damaged blood vessels. The device fabricated was found suitable in cardiovascular surgery. PLA stent made by braiding technology is shown in Figure 5.4.

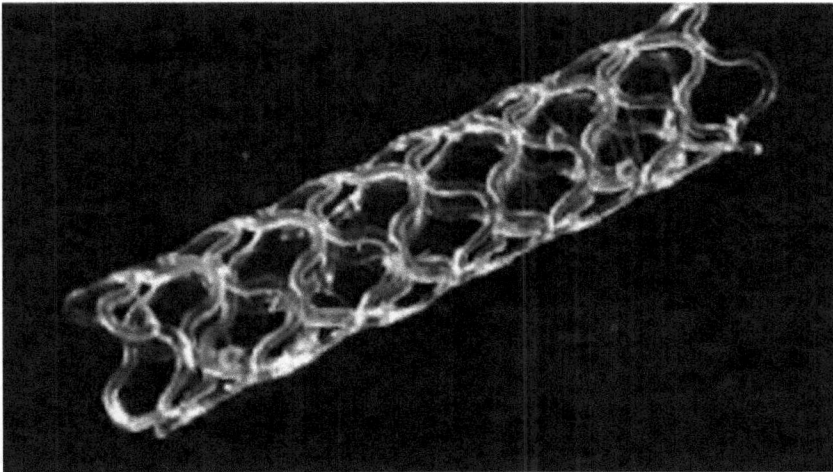

FIGURE 5.4   Implantable 3D printed PLA device.
Source: Reprinted with permission from Ref. [99]. © 2002 Elsevier.

5.4.1.4   DRUG DELIVERY

PLA and its composite materials are used as a drug delivery agent. The biodegradability of the biomaterial is an essential factor in drug release. Molecular weight, composition, device size along with physio-chemical factors for enhancing, or replacing biological functions influence the local delivery of drugs. It offers an alternative for treating damaged or missing tissues. Multi-filament yarns of PLA were developed by melt spinning technique and are used as bioabsorbable devices. PLA-based biodegradable materials involving a microcontroller system perform better as compared to traditional materials [101]. The filamentous carbon offers a potential scaffolding substrate to develop new tissues by virtue of which filamentous carbon-based PLA composite materials are used to treat tendon and ligament injuries. The properties and applications of these composites are significantly influenced by geometric structure and surface characteristics [102].

5.5 BIOLOGICAL PROPERTIES OF PLA

5.5.1 BIODEGRADATION

Degradation is a process that changes the overall material strength under controlled conditions. The widespread use of polymeric materials in the biomedical application is its ease of synthesizing devices and flexibility in altering the properties associated with mechanical strength and biodegradation. The improvement in the degradation behavior of polymers can be achieved by increasing the porosity and surface area to volume ratio. The two critical factors affecting the degradation rate are temperature and pH. With increasing temperature, the rate of biodegradation also increases. The biodegradation mechanism of PLA in the human body is via hydrolysis. Once the water molecule diffuses into the bulk of PLA, degradation starts simultaneously at the bulk and surface of the polymer leading to the formation of LA monomeric units as the degradation products. Due to

chain cleavage reaction, carboxylic groups are formed at the ends of the PLA chain, these carboxylic groups act as a catalyst for further degradation reaction of polymer inside the human body. As time progresses, the LA monomeric units diffuse out of polymer [103].

The implant particles that have been degraded must be nontoxic to the host body or environment [104, 105]. Microorganisms accomplish the degradation process by various enzymatic activities and cleavage of the bonds. Degradation of the polymer occurs in sequential order, biodegradation, biofragmentation, and mineralization [106].

5.5.2 TOXICITY AND BIOCOMPATIBILITY

The general criterion for choosing a material for its use as an orthopedic product is the efficiency to balance mechanical properties with degradation time as per application's requirements so that adequate strength remains until the bone is cured. The researchers focus on designing and developing an implant materials or implant devices that can induce predictable, controlled, guided, and rapid healing of interfacial tissues [107].

One of the major challenges that researchers face is developing an implant that can gradually degrade then distribute stress to the surrounding tissue, as it heals at a reasonable rate [108]. Even though, the above-mentioned synthetic polymers possibly will have excellent properties. There are complications when dealing with the specifications of other applications because their modified properties have resulted in variations in resorption time. For example, in animals, poly(lactide) hydrolyses to LA which is a natural product of muscle contraction. In the process of tricarboxylic acid cycle, the LA is further metabolized and then excreted as carbon dioxide and water [109].

PLA biomaterials have generally indicated adequate biocompatibility and lack of substantial toxicity, although some reductions in cell proliferation have been reported [110, 111]. However, it has been stated in the past study that PLA produces toxic solutions in vitro, possibly because of acidic degradation products [112]. Numerous other experiments have successfully demonstrated the in vivo biocompatibility properties of PLA. Since PLA has enjoyed effective clinical use of musculoskeletal tissues in the form of sutures, fixation devices, or replacement implants [113]. To prevent complications and surgical failure, the bone-healing time must be adjusted to the full deterioration of the polymer. It also involves the production of novel materials that can degrade at levels that meet the healing and load-bearing conditions.

Furthermost work in this field has taken advantage of combining two or more polymers to expand the synergistic and regulated properties of the resulting new material.

5.6 SUMMARY

In the present scenario, researchers have been pushed to find and design probable degradable biomaterials. The advanced techniques available for the synthesis of polymer bioimplants for all medical requirements have gained a wide response from research works to develop new bioimplants. The research advance in the past decade has seen a huge rise in the use of new polymeric materials and their composites to achieve better biocompatibility and degradability. Further, with the advancement in technology, computer-aided technology can be used to construct complicated bioimplants that can mimic biological tissues. These advancements have led to a deep understanding of how biological counterparts interact with the host tissues. The degradable PLA-based bioimplants were produced for orthopedic, CMF, cardiovascular, and drug delivery applications. The mechanical properties of the PLA-based implants can be monitored by varying the copolymers content. The scaffolds were to be tested and measured for toxicity, cell viability, and cell proliferation. The results agree with the research that

PLA-based biomaterials can be used as a successful biodegradable material in various medical applications. It is evident that several solutions will emerge and these solutions will have an impact on materials, increased longevity, improved health, and environment. The challenge being the homogeneity and functionality of the polymers compared to the biological systems. The synthetic scaffolds suffer from molecular weight distribution and chain composition. However, more fascinating and ground-breaking developments are necessary to counterpart biology and synthetic polymers.

KEYWORDS

- **biodegradable**
- **implants**
- **biomaterials**
- **composites**
- **gene carriers**
- **drug delivery**

REFERENCES

1. Ulery, B.D.; Nair, L.S.; Laurencin, C.T. Biomedical applications of biodegradable polymers. *J. Polym. Sci., Part B: Polym. Phys.* 2011, 49, 832–864.
2. Gross, R.A.; Kalra, B. Biodegradable polymers for the environment. *Science* 2002, 297, 803–807.
3. Chandra, R.; Rustgi, R. Biodegradable polymers. *Prog. Polym. Sci.* 1998, 23, 1273–1335.
4. Nair, L.S.; Laurencin, C.T. Biodegradable polymers as biomaterials. *Prog. Polym. Sci.* 2007, 32, 762–798.
5. Siracusa, V.; Rocculi, P.; Romani, S.; Dalla Rosa, M. Biodegradable polymers for food packaging: a review. *Trends Food Sci. Technol.* 2008, 19, 634–643.
6. Rodríguez, B.; Romero, A.; Soto, O.; Varorna, O. Biomaterials for orthopaedics. Applications of Engineering Mechanics in Medicine, GED, 2004, 1–26.
7. Seal, C.K.; Vince, K.; Hodgson, M.A. Biodegradable surgical implants based on magnesium alloys—a review of current research. *IOP Sci.* 2009, 4, 1–4.
8. Hermawan, H.; Ramdan, D.; Djuansjah, J.R.P. Metals for biomedical applications. *InTech*, 2011, 411–430. https://doi.org/10.5772/19033.
9. Hallab, N.; Merritt, K.; Jacobs, P.J. Metal sensitivity in patients with orthopaedic implants. *J. Bone Joint Surg.* 2006, 83-A, 427–436.
10. Reifenrath, J.; Bormann, D.; Lindenberg, A.M. Magnesium alloys as promising degradable implant materials in orthopaedic research. *InTech*, 2011, 94–106. https://doi.org/10.5772/14143
11. Rosson, J.; Egan, J.; Shearer, J.; Monro, P. Bone weakness after the removal of plates and screws. *J. Bone Joint Surg. Br.*, 1991, 72-B, 283–286.
12. Rosson, J.W.; Shearer, J.R. Refracture after the removal of plates from the fore arm. *J. Bone Joint Surg.* 1991, 73-B, 415–417.
13. Middleton, J.C.; Tipton, A.J. Synthetic biodegradable polymers as orthopaedic devices. *Biomaterials* 2000, 21, 2335–2346. http://dx.doi.org/10.1016/S0142-9612(00)00101-0.
14. Jacobs, J.J.; Gilbert, J.L.; Urban, R.M. Current concepts review corrosion of

metal orthopaedic implants. *J. Bone Joint Surg.* 1998, 80, 268–282.

15. Hansen, D.C. Metal corrosion in the human body: the ultimate bio-corrosion scenario. *Electrochem. Soc. Interfaces.* 2008, 17, 31–34.

16. Geringer, J.; Forest, B.; Combcade, P. Fretting-corrosion of materials used as orthopaedic implants. *Wear* 2005, 259, 943–951. http://dx.doi.org/10.1016/j.wear.2004.11.027.

17. Mitchell, A.; Shrotriya, P. Onset of nanoscale wear of metallic implant materials: influence of surface residual stresses and contact loads. *Wear* 2007, 263, 1117–1123. http://dx.doi.org/10.1016/j.wear.2007.01.068.

18. Farrar, D. Bioresorbable polymers in orthopaedics. *Med. Device Manuf. Technol.*, 2005, 1, 1–4.

19. Gunatillake, P.A.; Adhikari, R. Biodegradable synthetic polymers for tissue engineering. *Eur Cells Mater.* 2003, 5, 1–16.

20. Buddy, D.R.; Allan, S.H.; Fredrick, J.S.; Jack, E.S. Biomaterials Science. 2nd edn., Academic Press: Waltham, MA, 1996, 1–497.

21. Sravanithi, R. Preparation and characterization of poly(ὲ-caprolactone) PCL scaffolds for tissue engineering scaffolds. Department of Biotechnology and Medical Engineering, 2009, 1–59.

22. Ratner, Buddy D. Biomedical applications of synthetic polymers. Pergamon Press plc, Comprehensive Polymer Science, Oxford. 1989, 7, 201–247.

23. Okada, M. Chemical synthesis of biodegradable polymers. *Prog. Polym. Sci.* 2002, 27, 87–133.

24. Farng, E.; Sherman, O. Meniscal repair devices: a clinical and biomechanical literature review. *J. Arthrosc. Relat. Surg.* 2004, 20, 273–86.

25. Scycher, M. Scycher's handbook of polyurethanes. (CardioTech International, Inc.). CRC Press: Boca Raton, FL, USA, 1999.

26. Leong, K.W.; Brott, B.C. Langer R. Biodegradable polyanhydrides as drug carrier matrices. I: characterization, degradation and release characteristics. *J. Biomed. Mater. Res.* 1985, 19, 941–55.

27. Allcock, H.R. Chemistry and applications of polyphosphazenes. Wiley: New York, NY, 2003.

28. Lakshmi, S.; Katti, D.S.; Laurencin, C.T. Biodegradable polyphosphazenes for drug delivery applications. *Adv. Drug Deliv. Rev.* 2003, 55, 467–82.

29. Nair, L.S.; Lee, D.A.; Bender, J.D.; Barrett, E.W.; Greigh, Y.E.; Brown, P.W. Synthesis, characterization and osteocompatibility evaluations of novel alanine based polyphosphazenes. *J. Biomed. Mater. Res.* 2006, 76A, 206–213.

30. Nair, L.S.; Laurencin, C.T. Polymers as biomaterials for tissue engineering and controlled drug delivery. In: Lee K, Kaplan D, editors. Tissue engineering I. Advances in biochemical engineering/biotechnology. Review Series, Springer Verlag: Berlin, 2006; 47–90.

31. Sethuraman, S.; Nair, L.S.; El-Amin, S.; Farrar, R.; Nguyen, M.T.; Singh, A. In vivo biodegradability and biocompatibility evaluation of novel alanine ester based polyphosphazenes in a rat model. *J. Biomed. Mater. Res. A.* 2006, 77, 679–87.

32. Nair, L.S.; Lee, D.; Laurencin, C.T. Polyphosphazenes as novel biomaterials. In: Narasimhan A, Mallapragada A, editors. Handbook of biodegradable polymeric materials and applications. American Scientific Publication: Chicago, IL, USA, 2004; 277–306.

33. Nair, L.S.; Khan, Y.M.; Laurencin, C.T. Polyphosphazenes. In: Hollinger H, editor. An introduction to biomaterials. CRC publications: Boca Raton, FL, USA, 2006; 273–290.

34. Laurencin, C.T.; Nair, L.S. Polyphosphazene nanofibers for biomedical

applications: preliminary studies. In: Guceri S, Gogotsi YG, Kuznetsov V, editors. Nanoengineered nanofibrous materials. NATO-ASI Proceedings. Kluwer: Dordrecht, 2004; 281–300.

35. Kumbar, S.G.; Bhattacharyya, S.; Nukavarapu, S.P.; Khan, Y.M.; Nair, L.S.; Laurencin, C.T. In vitro and in vivo characterization of biodegradable poly(organophosphazenes) for biomedical applications. *J. Inorg. Organomet. Chem.* 2006, 16, 365–85.

36. Penczek, S.; Pretula, S.; Kalyzynski, K. Poly(alkylene phosphates): from synthetic models of biomacromolecules and biomembranes toward polymer-inorganic hybrids (mimicking biomineralization). *Biomacromolecules* 2005, 6, 547–551.

37. Penczek, S.; Pretula, J.; Kaluzynski, K. Models of biomacromolecules and other useful structures based on the poly(alkylene phosphate)chains. *Polym. J. Chem.* 2001, 75, 117–81.

38. Nair, L.S.; Laurencin, C.T. Biodegradable polymers as biomaterials. *Prog. Polym. Sci.* 2007, 32 (8), 762–798.

39. Eppley, B.L. Biomechanical testing of alloplastic PMMA cranioplasty materials. *J. Craniofac. Surg.* 2005, 16 (1), 140–143.

40. Lampin, M.; Warocquier-Clérout, R.; Legris, C.; Degrange, M.; Sigot-Luizard, M.F. Correlation between substratum roughness and wettability, cell adhesion, and cell migration. *J. Biomed. Mater. Res.* 1997, 36 (1), 99–108.

41. Ambrose, C.; Clanton, T. Bioabsorbable implants: review of clinical experience in orthopedic surgery. *Ann. Biomed. Eng.* 2004, 32 (1), 171–177.

42. Baro, M.; Sanchez, E.; Delgado, A.; Perera, A.; Evora, C. In vitro–in vivo characterization of gentamicin bone implants. *J. Control. Release.* 2002, 83 (3), 353–364.

43. Li, R.; Yao, D. Preparation of single poly(lactic acid) composites, *J. Appl. Polym. Sci.* 2008, 107 (5), 2909–2916.

44. Narayanan, G.; Vernekar, V.N.; Kuyinu, E.L.; Laurencin, C.T. Poly (lactic acid)-based biomaterials for orthopaedic regenerative engineering. *Adv. Drug Deliv. Rev.* 2016, 107, 247–276.

45. Shah, S.R.; Tatara, A.M.; Souza, R.N.; Mikos, A.G.; Kasper, F.K. Evolving strategies for preventing biofilm on implantable materials. *Mater. Today.* 2013, 16 (5), 177–182.

46. Lasprilla, A.J.; Martinez, G.A.; Lunelli, B.H.; Jardini, A.L.; Filho, R.M. Poly-lactic acid synthesis for application in biomedical devices—a review. *Biotechnol. Adv.* 2012, 30 (1), 321–328.

47. Migliaresi, C.; De Lollis, A.; Fambri, L.; Cohn, D. The effect of thermal history on the crystallinity of different molecular weight PLLA biodegradable polymers. *Clin. Mater.* 1991, 8 (1), 111–118.

48. Henton, D.E.; Gruber, P.; Lunt, J.; Randall, J. Natural fibers, biopolymers, and biocomposites. Taylor & Francis/ CRC Press: Boca Raton, FL, USA, 2005, 527–577.

49. Ray, S.S.; Okamoto, M. Biodegradable polylactide and its nanocomposites: opening a new dimension for plastics and composites. *Macromol. Rapid Commun.* 2003, 24 (14), 815–840.

50. Kovalevsky, G.; Barnhart, K. Norplant and other implantable contraceptives. *Clin. Obstet. Gynecol.* 2001, 44(1), 92–100.

51. Konda, R.K.; Arijit, B.; Abraham, J.D. Biodegradable polymers: medical applications. In: Encyclopedia of Polymer Science and Technology, Wiley, 2016, 1–22. https://doi.org/ 10.1002/0471440264.pst027.pub2

52. Walter, K.A.; Tamargo, R.; Olivi, A.; Burger, P. C.; Brem, H. Intratumoral

chemotherapy. *Neurosurgery* 1995, 37, 1129.

53. Montanari, L.; Costantini, M.; Sigoretti, E.C.; Valvo, L.; Santucci, M.; Bartolomei, M.; Fattibene, P.; Onori, S.; Faucitano, A.; Conti, B.; Genta, I. *J. Control. Release*. 1998, 56, 219.

54. Huang, S.J.; Ho, L.H. Biodegradable polymers derived from aminoacids. *Macromol. Symp*. 1999, 144, 7.

55. Leong, K.W.; Biodegradable polymers as drug delivery systems. *Polym. Control. Drug Del*. 1991, 127.

56. Hu, X.; Cebe, P.; Weiss, A.S.; Omenetto, F.; Kaplan, D.L. Protein-based composite materials. *Mater. Today*. 2012, 15, 208–215. http://dx.doi.org/10.1016/S1369-7021(12)70091-3.

57. Tashibana, M.; Yaita, A.; Tamiura, H.; Fukasawa, K.; Nagasue, N.; Nakanura, T. The use of chitin as a new absorbable suture material—an experimental study. *Jpn. J. Surg*. 1988, 18, 533–539.

58. Tabata, Y.; Ikada, Y. Synthesis of gelatin microspheres containing interferon. *Pharm. Res*. 1989, 6, 422–427.

59. Tabata, Y.; Ikada, Y. Macrophage activation through 37 phagocytosis of muramyl dipeptide encapsulated in gelatin microspheres. *J. Pharm. Pharmacol*. 1987, 39, 698–704.

60. Duarte, A.; Coelho, J.; Bordado, J.; Cidade, M.; Gil, M. Surgical adhesives: systematic review of the main types and development forecast. *Prog. Polym. Sci*. 2012, 37, 1031–1050.

61. Basu, A.; Kunduru, K.R.; Abtew, E.; Domb, A.J. Polysaccharide-based conjugates for biomedical applications. *Bioconj. Chem*. 2015, 26 (8), 1396–1412.

62. Salimi, H.; Aryanasab, F.; Banazadeh, A.R.; Shabanian, M.; Seidi, F. Designing syntheses of cellulose and starchderivatives with basic or cationic N-functions: part I—cellulose derivatives. *Polym. Adv. Technol*. 2016, 27, 5–32. doi: 10.1002/pat.3599.

63. Moon, R.J.; Martini, A.; Nairn, J.; Simonsen, J; Youngblood, J. Cellulose nanomaterials review: structure, properties and nanocomposites. *Chem. Soc. Rev*. 2011, 40, 3941–3994.

64. Khor, E. Chitin and chitosan as biomaterials: going forward based on lessons learnt. *J. Met. Mater. Miner*. 2005, 15(1), 69–72.

65. Heinze, T.; Koschella, A. Carboxymethyl ethers of cellulose and starch—a review. *Macromolec. Symp*. 2005, 223, 13–40.

66. Richard, Song.; Maxwell, Murphy.; Chenshuang, Li.; Kang, Ting.; Chia, Soo.; Zhong, Zheng. Current development of biodegradable polymeric materials for biomedical applications. *Drug Des. Devel. Ther*. 2018, 12, 3117.

67. Ulery, B.D., Lakshmi S.N.; Cato, T.L. Biomedical applications of biodegradable polymers. *J. Polym. Sci. B. Polym. Phys*. 2011, 49(12), 832–864.

68. Lakshmi S.N.; Cato, T.L. Biodegradable polymers as biomaterials. *Prog. Polym. Sci*. 2007, 32 (8-9), 762–798.

69. Guilherme, Fadel. Picheth.; Cleverton, Luiz. Pirich.; Maria, Rit.; Sierakowski.; Marco, Aurelio. Woehl.; Caroline, Novak. Sakakibara.; Clayton, Fernandes. De'Souza.; Andressa, Amado. Martin.; Renata, da Silva.; Rilton, Alves. de Freitas. Bacterial cellulose in biomedical applications: a review. *Int. J. Biol. Macromol*. 2017, 104, 97–106.

70. Drumright, R.E.; Gruber, P.R.; Henton, D.E. Polylactic acid technology. *Adv. Mater*. 2000, 12, 1841–1846.

71. Amini, A.R.; Wallace, J.S.; Nukavarapu, S.P. Short-term and long-term effects of orthopedic biodegradable implants. *J. Long Term Eff. Med. Implants*. 2011, 21, 93–122.

72. Miceli, E.; Kar, M.; Calderón, M. Interactions of organic nanoparticles with proteins in physiological conditions. *J. Mater. Chem. B*. 2017, 5, 4393–4440.

73. Fleige, E.; Quadir, M.A.; Haag, R. Stimuli-responsive polymeric nanocarriers for the controlled transport of active compounds: concepts and applications, *Adv. Drug Del. Rev.* 2012, 64, 866–884.

74. Cutright, D.E.; Hunsuck, E.E. Tissue reaction to the biodegradable polylactic acid suture, *Oral Surg. Oral Med. Oral Pathol.* 1971, 31, 134–139.

75. Weldon, C.B.; Tsui, J.H.; Shankarappa, S.A.; Nguyen, V.T.; Ma, M.; Anderson, D.G.; Kohane, D.S. Electrospun drug-eluting sutures for local anesthesia. *J. Control Release.* 2012, 161, 903–909.

76. Bostman, O.; Pihlajamaki, H. Clinical biocompatibility of biodegradable orthopaedic implants for internal fixation: a review. *Biomaterials.* 2000, 21, 2615–2621.

77. Kulkarni, R.K.; Pani, K.C.; Neuman, C.; Leonard, F. Polylactic acid for surgical implants. *Arch. Surg.* 1996, 93, 839–843.

78. Cutright, D.E.; Hunsuck, E.E. The repair of fractures of the orbital floor using biodegradable polylactic acid. *Oral Surg. Oral Med. Oral Pathol.* 1972, 33, 28–34.

79. Schwach, G.; Vert, M. In vitro and in vivo degradation of lactic acid-based interference screws used in cruciate ligament reconstruction. *Int. J. Biol. Macromol.* 1999, 25, 283–291.

80. Netti, P.A. editor. Biomedical Foams for Tissue Engineering Applications. 76. Woodhead Publishing Series in Biomaterials, Elsevier, 2014.

81. Cutright, D.E.; Hunsuck, E.E. The repair of fractures of the orbital floor using biodegradable polylactic acid. *Oral Surg. Oral Med. Oral Pathol.* 1972, 33, 28–34.

82. Schwach, E.E.; Vert, M. In vitro and in vivo degradation of lactic acid-based interference screws used in cruciate ligament reconstruction. *Int. J. Biol. Macromol.* 1999, 25, 283–291.

83. Tschakaloff, A.; Losken, H.W.; Vonoepen, R.; Michaeli, W.; Moritz, O.; Mooney, M.P.; Losken, A. Degradation kinetics of biodegradable Dl-polylactic acid biodegradable implants depending on the site of implantation. *Int. J. Oral Max. Surg.* 1994, 23, 443–445.

84. Ramzi A, Abd Alsaheb.; Azzam Aladdin.; Nor Zalina Othman.; Roslinda Abd Malek.; Ong Mei Leng.; Ramlan Aziz and Hesham A, El Enshasy. Recent applications of polylactic acid in pharmaceutical and medical industries. *J. Chem. Pharm. Res.* 2015. 7 (12), 51–63. doi:10.2174/138920021866617091 9170335

85. Chia-Tze Kao.; Chi-Chang Lin.; Yi-Wen Chen.; Chia-Hung Yeh.; Hsin-Yuan Fang.; Ming-You Shie. Poly(dopamine) coating of 3D printed poly(lactic acid) scaffolds for bone tissue engineering. *Mater. Sci. Eng. C.* 2015, 56, 165–173. doi: 10.1016/j.msec.2015.06.028.

86. Guduric, V.; Metz, C.; Siadous, R.; Bareille, R.; Levato, R.; Engel, E.; Fricain, J.C.; Devillard, R.; Luzanin, O.; Catros, S. Layer-by-layer bioassembly of cellularized polylactic acid porous membranes for bone tissue engineering. *J. Mater. Sci., Mater. Med.* 2017, 28 (5), 78. doi: 10.1007/s10856-017-5887-6.

87. Barber, F.A.; Elrod, B.F.; McGuire, D.A.; Paulos, L.E. Preliminary results of an absorbable interference screw. *Arthroscopy* 1995, 11, 537–548. doi: 10.1016/0749-8063(95)90129-9.

88. Matsue, Y.; Nakamura, T.; Suzki, S.; Iwasaki, R. Biodegradable pin fixation of osteochondral fragments of the knee. *Clin. Orthop. Relat. Res.* 1996, 322, 166–173.

89. Stahelin, A.C.; Weiler, A.; Rüfenacht, H.; Hoffmann, R.; Geissmann, A.; Feinstein, R. Clinical degradation andbiocompatibility of different

bioabsorbable interference screws: a report of six cases. *Arthroscopy* 1997, 13, 238–244. doi: 10.1016/S0749-8063 (97)90162-6.

90. Bostman, O.; Hirvensalo, E.; Vainionpaa, S.; Makela, A.; Vihtonen, K.; Tormala, P.; Rokkanen, P. Ankle fracturestreated using biodegradable internal fixation. *Clin. Orthop. Relat. Res.* 1989, 238, 195–203.

91. Lavery, A.; Higgins, K.R.; Ashry, H.R.; Athanasiou, K.A. Mechanical characteristics of poly-L-lactic acid absorbable screws and stainless steel screws in basilarosteotomies of the first metatarsal. *J. Foot. Ankle. Surg.* 1994, 33, 249–254.

92. Bucholz, R.W.; Henry, S.; Henley, M.B. Fixation withbioabsorbable screws for the treatment of fractures of the ankle. *J. Bone Joint Surg.* 1994, 76, 319–324.

93. Casteleyn, P.P.; Handelberg, F.; Haentjens, P. Biodegrad-able rods versus Kirschner wire fixation of wrist fractures. A randomised trial. *J. Bone Joint Surg.* 1992, 74, 858–861.

94. Hope, P.G.; Williamson, D.M.; Coates, C.J.; Cole, W.G. Biodegradable pin fixation of elbow fractures inchildren: a randomized trial. *J. Bone Joint Surg.,* 1991, 73, 965–968.

95. Saito, N.; Takaoka, K. New synthetic biodegradable polymers as BMP carriers for bone tissue engineering. *Biomaterials* 2003, 24, 2287–2293. doi: 10.1016/S0142-9612(03)00040-1.

96. Hamad, K.; Kaseem, M.; Yang, H.W.; Deri, F.; Ko, Y.G. Properties and medical applications of polylactic acid: a review. *eXPRESS Polym. Lett.* 2015, 9 (5), 435–455.

97. Murakami, N.; Saito, N.; Horiuchi, H.; Okada, T.; Nozaki, K.; Takaoka, K. Repair of segmental defects in rabbit humeri with titanium fiber mesh cylinders containing recombinant human bone morphogenetic protein-2 (rhBMP-2) and

a synthetic polymer. *J. Biomed. Mater. Res. A.* 2002, 62, 169–174. doi: 10.1002/jbm.10236.

98. Heller, J.; in Domb, A.J.; Kost, J.; Weiseman, D.M. eds., Handbook of biodegradable polymers, 7, Hardwood Academic Publishers: Amsterdam, 1997, 99.

99. Kaan, C. Yerit.; Georg, Enislidis.; Christian, Schopper.; Dritan, Turhani.; Felix, Wanschitz.; Arne, Wagner.; Franz, Watzinger.; Rolf, Ewers. Fixation of mandibular fractures with biodegradable plates and screws. *Oral Surg. Oral Med. Oral Pathol. Oral Radio.l Endod.* 2002, 94, 294–300.

100. Sang, Jin. Leea.; Ha, Hyeon. Joa.; Kyung, Seob. Lim.; Dohyung, Lim.; Soojin, Lee.; Jun, Hee. Lee.; Wan, Doo. Kim.; Myung, Ho. Jeong.; Joong, Yeon. Lim.; Keun, Kwon.; Youngmee, Jung.; Jun-Kyu, Park.; Su, A. Park. Heparin coating on 3D printed poly(L-lactic acid) biodegradable cardiovascular stent via mild surface modification approach for coronary artery implantation. *Chem. Eng. J.* 2019, 378, 122116

101. Vandorpe, J.; Schacht, E.; Dejardin, S.; Lemmouchi, Y. in Domb, A.J.; Kost, J.; Weiseman, D.M. eds., Handbook of Biodegradable Polymers. l, Hardwood Academic Publishers: Amsterdam, 1997, 161.

102. Silva, D.d.; Kaduri, M.; Poley, M.; Adir, O.; Krinsky, N.; Shainsky-Roitman, J.; Schroeder. A. Biocompatibility, biodegradation and excretion of polylactic acid (PLA) in medical implants and theranostic systems. *Chem. Eng. J.* 2018, 340, 9–14

103. Domb, A.J.; Amselem, S.; Maniar, M. Polymeric Bio-materials, S. Dumitriu, editor. Marcel Dekker: New York, 1994, 399.

104. Kost, J.; Weiseman, D.M. eds., Handbook of Biodegradable Polymers, Vol. 7, Hardwood Academic Publishers:

Amsterdam, the Netherlands, 1997; p. 135.

105. Han, J.H. Antimicrobial Food Packaging. *Food Technol.* 2000, 54 (4), 56–65.

106. Seneker, S.D.; Glass, J.E. Reaction parameter effects on substituent distributions in the heterogeneous synthesis of cellulose ethers. *Adv. Chem. Ser.* 1996, 248, 125–137.

107. Gardener, R.M.; Buchman, C.M.; Komark, R.;. Dorschel, D.; Boggs, C.; White, A.W, Compostability of cellulose acetate films. *J. Appl. Polym. Sci.* 1994, 52, 1477.

108. Bailey, W.J. Free radical ring-opening polymerization. *Polym. J. (Tokyo).* 1985, 17, 85.

109. Agboh, O.C.; Qin, Y. Chitin and chitosan fibers. *Polym. Adv. Technol.* 1997, 8, 355.

110. Filho, S.P.C.; Desbrieres, J. Chitin, chitosan and derivatives. *Nat. Polym. Agrofibers Based Compos.* 2000, 41, 2157–2168.

111. Jamshidian, M.; Tehrany, E.A.; Imran, M.; Jacquot, M.; Desobry, S. Poly-Lactic acid: production, applications, nanocomposites, and release studies. *Compr. Rev. Food Sci. Food Safety.* 2010, 9, 551–571.

112. Fukushima, K.; Hirata, M.; Kimura, Y. Synthesis and characterization of stereoblock poly(lactic acid)s with nonequivalent D/L sequence ratios. *Macromolecules* 2007, 40, 3049–3055.

113. Lehermeier, H.; Dorgan, J.; Way, J. Gas permeation properties of poly (lactic acid). *J. Membr. Sci.* 2001, 190, 243–251.

CHAPTER 6

METAL COMPLEXES-BASED CATALYSTS FOR OXIDATION REACTIONS AS NEW ALTERNATIVES FOR CATALYTIC PROCESSES IN THE PRODUCTION OF BIO-BASED POLYMERS

MIRELA-FERNANDA ZALTARIOV

Department of Inorganic Polymers, "Petru Poni" Institute of Macromolecular Chemistry, Aleea Grigore Ghica Voda, 41 A, 700487 Iasi, Romania

Corresponding author. E-mail: zaltariov.mirela@icmpp.ro

ABSTRACT

This chapter presents some results on the application of metal complexes as efficient catalysts in some oxidation reactions of alkane and alcohols by conventional and solvent-free microwave-assisted methods (in the absence of any added solvent, by using an environmentally acceptable oxidation agent *tert*-buthylhydroperoxide peroxides). The catalytic performance of metal complexes in oxidation reactions has been evaluated in the presence or absence of additives (at different concentrations and temperatures) and co-catalysts, which can modify the selectivity of the chosen substrates. The results highlight the great potential of metal complexes to address various energy-related problems that occur in industrial catalytic processes, mainly those related to the production of bio-based polymers from renewable sources.

6.1 INTRODUCTION

The most widespread compounds since they were discovered are plastics. They are the basis for the development of food packaging, medical equipment, clothing, and construction sources due to their high strength, transparency, low toxicity, resistance to corrosion, and, most important, very slow degradation [1]. Beside the special rise, with an increase of their production of at least 20-fold since 1960, the durability is the major problem that led to severe environmental pollution, affecting natural ecosystems, especially marine ones and quality of life. There are two ways to reduce landfill waste, gas emission and fossil fuel use, *recycling and reuse*, affording lower-in-value materials different in comparison with the original ones. The main problem is derived from the very low number of recycled polymers on a large scale. The most recycled are high-density polyethylene, polyethylene terephtalate (PET) (about 900,000 tons), polypropylene (PP), and polystyrene. They have a limited number of processing cycles, during which the properties (molecular weight and ductility) of the polymers are affected [1].

There are the following four approaches to polymer recycling:

1. Primary recycling: melting and remolding pure materials. It is the most used in recycling centers, the least expensive requiring cleaning processes and separation of the materials.
2. Mechanical recycling: reprocessing plastics by decontamination. The polymers undergo pretreatment of the contaminated plastics and pelletization before reprocessing.
3. Chemical recycling: obtaining of lower-molecular-weight species for reuse by the breakdown (chemolysis, thermal depolymerization) of the materials leading to monomers (complete reversion) or oligomers (incomplete reversions).
4. Energy recovery: incineration to recuperate the energy stored in chemical bonds. It is the least valuable, the polymers being destroyed during the process and at the same time is the high-energy costs [1].

At the European level, the recent processes route for plastic waste management is focused almost equally on landfill and for energy recovery. The recycling routes are based especially on the mechanical process to produce recycled materials in form of pellets.

However, a great interest is granted to the pyrolysis of waste plastics to obtain low-molecular-weight chains and oils for liquid fuels usually used in transport engine domains.

In thermochemical processing of waste plastics, the key role belongs to the catalysts. These are used to promote the targeted reactions, by reducing the temperature and thus, improving the efficiency. Also, they optimize the product distribution and selectivity, the quality and quantity of gaseous and liquid products, and yield appropriate properties to the conventional fuels (gasoline or diesel) [2, 3]. The most known catalysts are zeolite ZSM-5, Y-zeolite, and MCM-41, stable composites with various mesopore size, which are used in 1:1 or 1:2 ratios feedstock plastic: catalyst to produce hydrocarbons from C1 to C60, mainly C2 (ethene), C3 (propene), C4 (butane and butadiene), and C5–C9 (aliphatic and aromatic compounds) [3, 4].

The catalytic conversion of waste plastic to fuel provides a reasonable solution for waste management and generates alternative energy sources that will balance the reduction of fossil fuels.

However, the solution offered, a waste-reducing strategy is to replace the regular plastics with biodegradable ones, which can be metabolized by soil microorganisms, leading to disintegration of the materials without monomer/oligomer recovery, the effects on the environment being still unknown. So, this strategy provides part of the solution but is sustained by the exploration of chemically recyclable polymers in mild (low temperature) conditions by catalytic conversion. More than that, chemical recycling efficiency and selectivity can be established only by a judicious design of the monomers/polymers and catalysts, reducing the need for additives/compatibilizers [5].

Also, reducing the resources and energy by efficient waste recycling, optimizing the industrial processes, and prevention of environmental pollution are some of the principles of the green economy policy, with impact on sustainable chemistry [6].

In a green economy, three different strategies are proposed to be applied:

1. The renewable oil (the crude synthetic oil) and the "green" monomers for highly resource- and energy-effective polymer processing are obtained by biorefining of biomass and chemical conversion of carbon dioxide. These processes do not affect recognized recycling technologies.

2. Besides the green routes of polymers, biotechnologies and genetic engineering are involved in the production of bio-based polymers.

3. The activation and polymerization of CO_2.

Some basic concepts provide the use of bio-based products and renewable resources (solar power) to produce biomass as resources for biofuels and bioplastics (bio-based polymers and polymers considered biodegradable). "Bio" means in a broader sense water-soluble degradation products, inclusive of toxic metabolites, not only water and carbon dioxide [6].

The central idea is to use materials from biological sources such as starch, cellulose, sugars, proteins, fatty acids as feedstock, which are converted by bacterial consumption into suitable monomers for polymer production: D- and L-lactic acid, bio-1,4-butanedione, bio-ethylene, hydroxyalkanoic acids, bio-ethylene glycol, biopropylene, CO_2, etc. These monomers are actually applied to produce various bio-based plastics: polyethylene (PE), PP, PET, polylactic acid (PLA), polyhydroxyalkanoates, poly(propylene carbonate) (PPC), which are environmentally friendly and already available as disposable catering supplies, mainly in food packaging and agricultural films [7, 8].

6.2   OPPORTUNITIES FOR CATALYSIS

Concurrent with new bio-based materials, the other strategies (mechanical sorting, recycling, and decontamination), especially the depolymerization catalysts, must continue to be enhanced, especially that the bio-based products have high production costs and poor mechanical characteristics [7].

When chemical depolymerization processes are applied to bio-based starting materials, precious monomers can be recovered or biodegraded.

The development of cost-efficiently chemical methods (design of catalysts and targeted application of bio-based polymers) leads to economic and scientific opportunities.

An efficient catalyst should be as follows:

1. inexpensive,
2. accessible in the same way as the virgin material prices,
3. air and moisture stable as well as in the presence of other organic or inorganic impurities,
4. efficient in heterogeneous systems, and
5. highly selective and active on polymer-to-monomer conversion or monomer-to-polymer production [9].

In these attempts, several catalytic processes have been reported in which biomass was converted in useful ingredients, for example, CO_2 was used as "green" feedstock to generate new and established monomers in high yields and purities: carbon monoxide, methanol, formic acid, or formaldehyde production [8].

By this method, new polymers with excellent characteristics comparable or superior to those synthetic analogs can be obtained. The most representative example in this regard is the replacement of the terephthalic acid with the "bio" derivative analogue 2,5-furandicarboxylic acid in order to produce a high-performing fully bio-based PET exchange [10].

5-(hydroxymethyl)-furfural (HMF) is an important industrial feedstock that makes the transition from petroleum-based chemistry toward "green" and renewable resources. Its hydrogenation leads to the obtaining of high-energy biofuels (2,5-dimethylfuran, while its oxidation afford 2,5-furandicarboxylic acid (FDCA), a "green" platform for valuable polymers as a renewable alternative for the PET [10] (Figure 6.1).

The commonly used catalysts in the oxidation of HMF are noble metals: Pt, Pd, Ru, Au, which have been replaced with the transition-metal-based catalyst due to their high costs.

A selective oxidation of HMF was observed by the utilization of nickel and cobalt metal complexes of phosphorous and sulfur donor

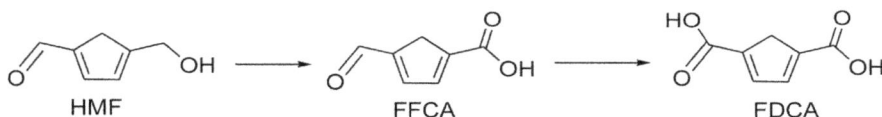

FIGURE 6.1   Reaction path for the oxidation of HMF to FDCA.

ligands or alloys, leading to high conversions and yields [11].

6.3   SALEN-TYPE-BASED CATALYSTS

6.3.1   SALEN METAL COMPLEXES AS CATALYSTS FOR THE PRODUCTION OF PLA

The development of novel and cheap nontoxic metal-based catalysts for biodegradable polymer production is an advantageous strategy for obtaining low costs polymers with improved biochemical properties. The ring-opening polymerization (ROP) catalyzed by metal complexes proved very effective in the control of the molecular weight and the structural conformation of polymers,

with a narrow polydispersity index (PDI) [12].

PLA has been widely investigated in the medical field (surgical sutures, drug delivery systems, tissue engineering) and packaging materials, because of its good biocompatibility and biodegradability. It is a hydrosoluble aliphatic polyester, which is usually obtained through a polycondensation reaction of lactic acid, often accompanied by some drawbacks (polydispersity of PLA, with low-molecular-weight, and high-molecular-weight fractions).

High-molecular-weight of PLA could be obtained by the ROP of D-, L-lactide or L-lactide, a reaction catalyzed by coordination compounds [12].

Literature data revealed that some metal complexes containing aryloxy moiety in the structure (imine aryloxy, salen-bis-aryloxy) exhibited excellent catalytic activity in the lactide ROP, without side reactions (transesterification of PLA), mainly due to the steric effect of the ligand [12].

Considering the applicability of these polymers, the catalysts should be biocompatible and nontoxic, with good stability. The most effective catalysts in ROP of lactide are metal complexes of calcium, magnesium, aluminum, zinc, and tin(II).

Besides these, titanium complexes proved a unique activity in producing PE with controlled molecular weights and structure, which drew attention to further evaluation of its potential for the bulk- and solution-phase polymerization of lactide.

An active catalyst for lactide requires a high ability to increase the electron density of the oxygen atoms (in an oxygen-metal bond of the catalyst) and also a strong attack of the carbonyl oxygen atom from lactide, by coordination of metal to oxygen. Utilization of the ligands having a large steric hindrance with metal ions leads to the inhibition of the side reactions, reducing the PDI of the molecular chains.

A D-,L-lactic acid ROP reaction by using a titanium-complex with a salen-ligand (Figure 6.2) as catalyst was performed under solvent-free conditions at a temperature of 160 °C for 16 h, obtaining PLA with a molecular weight of 9.31×10^4 g mol^{-1} [12].

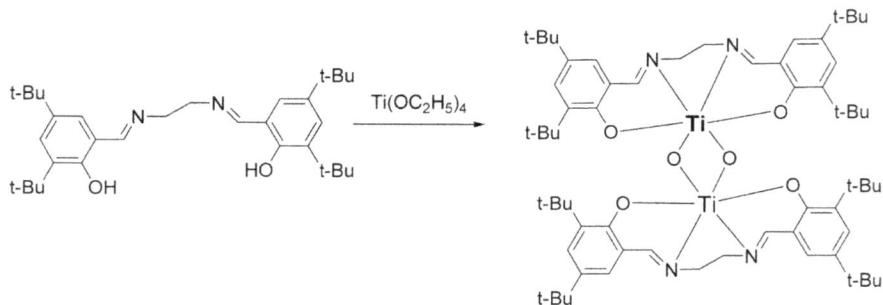

FIGURE 6.2   The chemical structure of the metal-based catalyst used in the ROP of lactic acid.

6.3.2   SALEN METAL COMPLEXES AS CATALYSTS FOR THE PRODUCTION OF POLYCARBONATES (PC)

Polycarbonate is one of the most commonly used engineering plastics

because of its greater chemical, physical, and mechanical properties. The carbonate bond in aliphatic PC is facile, this polymer being easily biodegradable. More than that, CO_2 included in their structure may be derived from a wide innovative biological sources.

CO_2 is an inexpensive, abundant, and nontoxic resource very suitable for industrial processes and applications (electronics, optical media, automotive, and medical industry) [13].

The recent studies evidenced the utilization of these renewable sources to make bio-based polymers (PC by copolymerization with bio-based epoxides), possessing excellent properties [10, 14].

In recent years, the coupling of CO_2 and epoxides to provide PC was studied by using metal complexes as catalysts with high selectivity (100% in cyclic carbonates with a salen Cr complex in the presence of nucleophiles (4-dimethyl aminopyridine (DMAP)) as cocatalysts).

The studies have been performed by using different substituted salen-M complexes (M = Cr, Al, Co) (Figure 6.3) in different systems with *cocatalysts*: *N-methylimidazole (N-MeIm)* [15], *tetra-n-butylammonium bromide (n-Bu4NBr), DMAP, and triethylamine (Et$_3$N)* with a TOF (turnover frequency) of 120 h^{-1} at 25 °C for propylene carbonate using Co complex and a TOF of 44 h^{-1} by using Cr complex, bis *(triphenylphophoranylidene)ammonium fluoride [PPN] F* in the presence of Co(III) complex with a high selectivity factor [16].

The most considerable advances using Co-salen complexes were the selective production of PPC by polymerization of rac-propylene oxide and carbon dioxide, with a selectivity >99% and 90%–99% carbonate linkage formation.

It is obvious from these studies that Schiff base complexes

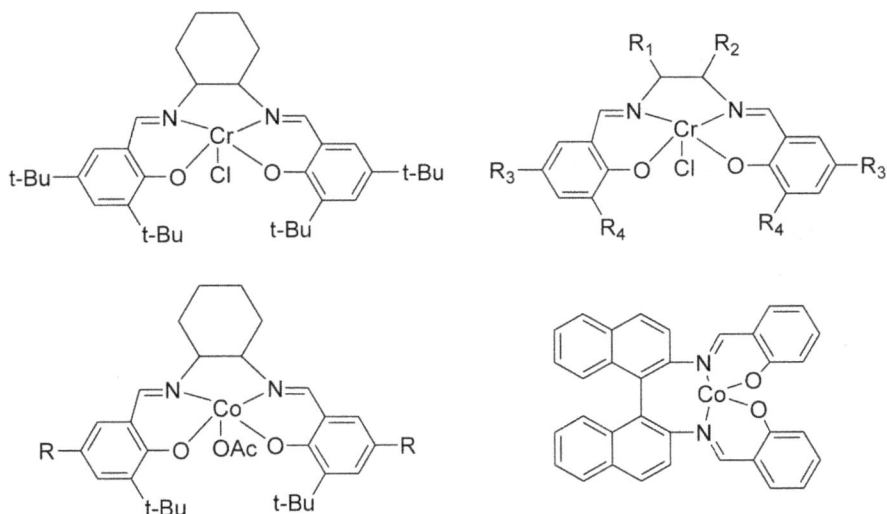

FIGURE 6.3 Some metal-based salen ligands with high selectivity in copolymerization between CO_2 and epoxides.

(especially Cr(III) and Co(III), Al(III) are less active) in the presence of a suitable cocatalyst are the most effective catalytic systems for the selective coupling of CO_2 with epoxides, proving a completely alternative for producing copolymers with high-molecular weights and low PDI [17].

6.3.3 METAL-BASED CATALYSTS FOR OXIDATION REACTIONS OF SIMPLE LIGNIN MODEL COMPOUNDS

There are three polymers in the nonfood biomass: cellulose (40%–80%), hemicellulose (15%–30%), and lignin (15%–30%). The conversion of lignin by oxidation and reduction reactions has been evaluated as a new approach toward more valuable chemicals. The hydrogenation products of lignin, phenol, alkane derivatives (cyclohexane), or aromatic compounds (xylene), are a promising source for new bio-based polymer products, while the oxidation ones can lead to aromatic and highly functionalized derivatives.

The oxidation of lignin was performed by using oxidants (hydrogen peroxide) and metal complexes as a catalyst in homogenous system. The results evidenced that, in the presence of cobalt salen complexes the aerobic oxidation of lignin (organosolv) produced quinones and other monomers, through a mechanism involving the breaking of C–C bonds in phenolic linkages via cobalt intermediate, that abstract a hydrogen atom from the substrate with the generation of phenoxy radical.

The simple lignin model compounds referring to 2-phenoxyethanol, 1-phneyl-2-phenoxyethanol, and 1,2-diphenyl-2-methoxyethanol (Figure 6.4) have been evaluated for a better understanding of the reactivity and selectivity of some catalysts: vanadium and copper complexes.

The oxidation reactions in the presence of vanadium complexes (Figure 6.5) have been performed in acetonitrile or pyridine solvents, in the presence of pinacol. The mechanism involves a C–C bond cleavage and formation of a mixture of acetone, methoxypropene, and vanadium complex, or a C–H bond cleavage to

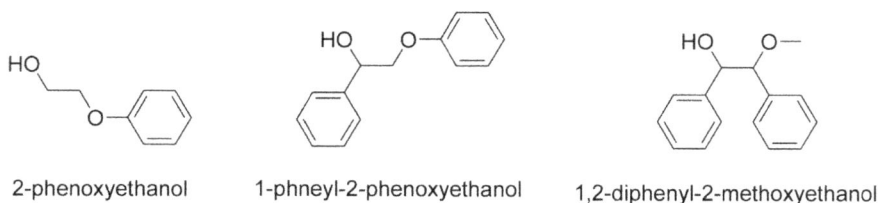

2-phenoxyethanol   1-phneyl-2-phenoxyethanol   1,2-diphenyl-2-methoxyethanol

FIGURE 6.4 The model derivatives of lignin are used as models in oxidation reactions.

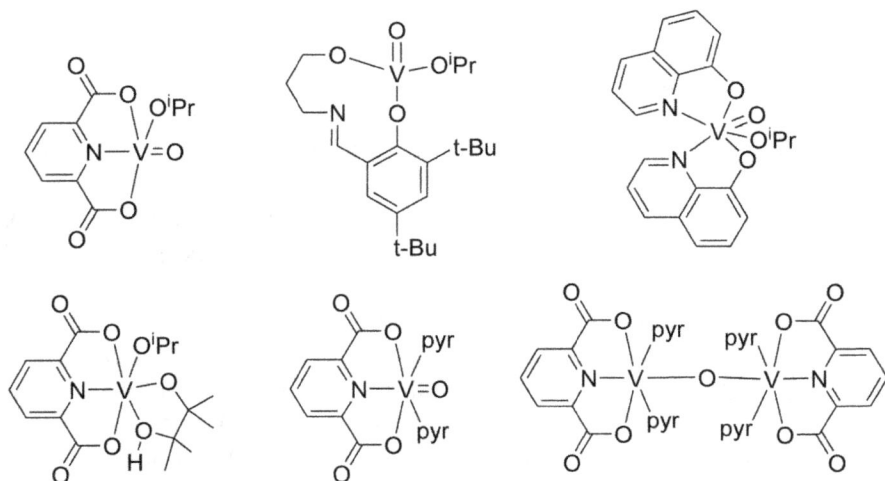

FIGURE 6.5 Vanadium complexes used in the oxidation reaction of lignin derivatives.

afford 2-phenoxyacetaldehyde and recovery of vanadium complex.

In general, vanadium complexes proved to be efficient catalysts of phenyl derivatives models affording mixtures of benzaldehyde, methanol, benzoic acid, or methylbenzoate (Figure 6.6). The catalyst showed a selectivity depending on the solvent used. When the reaction occurred in DMSO, the major products were benzaldehyde and methanol, while in pyridine, benzoic acid, and methyl benzoate were the major products.

The oxidation reaction with copper catalysts was performed also in aerobic conditions, in the presence of a strong base (potassium *tert*-butoxide or other alkoxide bases) which mediates the decomposition and other side reaction of the substrates. A catalytic system containing copper(I) chloride, TEMPO, and pyridine was used in oxidation, highlighting the role of the copper in the C–C bond cleavage. The oxidation of 1,2-diphenyl-2-methoxyethanol with the copper(I) catalytic system afford a mixture of benzaldehyde and methyl benzoate as major products. In this case, the mechanism of oxidation was by

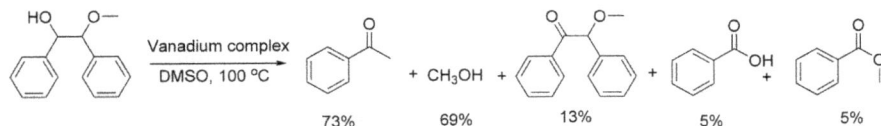

FIGURE 6.6 Oxidation products were obtained by using vanadium complexes catalysts and lignin derivatives.

direct C–C bond cleavage without ketone or aldehyde intermediates, as in the case of vanadium complexes. The disadvantage of this oxidation process was the instability of the copper salt so that new amounts were added to the system [18].

In this chapter, we propose to highlight the catalytic performance of new copper and cobalt metal complexes in oxidation reactions in order to provide an alternative to the catalytic systems mentioned in the literature and previously presented for obtaining new bio-based polymers. Our goal was directed toward the synthesis and catalytic evaluation of organic–inorganic structures by using flexible ligands (Schiff bases), particularly those containing trimethylsilyl tails in the structure. The reported data on the metal complexes with Schiff base ligands revealed a catalytic efficiency in the hydrocarboxylation of a variety of linear and cyclic alkanes to give carboxylic acids, the oxygenation of cyclohexane in mild conditions, the oxidation of 1-phenylethanol, microwave-assisted solvent-free oxidation of cyclohexane to cyclohexanol and cyclohexanone [19, 20].

6.4   EXPERIMENTAL

6.4.1   MATERIALS\

Trimethylsilyl-methyl-*p*-aminobenzoate was prepared according to [20]. 2,6-Diformyl-4-methylphenol (Polivalent-95), copper(II) chloride dihydrate (Aldrich, USA), cobalt chloride hexahydrate (Aldrich), triethylamine (Aldrich), diphenylamine (Fluka), pyrzinecarboxyic acid (Aldrich), TEMPO (Aldrich), cyclohexane (Fluka), *tert*-butyl hydroperoxide (TBHP) [(aq.) (Aldrich)], nitromethane (Aldrich), triphenylphosphine (Aldrich), trifluoroacetic acid (TFA) (Fluka), acetonitrile (MeCN, Aldrich), methanol (Chemical Company), dichloromethane (Chemical Company), and chloroform (Chemical Company) were used as received.

6.4.2   METHODS

Crystallographic measurements were carried out with an Oxford-Diffraction XCALIBUR E CCD diffractometer equipped having a graphite-monochromated Mo-$K\alpha$ radiation. The unit cell determination and data integration were carried out using the CrysAlis package of Oxford Diffraction [21]. The structures were solved by direct methods using Olex2 software [22] with the SHELXS structure solution program and refined by full-matrix least-squares on F2 with SHELXL-97 [23].

Gas chromatographic (GC) analyses were carried out with a FISONS Instruments GC 8000 series gas chromatograph with an FID detector and a capillary column (DB-WAX, column length: 30 m; internal diameter: 0.32 mm) (He as the carrier gas), using the Jasco-Borwin v.1.50 software.

6.4.3   CATALYTIC STUDIES

6.4.3.1   HYDROCARBOXY-LATION

Experiments have been performed by following a previously developed protocol [24–28]. Copper(II) complex was introduced in a stainless steel autoclave possessing a Teflon coated magnetic stirring bar. $K_2S_2O_8$, H_2O, and MeCN were added and the mixture was stirred al 60 °C for 4 h, then diethylether and cycloheptanone were added and energic stirred for another 10 min after that the organic layer was investigated by GC revealing the formation of the corresponding monocarboxylic acids as the dominant products.

6.4.3.2   PEROXIDATIVE OXIDATIONS

The oxidation of cyclohexane with aqueous H_2O_2 was carried out in the air in round-bottomed flasks using MeCN as solvent (up to 5.0 mL total volume). Catalyst (copper complex) and TFA were added, followed by cyclohexane and the reaction was started after the addition of hydrogen peroxide. The reaction mixture was stirred at 50 °C for 2 h and samples were analyzed after 5, 15, 30, 45, 60, and 120 min by GC using nitromethane as an internal standard, after the introduction of triphenylphosphine for reducing the cyclohexyl hydroperoxide to the corresponding alcohol.

6.4.3.3   MICROWAVE-ASSISTED SOLVENT-FREE PEROXIDATIVE OXIDATIONS

Catalytic experiments with 1-phenylethanol were performed with a focused Anton Paar Monowave 300 reactor. The alcohol, catalyst (copper complex), and a 70% aqueous solution of tBuOOH were introduced in a cylindric Pyrex tube that was sealed. This tube was then placed in the microwave reactor and the system was stirred under irradiation (5 or 20 W) at 80 °C or 120 °C for 0.5–3 h. After cooling to room temperature, benzaldehyde (internal standard) and MeCN were added. The obtained mixture was stirred for 10 min and then a sample was taken from the organic phase and analysed by GC using the internal standard method. The catalytic oxidations of cyclohexane were performed in sealed cylindrical Pyrex tubes, under focused microwave irradiation. The typical reaction conditions were the tube was charged with cyclohexane, catalyst (cobalt complex), and

TBHP. Diphenylamine was added to the reaction mixture. The tube was placed in the microwave reactor and the mixture was stirred (600 rpm) and irradiated (10 W) at 60 °C or 100 °C for 0.5–24 h. After the reaction, the mixture was allowed to cool to room temperature. Solution samples were analysed by GC after the addition of nitromethane (internal standard). For reactions performed under conventional heating, round-bottomed flasks equipped with reflux condensers in conventional oil baths in the air were used. The reagents were added to the flask and vigorously stirred at the desired temperature (60 °C–100 °C) for the established reaction time (up to 24 h).

6.5 RESULTS AND DISCUSSION

6.5.1 STRUCTURAL ANALYSIS OF METAL COMPLEXES

Cu(II) and Co(II) complexes have been prepared in a two-step reaction: (1) obtaining of the Schiff base ligand from amino and carbon precursors, 2,6-diformyl-4-methyl-phenol and trimethylsilyl-methyl-p-aminobenzoate, respectively, in a solvent mixture (methanol/chloroform) at relux for 2 h, followed by (2) the addition of metal salt, copper(II) chloride, and cobalt(II) chloride to the ligand and addition

of triethylamine for deprotonation according to Figure 6.7.

Slow evaporation of the solvent mixture led to the obtaining of crystalline products that were purified by chromatographic separation using methanol/dichloromethane (1:9) in the case of copper(II) complex and by recrystallization using a solvent mixture of methanol/chloroform/acetonitrile/diethyl ether in the case of the cobalt(II) complex.

Both compounds were separated in form of single crystals, their structure being established by X-ray diffraction analysis.

The central part of the Cu(II) complex consists of a tetranuclear (four copper atoms) core held together by a μ-oxido ligand. The coordination around copper(II) atoms is ensured by four nitrogen atoms and two phenolato oxygen atoms from two Schiff base ligands, linked together by a central μ-oxido ligand and four chloride atoms. The geometry of each copper(II) ion is square-planar (Figure 6.8) [20]. The X-ray structure of the cobalt(II) complex consists of two cobalt ions linked through two chloride atoms which act as bridging ligands and separate the two metal centers. The coordination around the cobalt atom is ensured by two N,O bidentate Schiff bases and two chloride atoms, in an octahedral geometry (Figure 6.8) [19].

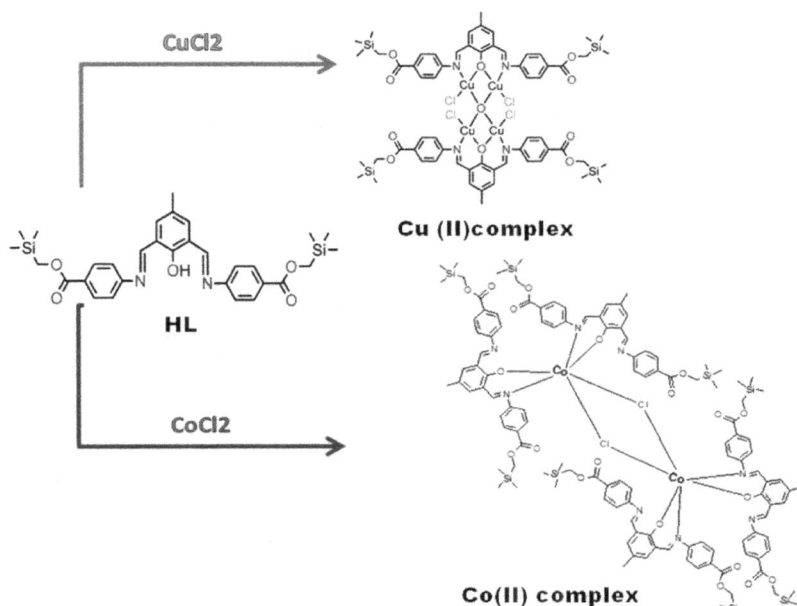

FIGURE 6.7 Schematic synthetic route to prepare copper(II) and cobalt(II) metal complexes.

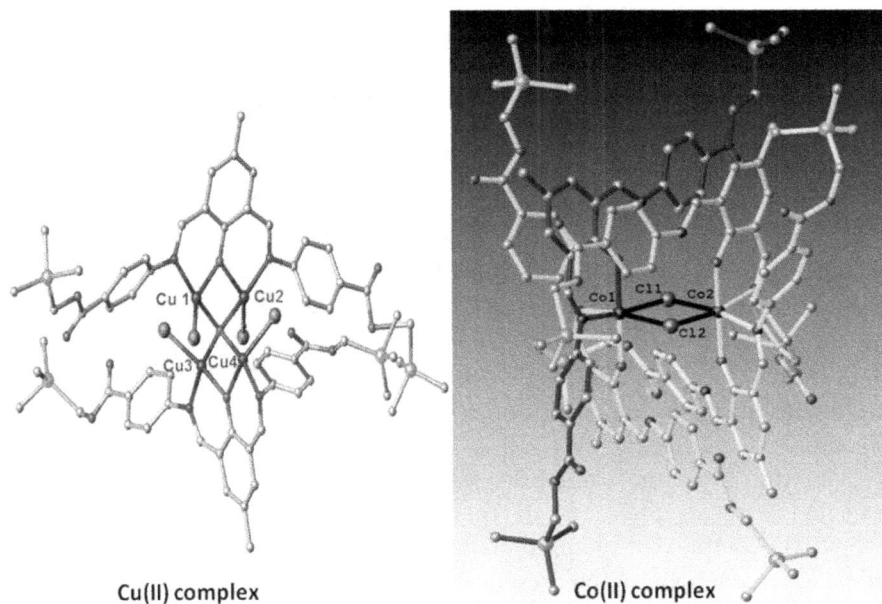

FIGURE 6.8 X-ray diffraction structure of the metal complexes.

6.5.2   CATALYTIC RESULTS

The copper(II) complex was tested as a catalyst precursor for hydrocarboxylation of a variety of linear and cyclic alkanes for obtaining carboxylic acids. The highest overall yield in the hydrocarboxylation of linear alkanes was 18.5%, while for cyclic alkanes was about 25.5%, proving the selectivity of the catalytic system on the secondary carbon atoms in the structure (Figure 6.9).

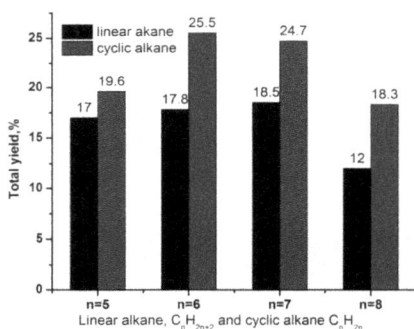

FIGURE 6.9   Total yield in the hydrocarboxylation of alkanes to carboxylic acids using copper(II) complex as catalyst.

The copper(II) complex also proved to be an efficient precatalyst in the peroxidative oxidation of cyclohexane in water/MeCN solvent mixture to afford a mixture of cyclohexanol and cyclohexanone [20].

The results of the catalytic test revealed a good efficiency of the copper(II) complex in the oxidation of cyclohexane both, in the presence and in the absence of TFA (cocatalyst). The total yield in the absence of the cocatalyst is 8% after 2 h, while the addition of the cocatalyst leads to a maximum yield of 13% after 1 h (Figure 6.10).

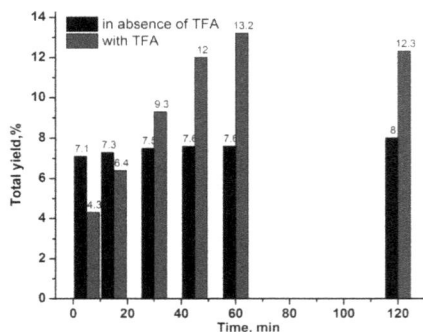

FIGURE 6.10   The total yield for the oxidation of cyclohexane with copper(II) complex.

The evaluation of the catalytic performance of the copper(II) complex toward the oxidation of 1-phenylethanol was performed by a solvent-free microwave-assisted method in the presence of different additives [20]. The acetophenone was the main product formed during the oxidation reaction, with a total yield of 76% after 30 min in the absence of the additives and 82% after 3 h. The addition of additives did not have a positive impact on the total yield. Thus, the presence of TFAs leads to a slow increase of the yield after 30 min (78%), while the addition of diphenylamine or TEMPO led to an inhibition of the catalytic reaction, the total yields being 4% and 6%, respectively (Figure 6.11).

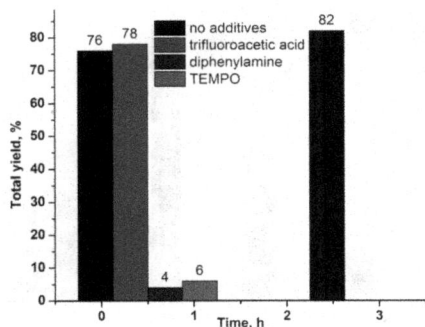

FIGURE 6.11 The total yield of the oxidation of 1-phenylethanol to acetophenone by using the copper(II) complex as catalyst.

The cobalt(II) complex was tested as a catalyst in a microwave-assisted method of solvent-free oxidation of cyclohexane to a mixture of cyclohexanol and cyclohexanone [19].

The influence of different additives was investigated: pyrazinecarboxylic acid and TEMPO at a concentration of 0.2% of catalyst. One can observe that the maximum total yield was obtained after 1.5 h in the absence of additives, while the addition of pyrazinecarboxylic acid has a moderate effect after 2 h. On contrary, the presence of TEMPO, as an oxidizing agent, has an inhibitory effect in the first hour of reaction and a relatively moderate effect after 2 h (Figure 6.12).

The concentration of the catalyst on the total transformation yield was also investigated. One can observe that a higher concentration of catalyst has a good impact on the yield, after 1.5 h being 52% as compared with 7% at a concentration of 10-fold smaller.

Also, the temperature increase has a positive impact on the catalytic oxidation, at 100 °C the total yield is 3-fold higher after 1.5 h (Figure 6.13).

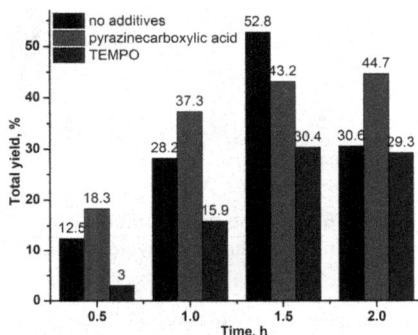

FIGURE 6.12 The catalytic results of the cobalt(II) complex on the microwave-assisted solvent-free oxidation of cyclohexane.

6.6 CONCLUSION

The present work highlighted the importance of different metal-based catalysts in the production of bio-based plastics. The development of novel and cheap nontoxic metal-based catalysts for biodegradable polymer production is an advantageous strategy for obtaining low costs polymers with improved biochemical properties. The ROP, the coupling of CO_2 and epoxides, or the oxidation of some simple lignin model compounds catalyzed by metal complexes proved to be very effective in the control of the molecular weight, selectivity, and reactivity of the catalysts. In addition, some new metal catalysts (copper(II)

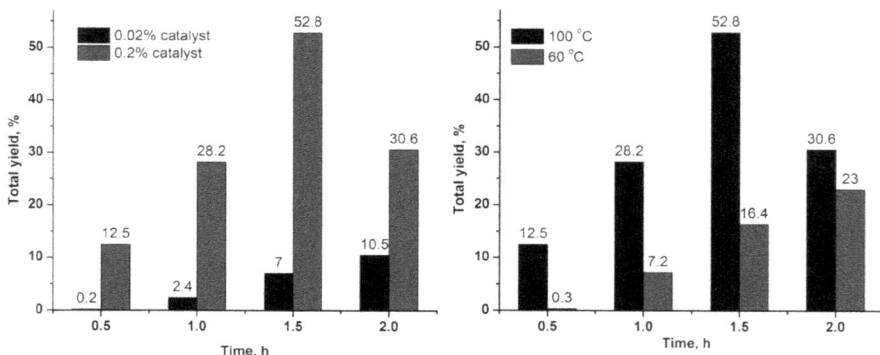

FIGURE 6.13   The concentration and tem-perature on the catalytic activity of cobalt(II) complex in the oxidation of cyclohexane.

and cobalt(II) complexes) have been proposed to be tested in oxidation processes, in different conditions, to provide new alternatives for catalytic processes in the production of bio-based polymers.

ACKNOWLEDGMENTS

This work was supported by a grant of Ministry of Research and Innovation, CNCS-UEFISCDI, project number PN-III-P1-1.1-PD-2016-1027 (Contract 5/2018).

KEYWORDS

- catalyst
- metal complexes
- oxidation reaction
- bio-based products

REFERENCES

1. Garcia, J. M.; Catalyst: Design Challenges for the Future of Plastics Recycling. *Chem* **2016;** 1:813–819.
2. Panda, A. K.; Alotaibi, A.; Kozhevnikov, I. V.; Shiju, N. R.; Pyrolysis of Plastics to Liquid Fuel Using Sulphated Zirconium Hydroxide Catalyst. *Waste Biomass. Valor.* **2019,** https://doi.org/10.1007/s12649-019-00841-4.
3. Budsaereechai, S.; Hunt, A. J.; Ngernyen, Y.; Catalytic Pyrolysis of Plastic Waste for the Production of Liquid Fuels for Engines. *RSC Adv.* **2019;** 9:5844–5857.
4. Miandad, R.; Rehan, M.; Barakat, M. A.; Aburiazaiza, A. S.; Khan, H.; Ismail, I. M. I.; Dhavamani, J.; Gardy, J.; Hassanpour, A.; Nizami, A.-S.; Catalytic Pyrolysis of Plastic Waste: Moving Toward Pyrolysis Based Biorefineries. *Front. Energy Res.* **2019;** 7:27, doi: 10.3389/fenrg.2019.00027.
5. Tang, X.; Chen, E. Y.-X.; Toward Infinitely Recyclable Plastics Derived from Renewable Cyclic Esters. *Chem* **2019;** 5:284–312.
6. Chen, G. Q.; Patel, M. K.; Plastics Derived from Biological Sources: Present and Future: A Technical and

Environmental Review. *Chem. Rev.* **2012;** 112:2082–2099.

7. Kulikowska, D.; Bernat, K.; Wojnowska-Baryla, I.; Effect of Bio-Based Products on Waste Management. *Sustainability* **2020;** 12:2088.

8. Mülhaupt, R.; Green Polymer Chemistry and Bio-based Plastics: Dreams and Reality. *Macromol. Chem. Phys.* **2013;** 214:159–174.

9. Rahimi, A.; Garcia, J.; Chemical Recycling of Waste Plastics for New Materials Production. *Nat. Rev. Chem.* **2017;** 1:0046.

10. Hillmyer, M. A.; The Promise of Plastics from Plants. *Science* **2017;** 358: 868–870.

11. Barwe, S.; Weidner, J.; Cychy, S.; Morales, D. M.; Dieckhçfer, S.; Hiltrop, D.; Masa, J.; Muhler, M.; Schuhmann, W.; Electrocatalytic Oxidation of 5-(Hydroxymethyl)furfural Using High-Surface-Area Nickel Boride. *Angew. Chem. Int. Ed.* **2018;** 57:11460–11464.

12. Li, X.; Yang, B.; Zheng, H.; Wu, P.; Zeng, G.; Synthesis and Characterization of Salen-Ti(IV) Complex and Application in the Controllable Polymerization of D, L-lactide. *PLoS One* **2015;** 13(8):e0201054.

13. Lambert, S.; Wagner, M.; Environmental Performance of Bio-Based and Biodegradable Plastics: The Road Ahead. *Chem. Soc. Rev.* **2017;** 46: 6855–6871.

14. Artham, T.; Doble, M.; Biodegradation of Aliphatic and Aromatic Polycarbonates. *Macromol. Biosci.* **2008;** 8:14–24.

15. Darensbourg, D. J.; Rodgers, J. L.; Fang, C. C.; The Copolymerization of Carbon Dioxide and [2-(3,4-Epoxycyclohexyl)Ethyl]Trimethoxysilane Catalyzed by (Salen)CrCl. Formation of a CO_2 Soluble Polycarbonate. *Inorg. Chem.* **2003;** 42(15):4498–500.

16. Darensbourg, D. J.; Mackiewicz, R. M.; Role of the Cocatalyst in the Copolymerization of CO_2 and Cyclohexene Oxide Utilizing Chromium Salen Complexes. *J. Am. Chem. Soc.* **2005;** 127:14026–14038.

17. Darensbourg, D. J.; Making Plastics from Carbon Dioxide: Salen Metal Complexes as Catalysts for the Production of Polycarbonates from Epoxides and CO_2. *Chem. Rev.* **2007;** 107:2388–2410.

18. Hanson, S. K.; Baker, R. T.; Knocking on Wood: Base Metal Complexes as Catalysts for Selective Oxidation of Lignin Models and Extracts. *Acc. Chem. Res.* **2015;** 48:2037–2048.

19. Zaltariov, M.-F.; Vieru, V.; Zalibera, M.; Cazacu, M.; Martins, N.M. R.; Martins, L. M. D. R. S.; Rapta, P.; Novitchi, G.; Shova, S.; Pombeiro, A. J. L.; Arion, V. B.; A bis(μ-Chlorido)-Bridged Cobalt(II) Complex With Silyl-Containing Schiff Base as a Catalyst Precursor in the Solvent-Free Oxidation of Cyclohexane. *Eur. J. Inorg. Chem.* **2017;** 2017:4324–4334.

20. Zaltariov, M. F.; Alexandru, M.; Cazacu, M.; Shova, S.; Novitchi, G.; Train, C.; Dobrov, A.; Kirillova, M. V.; Alegria, E. C. B. A.; Pombeiro, A. J. L.; Arion V. B.; Tetranuclear Copper(II) Complexes with Macrocyclic and Open-Chain Disiloxane Ligands as Catalyst Precursors for Hydrocarboxylation and Oxidation of Alkanes and 1-Phenylethanol. *Eur. J. Inorg. Chem.* **2014;** 2014:4946–4956.

21. CrysAlis RED, Oxford Diffraction Ltd. **2003,** version 1.171.36.32.

22. Dolomanov, O. V.; Bourhis, L. J.; Gildea, R. J.; Howard, J. A. K.; Puschmann, H. OLEX2: A Complete Structure Solution, Refinement and Analysis Program. *J. Appl. Crystallogr.* **2009;** 42:339–341.

23. Sheldrick G. M.; SHELXS. *Acta Crystallogr., Sect. A.* **2008;** 64:112–122.

24. Kirillova, M. V.; Kirillov, A. M.; Kuznetsov, M. L.; Silva, J. A. L.; Fraústo da Silva, J. J. R.; Pombeiro,

A. J. L.; Alkanes to Carboxylic Acids in Aqueous Medium: Metal-Free and Metal-Promoted Highly Efficient and Mild Conversions. *Chem. Commun.* **2009;** 17:2353–2355.

25. Kirillova, M. V.; Kirillov, A. M.; Pombeiro, A. J. L.; Metal-Free and Copper-Promoted Single-Pot Hydrocarboxylation of Cycloalkanes to Carboxylic Acids in Aqueous Medium. *Adv. Synth. Catal.* **2009;** 351:2936–2948.

26. Kirillova, M. V.; Kirillov, A. M.; Pombeiro, A. J. L.; Mild, Single-Pot Hydrocarboxylation of Gaseous Alkanes to Carboxylic Acids in Metal-Free and Copper-Promoted Aqueous Systems. *Chem. Eur. J.* **2010;** 16: 9485–9493.

27. Kirillova, M. V.; Kirillov, A. M.; Pombeiro, A. J. L.; Mild, Single-Pot Hydrocarboxylation of Linear C_5–C_9 Alkanes into Branched Monocarboxylic C_6–C_{10} Acids in Copper-Catalyzed Aqueous Systems. *Appl. Catal. A.* **2011;** 401:106–113.

28. Kirillov, A. M.; Kirillova, M. V.; Pombeiro, A. J. L.; Multicopper Complexes and Coordination Polymers for Mild Oxidative Functionalization of Alkanes. *Coord. Chem. Rev.* **2012;** 256:2741–2759.

PART II
Biodegradable Polymers and Sustainable Composites

CHAPTER 7

CURRENT TRENDS AND PERSPECTIVES IN BIODEGRADABLE POLYMERS

LUMINITA IOANA BURUIANA[1*], and CRISTIAN LOGIGAN[2]

[1]*"Petru Poni" Institute of Macromolecular Chemistry, 41A Grigore Ghica Voda Alley, 700487 Iasi, Romania*

[2]*Chemical Company SA, 14 Chemistry Bdv., 700293 Iasi, Romania*

Corresponding author. E-mail: buruiana.luminita@icmpp.ro

ABSTRACT

Recent trends in biodegradable polymers are focused on evolving new design strategies and technologies in order to obtain advanced polymers with improved performance. This chapter presents an up-to-date of the main types of biodegradable polymers and discusses their outstanding properties. Also, there are presented the most significant biomedical applications of biodegradable polymers, like implantable large or small devices, drug delivery vehicles, patches, or tissue engineering scaffolds. The recent technological advances of biodegradable polymers offer great promise toward achieving biodegradability with less pollutants and greenhouse emissions.

7.1 INTRODUCTION

Biodegradable and biocompatible polymers exhibit a huge scientific interest. The great progress in biotechnology has required specific features for biomaterials. In this context, new biodegradable polymers with special properties are demanding. Biodegradable polymers are produced from renewable sources that are completely biodegradable and mimic the properties of conventional polymers like polyethylene (PE), polypropylene, or polyethylene terephthalate (PET). Biodegradable polymers can be classified as natural or synthetic ones, according to their source of coming. Synthetic biodegradable polymers have more biomedical applications owing to their tailorable designs, from wound dressings or drug delivery vehicles to surgery and tissue engineering scaffolds. Also, they are used in nonmedical applications such as the controlled release of fertilizers and pesticides, in agriculture, for applications in the automotive industry [1–3], or in packaging technologies [4].

This work presents a comprehensive introduction to various types of synthetic biodegradable polymers with reactive groups and bioactive groups, their structure, and important applications domain properties. The focus is on advances in the past decade in functionalization and responsive strategies of biodegradable polymers and their medical applications.

7.2 CLASSIFICATION, PROPERTIES, AND APPLICATIONS

Considering the formation of biodegradable polymers, they can be divided into two main categories that include synthetic biodegradable polymers and natural biodegradable polymers.

7.2.1 SYNTHETIC BIODEGRADABLE POLYMERS

These types of polymers may be summarized, on the basis of their backbone, in polymers with hydrolyzable chains and polymers with a carbon backbone. From the first type, *polyesters* are the most encountered, especially poly(α-hydroxy acids), polylactones, polyorthoesters, and polyphosphoesters (PPEs). Poly(α-hydroxy acids) represent a class of biodegradable polymers that are frequently studied regarding medical applications. The synthesis of them is carried out by ring-opening polymerization or by polycondensation, depending on the used monomers. Poly(lactic acid) (PLA), poly(glycolic acid) (PGA), and their copolymers are more used. All of them are used in surgery as resorbable sutures and in tissue engineering [5]. The best example is PGA that has high cristallinity and good mechanical strength [6], properties that recommend this polymer for orthopedic application (named Biofix) [7]. Besides that, PGA is also applied in tissue engineering and drug delivery. The utilization of PGA has several drawbacks like higher degradation rates that induce an enhanced acidity of the neighboring tissues that can cause inflammation response [8–10]. In addition, lack of solubility and high cristallinity may cause some possible manufacturing problems.

PLA is commonly available in four optical active forms (stereoisomers). From these forms of lactides, poly(L-lactic acid) (PLLA) and poly(D,L-lactic acid) (PDLLA) are employed in the medical field. PLLA is used as tissue engineering scaffolds and in bone fixation (named bio-interference screw, bio-screw, or bio-anchor), while PDLLA is used especially in drug delivery [11]. Both polymers combined with PE glycol, PGA, and chitosan are applied to manufacture new biodegradable materials with requested properties by other interesting medical applications. In order to overcome the drawbacks of individual PLA and PLGA, copolymerization is developed poly(lactide-*co*-glycolide) (PLGA). The special characteristics of the copolymer make it a desired candidate for different biomedical applications, such as drug delivery, tissue engineering, or medical implants. The most used suture materials available on the market are Vicryl (90G/10 L), and Vicryl Rapide modified, and Polysorb and Purasorb, as well. A special feature of PLGA copolymer is its rapid degradation, which recommends it for delivery of drugs, vaccines, peptides, proteins, or macromolecules [12]. At the same time, this copolymer exhibits important bioadhesion and cell proliferation ability that makes it an exquisite candidate for tissue engineering applications [13].

Polycaprolactone (PCL)—a semicrystaline polymer—belongs to *polylactones* type of synthetic polymers. Some characteristic properties such as hydrophobicity and crystallinity cause slow degradation of the polymer. At the same time, a very important feature is represented by good permeability for drugs, which makes them suitable for sustained drug delivery. Copolymers like PLLA, PDLLA, PLGA lead to an increase of degradation kinetics in drug delivery [14]. Recent studies have focused more on tissue engineering applications and medical implants of the PCL composites against drug delivery. A new in vivo study shows that PCL composite scaffold works as an extracellular matrix for bone regeneration in the case of endochondral ossification process [15].

PPEs are synthesis originally polymers, prepared by ring-opening polymerization polycondensation and polyaddition [16]. These types of synthetic polymers exhibit important biodegradability and biocompatibility under physiological conditions. PACLIMER represents a device of drug delivery that is used in primary clinical trials for ovarian and lung cancer treatment. The most important applications of PPE and PPE composites are gene and drug delivery or tissue engineering [17].

Polycarbonates are polymers with linear backbone with a hydrolytically stable nature. Some examples of these polymer types include poly(ethylene carbonate) and poly(propylene carbonate)—thermoplastic polyesters of carbonic acid with aliphatic dihydroxy compounds.

Delivery of contraceptive steroids and antimalarial agents was studied from poly(dihydropyrans) matrices; in these directions, in vivo and in vitro tests were carried out [18]. 1,3-Propanediol and carbonic acid polymers show good biocompatible characteristics and were used for drug release applications [19]. At the same time, the low mechanical strength exhibited by this polymer is helpful for manufacturing of the soft tissue regeneration implants. Ethyl ester polycarbonate is other synthesis polymer used for biomedical applications, for example, bone fixation [20].

Polyanhydrides: the majority of the polyanhydrides are prepared by melt polycondensation. The most common monomers involved in this reaction are diacids; the final polymer is obtained by heating the prepolymer (formed by blending diacid anhydride with acetic anhydride) in a vacuum, in order to remove the acetic anhydride from the system. The most intense studied polyanhydride due to its special feature is polysebasic acid. Its characteristics include a good cristallinity, a melting point below 100 °C, and better solubility in halogenated solvents. Also, the polymer displays a rapid degradation property, which makes it suitable for drug delivery applications [21]. Aromatic polyanhydrides show a higher melting temperature, approximately 200 °C,

being insoluble in organic solvents. From them, the most known is 1,3-bis-carboxy phenoxy propane based polyanhydrides [22]. Other types of studied polyanhydrides are those that are based on fatty acids; they are hydrophobic, have a low melting point, and are soluble in organic solvents. Their hydrophobic nature determines bulk degradation. Several examples of polyanhydride and their uses are as follows:

1. Gliadel is a PCPP-SA polyanhydride applied for controlled drug delivery of chemotherapeutic agents to the brain in tumor treatment.

2. Septacin is a copolymer used for gentamicin delivery in the case of patients with osteomyelitis diseases.

At the same time, polyanhydrides are good delivery vehicles, but still exhibit low mechanical strength, which represents a drawback for applications in tissue regeneration. In order to overcome this disadvantage, the scientists obtained methacrylated polyanhydrides as injectable and crosslinked materials that have enhanced mechanical strength for tissue engineering applications [23, 24].

Besides the biodegradable synthesis polymers, mentioned above, some polymers are used in tissue engineering, named polyurethanes (PUs). Their preparation is based on the processing of raw materials and additives. PUs have important features, such as significant mechanical and physical properties, thermoplastic characteristics, good stability, and degradation. At the same time, taking into consideration their composition and the obtaining procedure, PU could be hydrophobic or hydrophilic [25]. PUs display suitable compatibility with blood elements, being biocompatible and bioresorbable; they have peculiar mechanical properties that can be tuned to the desired tissue [26, 27]. Initially, PU has been studied in medicine for vascular or bone grafts, artificial skin, and repair of articular cartilage [28–30].

Later, the research presents a PU complex appropriate for scaffold applications. These polyesters–urethanes are obtained especially from PCL, PLA, or PGA and their main characteristic must be excellent resorbability. This particular application requires some specific features of the material, for example, a porous structure with interconnected pores. Polymers must have a very good porosity (>90%) and an inherent pore dimension, in function of the application. In this context, some examples are presented below:

1. Scaffolds used for liver regeneration must have a pore diameter of maximum 20 μm to enable the hepatocytes growth.
2. The pore diameter for skin lesions should be in the 20–150

μm range, the best pore size for bone grafts is between 200 and 400 μm [31]. Furthermore, pores must be interconnected to allow cell and tissue ingrowth. All these properties depend, on the one hand, on the polymer used, and, on the other hand, on the preparation technique.

Solvent casting combined with particle percolating involves leaching out solid particles from the polymer solution. Scaffolds with high porosity (approximately 93%) and pore sizes up to 50 μm are obtained by this technique. Also, the shape and size of pores are determined by the shape and dimensions of the leachable particles [32]. Other parameters that influence the structure of the scaffolds are the amount of particles added and the initial concentration of the polymer solution. Moreover, the effect of the pore size on the scaffold mechanical properties must be considered. Thus, pores that have higher diameters are characterized by small porosity and increase mechanical strength. This preparation method has several advantages, the principal one being the easiness of fabrication without specific tools. But also has an important drawback that consists of the inconvenience of selecting the size of particles necessary to obtain high porosity of the material at the same time with keeping proper mechanical characteristics and material thickness (up to 3 mm).

Another important disadvantage is represented by organic solvents that must be removed from the system due to their toxicity. These types of PU scaffolds are found applications in soft tissue engineering, such as repair of coronary arteries and bones [33, 34].

Although melt molding is used usually for ceramics and metals preparation, it can be applied with fine results to PU. The preparation method includes patterns filled with porogen compounds and granulated or powdered polymers, which are heated above the polymer's glass transition temperature (T_g) at high pressure and raw materials connected to form a scaffold in the pattern shape. After pattern removing, the porogen agent is leached out, as in the method described above [35]. The main advantage of this method is the fast obtaining of polymer scaffolds with various shapes and sizes, which can be controlled by the porogen choice [36]. The melt molding method does not require an organic solvent to obtain scaffolds, as opposite to other methods [37]. Still, the method has some limitations like the pattern design and framework and the presence of the nonporous layer on the surface, thus being difficult to leach out the porogen. By this technique, PU and hydroxyapatite are combined for bone scaffold preparation [38, 39].

Scaffolds that include human cardiac progenitor cells (CPCs) represent a therapeutic challenge to treat heart regeneration after myocardial infarction. In this context, researchers focused on the preparation of square-grid scaffolds by melt extrusion additive, obtaining from a PU. Subsequently, the mixture is placed under plasma treatment for acrylic acid surface grafting/polymerization, and then, grafted with laminin-1 (PU-LN1) or gelatin (PU-G) by carbodiimide chemistry [40].

Laminin 1 is a cardiac niche component of the extracellular matrix and has a very important role in heart formation during the embryogenesis process. Gelatin represents a cell-adhesion protein that has the role of controlling the molecule functionalization. Although both PU-G and PU-LN1 scaffolds showed increased cell density in comparison with pristine PU scaffolds, LN1 functionalization is reflected in improved protection of CPCs from apoptosis and proliferation, stimulating their differentiation to cardiomyocytes, endothelial cells, and smooth muscle cells. Scaffolds obtain by this method are subcutaneously implanted in mice; it was observed a minimal inflammatory response and promoted angiogenesis. These tests evidenced the ability of PU-LN1 scaffolds to affect CPC behavior, being a key component of cardiac niche ECM. The possible application of PU-LN1 scaffolds is as in vitro models of myocardial microenvironment for studying CPC behavior or the implantable patches for tissue engineering.

Recently, a novel type of biodegradable PU having a very good elasticity was synthesized [41].

The obtaining process is based on the controlled crosslinking of poly(ester ether) triblock copolymer diols and PCL triols, using urethane chains. Thus, 3D porous scaffolds were prepared in situ, having ascertained geometry, controlled microstructures, and adjustable mechanical properties. The latest could be controlled by varying the ratio between the linear and branched polymer chains. The pore size and porosity of 3D porous scaffolds could be increased by reducing the solution concentration or using a porogen. While these scaffolds contain rigid PCL chains, they exhibit an outstanding elasticity and recoverability, which can be assigned to the triblock copolymer and crosslinking sites.

In vitro cell culture of 3T3 fibroblasts and MG63 osteoblast-like cells established the PU scaffolds biocompatibility. And, these scaffolds with different stiffness can activate different types of cell proliferation, as well. All properties mentioned above, recommend PU scaffolds suitable for different elastic tissue regeneration.

Other types of biodegradable and biocompatible PUs that have a great affinity for tissue regeneration were synthesized [42]. The obtaining process implies PCL (an aliphatic diisocyanate) and a biologically active compound (1,4:3,6-dianhydro-dsorbitol) as a chain extender. Thus, using a process that implies a combined salt leaching-phase inverse technique, 3D porous scaffolds were manufactured, which have a structure with controlled size pores. The pores size and geometry depend on solvent/nonsolvent used in the obtaining process and on the size of the solid porogen crystals, as well. From the solvents used, the best one is proved to be dimethylformamide (DMF). Other solvents were tried exhibited several drawbacks that are very important for the desired applications. In this context, by using dimethylsulfoxide or methyl-2-pyrrolidone, researchers did not achieve a scaffold with a homogenous pore structure. Although many pores were interconnected, some of them are still closed. That is why the prepared scaffolds display poor water permeability. The type of nonsolvent added in the DMF polymer solution has an important impact on the pore structure of the scaffolds [43]. The proper choice of solvent/nonsolvent mixture causes the obtaining of elastomeric PU scaffolds with regular interconnected pores, high water permeability, and 90% pore-to-volume ratio. Also, by loading these types of PU scaffolds with calcium phosphate salts (hydro-xyapatite or tricalcium phosphate), one can obtain a promising candidate for bone graft substitutes or for the repair of articular cartilage [44, 45].

An interesting type of PU scaffolds was synthesized, having a specific design, and could be adjusted for regenerated tissue. Thus, porous PU scaffolds using solvent casting/particulate leaching technique combined with thermally induced phase separation were prepared. The characterization of the scaffolds obtained (PPS) revealed a similar T_g with the pristine PU and a homogenous structure. The scaffold interaction with distilled water, saline, and phosphate-buffered saline solutions after three months of incubation showed the solutions stable character in the studied media. The microbiological tests performed did not point out any antimicrobial effect on the analyzed bacterial species (*Staphylococcus aureus, Pseudomonas aeruginosa,* and *Escherichia coli*). Then, the absence of inhibition zones around PPS and good adhesion showed by the bacterial species to the PPS surface indicate that the scaffold has no cell toxicity to the cells of the bacterial cultures. The cell culture studies exhibited that PPS is biocompatible with the 3T3 NIH cell line. In summary, all of these properties highlight the fact that the obtained PPS scaffolds may be a suitable material for soft tissue engineering, especially blood vessels [46].

7.2.1.1   CURRENT TRENDS IN PUs APPLICATIONS

Biodegradable PUs have specific physicochemical properties that make them suitable for regenerative medicine. Thus, depending on the tissue mechanical properties, rigid PU scaffolds are suitable for hard tissue replacement, and, on the contrary, soft PU scaffolds match better to soft tissue substitution.

7.2.1.1.1   Hard Tissue

Bone tissue applications are emerging issue research that implies the progress of alternative treatments to autografts for bone replacement in the case of orthopaedic lesions, bone tumors, pseudoarthrosis, and different applications in maxillofacial, craniofacial, reconstructive, neck, and head surgery [47, 48]. Bone tissue exhibits two different forms, compact bone and spongy bone. Compact bone is a dense tissue that forms a shell around the spongy bone; these ones in the mature stage have a lamellar or layered structure. Spongy bone is presented in the core of flat and irregular bones, being found in the epiphyses of the long bones. Spongy bone is comparable to a sponge in crosssection, yielding a divided appearance. Properties required for bone tissue applications are biocompatibility, bioactivity, osteoconduction, osteoinduction, and biodegradation [49]. Among biomaterials, metals, proteins, polysaccharides, or biocompatible synthetic polymers are the most used, due to their better mechanical properties, high biocompatibility, and reproducible physicochemical properties.

Elastomeric PUs are interesting polymers for bone graft substitutes because they avoid shear forces at the interface between bone and material, thus stimulating the proliferation of osteogenic cells and bone regeneration [50, 51].

Composites containing inorganic–organic phases were analyzed for reproducing bone composition. Scaffolds with the porous structure were prepared by polymer coagulation with a salt leaching technique, using a proper porogen (NaCl). These composite scaffolds present increase bioactivity emphasized by hydroxyapatite obtaining on the foam surfaces, immediately after dipping in simulated body fluid [52].

PU has also been used like fixation adhesive as another choice of the epoxy resins or cyanoacrylates. Epoxy resins have some application limits consisting of tissue necrosis due to polymerization heat, toxic effects, or weak bonding in wet conditions. On the other hand, cyanoacrylates are debatable due to the toxicity of some monomers or high infection rates [53].

For exceed, this drawback is a porous PU adhesive reinforced with hydroxyapatites that were manufactured [54].

A recent study in this direction is focused on improving the scaffold performance by incorporating the osteoinductive growth factor, bone morphogenetic protein-2 (BMP-2); this factor stimulates osteoblast differentiation and promotes bone formation. The BMP-2 mechanism implies bone formation by osteoblast differentiation, chemoattraction, and angiogenesis. Thus, the release of recombinant human BMP-2 from a reactive PU was evaluated, by using two approaches [55]. The first one is based on the direct incorporation of rhBMP-2 into the material before polymerization process, while, in the second one, it was loaded in the PLGA microparticles immersed in PU. In the first case, it was observed an important burst was released in the first 5 days, followed by a sustained release in the next 21 days. The encapsulation of rhBMP-2 with different sizes of PLGA microparticles causes the reducing burst release and improved sustained release. Also, PU could be tailoring for enhancing bone regeneration by bulk functionalization with bioactive proteins/peptides, increasing bone cell-adhesion and proliferation, or differentiation of collagen I and fibronectin [56].

7.2.1.1.2 Soft Tissue

This tissue category is referring to the tissue that connects, support, or surrounds other structures of the body besides bones. It includes tendons, ligaments, muscles, fascia, fibrous tissues, skin, nerves, and blood vessels. While hard tissues display small deformation, in soft tissue large deformation may occur. Mechanical properties of soft tissues can be simulated by some synthetic origin materials, like elastomers (PU is a genuine candidate for their substitution) [56]. There are several studies that recorded the PU used in muscle frame and vascular tissue engineering and nerve repair. To design scaffolds that simulate the mechanical properties of skeletal muscle, *Riboldi and coworkers* proposed the use of Degrapol™, a degradable block poly(ester urethane), obtained by electrospinning; thus, it is been realized a fibrous structure similar to native tissue [57]. Moreover, Degrapol™ were coated with collagen, fibronectin, and matrigel to simulate the native extracellular matrix. Hopefully, results were obtained with a low concentration of matrigel and collagen-coated Degrapol™ slides. In other studies, these authors changed some parameters for preparing oriented microfibrous scaffolds [58]. The main observation of performed tests showed that cells grown on oriented membranes displayed increase metabolic activity as compared to cells cultured on randomly oriented membranes. Still, the cell proliferation rate was similar for all scaffolds.

7.2.2 NATURAL BIODEGRADABLE POLYMERS

Natural polymers are biopolymers formed in nature during the growth cycles of the organisms. They are obtained by polymerization reactions of activated monomers catalyzed by enzyme, or growth chain, that are prepared in cells involving complex metabolic processes. In the following paragraphs, we will summarize the main classes of natural biodegradable polymers.

Polysaccharides are a class of polymeric materials of natural (animal, plant, and algal) origin formed by glycosidic linkages of monosaccharides. Depending on the nature of the monosaccharide unit, polysaccharides are classified as linear or branched-chain polysaccharides. Polysaccharides have different reactive functional groups in their chemical structure, such as hydroxyl, amino, and carboxylic acid groups, showing the possibility for chemical modification. Polysaccharides, having plant origin, are starch, cellulose, hemicellulose, hyaluronic, alginate, and guar gums, while the animal origin polysaccharides are chitin and chondroitin. Polysaccharides are found to be amorphous polymers. Research interests are highlighted on green polysaccharides for their abundance renewability, good biocompatibility and biodegradability, and nontoxicity. Natural biodegradable polymers are being used as binders in tablets, viscosity enhancers in liquids or emulsions, as a coating agent to eliminate the unpleasant taste of a drug and to improve drug stability, and to modify the amount and release rate of the drug [59]. Latter, polysaccharides are used as carriers for drug delivery in biomedical applications. Many research works are highlighted on linking polysaccharides and drugs in order to eliminate the side effects of the commonly used synthetic polymers [60].

Starch is formed in the chloroplasts of green leaves and amyloplasts of seeds, fruits, and tubers. Sources of starch include cereal grains like corn, wheat, sorghum, rice, and tubers or roots such as cassava, potato, tapioca, which are all sources of dietary carbohydrates [61]. Starch is employed in its native form or in the modified form. Native starch refers to starch in its natural, unmodified state, as extracted from its plant source, while modified starch is one in which certain properties have been modified for matching the desired applications [62]. The starch molecule consists of two major types of polymers, namely, amylose and amylopectin. In addition, starch also contains other noncarbohydrate components such as lipids, protein residues, and a small amount of minerals (calcium, magnesium, phosphorus, potassium, and sodium). The biodegradability and cost-friendly characteristics of starch make it an important raw material for many industrial applications. Still, native starches are limited in their direct application due to poor solubility in water and a strong tendency for decomposition and retrogradation. Other drawbacks are represented by high instability to temperature changes, pH, and shear forces [63].

Some issues may trouble the digestion degree of native starch and, hence, possible pharmaceutical application. These include amylopectin: amylose ratio, amylopectin chain length, degree of crystallinity, and intermolecular association in granules [64]. Modification typically affects all these properties, and the choice of modification can lead to customization and flexibility in starch use. Also, these modifications are done to develop specific properties such as solubility, hydrophobicity, thermal stability, amphiphilicity, mechanical strength, adhesion, and tolerance to high temperatures used in industrial applications. The main strenghts for which starch is investigated include its water adsorptive capacity, its behavior under agitation and high temperature, and the chemical modification of the molecule. For example, acetylated starch has some advantages as a fiber or film-forming polymer in comparison with pristine starch. Starch acetate has higher solubility compared to starch and is easier to cast into films from solvents. Also, the acetylation degree is slightly controlled by transesterification, enabling polymers to be produced with a range of hydrophobicities. Starch can also be modified with nonpolar groups, like fatty esters before the isocyanate reaction, to improve its reactivity degree [65].

Cellulose is the most abundant natural biopolymer [66]. Different natural fibers like cotton and plants have cellulose as their principal element. Cellulose is formed by long chains of anhydro-D-glucopyranose units (AGUs), each molecule having three hydroxyl groups in AGU, with the exception of the terminal ends. Cellulose is insoluble in water and most common solvents, due to the strong intramolecular and intermolecular hydrogen bonding between the chains [67]. Still, this natural polymer is used in a wide range of applications such as coatings, packing, paper, drug delivery, etc. Chemical modification of cellulose is performed to improve processability and to produce cellulose derivatives that can be tailored for specific applications. Thus, cellulose derivatives are reproducible, recyclable, and biocompatible, being used in

biomedical applications, such as blood purification membranes, drug release [68].

The most important cellulose derivatives are reaction products of one or more of the three hydroxyl groups, which are present in each glucopyranoside repeating unit, such as:

1. ethers: methylcellulose and hydroxyethyl cellulose,
2. esters: cellulose acetate and cellulose xanthate, which are used as soluble intermediates for processing cellulose into either fiber or film forms, during which the cellulose is regenerated by controlled hydrolysis, and
3. acetals especially the cyclic acetal formed between the C2 and C3 hydroxyl groups and butyraldehyde [69].

Chitin is a macromolecule found in the shells of crabs, lobsters, shrimps, and other insects, composed of 2-acetamide-2-deoxy-b-d-glucose through the b-(1-4)-glycoside linkage. Fibers of chitin were used in artificial skin substitutes (wound treatment) and absorbable sutures [70–72].

Chitin is an insoluble polymer, but its partly deacetylated form—*chitosan*—is soluble in water. These polymers are biodegradable and biocompatible, having also antimicrobial activities and the ability to chelate heavy metal ions. They were employed in the cosmetic industry due to the water-retaining characteristics and moisturizing properties. Several research works were made in medicine using chitin and chitosan as carriers in order to synthesize prodrugs soluble in water [73]. One of the most important applications of chitin derivatives is not only represented by drug carrier but also its use in absorbable sutures; that proves the lowest elongation property among the suture materials such as PGA, plain catgut, and chromic catgut [74].

Polypeptides are biomaterials composed of repeating amino acid units linked by a peptide bond. Polypeptides can display three-dimensional architectures, depending on their chemical composition. Their versatility, along with inherent biocompatibility and biological activity, makes them ideally proper for drug and gene transfer applications and in the development of tissue scaffolds [75, 76]. Polypeptides are obtained by sequential reactions of protected amino acids.

Polypeptides and proteins perform critical control and signaling functions in all tissues in the body and typically display high potency with low toxicity. Still, as pharmaceutical candidates, they present some limitations: short duration of action, limited receptor subtype selectivity, and low oral bioavailability. Current researches focused on antiviral medical direction and drug delivery approaches to answer each of these challenges [77].

Gelatin, an animal protein, is formed by 19 amino acids that are connected by peptide linkages. Gelatin can be hydrolyzed by different proteolytic enzymes to produce its amino acids or peptide components. The main gelatin sources are pig skin, bovine hide, pork, and cattle bones; the industrial use of gelatin obtained from nonmammalian species is growing in importance. Food, cosmetic, and pharmaceutical application of gelatin is based especially on its gel-forming properties [78]. Last research in the food industry, an enhancing number of new applications have been found for gelatin, including emulsifiers, foaming agents, biodegradable film-forming materials, and microencapsulating agents. Recent trends are focused on replacing synthetic agents with more natural ones. Thus, a large number of studies have discussed the enzymatic hydrolysis of collagen or gelatin for the production of bioactive peptides [79].

7.3 CONCLUSIONS AND FUTURE TRENDS AND PERSPECTIVES

Biodegradable polymers have attracted important interest all over the world in different domains such as surgery, pharmacology, agriculture, and the environment. Recent trends in this area indicate significant developments in terms of novel design strategies and engineering to provide advanced polymers with comparably good performance.

Though currently exists a wide range of biodegradable polymers that hold good potential, the idea of using well-characterized polymers (like PLGA or collagen) for biomedical applications was replaced with the utilization of newly evolved polymers that can match to certain niches (like inherent bioactivity of chitosan or DNA and RNA RNA related to phosphoesters). Furthermore, polymers combination constitutes the premise for the new materials that have desired features for specific applications. Also, later technological advances yield great promise toward obtaining biodegradability with less pollutants and greenhouse emissions.

On the other hand, future trends in the development of processing techniques will allow the obtaining of particles and scaffolds with complex architectures that can mimic the desired biological counterparts. Before that, it still needs a better understanding of how biomaterials interact with the host on cellular, tissue, organ, and systemic levels. Also, the researchers have to be more focused on improving the physiochemical performance or thermal properties of polymers, reducing their cost, and improving ease in production.

Consequently, there is a need to have a modern perspective on the design, properties, and functions of biodegradable polymers with a view to developing strategies for future developments.

ACKNOWLEDGMENT

The financial support of European Social Fund for Regional Development, Competitiveness Operational Programme Axis 1—Project "Petru Poni Institute of Macromolecular Chemistry—Interdisciplinary Pol for Smart Specialization through Research and Innovation and Technology Transfer in Bio(nano) polymeric Materials and (Eco)Technology" InoMatPol (ID P_36_570, Contract 142/10.10.2016, cod MySMIS:107464) is gratefully acknowledged.

KEYWORDS

- biodegradable
- synthetic polymers
- natural polymers
- medical applications

REFERENCES

1. Vroman, I.; Tighzert, L.; Biodegradable polymers. *Materials* **2009**, 2, 307–344.
2. Bastioli, C; Handbook of biodegradable polymers. Rapra Technology Ltd., Shawbury, Shrewsbury, Shropshire, **2005**.

3. Luckachan, G. E.; Pillai, C. K. S.; Biodegradable polymers—a review on recent trends and emerging perspectives. *J. Polym. Environ.* **2011**, 19, 637–676.

4. Siracusa, V.; Rocculi, P.; Romani, S.; Dalla Rosa, M.; Biodegradable polymers for food packaging: a review. *Trends Food Sci. Technol.* **2008**, 19, 634–643.

5. Verma, S.; Garkhal, K.; Mittal, A.; Kumar, N.; Biodegradable polymers in clinical use and clinical development. John Wiley and Sons, Inc.: Hoboken, NJ, **2011**, 565–629.

6. Jain, J. P.; Yenet Ayen, W.; Domb, A. J.; Kumar, N.; Biodegradable polymers in clinical use and clinical development. John Wiley and Sons, Inc.: Hoboken, NJ, **2011**, 1–58.

7. Ambrose, C.; Clanton, T. O.; Bioabsorbable implants: review of clinical experience in orthopedic surgery. *Annals Biomed. Eng.* **2004**, 32(1), 171–177.

8. Ceonzo, K.; Gaynor, A.; Shaffer, L.; Kojima, K.; Vacanti, C. A.; Stahl, G. L.; Polyglycolic acid-induced inflammation: role of hydrolysis and resulting complement activation. *Tissue Eng.* **2006,** 12, 301–308.

9. Pihlajamaki, H.; Salminen, S.; Laitinen, O.; Tynninen, O.; Bostman, O.; Long-term tissue response to bioabsorbable poly-l-lactide and metallic screws: an experimental study. *J. Orthop. Res.* **2006**, 24, 1597–1606.

10. Otto, J.; Binnebosel, M.; Pietsch, S.; Anurov, M.; Titkova, S.; Ottinger, A. P.; Jansen, M.; Rosch, R.; Kammer, D.; Klinge, U.; Large-pore PDS mesh compared to small-pore PG mesh. *J. Invest. Surg.* **2010**, 23, 190–196.

11. Gupta, A. P.; Kumar, V.; New emerging trends in synthetic biodegradable polymers- Polylactide: A critique. *Eur. Polym. J.* **2007**, 43, 4053–4074.

12. Jain, R. A.; The manufacturing techniques of various drug loaded biodegradable poly(lactídeo-co-glicolídeo) (PLGA) devices. *Biomaterials* **2000**, 21, 2475–2490.

13. Gentile, P.; Chiono, V.; Carmagnola, I.; Hatton, P.; An overview of poly(lactic-co-glycolic) acid (PLGA)-based biomaterials for bone tissue engineering. *Int. J. Mol. Sci.* **2014**, 15, 3640–3659.

14. Ito, Y.; Ochii, Y.; Fukushima, K.; Sugioka, N.; Takada, K.; Three-layered microcapsules as a long-term sustained release injection preparation. *Int. J. Pharm.* **2010**, 384, 53–59.

15. Xu, T.; Miszuk, J. M.; Zhao, Y.; Sun, H.; Fong, H.; Electrospun polycaprolactone 3D nanofibrous scaffold with interconnected and hierarchically structured pores for bone tissue engineering. *Adv. Healthc. Mater.* **2015**, 4, 2238–2246.

16. Nifantev, I. E.; Shlyakhtin, A. V.; Bagrov, V. V.; Komarov, P. D.; Kosarev, M. A.; Tavtorkin, A. N.; Minyaev, M. E.; Roznyatovsky, V. A.; Ivchenko, P. V.; Synthesis and ring-opening polymerization of glycidyl ethylene phosphate with a formation of linear and branched polyphosphates. *Mendeleev Commun.* 2018, 28, 155–157.

17. Molina, M.; Asadian-Birjand, M.; Balach, J.; Bergueiro, J.; Miceli, E.; Caldero ́n, M.; Stimuli-responsive nanogel composites and their application in nanomedicine. *Chem. Soc. Rev.* **2015**, 44, 6161–6186.

18. Grund, S.; Bauerand, M.; Fischer, D.; Polymers in drug delivery-state of the art and future trends. *Adv. Eng. Mater.* **2011**, 13, B61–B87.

19. Zhang, Z.; Grijpma, D. W.; Feijen, J.; Trimethylene carbonate-based polymers for controlled drug delivery applications. *J. Control. Release* **2006**, 116, e28–e29.

20. Papenburg, B. J.; Schuller-Ravoo, S.; M. Bolhuis-Versteeg, L. A.; Hartsuiker, L.; Grijpma, D.W.; Feijen, J.; Wessling, M.; Stamatialis, D.; Designing porosity and topography of poly(1,3-trimethylene

carbonate) scaffolds. *Acta Biomater.* **2005**, 5, 3281–3294.

21. Kumar, N.; Langer, R. S.; Domb, A. J.; Polyanhydrides: an overview. *Adv. Drug Del. Rev.* **2002**, 54, 889–910.

22. Kim, M. S.; Seo, K. S.; Seong, H. S.; Cho, S. H.; Lee, H. B.; Hong, K. D.; Kim, S. K.; Khang, G.; Synthesis and characterization of polyanhydride for local BCNU delivery carriers. *Biomed. Mater. Eng.* **2005**, 15, 229–238.

23. Weiner, A. A.; Shuck, D. M.; Bush, J. R.; Prasad Shastri, V.; *Biomaterials* **2007**, 28, 5259–5270.

24. Weiner, A. A.; Bock, E. A.; Gipson, M. E.; Shastri, V. P.; *Biomaterials* **2008**, 29, 2400–2407.

25. Kucinska-Lipka, J.; Gubanska, I.; Janik, H.; Polyurethanes modified with natural polymers for medical applications. Part I. Polyurethanes/chitosan and polyurethane/collagen. *Polymery* **2013**, 58, 37–43.

26. Gibas, I.; Janik, H.; Medical polyurethane with different hard segment content obtained from polycaprolactonediol, aliphatic diisocyanates and butanediol. *Polym. J. Appl. Chem.* **2009**, 53, 9–14.

27. Chen, Q.; Liang, S.; Thouas G.A.; Elastomeric biomaterials for tissue engineering. *Prog. Polym. Sci.* **2013**, 38, 584–671.

28. Chiono, V.; Sartori, S.; Rechichi, A.; Tonda-Turo, C.; Vozzi, G.; Vozzi, F.; D'Acunto, M.; Salvadori, C.; Dini, F.; Barsotti, G.; Carlucci, F.; Burchielli, S.; Nicolino, S.; Audisio, C.; Perroteau, I.; Giusti, P.; Ciardelli, G.; Poly(ester urethane) guides for peripheral nerve regeneration. *Macromol. Biosci.* **2011**, 11, 245–256.

29. Tetteh, G.; Khan, A.S.; Delaine-Smith, R.M.; Reilly, G.C.; Rehman, I.U.; Electrospun polyurethane/hydroxyapatite bioactive scaffolds for bone tissue engineering: the role of solvent and hydroxyapatite particles. *J. Mech. Behav. Biomed. Mater.* **2014**, 39, 95–110.

30. Kon, E.; Filardo, G.; Zaffagnini, S.; Di Martino, A.; Di Matteo, B.; Muccioli, G. M. M.; Busacca, M.; Marcacci, M.; Biodegradable polyurethane meniscal scaffold for isolated partial lesions or as combined procedure for knees with multiple comorbidities: clinical results at 2 years. *Knee Surg. Sports Traumatol. Arthrosc.* **2014**, 22, 128–134.

31. Burg, J. L.; Porter, S.; Kellam, J. F.; Biomaterial developments for bone tissue engineering. *Biomaterials* **2000**, 21, 2347–2359.

32. Zhu, N.; Chen, X.; Biofabrication tissue scaffolds. *Adv. Biomater. Sci. Biomed. Appl.* **2013**, 12, 315–328.

33. Rogers, L.; Said, S. S.; Meguanint, K.; The effects of fabrication strategies on 3d scaffold morphology, porosity, and vascular smooth muscle cell response. *J. Biomater. Tissue Eng.* **2013**, 3, 300–311.

34. Fare, S.; De Nardo, L.; De Cicco, S.; Jovenitti, M.; Tanzi, M. C.; Different processing methods to obtain porous structure in shape memory polymers. *Mater. Sci. Forum* **2007**, 539–543, 663–668.

35. De Nardo, L.; Bertoldi, S.; Tanzi, M. C.; Haugen, H. J.; Fare, S.; Shape memory polymer cellular solid design for medical applications. *Smart Mater. Struct.* **2011**, 20, 1–12.

36. Hou, Q.; Grijpma, D. W.; Feijen, J.; Porous polymeric structures for tissue engineering prepared by a coagulation, compression moulding and salt leaching technique. *Biomaterials* **2003**, 24, 1937–1947.

37. Haugen, H.; Ried, V.; Brunner, M.; Will, J.; Wintermantel, E.; Water as foaming agent for open cell polyurethane structures. *J. Mater. Sci. Mater. Med.* **2004**, 15, 343–346.

38. Leong, K. F.; Cheah, C. M.; Chua, C. K.; Solid freeform fabrication of three-dimensional scaffolds for engineering replacement tissues and organs. *Biomaterials* **2003**, 24, 2363–2378.

39. Bretcanu, O.; Samaille, C.; Boccaccini, A. R.; Simple methods to fabricate bioglass (R)- derived glass-ceramic scaffolds exhibiting porosity gradient. *J. Mater. Sci.* **2008**, 43, 4127–4134.
40. Boffito, M.; Di Meglio, F.; Mozetic, P.; Giannitelli, S. M.; Carmagnola, I.; Castaldo, C.; Nurzynska, D.; Sacco, A. M.; Miraglia, R.; Montagnani, S.; Vitale, N.; Brancaccio, M.; Tarone, G.; Basoli, F.; Rainer, A.; Trombetta, M.; Ciardelli, G.; Chiono, V.; Surface functionalization of polyurethane scaffolds mimicking the myocardial microenvironment to support cardiac primitive cells. *PLoS One* **2018**, 13, 1–21.
41. Mi, H. Y.; Jing, X.; Yilmaz, G.; Hagerty, B. S.; Enriquez, E.; Turng, L. S.; In situ synthesis of polyurethane scaffolds with tunable properties by controlled crosslinking of tri-block copolymer and polycaprolactone triol for tissue regeneration. *Chem. Eng. J.* **2018**, 348, 786–798.
42. Gorna, K.; Gogolewski, S.; Biodegradable porous polyurethane scaffolds for tissue repair and regeneration. *J. Biomed. Mater. Res. Part A.* **2006**, 79, 128–138.
43. Grad, S.; Kupcsik, L.; Gorna, K.; Gogolewski, S.; Alini, M.; The use of biodegradable polyurethane scaffolds for cartilage tissue engineering: potential and limitations. *Biomaterials* **2003**, 24, 5163–5171.
44. Lee, C. R.; Grad, S.; Gorna, K.; Gogolewski, S.; Goessl, A.; Alini, M.; Fibrin-polyurethane composites for articular cartilage tissue engineering: a preliminary analysis. *Tissue Eng.* **2005**, 11, 1562–1573.
45. Chia, S. L.; Gorna, K.; Gogolewski, S.; Alini, M.; Biodegradable elastomeric polyurethane membranes as chondrocytes carriers for cartilage repair. *Tissue Eng.* **2006**, 12, 1945–1953.
46. Kucinska-Lipka, J.; Gubanska, I.; Pokrywczynska, M.; Cieslinski, H.;

Filipowicz, N.; Drewa, T.; Janik, H.; Polyurethane porous scaffolds (PPS) for soft tissue regenerative medicine applications. *Polym. Bull.* **2017**, 75, 1957–1979.
47. A. R., Amini; C. T., Laurencin; S. P., Nukayarapu; *Crit. Rev. Biomed. Eng.* **2012**, 40, 363.
48. Ferrone, M. L.; Raut, C. P.; Modern surgical therapy: limb salvage and the role of amputation for extremity soft-tissue sarcomas. *Surg. Oncol. Clin. N. Am.* **2012**, 21, 201–213.
49. Li, X. M.; Wang, L.; Fan, Y. B.; Feng, Q. L.; Cui, F. Z.; Watari, F.; Nanostructured scaffolds for bone tissue engineering. *J. Biomed. Mater. Res., Part A.* **2013**, 101A, 2424–2435.
50. Huang, M. N.; Wang Y. L.; Luo, Y. F.; Biodegradable and bioactive porous polyurethanes scaffolds for bone tissue engineering. *J. Biomed. Sci. Eng.* **2009**, 2, 36–40.
51. Gorna, K.; Polowinski S.; Gogolewski, S.; Synthesis and characterization of biodegradable poly(ε-caprolactone urethane)s. I. Effect of the polyol molecular weight, catalyst, and chain extender on the molecular and physical characteristics. *J. Polym. Sci., Part A: Polym. Chem.* **2002**, 240, 156–170.
52. Heiss, C.; Kraus, R.; Schluckebier, D.; Stiller, A.; Wenisch, S.; Schnettler, R.; Bone adhesives in trauma and orthopedic surgery. *Eur. J. Trauma.* **2006**, 32, 141–148.
53. Petrie, E.; Cyanoacrylate adhesives in surgical applications. *Rev. Adhes. Adhes.* **2014**, 2, 253–309.
54. Schreader, K. J.; Bayer, I. S.; Milner, D. J.; Loth, E..; Jasiuk, I.; A polyurethane-based nanocomposite biocompatible bone adhesive. *J. Appl. Polym. Sci.* **2012**, 15, 4974–4982.
55. Li, B.; Yoshii, T.; Hafeman, A. E.; Nyman, J. S.; Wenke, J. C.; Guelcher, S. A.; The effects of rhBMP-2

released from biodegradable polyurethane/microsphere composite scaffolds on new bone formation in rat femora. *Biomaterials* **2009**, 30, 6768–6779.

56. Garcia, A. J.; Reyes, C. D.; Bio-adhesive surfaces to promote osteoblast differentiation and bone formation. *J. Dent. Res.* **2005**, 84, 407–413.

57. Riboldi, S. A.; Sampaolesi, M.; Neuenschwander, P.; Cossu, G.; Mantero, S.; Electrospun degradable polyesterurethane membranes: potential scaffolds for skeletal muscle tissue engineering. *Biomaterials* **2005**, 26, 4606–4615.

58. Riboldi, S. A.; Sadr, N.; Pigini, L.; Neuenschwander, P.; Simonet, M.; Mognol, P.; Sampaolesi, M.; Cossu, G.; Mantero, S.; Skeletal myogenesis on highly orientated microfibrous polyesterurethane scaffolds. *J. Biomed. Mater. Res.* **2008**, 84A, 1094–1101.

59. Boppana, R.; Kulkarni, R. V.; Setty, C. M.; Kalyane, N. V.; Carboxymethylcellulose—aluminum hydrogel microbeads for prolonged release of simvastatin. *ACTA Pharm. Sci.* **2010**, 52, 137–143.

60. Pushpamalar, V.; Langford, S. J.; Ahmad, M.; Hashim, K.; Lim, Y. Y.; Preparation of carboxymethyl sagopulp hydrogel from sagowaste by electron beam irradiation and swelling behavior in water and various pH media. *J. Appl. Polym. Sci.* **2013b**, 128, 451–459.

61. Pokhrel, S.; A review on introduction and applications of starch and its biodegradable polymers. *Int. J. Env.* **2015**, 4, 114–125.

62. Magallanes-Cruz, P. A.; Flores-Silva, P. C.; Bello-Perez, L. A.; Starch structure influences its digestibility: a review. *J. Food Sci.* **2017**, 82, 2016–2023.

63. Masina, N.; Choonara Y. E.; Kumar. P.; du Toit, L. C.; Govender, M.; Indermun, S.; Pillay, V.; A review of the chemical modification techniques of starch. *Carbohydr. Polym.* **2017**, 157, 1226–1236.

64. Alcázar-Alay, S. C.; Almeida Meireles, M. A.; Physicochemical properties, modifications and applications of starches from different botanical sources. *Food Sci. Technol.* **2015**, 35, 215–236.

65. Bashir, K.; Aggarwal, M.; Physicochemical, structural and functional properties of native and irradiated starch: a review. *J. Food Sci. Technol.* **2019**, 56, 513–523.

66. Bochek, A. M.; Effect of hydrogen bonding on cellulose solubility in aqueous and nonaqueous solvents. *Russ. J. Appl. Chem.* **2003**, 76, 1711–1719.

67. Gross, R. A.; Scholz, C.; Biopolymers from polysaccharides and agroproteins. American Chemical Society, Washington, DC, **2000**.

68. Akira, I.; Chemical modification of cellulose. In: Wood and Cellulosic Chemistry, Hon D. N-S., Shiraishi N., eds. Marcel Dekker: New York, NY, USA, **2001**, 599–626.

69. Ummartyotin, S.; Manuspiya, H.; A critical review on cellulose: from fundamental to an approach on sensor technology. *Renew. Sust. Energ. Rev.* **2015**, 41, 402–412.

70. Yang, T. L.; Chitin-based materials in tissue engineering: applications in soft tissue and epithelial organ. *Int. J. Mol. Sci.* **2011**, 12, 1936–1963.

71. Shigemasa, Y.; Minami, S.; Applications of chitin and chitosan for biomaterials. *Biotechnol. Gen. Eng. Rev.* **2013**, 13, 383–420.

72. Yussof, S. J.; Halim, A. S.; Saad, A. Z. M.; Jaafar, H.; Evaluation of the biocompatibility of a bilayer chitosan skin regenerating template, human skin allograft, and integra implants in rats. *Int. Scholarly Res. Net.* **2011**, 2011, 1–7.

73. Hosokawa, J.; Nishiyama, M.; Yoshihara, K.; Kubo, T.; Biodegradable film derived from chitosan and homogenized cellulose. *Ind. Eng. Chem. Res.* **1990**, 29, 800–805.

74. Bansal, V.; Sharma, P. K.; Sharma, N.; Pal, O. P.; Malviya, R.; Applications of chitosan and chitosan derivatives in drug delivery. *Adv. Biol. Res.* **2011**, 5, 28–537.

75. Gonzalez-Aramundiz, J. V.; Cordeiro, A. S.; Csaba, N.; de la Fuente, M.; Alonso, M. J.; Nanovaccines: nanocarriers for antigen delivery. *Biologie Aujourd'hui*, **2012**, 206, 249–261.

76. Grove, T. Z.; Forster, J.; Pimienta, G.; Dufresne, E.; Regan, L.; A modular approach to the design of protein-based smart gels. *Biopolymers* **2012**, 97, 508–517.

77. Wang, G.; Natural antimicrobial peptides as promising anti-HIV candidates. *Curr. Top. Pept. Protein Res.* **2016**, 13, 93–110.

78. Mariod, A. A.; Adam, H. F.; Review: gelatin, source, extraction and industrial applications. *Acta Sci. Pol., Technol. Aliment.* **2013**, 12, 135–147.

79. Nur Hanani, Z. A.; Roos, Y. H.; Kerry, J. P.; Use and application of gelatin as potential biodegradable packaging materials for food products. *Int. J. Biol. Macromol.* **2014**, 71, 94–102.

CHAPTER 8

SURFACE MODIFICATION AND ANALYSIS OF BIODEGRADABLE BIOPOLYMER MATERIALS FOR VARIOUS APPLICATIONS

ANDREEA IRINA BARZIC

"Petru Poni" Institute of Macromolecular Chemistry, Laboratory of Physical Chemistry of Polymers, 41A Grigore Ghica Voda Alley, 700487 Iasi, Romania

Corresponding author. E-mail: irina_cosutchi@yahoo.com

ABSTRACT

The surface properties of biodegradable polymers need adjustment through specific treatments in order to fulfill the criteria imposed by biomaterials. A good starting point for developing an efficient biocompatible material relies on treating the surface of a material that already displays good biofunctionality and bulk characteristics. In this way, its applicability is extended in various domains, such as bioengineering, microfluidics, textiles, bioelectronics, etc. Continued innovations in surface engineering led not only to novel processing routes of biodegradable polymers but also to new investigation techniques that emphasize the peculiarities created on their surfaces. Advancements regarding the processing and modification methodologies used in polymer science are described. Also, some relevant analysis tools for surface properties are shortly reviewed. The practical importance of surface modification of biodegradable biopolymers in food packaging, biosensors, and medicine is briefly presented, highlighting the future perspectives for the next generation of biomaterials with biodegradability and tuned surface features.

8.1   INTRODUCTION

Polymer materials characterized by biocompatibility and biodegradability are representing the best alternative for replacing plastics that severely damage the environment and implicitly the health of living beings [1–3]. Similar to common polymers, the biocompatible or eco-friendly ones need to satisfy certain standards for use in practice. Generally, the aimed properties can be accomplished by means of chemical structure and utilization of custom processing methodologies [4]. However, particular applications require not just the control of the bulk properties but also for greater performance, it is of paramount importance to adjust the surface features of the polymer coating or film [5, 6]. This is mainly imperative since polymers are known to display low surface energy, not to mention that the majority of plastics are hydrophobic [7–9]. As a consequence, these aspects diminish the possibility of macromolecular material to interact with other materials. For example, in the fabrication of an electronic device for technical uses or medical diagnosis, the polymer should adhere to other surfaces like metallic ones to ensure the stability and functionality of the device [10, 11]. Another example is referring to the case of bringing in contact the polymer to an organism or a biological medium (cell, blood, etc.) [6, 12]. In such situations, the response of the living body to the biopolymer layer is affected by its surface characteristics. If the biological response in regard to the introduced material is very strong then the polymer role will be significantly perturbed and it will be harder to perform the function. Polymer membranes for haemodialysis cannot function properly, if in the presence of blood a clot appears on the biomaterial surface so its surface characteristics must be modified [13, 14]. Contact lenses will negatively influence the cornea if the biopolymer surface is not wettable with tears [15]. Low adhesive capacity among a percutaneous device and the epidermis layer might determine infection in the free space found in between [6]. The yield of the immunoadsorbent element, which is decided by the ability of the material surface to interact with pathogenic compounds for their elimination, will quickly fail when protein adsorption and cell adhesion from the biological fluid environment hinder the immunoactive binding centers [6, 16, 17]. Biopolymers can be subjected to various types of surface treatments in order to render an antimicrobial characteristic [18–20].

All aforementioned examples clearly indicate that commonly employed materials do not entirely have the surface features necessary for biomaterials. A material having a less biocompatible surface can determine a powerful reaction from the biological medium in which the polymer is placed. Based on these aspects, it is presumable that the majority of conventional materials necessitate surface modification if they are to be utilized as biomaterials in various applications, such as medicine, bioengineering, food packaging industry, etc. It is rather odd that certain industrial polymeric materials were employed for a long time in the biomedical field lacking a surface treatment; they were just subjected to the purification procedure to expel toxic traces from the bulk of the compound [6].

Any biopolymer, which is intended to come in immediate or indirect contact with the human organism or biological fluids, must display outstanding properties both in bulk and surface; however, it is very scarce that a macromolecular compound with adequate bulk characteristics also has the surface features requisite for a biomaterial. So, based on such considerations, it is fundamental to apply a surface treatment for polymer materials prior to their use in medicine. Adaptation of surface characteristics is performed for at least two reasons: one is to confer biocompatibility to the material surface and the other to afford it the desired functionality (like specific physiological activity,

antimicrobial properties, cellular adhesion, etc.).

The next generation of biopolymers must be imperiously exposed to surface modification or alteration, imparting the pursued short- and long-term effects for efficient functionality. While augmenting the performance of the initial polymer, surface processing also expands the application scope of materials as a result of their tunable bulk features, including elasticity, optical clarity, tensile strength, and density. Surface-processed polymers have known progressive utility in numerous domains [21–29]. Figure 8.1 illustrates an overall picture concerning the practical importance of surface-treated biopolymers.

The wide spectrum of chemical and physical features for various families of known polymeric materials provides many possibilities to select from, so it is facile for researchers to elect and modify polymers for a targeted application. Among them, biodegradable and biocompatible polymers are receiving growing attention. These materials are imparted, based on their origin, into the following three classes [4]:

1. naturally synthesized renewable polymers,
2. synthetic polymers originated from renewable resources, and
3. synthetic polymers derived from petroleum-based resources.

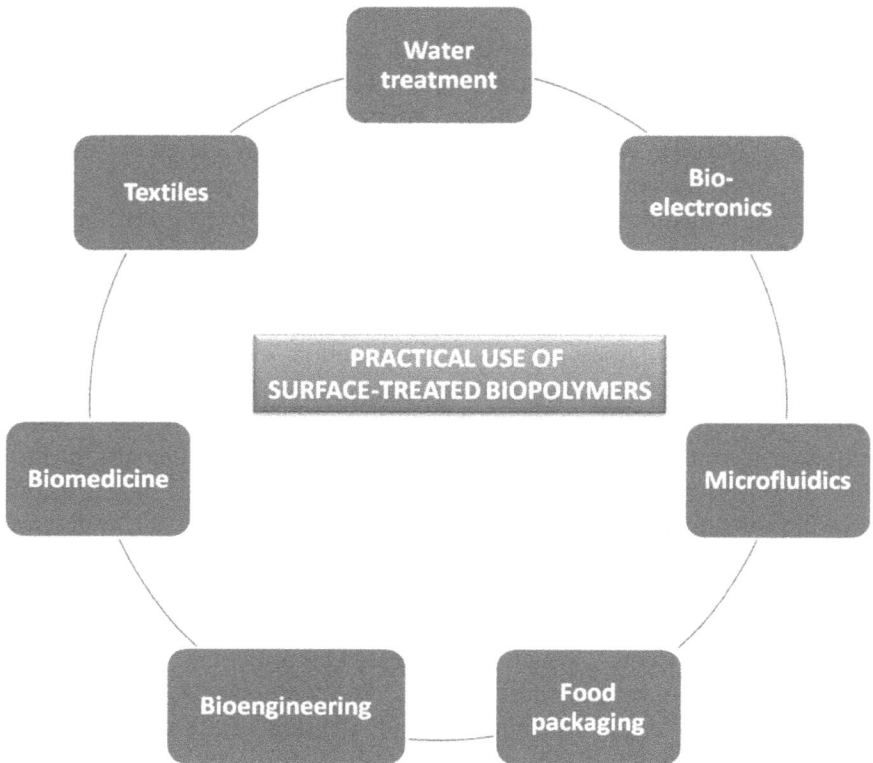

FIGURE 8.1 A schematic picture of the main applicative domains of the surface-treated biopolymer materials.

So, it is essential to present an outlook on various approaches of surface treatment and alteration (or processing) of biopolymers with a particular focus on their domain of utilization. The benefits, lifetime, and durability of created properties for the peculiar method are explored. In addition, the most used investigation procedures in the analysis of the peculiarities generated on the material surfaces are reviewed. The effort is to collate all the significant surface adjustment methods and elucidate their impact on the host polymer characteristics based on the pursued practical purpose.

In the following sections, the surface treatment techniques will be discussed that are widely used to adapt material surfaces, particularly thin biopolymer coatings or films, followed by some surface analysis methods and applicability in biosensors, food packaging, and biomedicine.

8.2   METHODS OF SURFACE MODIFICATION

Generally, surface modification of polymers can serve mainly three categories of purposes that include [7]:

1. surface functionalization,
2. surface cleaning or etching, and
3. surface deposition.

For a clearer image of the effects created by all three mentioned phenomena, they are sketched in Figure 8.2.

On a larger scale, surface modification or alteration of a polymer layer can be comprised in the following categories [30, 31]:

1. chemical procedures rely on chemical agents that "attack" the surface by creating novel functional groups on the material surface,

2. physical procedures involve mainly radiations to manufacture surface topographies with different scales (macro, micro, and nano), and

3. biological procedures are dealing with adsorption or seeding biological molecules of the polymer support.

The presented surface treatments can be accomplished through a broad range of processes, the most usual paths are to adjust and control the surface energy of the material, to enhance or reduce its adhesive character, roughness, absorbing, or releasing features.

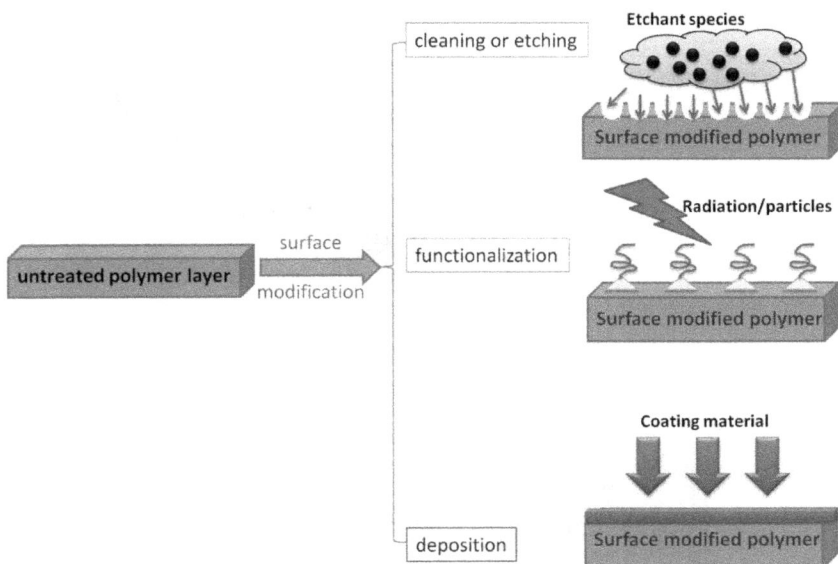

FIGURE 8.2   Representation of the main kinds of surface modification applied to biopolymer materials.

8.2.1 PHYSICAL PROCEDURES OF BIOPOLYMER SURFACE MODIFICATION

Surface alteration of polymer films or coatings by physical methods is a relatively facile approach, without much expense and scalable. Moreover, the application of such a technique does not harm the environment since it does not involve the utilization of chemical agents.

8.2.1.1 FLAME TREATMENTS

The method of flame treatment has the main goal to create introduce oxygen-containing polar groups (hydroxyls, carboxyls, or carbonyls) on the surface of hydrophobic plastics, like polypropylene, polyvinyl chloride, polyethylene (PE) terephthalate, or polystyrene [32–34]. With such treatment of surfaces, bond strength and durability is considerably enhanced. The procedure has a considerable influence on material wettability and adhesive abilities.

The setup designed for flame treatments is comprising a burner and a fuel tank, not complicated to construct and to work with. Figure 8.3 depicts the principle of flame surface modification used for

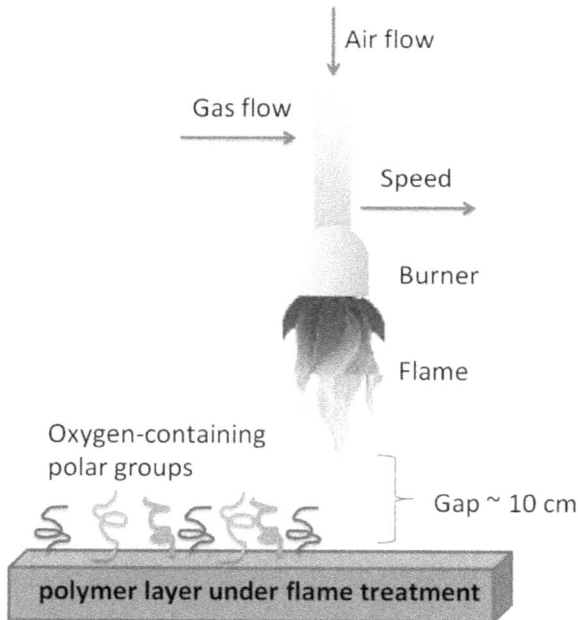

FIGURE 8.3 Illustration of general setup utilized in flame surface modification of the polymer materials.

polymer materials. This inexpensive technique is largely used in the packaging industry to modify the surface of blow-molded bottles and thick plastic coatings. Particular attention must be ascribed to control of flame temperature, contact time, flame composition (air-to-gas ratio), flow rate of the air-gas blend, and the space between flame and plastic material surface. All these factors are affecting the extent of the surface alteration [35]. Extreme caution must be done to avoid the contamination of modified material via chemical residual elements or partially burned compounds which can arise in the course of sample processing. Conversely, the material risks partially or entirely deteriorate the adhesion ability as a result of the occurrence of weak boundary layers at the interface. If the polymer is surface treated with a flame during an excessive period of time, then the sample could burn. For this reason, the time of surface modification using this method should be less than 1 s.

8.2.1.2 CORONA DISCHARGE AND PLASMA TREATMENTS

Plasma is considered the fourth state of matter and nowadays is encountered in numerous fields, such as displays, nuclear reactors, medical devices, propulsion turbines, and even the lighting industry [36]. There are several categories of plasma sources, and among them, the common ones are [37]:

1. gaseous plasma sources: radio frequency glow discharge, electron cyclotron resonance glow discharge, corona discharge, atmospheric arc glow discharge,
2. vacuum arc plasma sources, and
3. laser plasma sources.

The peculiarities of these plasma sources, namely, the electron temperature, ion temperature, electron density, and uniformity, are influenced by the plasma source and other experimental conditions (gas nature, applied voltage, pressure, reactor geometry, etc.).

Corona discharge is another type of plasma and it mainly represents an electrical discharge that is triggered by the ionization of a gas or fluid covering a conductor which is electrically charged [36]. The phenomenon occurs at atmospheric conditions (i.e., higher pressure) in contradistinction to cold plasma that needs a vacuum. Corona cab is described as a flux of charged particles, like electrons and ions that are accelerated in the presence of an electric field. This sort of plasma is produced when a space gap containing air or other gases is exposed to a certain magnitude of high voltage to provoke a chain reaction of high-speed particle bumps with neutral molecules generating in this way more ions. Therefore, the plastic sample is subjected to a discharge yield by high-frequency and high-voltage alternating current that action its wettability and roughness. So, corona discharge is often employed to treat the polymer foils with a positive impact on their adhesive characteristics.

Low temperature or cold plasma is made of highly energetic molecules (molecular segments, ions, radicals, and free electrons) that necessitate low pressure to ensure prevalent colliding with the polymer surface to cause major changes in surface properties. Upon application of the plasma-assisted procedure to a plastic one may notice that significant modifications in surface chemistry, wettability, and morphology take place expanding its use to a larger range of applications with the additional benefit that the characteristics of the bulk polymer are kept intact.

Plasma-supported surface alteration includes a range of techniques, which according to the literature [37] can be divided as follows:

1. Sputtering and cleaning species produced by inert or reactive gases reach the solid polymer surface and have sufficient energy to transfer to atoms on the superficial layer which in turn they detach from the surface. If the physical degradation of the bulk polymer is confined to the outermost shell, the process is known as sputtering [38]. So, the material is cleaned from surface impurities, producing volatile substances.

2. Etching is taking place when continuing with deeper misplacement of the plasma-altered material. This process involves the adsorption of energetic particles, followed by product occurrence, prior to product desorption [39].

3. Implantation introduces elements into the surface of the plastic foil in the absence of thermodynamic constraints [40].

4. Deposition refers to the process of placing though plasma of a thin layer of the desired compound on the polymer support [41]. The most used techniques included here are dual plasma deposition, plasma polymerization, plasma-grafting copolymerization, laser ablation, and plasma spraying [37].

Several kinds of surface reactions take place when a solid polymer surface is placed under plasma action [42–46]. All these have an important impact on several surface features such as optical, electrical, tribological, biocompatibility, surface energy, morphological, and textural [12, 37, 47–49]. Figure 8.4 depicts a scheme of synergistic interactions and processes occurring during plasma treatment of polymers, which affect the material evolution and performance. All illustrated processes are competitive but for a better understanding, they

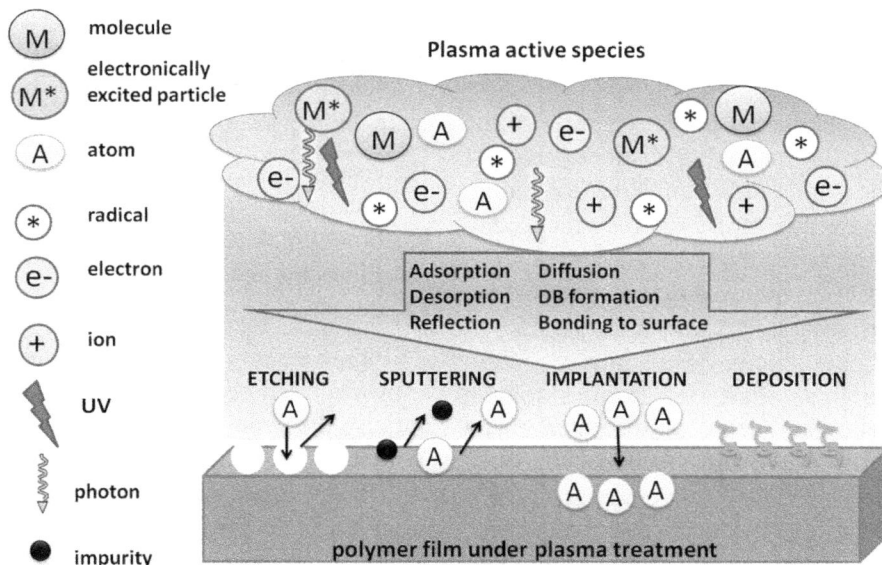

FIGURE 8.4 Schematic representation of the synergistic surface interaction processes occurring during plasma exposure of polymers.

are separately drawn. The presented image excludes the effect of reactor geometry on the plasma deposition procedure. A detailed investigation concerning this aspect can be found in the work of Whittle et al. [50], where an international consortium discusses data achieved on coatings processed in 14 distinct plasma reactors.

8.2.1.3 RADIATION EXPOSURE

Radiation can be considered a form of energy that arises from a source and moves through a medium or through space. Radiation is emitted from an atom when the electrons fall from a higher energy level to a lower

one. Based on the literature reports on polymer surface tuning using radiation sources, they might be classified based on their ionizing and coherence characteristics [51–53].

Ionizing radiations are generated by unstable atoms possessing an excess of energy or mass and sometimes both. Such radiations have high-energy radiations, thus they have the potential to extract electrons from the material resulting ions [54]. Conversely, nonionizing radiations are low-energy electromagnetic radiations that cannot remove electrons during impact with the material [55]. Both types of radiation can damage living organisms and might produce changes to the natural environment.

Among them, ionizing radiation is the most destructive since its very high energy.

Incoherent radiations refer to the fact that phases of distinct waves of the same radiation have a random phase difference. On the other hand, coherent radiation (i.e., laser) denotes that the phases of two waves are differing by a known constant [56].

Radiations can interact with polymer materials leading to the following three main phenomena (absorption, transmission, and reflection), which in turn, can generate a series of physicochemical reactions. Depending on the nature of incident radiation and experimental conditions (time of exposure, dose of irradiation, energy, etc.), the possible main categories of reactions are:

1. cross-linking,
2. ablation,
3. bond breaking,
4. functionalization, and
5. degradation or aging.

Figure 8.5 depicts not only a well-known classification of radiations but also the general effects produced by them on polymer materials.

The principle of radiation treatment is based on high-energy species, which are involved in the activation

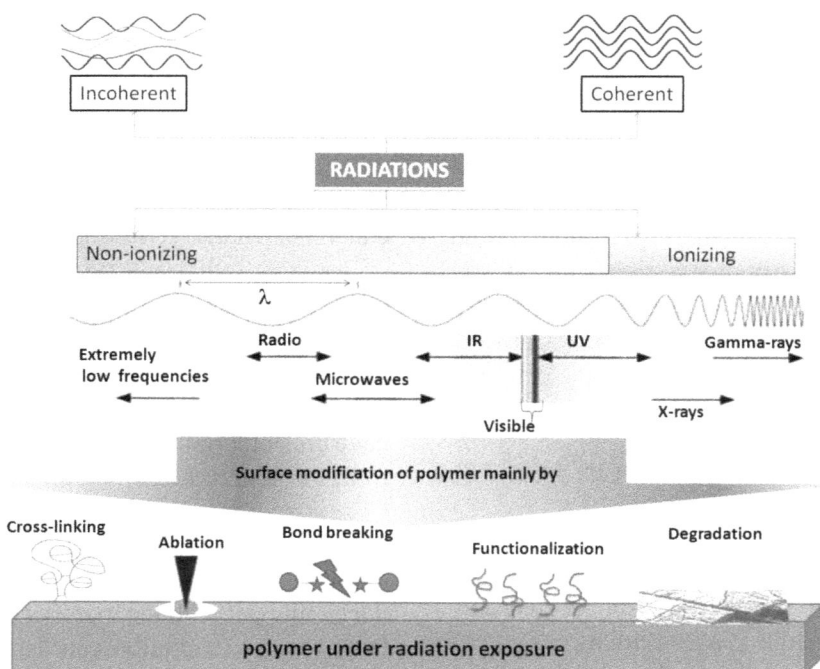

FIGURE 8.5   Representation of the types of radiations and their general effects created on the polymer surface.

of several sorts of chemical reactions. Some radiations (microwave and infrared) induce thermal effects on the materials, while others (ultraviolet, visible) also generate optical effects (photochemical reactions, electron excitation). Radiations with high energy (X-ray, gamma-rays) are able to disrupt atomic bonds and in this way, they have destructive effects on biomacromolecules, like DNA or RNA.

When discussing UV effects, it is important to mention that the chemical reactions under radiation exposure take place in the presence of an initiator or a sensitizer that absorbs the incident radiation. Thus, it is changed into an excited singlet state, which quickly is changed into a highly stable triplet state. In this state, the initiator is able to abstract atoms from the polymer film surface and therefore to produce active places for further chemical modification of the surface.

Gamma-ray treatments are employing moderate energy (100–300 kV) electrons to alter surface polymer features through a mechanism of free-radical formation creating active sites, which might react with oxygen and other reactive atoms. This is expected to positively impact adhesion and the mechanical characteristics of polymers [5].

Laser treatments of polymers mainly produce surface ablation and in specific conditions (fluence magnitude, number of pulses, polymer absorption features), a surface structuring can be noticed [57, 58]. If selecting a particular fluence range, upon treatment one may notice a carbon-enriched layer on the polymer film surface, the carbonized material is shown to be distinct from that which is deposited in the neighboring ablation crater [57].

8.2.1.4   PARTICLE BEAM TREATMENT

The particle beam method involves a flux of electrons or ions, which can produce changes in the surface of the polymer surface.

Electron beams are accelerated in an external field (under vacuum), so the high energy particles are determining the excitation or ionization of molecules encountered on the polymer superficial layers. These accelerated particles with low mass are traveling at the speed of light and they can pass through a thin metal window and reduce their energy. When reaching in the gaseous medium, they lead to free radicals that start chemical reactions. In this way, several functional species appear which can be further deposited on top of films of electron beam-modified polymers. The reactions produced by free radicals are affected by the acceleration voltage necessary for the electron flux, the features of the gas, the peculiarities of the polymer structure, as well as the external reaction conditions (temperature, pressure, etc.) [59]. An essential precaution relies on proper shielding during this procedure against X-rays arising from the inter- action of accelerated particles with the material. This surface-modifi- cation technique does not involve the use of photoinitiators since high-speed particles bombarding the surface create radicals are useful for initiation and propagation of other chemical reactions (polymerization, cross-linking, grafting, curing) [60]. Such surface adaptation procedure has a good impact on mechanical, thermal, wettability, permeability, printability, and antistatic behavior of the modified polymer coating.

In ion beam treatments, the parti- cles are speeded under vacuum and by application of a positive voltage. When particle beam reaches in the process chamber where the polymer film is placed, they bump with the atoms from the superficial macro- molecular chains and compel them to leave the structure, determining the appearance of several vacancies. Afterward, ions can be implanted in the newly formed spots on the polymer surface, generating a layer abundant in implanted material. Remarkable improvements in the polymer surface adhesivity, energy, printability, and sealability are noted after ion beam treatments [61]. The depth at which ions can infiltrate is dependent on their energy and mass. If the ions are not properly acceler- ated then they are not able to enter deeply in the solid polymer, while at sufficient high energy, these particles can penetrate up to micrometric level inside the sample. If accelerated ions are smaller (low mass) then they can enter deeply into the polymer. The characteristics of the polymeric material (i.e., molecular weight, structural ordering, cross-linking, etc.) also influence the infiltration depth of ions. Figure 8.6 shows a schematic picture of the setup for electron and ion beam surface treat- ment of polymer materials.

8.2.1.5 MECHANICAL TREATMENT

Surface features of polymers may be also adapted by mechanical treatments. Application of strong deformation forces can cause rear- rangement of polymer chains or even material displacement from the superficial layers of the polymer sample. Mechanical roughening of the surface can be achieved by using microrough surfaces to furbish polymer surfaces (Figure 8.7). The depth of the alteration is determined by the malleability of the material (textile, sandpaper, metal brush, etc.) placed in contact with the sample that needs treatment [62, 63]. If the desired surface alteration must be

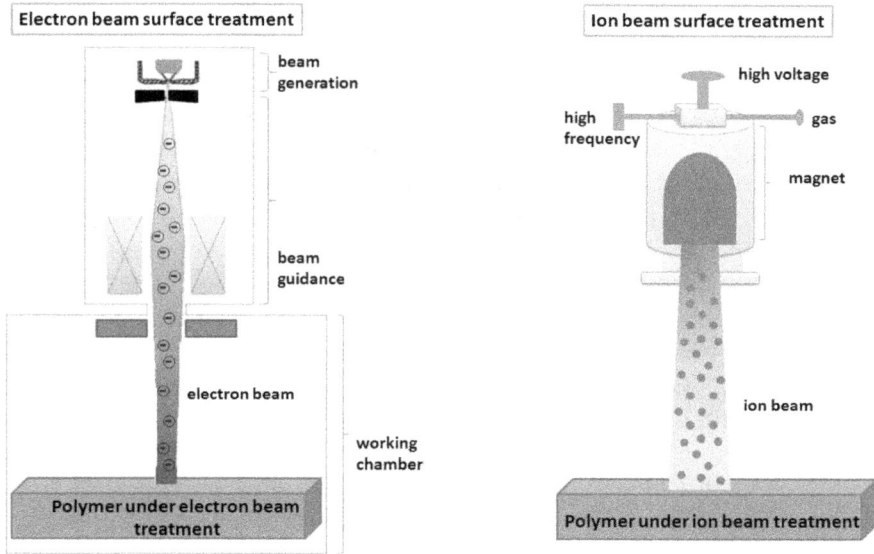

FIGURE 8.6 General principle of setup for electron and ion beam surface treatment of polymer materials.

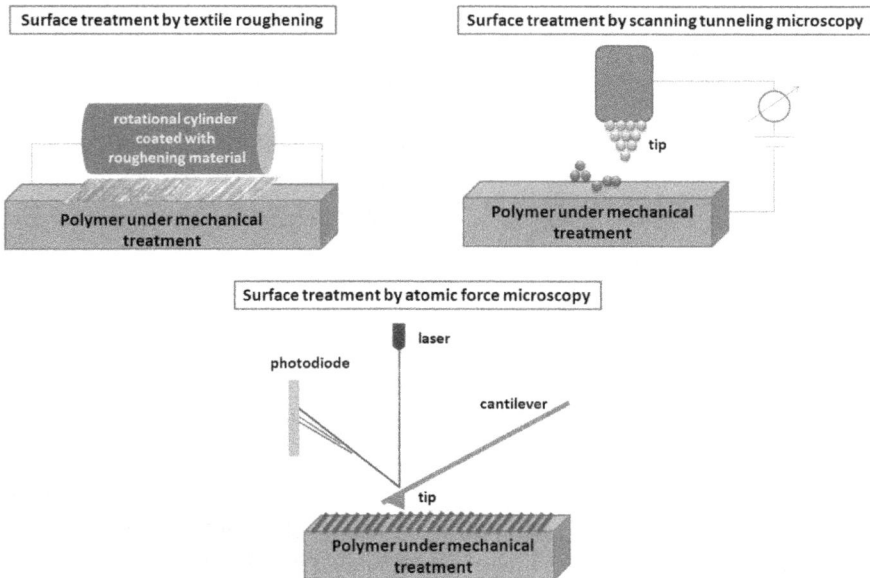

FIGURE 8.7 Schematic illustration of mechanical surface treatment of polymers by roughening, scanning tunneling microscopy, and dynamic plowing lithography.

made at a nanometric scale, one may use scanning tunneling microscopy or atomic force microscopy (Figure 8.7) [64–66]. In such situations, the homogeneity of the sample and its structural organization are essential factors in obtaining a uniform surface treatment.

8.2.2 CHEMICAL PROCEDURES OF BIOPOLYMER SURFACE MODIFICATION

Chemical modification approaches produce an alteration of the polymer surface without meaningfully impacting its bulk properties. The most common chemical procedures are as follows:

1. chemical vapor deposition,
2. grafting techniques,
3. self-assembled monolayers,
4. wet chemical etching (WCE), and
5. specific chemical treatments to modify groups from sample superficial layers (oxidization, sulfonation, chlorination, acetylation, and quaternization).

From the point of view of the interaction occurring at the surface, chemical modification of surface methods can be imparted as follows:

1. covalent: nonspecific oxidation, reduction reaction, addition reaction,
2. noncovalent: solvent coatings, surface active additives,

Such techniques are helpful for a situation where the majority of physical methods are not entirely adequate or inapt to render the needs of the pursued applications, particularly in the biomedical field, where the improved properties are achieved without changing the surface roughness. The majority of chemical surface adaptation approaches implicate wet procedures where the sample to be treated is dipped or coated/ sprayed with a chemical substance to outperform the initial surface features, as well as to remove debris and microbes, sterilizing the sample for biomedical purposes. WCE is adequate for increase wettability by localizing oxidized functional groups on the sample. The method is efficient owing to the deep penetration of the solvent into the polymer pores facilitating treatment of the material at a wide scale without many expenses.

Wet chemicals allow selective surface modification of materials and thus precise design for targeted applications. The procedure can be performed with a broad range of reagents to selectively modify polymers at a large scale, demanding a meticulous approach since the rate of reaction is influenced by the reagent reactivity, polymer sample composition, and period of treatment. Furthermore, additional steps like rinsing, washing, and drying are demanded prior to the further processing of the polymer, which invariably enhances the quantity of hazardous waste resulted during a single surface modification operation. Therefore, this surface processing method is good as long as the application is noninvasive and has low side effects, like proactive etching, and variations in the bulk crystalline phase of the polymer are not of main concern in the pursued application.

8.2.3 BIOLOGICAL PROCEDURES OF BIOPOLYMER SURFACE MODIFICATION

This third category of surface modification methods arises from the need to develop polymer surfaces with appropriate biopotential. The interactions occurring at the material interface with biological medium (cells, proteins, enzymes, peptides, ligands, lipids, DNA, RNA, blood, etc.) have to be adapted through specific techniques to ensure the desired evolution of the biological processes.

Among the attempts to tailor the surface biocompatible features of various polymeric materials, one may mention [30, 67]:

1. physical adsorption of molecules of biological nature,
2. self-cross-linking or self-assembling of biological molecules,
3. cell seeding and growth to confluence, and
4. chemical conjugation for immobilization of biological molecules.

As in the case of the above-described physical and chemical methods, biological surface processing determines alteration of the surface energy thus affecting the interfacial interactions. It was noted that certain types of polymer surfaces display similar free surface energy and a very distinct chemical nature [68, 69]. It is evident that surface tension is not a singular factor that impacts the bioconductive characteristics of the polymer coating. Lots of functional groups (OH, COOH, $C=O$, NH_2) are responsible for regulating the fate of attached biological molecules on polymer support [70]. For instance, the capacity of macrophages to produce giant, multinuclear cells on the superficial layers of a polymer hydrogel can be linked to the presence of some chemical groups [71]. Surface adaptation by biological means of synthetic polymers presenting adequate mechanical features with potential interfering particles similar to those of the extracellular matrix (ECM) leads to combined advantages of synthetic and natural materials. This enables to mimic stimulate the interaction at the biointerface with cells analogous to the interactions with some ligands of the ECM [72]. Apart from chemical functional groups, matrix proteins (fibrin, collagen, and fibronectin) could be mobilized on the synthetic polymer substrate. The basic idea of the biomimetic procedure was focused on resembling selected characteristics in the structure or functions of the natural extracellular microenvironment [73].

This group of approaches also refers to protein adsorption, enzyme immobilization, cell presowing, and others. Smaller biologically active compounds, like amino acid parts and integrin receptors of the cells, can be also included in polymer surface adaptation.

Biomacromolecules such as polysaccharides are adequate for increasing cellular adhesion in addition to adhesive peptides. The interfacial interactions are enhanced by the adsorption of homopolymers of some amino acids (i.e., polylysine). Immobilization of cellulosic polymers on synthetic polymeric surfaces influences cell adhesive abilities and over-the-surface functions.

8.3 SURFACE ANALYSIS AND APPLICABILITY OF SURFACE-TREATED POLYMER BIOMATERIALS

The modifications created by means of various methods should be closely examined to understand the relationship between surface adapted features and newly created properties and their impact on applications. The main surface investigation methods can be categorized as a function of the pursued property:

1. wettability: contact angle method,
2. topography: microscopy techniques (optical microscopy, atomic force microscopy, scanning electron microscopy, etc.),
3. adhesion: contact angle method, atomic force microscopy, peeling tests,
4. surface chemistry: infrared spectroscopy (FTIR), energy dispersive X-ray spectroscopy, X-ray photoelectron spectroscopy, Auger electron spectroscopy,
5. mechanical properties: atomic force microscopy,
6. electrical properties: electric force microscopy, and
7. magnetic properties: magnetic force microscopy.

In the following paragraphs, the advancements in surface-modified polymers in several applicative domains are reviewed, including the utility of some of the aforementioned surface analysis methods.

8.3.1 FOOD PACKAGING

Biopolymers with adapted surface features represent a good alternative for next-generation packaging materials. The recent trends revealed that petrochemical biostable polymers can be outperformed by biodegradable polymers packages particularly those from natural resources.

Siracusa et al. [74] subjected biodegradable poly(lactic acid) (PLA) to superficial treatments by metallization with pure aluminum and investigated the efficiency of the prepared materials from the point of view of their gas barrier behavior, thermal stability, and mechanical performance. They showed that the PLA-coated samples display a large variety in gas permeability properties, varying suddenly from a low (for air and C_2H_4) to a very high (N_2O and CO_2) permeable material.

Ke et al. [75] discussed the applicative potential in this field of polyhydroxyalkanoate (PHA) materials. The wet chemical approaches especially act on the ester bonds generating new functionalities. Alkaline treatments are useful for changing the polyhydroxybutyrate surface by inducing the formation of carboxylic groups that are compatible with the proliferation of human osteoblasts and impede the development of microorganisms, like S. aureus [75]. Chemical initiating-graft polymerization of vinyl monomers onto poly(hydroxybutyrate-co-hydroxyvalerate) led to the formation of diverse functional groups with an impact on wettability and functionality of the surface-engineered materials. Exposure of PHAs to gamma radiation generates chain scission and chain cross-linking reactions. The latter is mainly taking place in the amorphous regions or in the boundary zone from amorphous and crystalline parts of the material. Young's modulus presents a slight increase owing to the enhancement of crystallinity of the irradiated polymers [76]. Besides the mechanical properties of PHA packaging materials, it is very important to elucidate their UV resistance. The high UV barrier of this kind of biodegradable plastics is ideal to protect unsaturated lipid counterparts in food from the formation of radicals, therefore facilitating their quick spoilage [77].

Another procedure to design packaging materials aims to modify the surface of classical polymer foils with biodegradable macromolecules like chitosan [78, 79]. To augment the interactions at the interface of the materials, a corona or dielectric barrier discharge (DBD) plasma was applied. Theapsak et al. [78] exposed PE foils to DBD, which afterward were immersed in chitosan acetate solution. Atomic force microscopy indicated surface roughening induced by plasma, while Fourier transformed infrared spectroscopy (FTIR) and XPS investigations pointed out that the PE surface became rich in oxygen-containing functional groups. Chitosan-coated polymer films presented good antibacterial activity against E. coli and S. aureus, which is compatible for food packaging purposes. In the same trend, Panaitescu et al. [80] modified poly(3-hydroxybutyrate) by reinforcing with biodegradable nanocellulose and adapting surface properties by plasma irradiation or ZnO nanoparticle plasma coating. Plasma treatment maintained the thermal stability and mechanical resistance regardless of the added percent of nanoparticles, while augments the antibacterial behavior against *E. coli* and *S. aureus*. In addition, ZnO plasma-assisted coating showed that enabled a full inhibition growth of *S. aureus*.

Cellulose materials can be surface treated to achieve food packaging materials [81, 82]. Graft copolymerization onto cellulose membranes is suitable for binding silver nanoparticles. The entrapment of the antibacterial nanoagent was demonstrated by transmission electron microscopy and selected area electron diffraction. The new cellulosic material displays show strong antibacterial properties and are good for food science purposes [82]. Ahmadi et al. [83] used carboxymethyl cellulose foils and altered the surface properties by UV-induced and chemical cross-linking reactions. In both cases, the surface wettability, water barrier, and mechanical properties were changed and more obviously for photo-cross-linked samples. Surface morphology studies were done by using scanning electron microscopy revealed that microcracks were eliminated on the neat polymer after the cross-linking procedure. The cytotoxicity tests highlighted that the cellulosic materials correspond to food regulations criteria. Dairi et al. [84] prepared multicomponent systems based on cellulose acetate for packaging applications. Cellulose acetate was mixed with triethyl citrate in solution in which was subsequently introduced thymo (Th) additive. The system was further doped with silver nanoparticles/gelatin-modified montmorillonite (AgM). Increasing amounts of these additives into cellulosic matrix changes its surface properties. Also, UV barrier and oxygen barrier properties were more efficient, particularly when clay was prevalent in the composition. Antioxidant and antibacterial investigations indicated that these materials are promising active packaging products. In recent work, Sengupta and Han [85] presented some approaches to create special surfaces of biopolymer packaging products inspired by biomimetics. A particular case is that of superhydrophobicity, which is reflected in the combined effect of surface roughness and hydrophobicity to determine water expelling after contact with the polymer. This is useful for avoiding the development of microorganisms on the packaging material and future perspectives in this field should take nature models applied to the biodegradable polymer to accomplish products with reduced impact on the environment.

8.3.2   BIOMEDICINE

The ability to bind a substance to the surface of a polymer while keeping its bulk features is an excellent approach for designing products for pharmaceuticals or bioengineering.

Given the inert surface of polymers, biomedical applications require reconditioning of their surface energy, and topography in order to attain the desired interaction with biological molecules. Surface treatments should have an influence mainly on the topmost layer of the surface. There are several paths to alter the surface of polymers to enhance their functionality for the following purposes [6]:

1. development of lubricious surface materials,
2. designing of blood-compatible materials,
3. preparation physiologically active surface, and
4. fabrication of tissue-bonding surfaces.

All these aspects will be discussed below in order to understand what surface treatment is more suitable for each bioapplication.

Polymers having lubricious surfaces are highly desirable in the production of tubular medical devices (catheters, cannulae, endoscopes, and cytoscopes), which are introduced in the organism orifices open to the outside. During such medical procedures, the plastic material will endure frictional resistance in contact with the mucous membrane and thus will damage it. Consequently, lubrication is essential. Such a surface can be achieved by graft polymerization of nonionic aqueous-soluble reactants on the polymeric tube. The lowering of the coefficient of friction between a polymeric surface and a glass in water is attained when the surface was graft polymerized with dimethyl acrylamide [6]. A large amount of water content in the superficial grafter layer led to the slipperiness of the device surface. Fast squeezing-out of water from grafted hydrogel sheet by low pressure renders the impression of slipperiness. Zhou et al. [86] showed that biodegradable polyurethanes can reach advanced hydrophilic and lubricious properties upon surface grafting with poly(acrylic acid). Attenuated total reflection Fourier transformed infrared spectroscopy (ATR-FTIR) combined with microscopy techniques analyses supported good hydrophilic and lubricious performance, particularly after pretreatment with ozone. A recent report by Zhang et al. [87] presented a versatile route for the surface alteration of polymer hydrogels via Cu^0-mediated controlled radical polymerization. When a Cu^0 plate is brought in the vicinity of initiator-bearing hydrogel surfaces and in the attended by ligand and monomer, it starts to quickly consume dissolved oxygen present in the reaction system, and afterward, it acts as a source of catalyst. In this way, it determines the fast growth of hydrogel-bound macromolecular chains. Three sorts of functional surfaces were obtained through this procedure, namely, a protective hydrogel having a hydrophobic surface, a lubricious hydrogel, and, finally, a hydrogel with thermally controlled frictional characteristics.

Blood compatible surfaces represent another important goal in the surface engineering of biopolymers. When designing a blood-contacting biomedical device, the interfacial interactions must not interfere with blood flow, should not produce platelet aggregation and fibrin network formation. Biopolymers, which are brought in direct contact with blood, are utilized for short-term and long-term applications. In practical situations, when the plastic material is surrounded by fresh blood for a couple of hours as needed for extracorporeal circulation, then an anticoagulant is administered to the blood to avoid coagulation. For the cases that require longer times of interactions such as the case of plastic containers for clinical tests or transfusions, chelating agents are introduced to the biological fluid to remove calcium ions. If the blood-contacting device needs to operate for a larger time (weeks), slow deliverance of anticoagulant from the device or immobilization of fibrolytic enzymes on the surface is done. Such treatment is often employed for the catheters that are involved for intravenous hyperalimentation. No effective approaches of surface adaptation are reported by

the polymer surfaces in permanent interaction with flowing blood. The medical devices for such long-term uses refer to artificial heart or valves.

Morimoto et al. [88] have prepared polyurethane materials with improved surface hemocompatibility via a chemical procedure of the treatment with 2-methacryloyloxyethyl phosphorylcholine by creating a semi-interpenetrating polymer network. It was proved that these materials prevented platelet adhesion from being suitable for the biomedical field. The immobilization of bioactive agents on the polymer surface allows drugs to freely interact with the enzymes, proteins, or receptors present at the blood/polymer interface. The benefit would rely on the permanent effect of the drug thus rendering its pharmacological role without affecting the systemic circulation [89]. Polymer surfaces can be modified with several biomedical substances to ensure hemocompatibility: heparin (avoids fibrin occurrence), prostaglandins (influences platelet activation), and PGE1-heparin conjugate. Yang and Lin [90] modified the surface of polyacrylonitrile by grafting with chitooligosaccharides followed by the immobilization of heparin. Lan et al. [91] subjected to sulfonation a nanofibrous cellulose acetate for creating a heparin-rich surface. Chandy and Sharma [92] discussed the antithrombotic processes on albuminate polymer having prostaglandin attached to their surface. More detailed aspects of immobilization of such drugs on various surfaces can be found in the literature [93, 94].

Design of physiologically active polymer surfaces is desirable for applications where physiologically active functions like molecular recognition and catalysis are needed. For such purposes, macromolecules are mainly employed since many synthetic polymers do not possess the adequate physiological activity as compared to naturally formed biopolymers. The enzymatic activity can be accomplished by covalent immobilization of such molecules onto polymer surfaces with considerable impact on the future manufacturing of biosensors, bioreactors, and artificial cells [95, 96]. A great issue in enzyme immobilization arises from the lowering of enzymatic activity during the binding procedure. To solve this problem, space is introduced between the polymer surface and the enzyme molecule. Flexible spacers are not ideal since they reduce the stability of the enzyme [6]. Other applications related to physiologically active polymer surfaces are concerning immunoadsorbent surfaces. These are important in assisting the metabolic functions of the organism like those exerted by the liver. Activated carbons are useful for removing toxins from the blood. Since their

elimination via physical adsorption to the carbon surface is narrowed to certain cases, more selective adsorbents were investigated, mostly immobilizing peptides, polypeptides, or polysaccharides on the adsorbents [6, 97]. Some of them present the benefit of the biospecific antigen–antibody interaction. Furthermore, immunoadsorbents that can be able to selectively link the toxins of autoimmune diseases were intensely investigated. The toxin of such diseases is believed to be the antibody of patients. In the case of systemic lupus erythematosus, a treatment alternative could be based on immobilization of DNA on molecules on an ultrafine PE terephthalate fiber by grafting of polyacrylic acid. The latter was chosen since it presents a very big specific area in regard to the conventional microbeads employed as adsorbents. This method determined immunoglobulin G elimination from the serum of a rat suffering from this disease [6].

Altering polymer surfaces for tissue-bonding purposes is a hot topic of actual interest in polymer science and biomedicine [98–111]. A huge variety of polymer scaffolds have been prepared for both soft and hard tissues. Most relevant studies in the past two decades on this topic are briefly discussed. Zhu et al. [98] adapted surface features of polycaprolactone via aminolysis creating a surface rich in NH_2 groups as proved by spectroscopy methods. The existence of pores with big depth on polymer membrane allowed a reaction to occur as deep as 50 μm. Biomolecules like gelatin, chitosan, or collagen could be further attached on altered polymer surfaces in the presence of a cross-linking agent, as showed by XPS data. The changes in contact angle and wettability could be correlated with the good cytocompatibility results. Elvira et al. [99] used plasma- and chemical-induced graft polymerization to change the properties of starch scaffolds. FTIR, XPS, and contact angle measurements revealed that the biodegradable polymer coating was chemically modified. Preliminary cell adhesion and proliferation analyses indicated outstanding improvement in regard to initial nonsurface adapted starch-based scaffolds. Hsieh [100] grafted chitosan surface with biodegradable poly(vinyl alcohol) chains as supported by FTIR study. Scanning electron microscopy imaging revealed that the prepared material presents a porous tridimensional architecture. In vitro tests demonstrated excellent cell proliferation and growth in the biopolymer scaffold, which is susceptible to enzymatic degradation. Li et al. [101] reviewed the most suitable surface treatment methods for processing biopolymers that are used in vascular scaffolds. Surface adaptation procedures

mainly refer to the attachment of gelatin, heparin, peptides, and other biomacromolecules on hydrophilic surfaces like that of poly(ethylene glycol) and zwitterion polymers. Liu et al. [102] improved collagen surface features by chemical modification with chondroitin sulfate. The proposed surface grafting procedure led to the enhancement of moisture capacity and mechanical properties. The new biopolymer scaffold seems to be suitable for the proliferation of human corneal epithelial cells as proved by in vitro tests. Daskalova et al. [103] prepared chitosan scaffolds by laser-assisted surface modification. Scanning electron microscopy analysis indicated that for more pulses, the material surface has a degree of microporosity. The generated surface texturing allowed us to strengthen and accelerate cell proliferation and tissue organization. Courtenay et al. [104] changed the surface of bacterial cellulose layers by chemical means by reacting with glycidyltrimethylammonium chloride or by oxidation with sodium hypochlorite, without degrading the mechanical performance. In this way, positive or negative charges are introduced in the superficial layers of the scaffold. The cationic cellulose coatings augmented cell attachment by 70% in regard to the pristine polymer. The oxidized cellulose films present a small degree of cell attachment almost similar to initial cellulose. Nedela et al. [105] discussed the most employed surface modification procedures applied to polymers for biomedicine, namely, laser irradiation, ion implantation, plasma exposure, and grafting. They show that depending on the polymer substrate properties, these techniques can lead to periodic patterns on the surface, thus enabling a controlled growth and orientation of cells on such scaffolds. They point out that besides the development of new polymer materials, it is also useful to combine surface techniques for achieving scaffolds with unique properties, having a great impact on tissue replacements or organ reconstruction. Jaganathan and Prasat [106] used UV irradiation to modify the surface of metallocene PE for tissue-bonding purposes. Microscopy imaging allowed observation of crests and troughs, which illustrates the increase of surface roughness of polymer substrates provoked by UV etching. The contact angle decreased as a result of better hydrophilicity upon radiation exposure. Biological measurements indicated good blood compatibility and proliferation of seeded 3T3 cell population. In the work, Sengupta and Prasad [107, 108] transformed a hydrophobic polyetherimide surface into a hydrophilic one by layer-by-layer assembly technique implicating gold nanoparticles and arginine. The latter has the role of a sticky protein that facilitates viable cell accommodation. The murine fibroblasts present higher adherence and accommodation on the arginine-modified polymer films in comparison with other commercial tissue culture plates. Teixeira [109] and Ino [110] prepared scaffolds based on poly(vinyl alcohol) presenting considerable potential for skin, bone, cartilage, vascular, neural, and corneal tissue regeneration. Glow discharge can be used to functionalize this biodegradable polymer. XPS and FTIR studies show that microwave plasma led to the formation of carbonyl and nitrogen groups, while the basic structure of polymer remains unaltered. Plasma surface processing generates the augmentation of the surface wettability without relevant variation of roughness. Surface treated films enabled the development of mouse fibroblasts and human endothelial cells. These data pointed out that the grafting can be considered as being stable upon rehydration and led to good cell adhesive properties. Therefore, plasma amination of biodegradable polymers represents a good route for designing synthetic scaffolds for tissue engineering. In a more recent paper, Asadian et al. [111] present the technique for fabrication of scaffolds by nonthermal plasma surface adaptation of nanofibrous supports based on biodegradable polyesters. The positive influence of plasma activation and polymerization treatments on the nanofiber features such as wettability, surface chemistry, cell adhesion, migration, proliferation, and protein grafting is reviewed for various biodegradable polymers.

8.3.3   BIOSENSORS

Polymers with controlled surface characteristics can be also used in biosensor applications. Some relevant studies in this research area are shortly described. Reimhult and Höök [112] highlight the importance of texturizing molecular films for better controlling of the interactions of biological molecules with nanoscale biosensor surfaces. It is essential to take into consideration that the performance of the biosensor is significantly affected by the substrate material properties and its meticulous surface alteration for targeted applications. So, first one should comprehend the principles of a biosensor to recognize the substrate material and its surface alteration chemistry. The mandatory surface adaptation for the binding of biomolecules without damaging their bioactivity is vital for sensitive detection. Numerous surface processing technologies are developed in agreement with the kind of the substrate material. Such techniques of surface alteration of materials employed as platforms for biosensor purposes are electrochemical modification,

covalent modification, and physical adsorption [113]. Heinemeyer et al. [114] proposed a surface modification methodology to fabricate biosensor chips based on chemically sensitive polymers. They used chemical and plasma activation of some polymers for immobilization of biomolecules that are prepared for application in surface acoustic wave biosensors. Alemán and collaborators [115] attempted to enhance the biodetection process by polymer surface modification and/or functionalization through plasma exposure. Processes occurring at the surface like etching, activating, or crosslinking are ideal for creating electrocatalytic species that are capable to encourage the oxidation of certain bioanalytes and/or gas traces that are incompatible with human health. Some advances regarding sensors used in biomedical field applications (e.g., neurotransmitter, temperature, pH, pressure, and glucose sensors) are reported. In particular, sensors made for biomolecules and living systems are focused on the recognition of biochemical molecules, such as simple electrolytic salts, neurotransmitters, and enzymes [116]. Among these devices, special attention is ascribed to the organic electrochemical transistor, which was utilized for enzymatic sensing involving dedoping mechanisms and/or field-effect transistor (FET) principles [117]. In addition, the application of plasma technology enhances FET reaction of a flexible parylene support under the action of simple electrolytes [118]. On the other hand, biosensors for glucose monitoring and detection of other physiologically relevant chemicals are essential for treatment of brain diseases. Classical glucose tracking devices rely on amperometric detection performs specific recognition by interacting with an enzyme known as glucose oxidase (GOx). The latter is responsible for catalyzing the oxidation reaction of glucose to gluconolactone [115]. Maekawa et al. [119] projected a better biosensor for glucose by binding an enzyme on a poly(dimethylsiloxane) coating expose to an oxygen glow discharge. The surface treatment of polymer helped to overcome issues mainly encountered in enzyme-based sensors, like low reproducibility, tricky immobilization processes, and elevated costs, while making easier the integration of the GOx. The designed biosensor is able to measure glucose in the concentration range of 0.02 and 1.8 mM and does not present errors determined by interferents

8.4 CONCLUSION

The inert character of most polymer surfaces demands the utilization of surface adaptation procedures in order to accomplish the pursued properties imposed by various applications, particularly those of biological nature. Physical, chemical, and biological surface alteration approaches used in polymer science are reviewed. A short enumeration of the most relevant analysis techniques useful for emphasizing the peculiarities produced on biopolymer surface is made. The practical implication of surface treats biodegradable polymers. The applicability of several categories of such biomaterials with surface features adapted through one of the aforementioned methods is discussed, especially for food packaging, biomedicine, and biosensors.

Despite plenty of positive progress in the field of surface engineered polymers for biomedicine, future work should concern the following aspects, namely:

1. overcoming aspects of hydrophobic recovery of plasma-treated polymers,
2. deeper investigation of induced effects on the metabolic activity of the seeded cells on nanofibrous scaffolds,
3. clarification of the discrepancies reported between the in-vitro and in-vivo performance of surface adapted scaffolds, and
4. closer examination of the correlation between wettability, the regularity of surface features, and local mechanical properties of polymer support with cell proliferation.

ACKNOWLEDGMENT

This work is dedicated to the 71st Anniversary of "Petru Poni" Institute of Macromolecular Chemistry of Romanian Academy of Sciences.

KEYWORDS

- biodegradable polymer
- surface modification
- surface analysis
- interface
- applications

REFERENCES

1. Serwańska-Leja, K.; Lewandowicz, G.; Polymer biodegradation and biodegradable polymers—a review. *Pol. J. Environ. Stud.* **2010**, *19*, 255–266.
2. Wróblewska-Krepsztul,J.;Rydzkowski, T.; Borowski, G.; Szczypiński, M.; Klepka T.; Thakur, V.V.K.; Recent progress in biodegradable polymers and nanocomposite-based packaging materials for sustainable environment. *Int. J. Polym. Anal. Charact.* **2018**, *23*, 383–395.
3. Zeng, S.H.; Duan, P.P.; Shen, M. X.; Xue, Y.J.; Wang, Z. Y.; Preparation and degradation mechanisms of biodegradable polymer: a review. *IOP Conf. Ser.: Mater. Sci. Eng.* **2016**, *137*, 012003.
4. Ashter, S.; Processing biodegradable polymers. In: *Introduction to Bioplastics Engineering*. William Andrew–Elsevier: Oxford, **2016**, 179–209.
5. Ozdemir, M.; Yurteri, C.U.; Sadikoglu, H.; Physical polymer surface modification methods and applications in food packaging polymers. *Crit. Rev. Food Sci. Nutr.* **1999**, *39*, 457–477.
6. Ikada, Y.; Surface modification medical applications. *Biomaterials* **1994**, *15*, 725–736.
7. Nemani, S.K.; Annavarapu, R.K.; Mohammadian, B.; Raiyan, A.; Heil, J.; Haque, M.A.; Abdelaal, A.; Sojoudi, H.; Surface modification of polymers: methods and applications. *Adv. Mater. Interfaces.* **2018**, 1801247.
8. Yu, L.Y.; Zhu, B.; Cai, X.; Wang, Y.W.; Han, R.H.; Li, Y.W.; Review of polymer surface modification method. *Mater. Sci. Forum* **2016**, *852*, 636–631.
9. Thakur, M.K.; Rana, A.K.; Liping, Y.; Singha, A.S.; Thakur, V.K.; Surface modification of biopolymers: an overview. In: *Surface Modification of Biopolymers*. Thakur, V.K.; Singha, A.S.; Eds.; Wiley: Hoboken, NJ, USA, **2015**, 1–19.
10. Behl, M.; Seekamp, J.; Zankovych, S.; Torres, S.; Zentel, R.; Ahopelto, J.; Towards plastic electronics: patterning semiconducting polymers by nanoimprint lithography. *Adv. Mater.* **2002**, *14*, 588–591.
11. Soroceanu, M.; Barzic, A.I.; Stoica, I.; Sacarescu, L.; Ioanid, E.G.; Harabagiu, V.; Plasma effect on polyhydrosilane/metal interfacial adhesion/cohesion interactions. *Int. J. Adhes. Adhes.* **2017**, *74*, 131–136.
12. Popovici, D.; Barzic, A.I.; Stoica, I.; Butnaru, M.; Ioanid, G. E.; Vlad, S.; Hulubei, C.; Bruma, M.; Plasma modification of surface wettability and morphology for optimization of the interactions involved in blood constituents spreading on some novel copolyimide films. *Plasma Chem. Plasma Process.* **2012**, 32, 781–799.

13. Lin, W.-C.; Liu, T.-Y.; Yang, M.-C.; Hemocompatibility of polyacrylonitrile dialysis membrane immobilized with chitosan and heparin conjugate. *Biomaterials* **2004**, *25*, 1947–1957.

14. Alibeik, S.; Sask, K.N.; Blood compatible polymers. In: *Functional Biopolymers*. Mazumder, M.A.J.; Sheardown, H.; Al-Ahmed, A.; Eds.; Springer: Cham, **2019**, 149–189.

15. Deng, X.M.; Castillo, E.J.; Anderson, J.M.; Surface modification of soft contact lenses: silanization, wettability and lysozyme adsorption studies. *Biomaterials* **1986**, *7*, 247–251.

16. Hlady, V.V.; Buijs, J.; Protein adsorption on solid surfaces. *Curr. Opin. Biotechnol.* **1996**, *7*, 72–77.

17. Holubnycha, V.; Kalinkevich, O.; Ivashchenko, O.; Pogorielov, M.; Antibacterial activity of in situ prepared chitosan/silver nanoparticles solution against methicillin-resistant strains of *Staphylococcus aureus*. *Nanoscale Res. Lett.* **2018**, *13*, 71.

18. Bespalova, Y.; Kwon, D.; Vasanthan, N.; Surface modification and antimicrobial properties of cellulose nanocrystals. *Appl. Polym. Sci.* **2017**, *134*, 44789.

19. Amirsoleimani, M.; Khalilzadeh, M.A.; Sadeghifar, F.; Sadeghifar, H.; Surface modification of nanosatrch using nano silver: a potential antibacterial for food package coating. *J. Food Sci. Technol.* **2018**, *55*, 899–904.

20. Greenhalgh, R.; Dempsey-Hibbert, N. C.; Whitehead, K. A.; Antimicrobial strategies to reduce polymer biomaterial infections and their economic implications and considerations. *Int. Biodeter. Biodegr.* **2019**, *136*, 1–14.

21. Tu, Q.; Wang, J.-C.; Zhang, Y; Liu, R.; Liu, W.; Ren, L.; Shaofei, S.; Xu, J.; Zhao, L.; Wang, J.; Surface modification of poly(dimethylsiloxane) and its applications in microfluidics-based

22. Jocic, D.; Smart textile materials by surface modification with biopolymeric systems. *Res. J. Text. Apparel* **2008**, *12*, 58–65.

23. Feron, K.; Lim, R.; Sherwood, C.; Keynes, A.;, Brichta, A.; Dastoor, P.C.; Organic bioelectronics: materials and biocompatibility. *Int. J. Mol. Sci.* **2018**, *19*, 2382.

24. Bose, S.; Robertson, S.F.; Bandyopadhyay, A.; Surface modification of biomaterials and biomedical devices using additive manufacturing. *Acta Biomater.* **2018**, *66*, 6–22.

25. Vartiainen, J.; Vähä-Nissi, M.; Harlin, Ali.; Biopolymer films and coatings in packaging applications—a review of recent developments. *Mater. Sci. Appl.* **2014**, *5*, 708–718.

26. Khwaldia, K.; Arab-Tehrany, E.; Desobry, S.; Biopolymer coatings on paper packaging materials. *Compr. Rev. Food Sci. Food Saf.* **2010**, *9*, 82–91.

27. Saqib, J.; Aljundi, I. H.; Membrane fouling and modification using surface treatment and layer-by-layer assembly of polyelectrolytes: state-of-the-art review. *Water Process. Eng.* **2016**, *11*, 68–87.

28. Stoica, I.; Barzic, A.I.; Butnaru, M.; Doroftei, F.; Hulubei, C.; Surface topography effect on fibroblasts population on epiclon-based polyimide films. *J. Adhes. Sci. Technol.* **2015**, *29*, 2190–2207.

29. Tejero, R.; Anitua, E.; Orive, G.; Toward the biomimetic implant surface: biopolymers on titanium-based implants for bone regeneration. *Progr. Polym. Sci.* **2014**, *39*, 1406–1447.

30. Hoffman, A. S.; Surface modification of polymers: physical, chemical, mechanical and biological methods. *Macromol. Symp.* **1996**, *101*, 443–454.

31. Fabbri, P.; Messori, M.; Surface modification of polymers: chemical, physical, and biological routes. In:

Modification of Polymer Properties. Jasso-Gastinel, C.F..; Kenny, J. M.; Eds.; William Andrew–Elsevier: Oxford, **2017**, 109–130.

32. Briggs, D.; Brewis, D. M.; Konieczko, M.B.; X-ray photoelectron spectroscopy studies of polymer surfaces. III. Flame treatment of polyethylene. *J. Mater. Sci.* **1979**, *14*, 1344–1348.

33. Garbassi, F.; Occhiello, E; Polato, F.; Surface effect of flame treatments on polypropylene. *J. Mater. Sci.* **1987**, *22*, 207–212.

34. Pascoe, R. D., O'Connell, B.; Flame treatment for the selective wetting and separation of PVC and PET. *Waste Manag.* **2003**, *23*, 845–850.

35. Sutherland, I.; Brewis, D. M.; Heath, R. J.; Sheng, E.; Modification of polypropylene surfaces by flame treatment. *Surf. Interface Anal.* **1991**, *17*, 507–510.

36. Ebnesajjad, S.; Surface treatment of polyvinyl fluoride films and coatings. In: *Polyvinyl Fluoride Technology and Applications of PVF—A Volume in Plastics Design Library.* Ebnesajjad, S.; Ed.; William Andrew Publishing: Norwich, **2013**, 193–212.

37. Chu, P. K.; Chen, J. Y.; Wang, L. P.; Huang, N.; Plasma-surface modification of biomaterials. *Mater. Sci. Eng. R Rep.* **2002**, *36*, 143–206.

38. Chapman, B.; *Glow Discharge Processes: SPUTTERING and Plasma Etching.* Wiley: Hoboken, NJ, **1980**.

39. Coburn, J. W.; Winters, H. F.; Plasma etching—a discussion of mechanisms. *J. Vac. Sci. Technol.* **1979**, 16, 391–403.

40. Han, S.; Lee, Y.; Kim, H.; Kim, G.H.; Lee, J.; Yoon, J.H.; Kim, G.; Polymer surface modification by plasma source ion implantation. *Surf. Coat. Technol.* **1997**, *93*, 261–264.

41. De Geyter, N.; Morent, R.; Cold plasma surface modification of biodegradable polymer biomaterials. In:

Biomaterials for Bone Regeneration. Novel Techniques and Applications. Dubruel, P.; Van Vlierberghe, S.; Eds.; Woodhead Publishing: Cambridge, **2014**, 202–224.

42. Wirth, B. D.; Hammond, K. D.; Krasheninnikov, S. I.; Maroudas, D.; Challenges and opportunities of modeling plasma–surface interactions in tungsten using high-performance computing. *J. Nuclear Mater.* **2015**, *463*, 30–38.

43. Sodhi, R.N.S.; Application of surface analytical and modification techniques to biomaterials research. *J. Electr. Spectrosc. Phenom.* **1996**, *81*, 269–284.

44. Oehrlein, G. S.; Plasma-polymer interactions: a review of progress in understanding polymer resist mask durability during plasma etching for nanoscale fabrication. *J. Vacuum Sci. Technol. B.* **2011**, *29*, 010801.

45. Akishev, Y.; Grushin, M.; Dyatko, N.; Kochetov, I.; Napartovich, A.; Trushkin, N.; Duc, T.; Descours, S.; Studies on cold plasma–polymer surface interaction by example of PP- and PET-films. *J. Phys. D: Appl. Phys.* **2008**, *41*, 235203.

46. Fisher, E.R.; A review of plasma-surface interactions during processing of polymeric materials measured using the IRIS technique. *Plasma Proc. Polym.* **2004**, *1*, 13–27.

47. Stoica, I.; Aflori, M.; Ioanid, E.; Hulubei, C.; Effect of oxygen plasma treatment and gold sputtering on morphological and local mechanical properties of copolyimide/gold micropatterned structures. *Surf. Interface Anal.* **2018**, *50*, 154–162.

48. Ioan, S.; Cosutchi, A.I.; Hulubei, C.; Macocinschi, D.; Ioanid, E.G.; Surface and interfacial properties of poly(amic acid)s and polyimides. *Polym. Eng. Sci.* **2007**, *47*, 381–389.

49. Popovici, D.; Vlad, S.; Stoica, I.; Vasilescu, D. S.; Evaluation of bio- and hemo-compatibility of two partially

aliphatic copolyimides modified by plasma and chemical treatment. *UPB Sci. Bull. Ser. B: Chem. Mater. Sci.* **2016**, *78*, 149–160.

50. Whittle, J. D.; Short, R. D.; Steele, D. A.; Bradley, J. W.; Bryant, P. M.; Jan, F.; Biederman, H.; Serov, A. A.; Choukurov, A.; Hook, A. L. Variability in plasma polymerization processes–an international Round-Robin study. *Plasma Process. Polym.* **2013**, *10*, 767–778.

51. Valentin J. The recommendations of the international commission on radiological protection, *Ann. ICRP.* **2007**, *37*(2–4), 1–332.

52. Hamada, N.; Fujimichi, Y.; Classification of radiation effects for dose limitation purposes: history, current situation and future prospects. *J. Radiat. Res.* **2014**, *55*, 629–640.

53. Sporea, D.; Sporea, A.; Radiation effects in optical materials and photonic devices. In: *Radiation Effects in Materials.* Monteiro, W.A.; Ed.; InTech: Rijeka, **2016**, 37–67.

54. Ferry, M.; Ngono-Ravache, Y.; Aymes-chodur, C.; Clochard, M.C.; Coqueret, X.; Cortella, L.; Pellizzi, E.; Rouif, S.; Esnouf, S.; Ionizing radiation effects in polymers. In: *Reference Module in Materials Science and Materials Engineering*, Elsevier: Amsterdam, **2016**, 1–28.

55. Ng, K. H.; Non-ionizing radiations-sources, biological effects, emissions and exposures. *Proceedings of the International Conference on Non-Ionizing Radiation at UNITEN* **2003**, 1–16.

56. Meyerhofer, D.; Incoherent radiation and its applications (Visible, UV, X rays). In: *VLSI Electronics Microstructure Science.* Einspruch, N.G.; Cohen, S.S.; Singh, R.N.; Eds.; Elsevier: London, **1989**, 439–476.

57. Barzic, A. I.; Stoica, I.; Popovici, D.; Ursu, C.; Gradinaru, L. M.; Hulubei,

C.; Physico-chemical insights on tuning the morphology of a photosensitive polyimide by UV laser irradiation. *Mater. Plast.* **2013**, *50*, 88–92.

58. Bujak, K.; Sava, I.; Stoica, I.; Tiron, V.; Topala, I.; Weglowski, R.; Schab-Balcerzak, E.; Konieczkowska, J.; Photoinduced properties of "T-type" polyimides with azobenzene or azopyridine moieties. *Eur. Polym. J.* **2020**, *126*, 109563.

59. Schiller, S.; Heisig, U.; Panzer, S.; Electron beam radiation techniques. In: *Electron Beam Technology.* Wiley: New York, NY, USA, **1982**, 463–500.

60. Garbassi, F.; Morra, M.; Occhiello, E.; Physical modifications. In: *Polymer Surfaces from Physics to Technology*, Wiley: Chichester, **1994**, 223–241.

61. Kondyurin, A.; Bilek, A.; *Ion Beam Treatment of Polymers*, 2nd ed. Elsevier: Amsterdam, **2014**.

62. Stoica, I.; Barzic, A. I.; Hulubei, C.; The impact of rubbing fabric type on surface roughness and tribological properties of some semi-alicyclic polyimides evaluated from atomic force measurements. *Appl. Surf. Sci.* **2013**, *268*, 442–449.

63. Nechifor, C.-D.; Postolache, M.; Albu, R.M.; Barzic, A.I.; Dorohoi, D.; Induced birefringence of rubbed and stretched polyvinyl alcohol foils as alignment layers for nematic molecules. *Polym. Adv. Technol.* **2019**, *30*, 2143–2152.

64. Stoica, I.; Barzic, A. I.; Hulubei, C.; Fabrication of nanochannels on polyimide films using dynamic plowing lithography. *Appl. Surf. Sci.* **2017**, *426*, 307–314.

65. Albrecht, T.; Dovek, M.; Lang, C.; Grutter, P.; Quate, C.; Kuan, S.; Frank, C.; Pease, F.; Imaging and modification of polymers by scanning tunneling and atomic force microscopy. *J. Appl. Phys.* **1988**, *64*, 1178–1184.

66. Nyffenegger, R. M.; Penner, R. M.; Nanometer-scale surface modification

using the scanning probe microscope: progress since 1991. *Chem. Rev.* **1997**, *97*, 1195–1230.

67. Ratner, B.D.; Surface modification of polymers: chemical, biological and surface analytical challenges. *Biosens. Bioelectron.* **1995**, *10*, 797–804.

68. Lydon, M.J.; Minett, T.W.; Tighe, B.J.; Cellular interactions with synthetic polymer surfaces in culture. *Biomaterials* **1985**, *6*, 396–402.

69. Albu, R. M.; Hulubei, C.; Stoica, I.; Barzic, A. I.; Semi-alicyclic polyimides as potential membrane oxygenators: rheological implications on film processing, morphology and blood compatibility. *eXPRESS Polym. Lett.* **2019**, *13*, 349–364.

70. Bhattacharyya, D.; Xu, H.; Deshmukh, R.R.; Timmons, R.B.; Nguyen, K.T.; Surface chemistry and polymer film thickness effects on endothelial cell adhesion and proliferation. *J. Biomed. Mater. Res. A.* **2010**, *94*, 640–648.

71. Gölander, G.; Lassen, B.; Nilsson, K.; Nilsson, U.; RF-plasma-modified polystyrene surfaces for studying complement activation. *J. Biomater. Sci. Polym. Edn.* **1992**, *4*, 25–30.

72. West, J. L.; Biofuntcional polymers. In: *Encyclopedia of Biomaterials and Biomedical Engineering.* Wnek, G.E.; Bowlin, G.L.; Eds.; CRC Press: New York, NY, USA, **2007**, 89–95.

73. Peña, B.; Martinelli, V.; JeongM.; Bosi, S.; Lapasin, R.; Taylor, M.R.G.; Long, C.S.; Shandas, R.; Park, D.; Mestroni, L.; Biomimetic polymers for cardiac tissue engineering. *Biomacromolecules* **2016**, *17*, 1593–1601.

74. Siracusa, V.; Dalla Rosa, M.; Iordanskii, A.; Performance of poly(lactic acid) surface modified films for food packaging application. *Materials* **2017**, *10*, 850.

75. Ke, Y.; Liu, C.; Zhang, X.; Xiao, M.; Wu, Gang.; Surface modification of polyhydroxyalkanoates toward enhancing cell compatibility and antibacterial activity. *Macromol. Mater. Eng.* **2017**, *302*, 1700258.

76. Yang, H.-L.; Liu, J.-J.; Thermal analysis of poly(3-hydroxybutyrate-*co*-3-hydroxyvalerate) irradiated under vacuum. *Polym. Int.* **2004**, *53*, 1677–1681.

77. Siracusa, V.; Rocculi, P.; Romani, S.; Rosa, M. D.; Biodegradable polymers for food packaging: a review. *Trends Food Sci. Tech.* **2008**, *19*, 634–643.

78. Theapsak, S.; Watthanaphanit, A.; Rujiravanit, R.; Preparation of chitosancoated polyethylene packaging films by DBD plasma treatment. *ACS Appl. Mater. Interface* **2012**, *4*, 2474–2482.

79. Munteanu, B.; Stoleru, E.; Zemljič, L.; Sdrobis (Irimia), A.M.; Pricope, G.; Vasile, C.; Chitosan coatings applied to polyethylene surface to obtain food-packaging materials. *Cell Chem. Technol.* **2014**, *48*, 565–575.

80. Panaitescu, D. M.; Ionita, E. R.; Nicolae, C. A.; Gabor, A. R.; Ionita, M.D.; Trusca, R.; Lixandru, B.-E.; Codita; I., Dinescu, G.; Poly(3-hydroxybutyrate) modified by nanocellulose and plasma treatment for packaging applications. *Polymers (Basel)* **2018**, *10*, 1249.

81. Khan, A.; Huq, T.; Khan, R.A.; Riedl, B.; Lacroix, M.; Nanocellulose-based composites and bioactive agents for food packaging. *Crit. Rev. Food Sci. Nutr.* **2014**, *54*, 163–174.

82. Tankhiwale, R.; Bajpai, S. K.; Graft copolymerization onto cellulose-based filter paper and its further development as silver nanoparticles loaded antibacterial food-packaging material. *Colloids Surfaces B.* **2009**, *69*, 164–168.

83. Ahmadi, S.; Seif, A.; Rajabzadeh, G.; Carboxymethyl cellulose film modification through surface photo-crosslinking and chemical crosslinking for food packaging applications. *Food Hydrocoll.* **2016**, *61*, 378–389.

84. Dairi, N.; Hafida, F.; Ramos, M.; Garrigós, M.; Cellulose acetate/AgNPs-organoclay and/or thymol nano-biocomposite films with combined antimicrobial/antioxidant properties for active food packaging use. *Int. J. Biol. Macromol.* **2019**, *121*, 508–523.

85. Sengupta, T.; Han, J. H.; Packaging and biopolymer materials. In: *Surface Chemistry of Food, Packaging, and Biopolymer Materials. Innovations in Food Packaging.* Elsevier: London, **2014**, 51–86.

86. Zhou, X.; Zhang, T.; Jiang, X.; Gu, N.; The surface modification of medical polyurethane to improve the hydrophilicity and lubricity: the effect of pretreatment. *J. Appl. Polym. Sci.* **2010**, *116*, 1284–1290.

87. Zhang, K.; Yan, W.; Simic, R.; Benetti, E.M.; Spencer, N.D.; Versatile surface modification of hydrogels by surface-initiated, Cu^0-mediated controlled radical polymerization. *ACS Appl. Mater. Interface* **2020**, *12*, 6761–6767.

88. Morimoto, N.; Iwasaki, Y.; Naka-bayashi, N.; Ishihara K.; Physical properties and blood compatibility of surface-modified segmented polyurethane by semi-interpenetrating polymer networks with a phospholipid polymer. *Biomaterials* **2002**, *23*, 4881–4887.

89. Jacobs, H..; Grainger, D.; Okano, T.; Kim, S. W.; Surface modification for improved blood compatibility. *Artif. Organs* **1988**, *12*, 506–507.

90. Yang, M.-C.; Lin, W.-C.; Surface modification and blood compatibility of polyacrylonitrile membrane with immobilized chitosan–heparin conjugate. *J. Polym. Res.* **2002**, *9*, 201–206.

91. Lan, P.; Wang, W.; Cao, J.; Study on heparin-like surface based on nanofibrous membrane of cellulose acetate and its blood compatibility. *Appl. Mech. Mater.* **2012**, *159*, 317–321.

92. Chandy, T.; Sharma, C.P.; The anti-thrombotic effect of prostaglandin E1 immobilized on albuminated polymer matrix. *J. Biomed. Mater. Res.* **1984**, *18*, 1115–1124.

93. Murugesan, S.; Xie, J.; Linhardt, R.J.; Immobilization of heparin: approaches and applications. *Curr. Top. Med. Chem.* **2008**, *8*, 80–100.

94. Jacobs, H.; Okano, T.; Lin, J. Y.; Kim, S. W.; PGE1—heparin conjugate releasing polymers. *J. Controlled Release* **1985**, *2*, 313–319.

95. Hoffmann, C.; Modification of polymer surfaces to enhance enzyme activity and stability. Technical University of Denmark, Kgs. Lyngby, **2017**.

96. Fischer-Colbrie, G.; Heumann, S.; Guebitz1, G.; Enzymes for Polymer Surface Modification. In: *Modified Fibers with Medical and Specialty Applications.* Edwards, J.V.; Buschle-Diller, G.; Goheen, S.C.; Eds.;. Springer: Dordrecht, **2006**, 181–189.

97. Lett, E.; Gangloff, S.; Zimmermann, M.; Wachsmann, D.; Klein, J.P.; Immunogenicity of polysaccharides conjugated to peptides containing T- and B-cell epitopes. *Infect Immun.* **1994**, *62*, 785–792.

98. Zhu, Y.; Gao, C.; Liu, X.; Shen, J.; Surface modification of polycaprolactone membrane via aminolysis and biomacromolecule immobilization for promoting cytocompatibility of human endothelial cells. *Biomacromolecules* **2002**, *3*, 1312–131.

99. Elvira, C.; Yi, F.; Azevedo, M.C.; Rebouta, L.; Cunha, A.M.; Roman, J.S.; Reis, R.L.; Plasma- and chemical-induced graft polymerization on the surface of starch-based biomaterials aimed at improving cell adhesion and proliferation. *J. Mater. Sci. Mater. Med.* **2003**, *14*, 187–194

100. Hsieh, W.C.; Liau, J.J.; Li, Y.J.; Characterization and cell culture of a grafted

chitosan scaffold for tissue engineering. *Int. J. Polym. Sci.* **2015**, *2015*, 1–7.

101. Li, Z.K.; Wu, Z.S.; Lu, T.; Yuan, H.Y.; Tang, H.; Tang, Z.J.; Tan, L.; Wang, B.; Yan, S.-M.; Materials and surface modification for tissue engineered vascular scaffolds. *J. Biomater. Sci. Polym. Ed.* **2016**, *27*, 1534–1552.

102. Liu, Y.; Lv, H.; Ren, L.; Xue, G.; Wang, Y.; Improving the moisturizing properties of collagen film by surface grafting of chondroitin sulfate for corneal tissue engineering. *J. Biomater. Sci. Polym. Ed.* **2016**, *27*, 785–772.

103. Daskalova, A.; Bliznakova, I.; Zhelyazkova, A.; Trifonov, A.; Angelova, L.; Avramov, L.; Buchvarov, I.; Fabrication of chitosan scaffolds via laser assisted modification. *J. Phys. Technol.* **2017**, *1*, 8–12.

104. Courtenay, J. C.; Johns, M. A.; Galembeck, F.; Deneke, C.; Lanzoni, E. M.; Costa, C. A.; Scott, J.L.; Sharma, R.I.; Surface modified cellulose scaffolds for tissue engineering. *Cellulose* **2017**, *24*, 253–267.

105. Nedela, O.; Slepicka, P.; Švorcík, V.; Surface modification of polymer substrates for biomedical applications. *Materials* **2017**, *10*, 1115.

106. Jaganathan, S. K.; Prasat, M. M.; UV induced surface modification on improving the cytocompatibility of metallocene polyethylene. *An. Acad. Bras. Ciênc.* **2018**, *90*, 195–204.

107. Sengupta, P.; Prasad, B.L.V.; Surface modification of polymers for tissue engineering applications: arginine acts as a sticky protein equivalent for viable cell accommodation. *ACS Omega* **2018**, *3*, 4242–4251.

108. Sengupta, P.; Prasad, B.L.V.; Surface modification of polymeric scaffolds for tissue engineering applications. *Regen. Eng. Transl. Med.* **2018**, *4*, 75–91.

109. Teixeira, M.A.; Amorim, M.T.P.; Felgueiras, H.P.; Poly(vinyl alcohol)-based nanofibrous electrospun scaffolds for tissue engineering applications. *Polymers (Basel)* **2019**, *12*, 7.

110. Ino, J. M.; Chevallier, P.; Letourneur, D.; Mantovani, D.; Le Visage, C.; Plasma functionalization of poly(vinyl alcohol) hydrogel for cell adhesion enhancement. *Biomater.* **2013**, *3*, e25414.

111. Asadian, M.; Chan, K. V.; Norouzi, M.; Grande, S.; Cools P.; Morent, R.; De Geyter, N.; Fabrication and plasma modification of nanofibrous tissue engineering scaffolds. *Nanomaterials* **2020**, *10*, 119.

112. Reimhult, E.; Höök, F.; Design of Surface modifications for nanoscale sensor applications. *Sensors* **2015**, *15*, 1635–1675.

113. Sonawane, M.D.; Nimse, S.B.; Surface modification chemistries of materials used in diagnostic platforms with biomolecules. *J. Chem.* **2016**, *2016*, 1–19.

114. Heinemeyer, C.; van der Loh, M.; Wagner, M.-N.; Länge, K.; Surface modification procedure for biosensor chips made of chemically sensitive polymers. *Procedia Technol.* **2017**, *27*, 165–166.

115. Alemán, C.; Fabregat, G.; Armelin, E.; Buendía, J. J.; Llorca, J.; Plasma surface modification of polymers for sensor applications. *J. Mater. Chem. B.* **2018**, *6*, 6515–6533.

116. Moon, J. M.; Thapliyal, N.; Hussain, K. K.; Goyal, R. N.; Shim, Y. B.; Conducting polymer-based electrochemical biosensors for neurotransmitters: a review. *Biosens Bioelectron.* **2018**, *102*, 540–552.

117. Bernards, D. A.; Macaya, D. J.; Nikolou, M.; DeFranco, J. A.; Takamatsu, S.; Malliaras, G. G.; Enzymatic sensing with organic electrochemical

transistors. *J. Mater. Chem.* **2008**, *18*, 116–120.

118. Werkmeister, F.; Nickel, B.; Towards flexible organic thin film transistors (OTFTs) for biosensing. *J. Mater. Chem. B* **2013**, *1*, 3830–3835.

119. Maekawa, E.; Kitano, N.; Tasukawa, T.; Mizutani, F.; Plasma surface modification of polymers for sensor applications. *Electrochem. (Tokyo)* **2010**, *77*, 319–321.

CHAPTER 9

GALACTOMANNAN: A BIODEGRADABLE POLYMER USED FOR BIO-BASED EDIBLE COATING AND FILM MATERIALS

GUILLERMO CRISTIAN G. MARTÍNEZ-AVILA[1,2],
MIGUEL A. AGUILAR-GONZALEZ[3],
MÓNICA L. CHÁVEZ-GONZALEZ[1], DEEPAK KUMAR VERMA[4],
HAROON KHAN[5], and CRISTOBAL N. AGUILAR[1*]

[1]Bioprocesses and Bioproducts Research Group,
Food Research Department, School of Chemistry,
Autonomous University of Coahuila, Saltillo, 25280 Coahuila, Mexico

[2]Laboratory of Chemistry and Biochemistry, School of Agronomy,
Autonomous University of Nuevo Leon, General Escobedo,
C.P. 66050 Nuevo León, Mexico

[3]Center for Research and Advanced Studies of the National
Polytechnic Institute (CINVESTAV-IPN) Unit-Saltillo,
C.P. 25900 Ramos Arispe, Mexico

[4]Agricultural and Food Engineering Department, Indian Institute of
Technology Kharagpur, Kharagpur 721302, India

[5]Department of Pharmacy, Abdul Wali Khan University,
Mardan 23200, Pakistan

*Corresponding author. E-mail: cristobal.aguilar@uadec.edu.mx

ABSTRACT

The use of edible coatings and film to prolong the quality and shelf life of fruit and vegetables has been a strategy that has been popularized because of its advantages as a sustainable food preservation technology system, creating a gas barrier and producing a modified atmosphere around fruit and vegetables. This atmosphere reduces the availability of O_2 and increases the concentration of CO_2; thus reducing the breathing rate and the loss of water, thereby increasing the shelf life of the product. One of the advantages of using this technology is that these materials can serve as vehicles for other ingredients with a different specific purpose, so that, for example, antimicrobial agents, flavorings, antioxidants, and pigments have been incorporated into the formulations. Biodegradable polymers and other bio-based packaging materials are among the materials used to make edible casings and film. Galactomannans are a type of biopolymer that has unique properties for this purpose. This chapter analyzes the use of galactomannans as an attractive biodegradable polymer used for bio-based edible coating and film materials.

9.1   INTRODUCTION

Edible coatings and film are an alternative to commercial bio-based packaging materials used in food products because, from an environmental point of view, they are thought to be less expensive than plastics, so that their use for this purpose would significantly reduce the packaging waste associated with fresh and processed food [8]. Generally, edible coatings and film are based on a multicomponent system that is based

on a structural agent, a plasticizer, and a hydrophobic agent, and may or may not have an antimicrobial agent. The thickening and gelling characteristics of some polysaccharides give them the ability to interact with other polymeric materials, including other polysaccharides, proteins, and fats, providing synergy between these materials, thus modifying their physicochemical properties. Such interactions have been extensively studied using a wide range of polysaccharides, such as alginates, pectin, starch, and others. In recent years, however, special attention has been paid to the type of polysaccharides that are found in the endosperm of certain legumes, known as galactomannans, which are a group of neutral heteropolysaccharides that perform different functions within the legume tissue of the plant, since they contribute to the accumulation of water during the growth of the plant, in addition to the polysaccharide reserve [16].

A biopolymer called galactomannan is a special polysaccharide formed by a chain of mannose units with branches made up of galactose units. They are the components of some seeds and oilseeds, as well as the cell walls of filamentous fungi. Galactomanans are used in foods such as thickeners and stabilizers, particularly in the dairy sector, in the preservation of fruit and salad dressings, and as flour for the formulation of galactomannan-based foods.

9.2 EDIBLE COATINGS AND FILMS

The edible coating is a type of thin layer of edible material that coats a food product, while the edible film is a thin layer of preformed edible polymer (polysaccharide, lipid, or protein) that, once formed, is placed on or between food components. The main difference between the edible coating and the edible film is that the edible coating is applied in a liquid form to the food, usually by immersing the product in a solution formed by a structural matrix, either based on carbohydrates, proteins, lipids, or a mixture of several components. However, the edible film is first molded as solid leaves that are then applied as a wrapper in a food product or used as a biodegradable packaging for their packaging [22].

Edible films and coatings are based on a multicomponent system where each of the compounds plays a specific role in these formulations. Generally, edible films and coatings are formulated on the basis of the following points, they are as follows:

1. A structural agent such as a hydrocolloid (polysaccharides or proteins) is responsible for providing structural support for the formulation in addition to acting as an emulsifying agent.

2. A plasticizing agent (such as glycerol, sorbitol) provides flexibility to the polymer matrix and therefore makes it more flexible.
3. A hydrophobic agent (such as oils or waxes) responsible for reducing the loss of moisture from the food in which the coating is applied.
4. Finally, edible films and covers may include an antimicrobial agent (synthetic or natural) responsible for the prevention of microbial contamination.

9.3 EDIBLE COATINGS AND BIOACTIVE COMPOUNDS

In order to meet the growing consumer demand for the use of chemicals in minimally processed fruit and vegetables, more attention has been paid to the production of natural antimicrobial compounds [49, 59]. The process of solid-state fermentation is one of the options for obtaining such compounds since the use of this technique can be obtained by monomeric compounds and thus increases their effectiveness. Polyphenolic compounds derived from fruit and vegetable residues, organic acids present in fruits, and some phytoalexins present in plants are among the antimicrobial compounds that are obtained by FMS [7]. Polyphenolic compounds are natural antioxidants that are derived mainly from lignocellulosic residues. These compounds have been used in the formulation of edible films and casings, which have been shown to help prolong the shelf life of certain foods. Polyphenolic compounds have been reported to improve the stability of food lipids and thus help prevent loss of nutritional and sensory quality of such products [31, 50, 54, 63].

Edible polysaccharide-based coatings and natural waxes may also serve as a vehicle for antimicrobial and/or antioxidant compounds to maintain a high concentration of preservatives on the surface of the food. Polymeric and monomeric polyphenolic compounds such as tannins and galactomannan, ferulic and p-cumaric acid have been associated with inhibition of the growth of certain microorganisms such as *Aspergillus flavus*, *A. ochraceus*, *A. parasiticus*, *A. versicolor*, *Bacillus cereus*, *Clostridium botulinum*, *Escherichia coli*, *Listeria monocytogenes*, *Penicillium urticae*, *Staphylococcus aureus*, and *Vibrio parahaemolyticus* [7]. Puupponen et al. [79] shown that myricetin compound (a type of flavonoid) inhibited the growth of lactic acid bacteria from human gastrointestinal flora. Arabic gum, candelilla wax, and ellagic acid edible films have been successfully used to reduce the adverse effects caused by the overcrowding of *Collecotrichum overcameins* of Hass avocado and have therefore significantly improved their appearance [77]. In addition,

antioxidants could be used in combination with process technologies such as heat treatment, modified and controlled atmospheres, edible coatings and film, gamma radiation, and electrical pulses to prevent darkening of the fruit. Saucedo-Pompa et al. [78] evaluated the addition of gallic acid and ellagic acid to candelilla wax/Arabic gum-based coatings in order to study the effect of the addition of natural antioxidants on the quality of life of cut fruits (avocado, banana, and apple). The authors reported that the formulation of bioactive compounds helps to reduce color changes in fruit. These results were consistent with those reported by Ponce et al. (2008) who demonstrated the great advantage of using oleoresin-rich chitosan coatings with high antioxidant activity to prevent darkening reactions that often result in a loss of quality in fruit and vegetables.

9.4   POLYSACCHARIDES: AS A STRUCTURAL BASE OF EDIBLE COATINGS

First, a solution that may constitute a structural matrix with sufficient cohesion is needed for the formulation of an edible coating. When lipids, proteins, and polysaccharides that may interact physically and/or chemically are combined, coatings with improved properties can be obtained [59]. Hydrocolloid coatings are excellent as an obstacle to the diffusion of O_2 and CO_2. Most of these biomaterials also have desirable mechanical and structural properties, which make them useful for improving the structural integrity of fragile products. The hydrocolloids used in the production of coatings are classified according to their composition, molecular load, and solubility in water [8].

Generally, polysaccharide-based coatings have a high water vapor permeability due to their hydrophilic nature. Water-soluble polysaccharides are long-chain, linear, or branched polymers that have glycosidic units which, on average, have three hydroxyl groups, so that they form hydrogen bridges with water, and therefore, polysaccharide particles can take molecules out of the water, swell with them, and soluble completely or partially, giving a thickening and/or gelling character to the aqueous phase [61]. Some scientific reports have described the use of polysaccharides called galactomannans in film design and edible coatings, with particular emphasis on discussing the most important characteristics of these polysaccharides, as well as the physical, chemical, thermal and mechanical properties of the film, and edible polymer-based coatings. In addition, they describe the properties of the transport of moisture and the phenomena of the exchange

of gasses such as CO_2 and O_2 that occur in such materials [11, 22].

9.5 GALACTOMANNANS

Galactomannans have the main chain of mannose units joined by glycosidic bonds $\beta(1\rightarrow4)$ that are replaced by simple galactose units attached to it by means of $\alpha(1\rightarrow4)$ glycosidic bonds. [9]. The typical chemical structure of the biopolymer "galactomannans" is shown in Figure 9.1. These are hydrocolloids with unique chemical and functional properties. These compounds are identified as the main reserve compounds in the various endosperm seeds. The physicochemical and functional properties of these polymers depend mainly on the mannose/galactose ratio that they have within their structure, which depends essentially on the source and extraction method used for their recovery [11]. Some extractants, such as acetic acid solutions [23], sodium hydroxide [16], and sulfuric acid [19], have been reported to be used to extract galactomanides from various plant sources, including mesquite (*Prosopis* spp.) and to carry out functionality and characterization studies. However, this type of polysaccharide can be extracted by simpler and more environmentally sound methods for use in the biomedical and food industries.

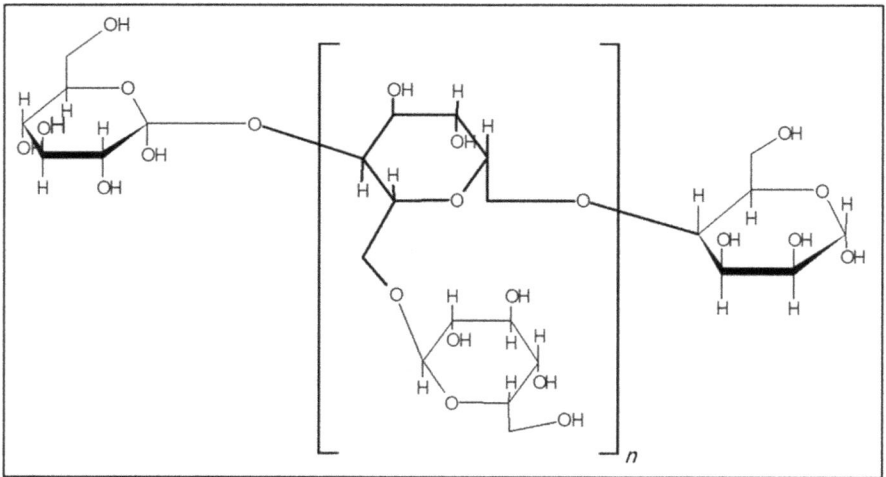

FIGURE 9.1   Typical chemical structure of galactomannan.

Some mesquite species have been reported as potential sources of galactomannan extraction that have been physicochemically characterized. However, research has focused on species such as *P. juliflora, P. chilensis, P. pallida,* and *P. flexuosa,* which are widely distributed throughout South America [33]. Glandulosa is another type of mesquite that grows naturally in Northwestern Mexico, which is also very resistant to drought and has a good nitrogen-fixing effect [16]. Traditionally, mesquite pods are used as a nutritional supplement and for the production of alcoholic and nonalcoholic drinks due to their high sugar content [56]. In addition, mesquite seeds could be used as an alternative source of galactomannans for possible industrial applications.

Due to their unique characteristics and lack of toxicity, galactomannans have been used for the production of hypocaloric cheeses [32] as well as for the preparation of edible coatings to extend the shelf life of some tropical fruits [11].

9.6 EXTRACTION AND PURIFICATION OF MESQUITE SEED GALACTOMANNANS

The first step in extracting the desired natural products from the raw materials is the extraction [17, 73]. Methods of extraction include solvent extraction, distillation method, pressing, and sublimation in accordance with the extraction principle. The most commonly used method is solvent extraction [73, 74]. The extraction of mesquite seed galactomannans is generally carried out in accordance with the methodology previously reported by Sciarini et al. [62] or with some minor modifications [44]. Briefly, the seeds are stirred with water at a ratio of 1:25 (w/v) to 100 °C for 1 h. After the sample is cooled to room temperature, it is centrifuged at 10,000 rpm for 10 min. The supernatant is then collected and the material is resuspended in hot water and the process is repeated twice. Subsequently, all supernatants are mixed and the polysaccharide is precipitated by two volumes of 96% ethanol. After the excess ethanol has been removed, the polymer is frozen and sprayed into a mortar.

Purification is the process by which the target natural products of interest (compound of other potentially structurally related compounds or contaminants) are separated or extracted from the sample [73, 74]. For the purification of the biopolymer, the technique previously reported by Chaires-Martínez et al. [16] is used in our laboratory (BBG-DIA/UAdeC) with some minor modifications. Solutions are prepared with 1% of the alcohol material obtained from extraction. The solutions are centrifuged at 10,000 rpm by 15 min at 4 °C, and the supernatant is precipitated by two volumes of 96% ethanol. Galactomannans are separated by filtration and are frozen. Finally, the material is polished in a mortar and passed through a 0.125 mm mesh sieve in order to have the same particle size and to facilitate the completion of the next stages.

9.7 CHEMICAL ANALYSIS AND FUNCTIONAL PROPERTIES

The moisture, ash, lipid, and protein contents are determined in accordance with the official procedures of the Association of Official Analytical Chemists [5]. The analysis of glycosyl composition is determined by gas chromatography combined with mass spectrometry (CG/EM) of O-trimethylsilyl (TMS) derivatives of methylglycoside monosaccharides produced from the sample by acid methanolysis.

Approximately 500 g of the sample is stored in the test tube. Twenty grams of inositol were added as an internal pattern. The sample is frozen and then proceeds for further process in freeze-dried. Methylglycoside is prepared from a dry sample by means of a methanol analysis in HCl 1 M in methanol at 80 °C (18 h), followed by N-acetylation with pyridine and acetic anhydride/methanol (for the detection of amino sugars). Samples are O-trimethylisylated at 80 °C for 30 min, following the methods reported by Merkle and Poppe [46] and York et al. [75]. The CG/EM analysis of methylglycosides is performed on a CG interfaced with MSD using the

Supelco®EC-1 capillary column of molten silica (30 mm × 0.25 mm in diameter). Functional characterization of galactomannans can be evaluated using the methodology previously reported by Sciarini et al. [62] with some minor modifications for solubility tests, emulsification capacity, and emulsion stability [44] including water absorption capacity, water-solubility, emulsification capacity, macromolecular characterization to evaluate water absorption capacity.

9.8 APPLICATIONS AND POTENTIAL USES

Several applications of galactomannan have demonstrated the great versatility of the functions of this biopolymer. The main applications in the agricultural and food sectors have been shown in Table 9.1(A). Use as a carrier of bioactive molecules, as a component of edible coatings and film, and as a gelling agent, among others, have been reported. Table 9.1(B) describes the main applications of galactomannans in the design of new materials and nanotechnology. In this regard, it is important to note the most recent publications which report the use of galactomannan as a versatile material for various purposes. A number of pharmacological and biotechnological applications of galactomannans has been shown in Table 9.1(C).

9.9 CONCLUDING REMARKS

It is important to bear in mind that in the food packaging industry, there are constant science and technology advances aimed at creating sustainable food systems that are more successful in maintaining quality and increasing the quality, marketability, and appeal of foods. One of the key developments in the food industry is the use of sustainable sources for biologic packaging materials, such as hydrocolloids. Edible coatings and film are one of the potential technological strategies to enhance the storage of food by improving current packaging technology, promoting food safety, and safety and protecting food from environmental factors. Galactomannans are an excellent option for the production of edible coatings and films. The most important features of this biomaterial have been discussed, and potential applications are considered for the near future tailored edible packaging of bio-based materials that can be applied to selected foods, substituting nonbiodegradable, or nonedible polymers.

ACKNOWLEDGMENTS

We thank CONACYT for the financial support provided by the PEI-CONACYT programme.

TABLE 9.1   Application of Galactomannans for Various Agricultural, Food and Biomedical Purposes

Potential Application	References
A) In agricultural and food industry	
Hydrogel as a potential water-retaining agent in agriculture	[39]
New natural adjuvant reduces the amount of copper necessary to control downy mildew of grapevine	[35]
Promoting of root nodulation of Rhizobium inoculants	[70]
Nutritional and antioxidant potential of galactomannan flour	[53]
As a source of mannooligosaccharides	[55]
Environment-friendly materials	[30]
Improving of rheological, pasting, and physical properties of water chestnut starch	[36]
Edible films and coatings to extend the shelf life of cheese and meat	[12, 13]
Formulation of new edible films and coatings	[14, 26, 28, 47, 58]
Microbial control and reduction of pathogens	[45, 68]
Edible coating and films to extend the shelf life quality of fruit and vegetables	[1, 11, 38]
B) In designing of new materials for nanotechnology	
As a component of hydrogel with reductive stimuli-responsive degradable properties	[25]
Functionalization of hydroxyapatite with excellent drug-loading properties	[72]
Nanofibrous mats seeded with galactomannan for wound healing applications	[34]
Promising biomaterial for films fabrication	[6]
Biocompatible nanoparticles	[51]
Composites as potential materials for several applications	[64]
Novel nanobiocomposite hydrogels	[48]
Biocomposites aerogels	[27]
Chemo-enzymatically oxidized galactomannans as novel polymeric biomaterials	[60]
Promising starting material for diverse applications	[71]
Binding agents for silicon anodes in Li-ion batteries	[21]
Chemically grafted biopolymer composites in the printing of cotton fabric	[68]
New biomaterials with novel properties and applications, including as a delivery system	[65, 67]

TABLE 9.1   *(Continued)*

Potential Application	References
C) In pharmaceutical and biotechnological industries	
Improvement of physical and chemical properties of hyaluronic acid	[43]
Injectable hydrogel as a suitable platform for cell culture	[40]
A new gelling agent and rheology modifier in cosmetics	[57]
Treatment of diseases	[37]
Galactomannan as anticancer compound	[76]
Galactomannan hydrolysates as anti-inflammatory factor	[18]
Prebiotic application and its fermentation by the probiotic strain	[42]
Synthetic glycopeptides	[29]
Aerogels to delivery systems of antimicrobial peptides and enzymes	[10]
Biodegradable hydrogel for pharmaceutical application	[52]
Biomarker applications in diagnostics of fungal infections	[41]
Biotechnological applications for immobilization of biomolecules and bioactive compounds	[2–4]
Biomaterials for tissue engineering applications	[66]
Galactomannans as drug delivery systems	[15, 20, 24]

KEYWORDS

- galactomannans
- edible coating and film
- biodegradable polymer

REFERENCES

1. Aguiar, R. P., Miranda, M. R. A., Lima, Á. M., Mosca, J. L., Moreira, R. A., & Enéas-Filho, J. (2011). Effect of a galactomannan coating on mango post-harvest physicochemical quality parameters and physiology. *Fruits* 66(4), 269–278.

2. Albuquerque, P. B., Cerqueira, M. A., Vicente, A. A., Teixeira, J. A., & Carneiro-da-Cunha, M. G. (2017). Immobilization of bioactive compounds in *Cassia grandis* galactomannan-based films: influence on physicochemical properties. *International Journal of Biological Macromolecules*. 96, 727–735.

3. Albuquerque, P. B., Coelho, L. C., Correia, M. T., Teixeira, J. A., & Carneiro-da-Cunha, M. G. (2016). Biotechnological applications of galactomannan matrices: emphasis on immobilization of biomolecules. *Advances in Research*. 6, 1–17.

4. Albuquerque, P. B., Silva, C. S., Soares, P. A., Barros Jr, W., Correia, M. T., Coelho, L. C., Teixeira, J. A., & Carneiro-da-Cunha, M.G., (2016). Investigating a galactomannan gel obtained from *Cassia grandis* seeds as immobilizing matrix for Cramoll lectin. *International Journal of Biological Macromolecules*. 86, 454–461.

5. AOAC. (1992). Official Methods for Analysis of the Association of Official Analytical Chemist. William Horwitz, Ed. AOAC: México.

6. Batista, M. J., Ávila, A. F., Franca, A. S., & Oliveira, L. S. (2020). Polysaccharide-rich fraction of spent coffee grounds as promising biomaterial for films fabrication. *Carbohydrate Polymers*. 233, 115851.

7. Beuchat, L R. (2001). Chapter 11: Control of food-borne pathogens and spoilage microorganisms by naturally occurring antimicrobials. In: Microbial Food Contamination. Wilson, C. L., Droby, S. Eds. CRC Press: London. 149–169.

8. Bósquez-Molina. (2003). Elaboración de recubrimientos comestibles formulados con goma de mezquite y cera de candelilla para reducir la cinética de deterioro en fresco del limón persa (Citrus latifolia Tanaka). Tesis profesional. UAM. México, D.F.

9. Buckeridge, M. S. (2010). Seed cell wall storage polysaccharides: models to understand cell wall biosynthesis and degradation. *Plant Physiology.* 154, 1017–1023.

10. Campia, P., Ponzini, E., Rossi, B., Farris, S., Silvetti, T., Merlini, L., Brasca, M., Grandori, R., & Galante, Y.M., (2017). Aerogels of enzymatically oxidized galactomannans from leguminous plants: versatile delivery systems of antimicrobial peptides and enzymes. *Carbohydrate Polymers*. 158, 102–111.

11. Cerqueira, M. A., Lima, A. M., Texeira, J. A., Moreira, R. A., Vicente, A. A. (2009). Suitability of novel galactomannans as edible coatings for tropical fruits. *Journal of Food Engineering*, 94, 372–378.

12. Cerqueira, M. A., Bourbon, A. I., Pinheiro, A. C., Martins, J. T., Souza, B. W. S., Teixeira, J. A., & Vicente,

A. A. (2011). Galactomannans use in the development of edible films/coatings for food applications. *Trends in Food Science & Technology*. 22(12), 662–671.

13. Cerqueira, M. A., Sousa-Gallagher, M. J., Macedo, I., Rodriguez-Aguilera, R., Souza, B. W., Teixeira, J. A., & Vicente, A. A. (2010a). Use of galactomannan edible coating application and storage temperature for prolonging shelf-life of "Regional" cheese. *Journal of Food Engineering*. 97(1), 87–94.

14. Cerqueira, M. A., Souza, B. W., Martins, J. T., Teixeira, J. A., & Vicente, A. A. (2010b). Seed extracts of *Gleditsia triacanthos*: functional properties evaluation and incorporation into galactomannan films. *Food Research International*, 43(8), 2031–2038.

15. Cerqueira, M. A., Pinheiro, A. C., Pastrana, L. M., Vicente, A. A. (2019). Amphiphilic modified galactomannan as a novel potential carrier for hydrophobic compounds. *Frontiers in Sustainable Food Systems*, 3, 17.

16. Chaires-Martínez, L., Salazar-Montoya, J. A., Ramos-Ramírez, E. G. (2008). Physicochemical and functional characterization of the galactomannan obtained from mesquite seeds (*Prosopis pallida*). *European Food Research and Technology*, 227, 1669–1676.

17. Chávez-González, M. L.; Sepúlveda, L.; Verma, D. K., Luna-García, H. A.; Rodríguez-Durán, L. V.; Ilina, A.; and Aguilar, C. N. (2020). Conventional and emerging extraction processes of flavonoids. *Processes* 8, 434. doi: 10.3390/pr8040434.

18. Chen, W.-L., Chen, H.-L., Guo, G.-W., Huang, Y-C., Chen, C-Y., Tsai, Y., Huang, K.-F., Yang, C.-H. (2018). Locust bean gum galactomannan hydrolyzed by thermostable β-D-mannanase may reduce the secretion of pro-inflammatory factors and

the release of granule constituents. *International Journal of Biological Macromolecules*. 114, 181–186.

19. Dakia P. A., Bleckerb C, Roberta C, Watheleta B, Paquota M. (2008). Composition and physicochemical properties of locust bean gum extracted from whole seeds by acid or water dehulling pre-treatment. *Food Hydrocolloids*, 22, 807–818.

20. de Almeida, R. R., Magalhães, H. S., de Souza, J. R. R., Trevisan, M. T. S., Vieira, Í. G. P., Feitosa, J. P. A., Araújo, T. G., Ricardo, N. M. P. S. (2015). Exploring the potential of *Dimorphandra gardneriana* galactomannans as drug delivery systems. *Industrial Crops and Products*. 69, 284–289.

21. Dufficy, M. K., Khan, S. A., Fedkiw, P. S. (2015). Galactomannan binding agents for silicon anodes in Li-ion batteries. *Journal of Materials Chemistry* A. 3(22), 12023–12030.

22. Falguera, V., Quintero, J. P., Jiménez, A., Muñoz, J. A., Ibarz, A. (2011). Edible films and coatings: structures, active functions and trends in their use. *Trends in Food Science and Technology*. 22, 292–303.

23. Figueiró, S. D., Góes, J. C., Moreira, R. A., Sombra, A. S. B. (2004). On the physico-chemical and dielectric properties of glutaraldehyde crosslinked galactomannan–collagen films. *Carbohydrate Polymers*. 56, 313–320.

24. Galante, Y. M., Merlini, L., Silvetti, T., Campia, P., Rossi, B., Viani, F., Brasca, M. (2018). Enzyme oxidation of plant galactomannans yielding biomaterials with novel properties and applications, including as delivery systems. *Applied Microbiology and Biotechnology*. 102 (11), 4687–4702.

25. Gao, Y., Zong, S., Huang, Y., Yang, N., Wen, H., Jiang, J., Duan, J. (2020). Preparation and properties of a highly elastic galactomannan- poly(acrylamide- N,

N-bis (acryloyl) cysteamine) hydrogel with reductive stimuli-responsive degradable properties. *Carbohydrate Polymers*. 231, 115690.

26. Germano, T. A., Aguiar, R. P., Bastos, M. S. R., Moreira, R. A., Ayala-Zavala, J. F., & de Miranda, M. R. A. (2019). Galactomannan-carnauba wax coating improves the antioxidant status and reduces chilling injury of *Paluma guava*. *Postharvest Biology and Technology*. 149, 9–17.

27. Ghafar, A., Gurikov, P., Subrahmanyam, R., Parikka, K., Tenkanen, M., Smirnova, I., Mikkonen, K. S. (2017). Mesoporous guar galactomannan based biocomposite aerogels through enzymatic crosslinking. *Composites Part A: Applied Science and Manufacturing*. 94, 93–103.

28. González, A., Barrera, G. N., Galimberti, P. I., Ribotta, P. D., Alvarez Igarzabal, C. I. (2019). Development of edible films prepared by soy protein and the galactomannan fraction extracted from *Gleditsia triacanthos* (Fabaceae) seed. *Food Hydrocolloids*. 97, 105227.

29. Hammura, K., Ishikawa, A., Ravi Kumar, H.V., Miyoshi, R., Yokoi, Y., Tanaka, M., Hinou, H., Nishimura, S.-I. (2018). Synthetic glycopeptides allow for the quantitation of scarce nonfucosylated IgG Fc N-Glycans of therapeutic antibody. *ACS Medicinal Chemistry Letters*. 9(9), 889–894.

30. Hebeish, A., Rekaby, M., Abd El-Thalouth, J. I. (2016). A benign strategy for synthesizing environment-friendly textile prints using guar acetate and natural dye. *International Journal of Pharmaceutical Sciences Review and Research*. 40(1), 34, 173–181.

31. Hemeda H, Klein B. (1990). Effects of naturally occurring antioxidants on peroxidase activity of vegetable extracts. *Journal of Food Science*. 55, 184–192.

32. Hernández-Tinoco, A., Ramos-Ramírez, E. G., Falcony-Guajardo, C., Salazar-Montoya, J. A. (2004). Rheometry and scanning electron microscopy study of casein curds added with mesquite seed gum and soy proteins. *Latin American Applied Research.* 34, 195–202.

33. Ibañez, M. C., Ferrero, C. (2003). Extraction and characterization of the hydrocolloid from *Prosopis flexuosa* DC seeds. *Food Research International.* 36, 455–460.

34. Kalachaveedu, M., Jenifer, P., Pandian, R., Arumugam, G. (2020). Fabrication and characterization of herbal drug enriched Guar galactomannan based nanofibrous mats seeded with GMSC's for wound healing applications. *International Journal of Biological Macromolecules.* 148, 737–749.

35. Lahoz, E., Tarantino, P., Mormile, P., Malinconico, M., Immirzi, B., Cermola, M., Carrieri, R. (2018). Evaluation of a new natural adjuvant obtained from locust bean gum to reduce the amount of copper necessary to control downy mildew of grapevine. *Journal of Plant Diseases and Protection.* 125(3), 287–296.

36. Lee, Y., Chang, Y.H. (2015). Effects of galactomannan addition on rheological, pasting and physical properties of water chestnut starch. *Journal of Texture Studies.* 46(1), 58–66.

37. Li, P. C., Yang, X. S., Li, W. Y. (2019). Galactomannan testing in the treatment of autoimmune disease combined with invasive fungal disease. *Journal of Biological Regulators and Homeostatic Agents.* 33(1), 139–144

38. Lima, Á. M., Cerqueira, M. A., Souza, B. W., Santos, E. C. M., Teixeira, J. A., Moreira, R. A., & Vicente, A. A. (2010). New edible coatings composed of galactomannans and collagen blends to improve the postharvest quality of fruits–influence on fruits gas transfer rate. *Journal of Food Engineering,* 97(1), 101–109.

39. Liu, C., Lei, F., Li, P., Jiang, J., Wang, K. (2020). Borax crosslinked fenugreek galactomannan hydrogel as potential water-retaining agent in agriculture. *Carbohydrate Polymers.* 236, 116100.

40. Lucas de Lima, E., Fittipaldi Vasconcelos, N., da Silva Maciel, J., Andrade, F. K., Silveira Vieira, R., Andrade Feitosa, J.P. (2020). Injectable hydrogel based on dialdehyde galactomannan and N-succinyl chitosan: a suitable platform for cell culture. *Journal of Materials Science: Materials in Medicine.* 31(1), 5.

41. Maertens, J., Lagrou, K. (2017). Biomarker applications in diagnostics of fungal infections. Immunogenetics of Fungal Diseases. Springer: Berlin, 173–186

42. Majeed, M., Majeed, S., Nagabhushanam, K., Arumugam, S., Natarajan, S., Beede, K., Ali, F. (2018). Galactomannan from *Trigonella foenum-graecum* L. seed: prebiotic application and its fermentation by the probiotic *Bacillus* coagulans strain MTCC 5856. Food Science And Nutrition. 6(3), 666–673.

43. Martin, A. A., Sassaki, G. L., Sierakowski, M. R. (2020). Effect of adding galactomannans on some physical and chemical properties of hyaluronic acid. *International Journal of Biological Macromolecules.* 144, 527–535.

44. Martínez-Ávila, G. C. G., Hernández-Almanza, A. Y., Sousa, F. D., Moreira, R., Gutierrez-Sanchez, G., & Aguilar, C. N. (2014). Macromolecular and functional properties of galactomannan from mesquite seed (*Prosopis glandulosa*). *Carbohydrate polymers.* 102, 928–931.

45. Martins, J. T., Cerqueira, M. A., Souza, B. W., Carmo Avides, M. D., & Vicente, A. A. (2010). Shelf life extension of

ricotta cheese using coatings of galacto-mannans from nonconventional sources incorporating nisin against *Listeria monocytogenes*. *Journal of Agricultural and Food Chemistry*. 58(3), 1884–1891.

46. Merkle, R. K., Poppe, I. (1994). Carbohydrate composition anlysis of glycoconjugates by gas-liquid chroma-tography/mass spectrometry. *Methods Enzymol*. 230, 1–15

47. Naeem, A., Abbas, T., Ali, T. M., Hasnain, A. (2018). Effect of guar gum coatings containing essential oils on shelf life and nutritional quality of green-unripe mangoes during low temperature storage. *International Journal of Biological Macromolecules*. 113, 403–410.

48. Oleyaei, S. A., Razavi, S. M. A., Mikkonen, K. S. (2018). Novel nano-biocomposite hydrogels based on sage seed gum-laponite: physico-chemical and rheological characterization. *Carbo-hydrate Polymers*. 192, 282–290.

49. Oya, A., Yoshida, S., Alcañiz-Monge, J., Linares-Solano, A. (1996). Prepa-ration properties of an antibacterial activated carbon fiber containing mesopores. *Carbon*. 34, 53–57.

50. Ozcan, M. (2003). Antioxidant activi-ties of rosemary, sage and sumac extract and their combinations on stability of natural peanut oil. *Journal of Medicinal Food*. 6, 267–270.

51. Padinjarathil, H., Joseph, M. M., Unni-krishnan, B.S., Preethi, G. U., Shiji, R., Archana, M. G., Maya, S., H. P., Sreelekha, T. T. (2018). Galactomannan endowed biogenic silver nanoparticles exposed enhanced cancer cytotoxicity with excellent biocompatibility. *Inter-national Journal of Biological Macro-molecules*. 118, 1174–1182.

52. Pandit, B. (2017). Biodegradable guar gum based hydrogel for pharmaceutical application. *Current Chemical Biology*. 11(1), 3–9.

53. Petkova, N., Petrova, I., Ivanov, I., Mihov, R., Hadjikinova, R., Ognyanov, M., Nikolova, V. (2017). Nutritional and antioxidant potential of carob (*Ceratonia siliqua*) flour and evaluation of functional properties of its polysac-charide fraction. *Journal of Pharma-ceutical Sciences and Research*. 9(11), 2189–2195.

54. Ponce A, Del Valle C, Roura S. (2004). Shelf life of leafy vegetables treated with natural essential oils. *Journal of Food Science*. 69, 50–56.

55. Pradeep G. C, Cho, S. S., Choi, Y. H., Choi, Y. S., Jee, J-P., Seong, C.N., Yoo, J. C. (2016). An extremely alka-line mannanase from *Streptomyces* sp. CS428 hydrolyzes galactomannan producing series of mannooligosaccha-rides. *World Journal of Microbiology and Biotechnology*. 32(5), 84

56. Prokopiuk, D., Cruz, G., Grados, N., Garro, O., Chiralt, A. (2000). Estudio comparativo entre frutos de *Prosopis alba* y *Prosopis pallida*. *MULTE-QUINA Latin American Journal of Natural Resources*. 9, 35–45.

57. Rigano, L., Deola, M., Zaccariotto, F., Colleoni, T., Lionetti, N. (2019). A new gelling agent and rheology modifier in cosmetics: Caesalpinia spinosa gum. *Cosmetics*. 6(2), 34.

58. Rodriguez-Canto, W., Cerqueira, M. A., Chel-Guerrero, L., Pastrana, L. M., & Aguilar-Vega, M. (2020). Delonix regia galactomannan-based edible films: effect of molecular weight and k-carra-geenan on physicochemical properties. *Food Hydrocolloids*. 103, 105632.

59. Rojas-Graü, M. A., Tapia, M. S., Rodrí-guez, F. J., Carmona, A. J., Martín-Belloso, O. (2007). Alginate and gellan based edible coatings as support of anti-browning agents applied on fresh-cut Fuji apple. *Food Hydrocol*. 21, 118–127.

60. Rossi, B., Ponzini, E., Merlini, L., Grandori, R., Galante, Y. M. (2017).

Characterization of aerogels from chemo-enzymatically oxidized galactomannans as novel polymeric biomaterials. *European Polymer Journal.* 93, 347–357.

61. Ruiz-Hernández F. (2009). Aplicación de películas comestibles a base de quitosano y mucílago de nopal en fresa (*Fragaria ananassa*) almacenada en refrigeración. Tesis Maestría. Ciencia de Alimentos. Departamento de Ingeniería Química y Alimentos, Escuela de Ingeniería, Universidad de las Américas, Puebla.

62. Sciarini, L. S., Maldonado, F., Ribotta, P.D., Pérez, G. T., León, A. E. (2009). Chemical composition and functional properties of *Gleditsia triacanthos* gum. *Food Hydrocol.* 23, 306–313.

63. Sebranek, J. (2004). Antioxidant effectiveness of natural rosemary extract in pork sausage. IOWA State University, Animal Industry Report. 1–2.

64. Sharma, G., Sharma, S., Kumar, A., Al-Muhtaseb, A. H., Naushad, M., Ghfar, A. A., Mola, G. T., Stadler, F. J. (2018). Guar gum and its composites as potential materials for diverse applications: A review. *Carbohydrate Polymers.* 199, 534–545.

65. Sharma, P., Gupta, S., Soni, P.L., Kumar, V. (2020). Ce(IV)-ion initiated grafting of 1,3 galactomannan biopolymer with acrylonitrile. *Journal of Macromolecular Science, Part A*, 57(7), 519–530.

66. Siqueira, N. M., Paiva, B., Camassola, M., Rosenthal-Kim, E. Q., Garcia, K. C., Dos Santos, F. P., Soares, R. M. D. (2015). Gelatin and galactomannan-based scaffolds: characterization and potential for tissue engineering applications. *Carbohydrate Polymers.* 133, 8–18.

67. Soares, P. A. G., C De Seixas, J. R. P., Albuquerque, P. B. S., Santos, G. R. C., Mourão, P. A. S., Barros Jr., B., Correia, M. T. S., Carneiro-Da-Cunha,

M. G. (2015). Development and characterization of a new hydrogel based on galactomannan and κ-carrageenan. *Carbohydrate Polymers.* 134, 673–679.

68. Tarigan, J. B., Nainggolan, I., & Kaban, J. (2018). The physicochemical and antibacterial properties of galactomannan edible film of *Arenga pinnata* incorporated with *Zingiber officinale* essential oil. *Asian Journal of Pharmaceutical and Clinical Research.* 11(12), 138–142.

69. Teli, M. D., Rangi, A., & Valia, S. P. (2015). Application of chemically grafted biopolymer composites in printing of cotton fabric. *Journal of Landscape Ecology.* 12(5), 36–40.

70. Thapa, S., Adams, C. B., & Trostle, C. (2018). Root nodulation in guar: effects of soils, *Rhizobium* inoculants, and guar varieties in a controlled environment. *Industrial Crops and Products.* 120, 198–202.

71. Thombare, N., Jha, U., Mishra, S., & Siddiqui, M. Z. (2016). Guar gum as a promising starting material for diverse applications: A review. *International Journal of Biological Macromolecules.* 88, 361–372.

72. Tian, C., Xu, P., Jiang, J., & Han, C. (2020). Preparation and drug-delivery study of functionalized hydroxyapatite based on natural polysaccharide gums with excellent drug-loading properties. *Journal of Dispersion Science and Technology.* Article in Press.

73. Verma, D. K., & Srivastav, P. P. (2020a). Extraction, Identification and quantification methods of rice aroma compounds with emphasis on 2-Acetyl-1-Pyrroline (2-AP) and its relation with rice quality: a comprehensive review. *Food Reviews International.* https://doi.org/10.1080/87559129.2020.1720231.

74. Verma, D. K., & Srivastav, P. P. (2020b). A paradigm of volatile aroma compounds in rice and their product with extraction and identification methods: a

comprehensive review. *Food Research International.* 130: 1–33. https://doi.org/10.1016/j.foodres. 2019.108924

75. York, W. S., Darvill, A. G., McNeil, M., Stevenson, T. T., & Albersheim, P. (1986). Isolation and characterization of plant cell walls and cell wall components. *Method in Enzymol.* 118, 3–40.

76. Zhou, M., Yang, L., Yang, S., Zhao, F., Xu, L., & Yong, Q. (2018). Isolation, characterization and in vitro anticancer activity of an aqueous galactomannan from the seed of *Sesbania cannabina*. *International Journal of Biological Macromolecules.* 113, 1241–1247.

77. Saucedo-Pompa, S., Rojas-Molina, R., Aguilera-Carbó, A. F., Saenz-Galindo, A., de La Garza, H., Jasso-Cantú, D.,

& Aguilar, C. N. (2009). Edible film based on candelilla wax to improve the shelf life and quality of avocado. *Food Research International*, 42(4), 511–515.

78. Saucedo-Pompa, S., Jasso-Cantu, D., Ventura-Sobrevilla, J., Saenz-Galindo, A., Rodriguez-Herrera, R., & Aguilar, C. N. (2007). Effect of candelilla wax with natural antioxidants on the shelf life quality of fresh-cut fruits. *Journal of Food Quality*, 30(5), 823–836.

79. Puupponen-Pimiä, R., Nohynek, L., Meier, C., Kähkönen, M., Heinonen, M., Hopia, A., & Oksman-Caldentey, K. M. (2001). Antimicrobial properties of phenolic compounds from berries. *Journal of Applied Microbiology*, 90(4), 494–507.

CHAPTER 10

POLY(lactic acid)-BASED MATERIALS: FOOD PACKAGING APPLICATION AND BIODEGRADABILITY EVALUATION

ELENA STOLERU[1,2]

[1]*"Petru Poni" Institute of Macromolecular Chemistry, Physical Chemistry of Polymers Laboratory, 41A Grigore Ghica Voda Alley, 700487 Iasi, Romania*

[2]*"Alexandru Ioan Cuza" University of Iasi, Faculty of Chemistry, 11 Carol I Blvd, 700506 Iasi, Romania*

Corresponding author. E-mail: elena.paslaru@icmpp.ro.

ABSTRACT

The lack of degradability and xenobiotic nature of the synthetic polymeric materials leads to high levels of environmental pollution and health hazards. The recalcitrant nature of plastic is a matter of huge concern, and hence new challenges came in front in response to plastic degradation. With the increasing demand for plastics and rising pressure for their safe disposal, biodegradable plastics and plastic biodegradation gained a lot of attention in recent years. The application of biopolymers is limited to short shelf-life products, but they have the huge potential to govern the packaging sector in the coming years. Poly(lactic acid) (PLA) is among the most important commercially available thermoplastic polyesters that are bio-based and biodegradable. It offers a sustainable alternative for food packaging across a wide range of potential product-based applications as per consumers' demand and market trends. The analysis of biochemical processes involved in PLA biodegradation is a key factor for exploring the high efficient methods of PLA degradation in natural environments. This chapter aims to present the latest developments and challenges in the field of PLA-based materials applied in the food packaging sector and to evaluate in a critical manner the way in which such materials are studied for their biodegradability and/or biocompostability in specific biological environments. Furthermore, the end of life options and life cycle assessment studies for PLA reported in recent literature are being discussed.

10.1 INTRODUCTION

Nowadays, petroleum-based plastic packaging is extensively used in preserving various food products, but the magnitude of their application and the consequent environmental repercussions have forced many countries to put strict restrictions on their usage and necessitate the application of alternatives in our daily life. In this regard, developing biodegradable polymers from renewable sources is highly desirable for food preservation and packaging, provided it would be as effective as the currently used plastics in protecting food against microbial contamination, physical damage, and chemical reactions (e.g., oxidation) [1]. Polymers with ester bonds in the macromolecular backbone are a particular interesting group of polymers in the family of synthetic biodegradable plastics. A number of aliphatic polyesters or copolyesters based on hydroxyacids, diacids and diols, lactones or lactides are reported to be completely biodegradable [2],

ester-linkages being generally easy to hydrolyze [3–5]. These features have attracted the attention to this class of polymers to be used in developing sustainable and biodegradable packaging materials. From this group of bio-based and biodegradable polymers, the most commonly used in food packaging applications is poly(lactic acid) (PLA).

10.2 GENERALITIES ON PLA BIOPOLYMER

PLA is a bio-based polymer, commonly derived from corn, sugar cane, or other plant residues. It is an aliphatic polyester primarily produced by industrial polycondensation of lactic acid (2-hydroxy propionic acid) and/or ring-opening polymerization of lactide [6]. PLA exists in L- and D-form, which are optical isomers. PLA with a large amount of L-form isomer is highly crystalline. In general, the crystallinity and biodegradability depend on the content of the D-form isomer [7]. There are two main methods to produce lactic acid, namely, bacterial fermentation of carbohydrates and chemical synthesis [8]. The bacterial fermentation method is preferred by the major producers of PLA (e.g., NatureWorks LLC and Corbion®), mainly because the chemical synthesis route is economically unviable [9]. Melt processing is the main technique used for the mass production of PLA products for the medical, textile, plasticulture, and packaging industries [6]. PLA has been receiving much attention in the last decades due to its biodegradability in the human body as well as in the soil, biocompatibility, environmentally friendly characteristics, and nontoxicity [10, 11], and its applications were initially oriented toward the manufacture of medical grade sutures, implants, and controlled drug release applications mainly because of its high costs. Actually, PLA has many potential uses including packaging and textile industries [10–14]. Even if the demand for PLA is expanding, it is facing a high selling price industrially (mainly because of high-cost production of its monomer, lactic acid) as such it still cannot compete with fossil fuel-based plastics [15]. Due to its higher cost, the initial focus of PLA as a packaging material has been in high-value films, rigid thermoforms, food and beverage containers, and coated papers [16]. Recently, the production cost of PLA has been lowered by using modern and emerging production technologies, thus the applicability range in packaging was broadened. Early economic studies showed that PLA tends to become an economically feasible material to be used as a packaging polymer [7, 17, 18]. In this context, PLA is frequently used in combination with other bio-based and or biodegradable polymers to improve stiffness and strength and to reduce costs [19].

PLA biopolymer is approved for use in food packaging, including direct contact applications with its classification being generally recognized as safe, by the US Food and Drug Administration, and it is authorized by the European Commission [20]. Currently, PLA is being used as a food packaging polymer for short shelf-life products [16]. The key selling feature of PLA is its biodegradability under industrial composting conditions [21] and by providing consumers with extra end-user benefits such as avoiding paying the "green tax" or meeting environmental regulations [16]. However, its true biodegradability when used in commercial applications is disputed [22, 23], and life cycle assessments (LCAs) have shown that the overall sustainability benefits of PLA compared with conventional plastics are uncertain [24, 25].

Even if the mass production of bio-based polymers presents an increasing trend is still far to be competitive with petroleum-based plastics. For instance, in 2019, the total production volume of bio-based polymers was 3.8 million tonnes, which is 1% of the production volume of fossil-based polymers and about 3% more than in 2018. This compound annual growth rate

(CAGR) is expected to continue until 2024. A total production volume of around 0.3 million tonnes represents the production of PLA [26]. Nearly 50% of the produced PLA is used for packaging [27, 28]. The major PLA producers are NatureWorks, WeforYou, Evonik, and Total Corbion [29].

10.3   PLA-BASED MATERIALS IN FOOD PACKAGING

With multiple initiatives to keep food waste away from landfills, packaging systems that are compatible with the alternative end of life scenario (e.g., composting, biodigestion) are required [30]. Since the 1990s, researchers have concentrated their efforts toward using PLA in food packaging applications [31].

10.3.1   PLA PROPERTIES FOR FOOD PACKAGING

The good properties of PLA that have attracted a lot of attention in this domain are high transparency, biodegradability, large-scale production, high tensile strength, and reasonable barrier properties for some applications [32]. However, poor toughness, high brittleness with <10% elongation at break, slow degradation rate, and its high cost (2.8 times more expensive than polypropylene (PP) and 3.2 times more expensive than high-density polyethylene) are its

shortcomings, which still limits is application as packaging material [10, 30, 31]. It is important to note that some features of PLA can be a drawback in one application and a benefit in another, for example, the low water vapor barrier of the bio-based PLA is a disadvantage for a water bottle but an advantage in (breathable) packaging of vegetables and fruits [19]. Even though there are many limitations due to its material properties, there exist potential routes to resolve these shortcomings such as copolymerization, blending, plasticization modification, or the addition of reinforcing phases (fibers or fillers and additives), etc. [34–37]. Blending PLA with various soft and tough polymers can enhance mechanical properties and biodegradability. Various studies [37–40] had shown that the addition of nanoparticles (NPs) such as organo-modified montmorillonite, nano-silica, and silver NP into the PLA-based films may lead to improved thermal stability and tensile strength properties. NPs could also offer some new features to the PLA–based packaging films including antimicrobial activity, an increase of film's water barrier property, and mechanical strength. Besides NPs, other additives have been tested in the PLA-based eco-friendly food packaging. The addition of plant-derived essential oils has shown to lead to the development of the packaging materials with improved food preservation [41-44].

10.3.2 COMMERCIAL PLA-BASED FOOD PACKAGING

In the last two decades, PLA has been used in a wide range of primary packaging applications including oriented and flexible films, extruded and/or thermoformed packages suitable for common applications, such as food and beverage containers, cups, overwrap, blister packages, as well as coated paper and board [45]. Because PLA films are highly transparent, they started to be used in packaging materials as a homo-material (e.g., bi-axially oriented PLA (BOPLA) films) and as laminates (barrier films). PLA films have similar flexibility to polyethylene terephthalate (PET) and cellophane films but are not as flexible as LDPE films, thus can be sensitive to tearing. BOPLA films (e.g., Tafhleef Industries' NATIVIA™) can successfully replace the widely used bi-axially oriented PP in the packaging of cut fresh vegetables and fruit. In this application, the higher water vapor permeability of BOPLA is an advantage, yet the films are supplementarily perforated to reach the required level of water permeation [19, 45]. BOPLA films as flexible packaging have extended to fresh products and bakery goods, frozen foods, snacks and confectionary packaging, cookies and cereal and nutrition bars, and for lidding films. However, various shortcomings still must be addressed, such as noise reduction, barrier properties, shrink films, and thinner films down to 15 µm [46].

PLA rigid containers (trays) are highly comparable to polystyrene (PS) trays and can be used to pack fruits and vegetables [47]. In this application, barrier properties are not so important. PLA thermoformed cups are also very suitable to replace PS in dairy product cups. One of the first commercial applications of PLA in food packaging was as a thermoformed PLA composite cup for organic yoghurt. This product was introduced with technical success [48] on the German market in the late 1990s by Danone (French world-leading food company). However, the market introduction failed, and the product was subsequently withdrawn from the market. Later, in 2011, Danone has switched to Ingeo™ PLA from NatureWorks for its Activia brand yogurt in Germany. Danone was the first company to switch to PLA packaging for a leading yogurt product in Europe and Stonyfield Farm, Londonderry, NH, was the first in the United States (few months earlier, in October 2010). The new PLA cups, replaced the previous packaging made from high-impact PS, resulting in 48% less greenhouse gas emissions [49]. The use of PLA for the packaging of organic yoghurt as well as milk was also supported by Valio Ltd (a

Finnish dairy company), but with some reservations due to the still higher cost [50]. Moreover, a Danish dairy company has replaced the yogurt cups, traditionally made from high-impact PS, with biodegradable PLA cups [51].

The development of higher heat-resistant and high-impact PLA allows it to be used as a renewable replacement for PS and PP in high-heat thermoformable applications, such as microwaveable frozen food trays, hot food take-out containers, and hot beverage cups lids. [46]. A commercial example for beverage cups is a corn-based PLA cup for cold drinks produced by Biocorp (USA), a manufacturer of compostable and biodegradable materials [52]. Thermoformed PLA cups are also used for packaging fresh salads by McDonald's (the world's largest fast-food chain). The Italian supermarket chain, IPER, is using PLA trays and films to pack fresh foods and pasta. Other commercial examples include the use of PLA for the production of lunch boxes and fresh food packaging [53] and containers for the packaging of bottled water and juices [54]. PLA has a high water vapor and gas permeability is limited to be applied solely as the packaging of water sensitive products that will be stored over longer periods [55]. This can be overcome by using SiOx or AlOx technologies, obtaining composites [35], or used in laminates films with barrier materials, while still maintaining composability. These methods were used by Nature-Works and Metalvuoto to develop a new generation of high barrier PLA-based flexible substrates designed to keep processed foods fresh on store shelves [19, 56].

Even if PLA can be used to obtain bottle-shaped containers, having a very similar look and feel as PET bottles, the barrier properties of PLA are not sufficient to replace PET in long shelf-life applications, thus the producers do not actively promote this type of application. Some attempts to improve these properties of PLA are by using SiOx technology, which was reported by KHS company (Germany) [57], but because the technology is rather expensive and has not been implemented for PLA bottles [19].

As is evident from the state of the art, actual commercial applications of PLA-based food packaging are few, one of the main reasons being its high costs.

10.3.3   PLA-BASED MATERIALS IN ACTIVE FOOD PACKAGING

The increasing demand for healthier and more nutritional food has prompted the development of new conservation and packaging technologies. Active/intelligent packaging attempts to strengthen or take advantage of interactions in the food/packaging/environment system, acting in a coordinated manner to improve the healthiness and quality of the packaged food and increase its shelf-life [58]. The wide diversity of active packaging systems comprise additives with a multitude of active functions, namely, absorbing/scavenging properties (e.g., oxygen, carbon dioxide, ethylene, moisture, flavors, taints, and UV light); releasing/emitting properties (e.g., ethanol, carbon dioxide, antioxidants, preservatives, sulphur dioxide, and flavors); removing properties (catalyzing food component removal: lactose, cholesterol); and temperature, microbial, and quality control [59]. It can be obtained by the incorporation of active compounds into matrices generally used in existing packaging materials, or by the application of coatings with the mentioned functionality through physical or chemical surface modification [60].

Based on its intrinsic properties, PLA can be used as a matrix to develop sustainable active food packaging. To this end, PLA could provide a controlled release system where the slow release of the active agents from the PLA polymer to the wrapped/packed food could offer long-term protection against food spoilage [1]. The active agents used for active packaging are following the same trend as the packaging materials toward natural-based and eco-friendly alternatives. Natural additives not only act as reinforcing agents but they also improve barrier and wetting properties, as well as conferring antioxidant and antimicrobial properties, rendering the PLA matrix and active packaging material [61–64].

Active PLA films were developed by Safei and Azad [65] through the incorporation of different amounts of propolis extract (PE), as an active agent, and tested on dry meat sausage. Furthermore, incorporation of PEG/CaCO$_3$ into PLA/PE films remarkably improved the antimicrobial activity of films, enhanced flexibility and stiffness of polymers, and reduced their tensile strength. Heydari-Majd et al. [1] have improved the efficacy of a PLA packaging film, by extending the shelf-life of *O. ruber* fish fillets in the refrigerator, by supplementing it with ZnO NPs, and two essential oils *Zataria multiflora* Boiss. (ZEO) and *Menta piperita* L. (MEO).

With the addition of fennel (FEN) oil, biodegradable films based on PLA and polyhydroxybutyrate (PHB) was developed to possess the antimicrobial ability and were used on the preservation of oysters. The study suggested that the PLA-PHB-FEN film could prolong the shelf-life of oysters for 2–3 days while maintaining their quality compared to ethanol vinyl alcohol (EVOH) film [66].

10.4 BIODEGRADABILITY ASPECTS

Polymer degradation is induced by a range of factors in nature, such as oxidation, photodegradation, thermolysis, hydrolysis, biodegradation, or enzymolysis [67]. In general, the biodegradability of polymers is predetermined by their chemical and/or physical structure most of them having hydrolyzable or oxidizable bonds in the polymer backbone. The accessibility of these bonds for enzymes and sufficient flexibility of macromolecules are subsequent requirements for biodegradation [2].

Poly(L-lactic acid) degradation occurs mainly through the scission of ester bonds. PLA degrades biologically into lactic acid, a product of the carbohydrate metabolism, and its importance as a substitute for the nondegradable thermoplastics is an attractive feature [33]. Although PLA is a biodegradable polymer, the complete disappearance of PLA in a natural environment may take several years [68]. The biodegradability of PLA depends on the environment to which it is exposed and can be degraded in both aerobic and anaerobic conditions [69].

In human or animal bodies, it is assumed that PLA is initially degraded by hydrolysis and the soluble oligomers formed are metabolized by cells [70]. Upon disposal in the environment (e.g., soil, water),

the degradation process and its duration and mechanism strongly depend on the environmental conditions including heat, humidity, pH, oxygen, microbes, and so on. PLA requires a temperature around 60 °C (its glass transition temperature) which is not reached in normal soil conditions [71]. PLA decays best in environments with elevated temperatures because rapid chemical hydrolysis occurs and has a rich presence of microorganisms. But even in such conditions, PLA will start to show signs of biodegradation approximately in 6 months [72]. Nevertheless, an important feature of PLA is that its biodegradation will not pollute the environment [73].

10.4.1 BIODEGRADATION STAGES

PLA biodegradation is a complex process that occurs in three phases [74]: biodeterioration, biofragmentation, and assimilation. Biodeterioration encompasses the physical and chemical modifications of PLA properties after a microbial community adheres to the material surface as a biofilm; biofragmentation is the cleavage of PLA into oligomers, dimers, or monomers by extracellular hydrolytic enzymes; assimilation involves the simple molecules resulting from biofragmentation being transported to the cytoplasm of microbial cells and catabolized [75].

Figure 10.1 presented schematically the biochemical processes involved during microbial degradation of PLA [73]. PLA-degrading microorganisms first excrete extracellular depolymerase of PLA generally stimulated by proteins (e.g., elastin, gelatin, silk fibroin), peptides, and aminoacids—the so-called inducers [76, 77]. The depolymerase attacks the intramolecular ester links of PLA, which result in the production of oligomers, dimers, and monomers. Subsequently, the low molecular weight compounds are taken over by the microbial membranes and

are decomposed by the intercellular enzymes into CO_2, H_2O, or CH_4 [70, 71, 78].

As reported in different studies [70, 79], a variety of factors affect the microbial degradation of PLA-based materials, such as polymer properties (molecular weight and molecular weight distribution, glass transition temperature (T_g), melting temperature (T_m), crystallinity, and modulus of elasticity), abiotic processes (oxidative, thermal, chemical, or photodegradation), and characteristics of the biological material used. The shape of PLA materials did not show any

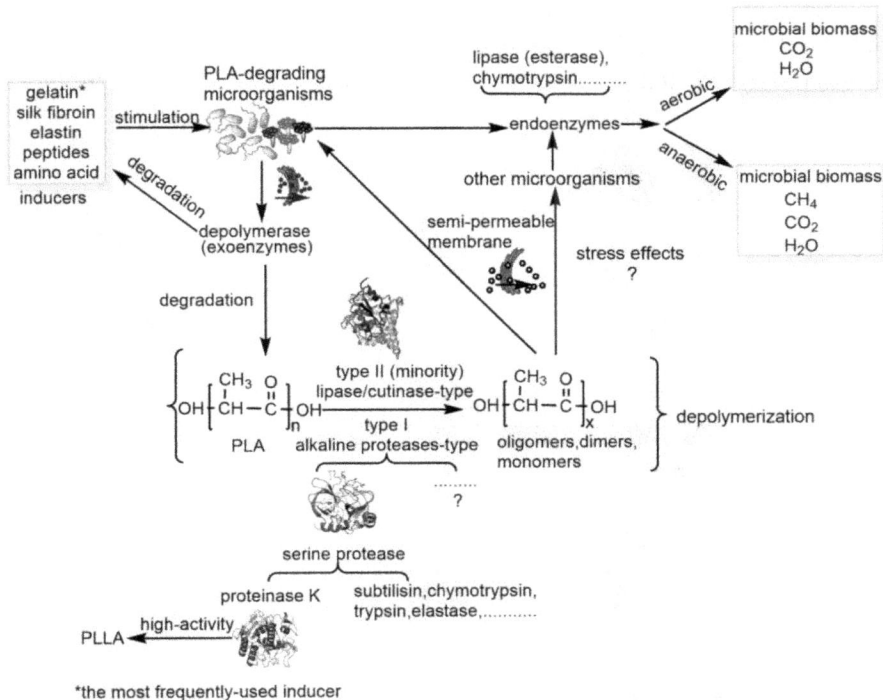

*the most frequently-used inducer

FIGURE 10.1 Schematic representation of the biochemical processes involved in PLA degradation. Reprinted with permission from [73], Copyright 2020 Elsevier.

significant difference since degradation occurs simultaneously on the surface and the bulk. As reported by Castro-Aguirre et al. [79], the abiotic step or hydrolysis is the main contribution to the degradation process of PLA in the early stage of degradation. Therefore, it is a limiting factor for the subsequent biodegradation of PLA, knowing that the evolution of the process in the initial stage (lag phase) is the time determining step for degradation [80].

In general, the hydrolytic degradation of PLA-based solid polymer matrices can proceed through under two different mechanisms [81, 82]: (1) surface or heterogeneous reactions and (2) bulk or homogeneous erosion—Figure 10.2.

The degradation behavior of PLA is greatly influenced by its average molecular weight, namely, PLA samples with high molecular weight are degraded at a slower rate than those with low molecular weights. Hence, the high molecular weight PLA needs to undergo an initial chemical hydrolysis to reduce its molecular weight prior to microbial mineralization [69, 83]. Moreover, also the morphology of PLA greatly affects their biodegradation rates. The degree of crystallinity is a crucial factor affecting biodegradability since enzymes mainly attack the amorphous domains of a polymer. The molecules in the amorphous region are loosely packed, and thus make

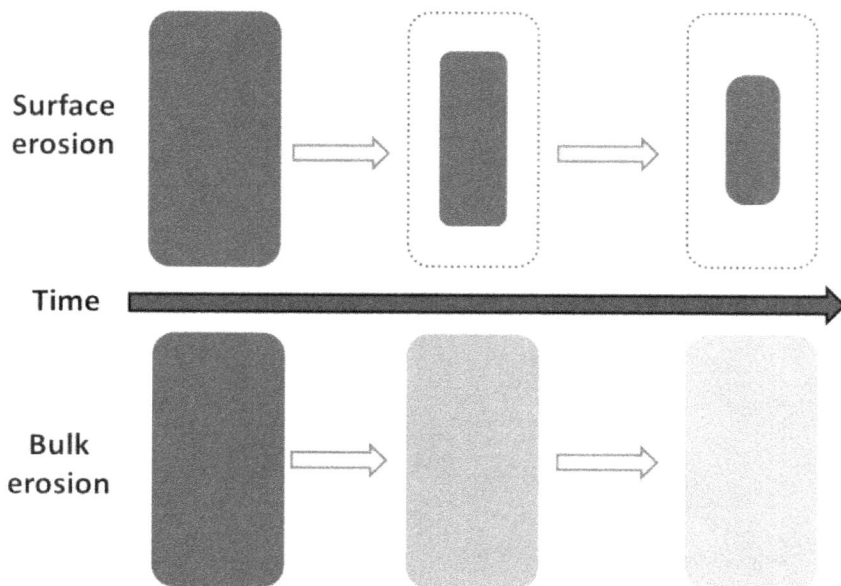

FIGURE 10.2   Mechanisms of hydrolytic degradation in PLA.

it more susceptible to degradation. The crystalline part of the polymers is more resistant than the amorphous region [84].

10.4.2 PLA-DEGRADING MICROORGANISMS

Biodiversity and occurrence of PLA-degrading microorganisms vary depending on the environment, such as soil, sea, compost, activated sludge, etc [84]. Compared to other biodegradable plastics (e.g., poly-ε-caprolactone or PHB), PLA materials are less sensitive to attacks of microbial in the native soil environment, mainly because PLA-degrading microorganisms are not widely distributed in natural soils and present at very low percentages [75]. Hence, the studies on pure isolation of PLA-degrading microorganisms have been increased in recent years. Researchers have primarily focused their attention on the isolation of thermophilic microorganisms because PLA undergoes rapid chemical hydrolysis at elevated temperatures, which facilitates the microbial mineralization of PLA. Tokiwa and Calabia [70] concluded that most of the PLA-degrading microorganisms phylogenetically belong to the family of *Pseudonocardiaceae* and related genera, such as *Amycolatopsis, Lentzea*, etc., in which proteinous

materials promote the production of the PLA-degrading enzyme. Nowadays, some types of microbes that are able to degrade PLA have been isolated from soil or water [5, 69, 71, 85].

The study elaborated in 1981 by William [86] was firstly reporting that PLA can be degraded by proteinase K, a serine protease produced by the mold strain *Tritirachium album* (*T. album*). Since then, the research in the field of PLA biodegradation has intensified with a more pronounced emphasis in the last decade on bacteria and fungi able to degrade the biopolymer. Table 10.1 listed the main PLA-degrading microorganisms reported in the literature until now. They mostly are actinomycetes (*Amycolatopsis, Saccharothrix, Lentzea, Kibdelosporangium, Streptoalloteichus, Pseudonocardiaceae*) which accelerate PLA biodegradation in the natural soil microcosms [87–90], bacteria (*Bacillus, Pseudomonas, Stenotrophomonas*, etc.) [91–94] and various fungi (*T. album, Chaetomium globosum, Phanerochaete chrysosporium*, etc.) [71, 73, 95–97].

To increase the biodegradation rate of PLA, different studies have focused on the addition of natural fillers (starch, kenaf bast fibers, chitosan) into a polymeric matrix to obtain composite materials or by using different predegradation methods (e.g., irradiation) [33, 35,

TABLE 10.1   Main PLA-degrading Microorganisms and Their Sources Reported in Literature

Microorganism	Substrate	Sample Source	Reference
Actinomycete			
Amycolatopsis SNC strain	PLA	Soil	[75]
Amycolatopsis strain HT-32, No.3118, KT-s-9	PLLA	Soil	[78]
Amycolatopsis strain K104-1	PLLA	Soil	[87]
Amycolatopsis strain 41	PLLA	Soil	[98]
Amycolatopsis orientalis	PLLA	IFO12362[a]	[99]
Amycolatopsis thailandensis	PLA07 PLLA	Soil	[100]
Amycolatopsis strain SCM_MK2-4	PLLA	Soil	[101]
Saccharothrix waywayandensis	PLLA	JCM 9114[b]	[102]
Kibdelosporangium aridum	PLLA	JCM 7912[b]	[103]
Actinomadura strain T16-1	PLLA	Soil	[104]
Laceyella sacchari LP175	PLLA	Soil	[105]
Pseudonocardia alni AS4.1531(T)	PLLA	Soil	[106]
Pseudonocardia sp. RM423	PLLA	KU[c]	[88]
Bacteria			
Bacillus brevis	PLLA	Soil	[107]
Bacillus stearothermophilus	PDLA	Soil	[108]
Bacillus smithii strain PL21	PLLA	Garbage	[109]
Bacillus licheniformis	PLLA	Compost	[110]
Paenibacillus amylolyticus strain TB-13	PDLLA	Soil	[111]
Alcaligenes sp.	PLLA	MS[d]	[112]
Geobacillus thermocatenulatus	PLLA	Soil	[92]
Pseudomonas sp. *strain DS04-T*	PLLA	Activated sludge	[113]
Pseudomonas geniculate WS3	PLA	Bacto™ Soytone	[114]
Stenotrophomonas maltophilia LB 2-3	PLLA	Soil	[115]
Pseudomonas tamsuii TKU015	PLA	Soil	[94]
Alcanivorax borkumensis ABO2449	PDLLA	–	[93]
Rhodopseudomonas palustris RPA1511	PDLLA	–	[93]
Fungus			
T. album ATCC 22563	PLLA	SCC[e]	[95]
Cryptococcus sp. *strain S-2*	PDLA	Wastewater	[96]
Aspergillus oryzae RIB40	PDLLA	Fermentation	[97]
Trichoderma viride	PLLA	SIHB[f]	[85]
Phanerochaete chrysosporium	PLA	SIHB[f]	[5]
Chaetomium globosum	PLA	SIHB[f]	[71]

Notes: [a]Institute for Fermentation, Osaka; [b]Japan Collection of Microorganisms; [c]Kasetsart University; [d]Meito Sangyo (Tokyo, Japan); [e]Sigma Chemical Co.; [f]Scientific Institute of Health Belgium. Adapted from [73].

116, 117]. Moreover, using plasticizers or blending PLA with more flexible polymers can decrease the glass transition of the material, favoring biodegradation [80, 118, 119]. It was found that the degradation rate of PLA was lower than that of PLA-based biocomposites with coir (CF)/pineapple leaf fibers (PALF) (CF/PALF/PLA) after 250 h of accelerated weathering. Biocomposites degrade after weathering through photoradiation, thermal degradation, oxidation, and hydrolysis. Water enhances the rate of degradation through the swelling of the fiber, which leads to further light dispersion. The soil burial tests imply good biodegradability of the CF/PALF/PLA biocomposites. The percentage weight loss in all the biocomposites was linearly related to the number of days of soil burial [35].

Chitosan, added especially in high amounts, increases the hydrophilicity of PLA composites, favoring moisture uptake, and the attack of microorganisms in soil burial for 150 days [117].

The PLA/ethylene vinyl acetate (EVA) copolymer blends biodegradation under different conditions of the agricultural soil in Vietnam (and mud) were study by Cong et al. [33]. It was revealed that PLA/EVA blends were easier to be decomposed under Vietnamese agricultural conditions than pristine PLA and were discovered some novel strains in Vietnamese agricultural environment, belonging to the *Rhizobium* sp. and *Alpha proteobacterium* species, which were responsible for the biodegradation of PLA-alike polymers.

Biodegradation of PLA can be achieved both aerobic and anaerobic, depending on the environment in which it is being degraded. The anaerobic degradation of PLA is less studied than the aerobic biodegradation, most likely because its degradation is slow at low temperatures, even though, anaerobic conditions are of interest in landfill, marine disposal, and anaerobic digestion systems [120].

10.4.3 AEROBIC BIODEGRADATION

During aerobic biodegradation, microorganisms use the polymer as a source of carbon for growth [79]. Under aerobic conditions, the biodegradation of the polymer forms CO_2 and H_2O in addition to the cellular biomass of microorganisms. Aerobic microorganisms use oxidative reactions to degrade the organic material. The aerobic process is more efficient than the anaerobic process through energy production as less energy is produced in anaerobic processes because of the lack of O_2, which serves as an electron acceptor, and this is more efficient in comparison to CO_2 and SO_4^{2-} [121, 122].

10.4.4 ANAEROBIC BIODEG-RADATION

Anaerobic digestion is a multistage process involving a complex population of microorganisms. Degradation by anaerobic bacteria takes place by reductive types of reactions [123]. The main product of anaerobic digestion is biogas, for example, a mix of methane, carbon dioxide, and trace amounts of nitrogen, ammonium, water vapor, and hydrogen sulphide [124]. Hence, anaerobic digestion converts biodegradable materials into energy-rich biogas. The anaerobic fermentation plant is a nearly closed system as compared to the more stenchful aerobic one with a shorter processing time and produces CH_4 as an energy source [125]. In anaerobic digestion, the decomposition of organic particles is usually a limiting step and this process is assumed to be surface limited [124]. The biodegradability of PLA films under thermophilic anaerobic conditions ranges from 86% to 99% depending on the specimens' size [126, 127]. Few studies have been reported until now on the behavior of PLA-based materials under anaerobic conditions. Iwańczuk et al. [124] have noticed that PLA exhibited a little change in properties after immersion in the anaerobic sludge.

Different analytical techniques have been used to evaluate biodegradation of polymers in composting using a direct or an indirect approach. Even though techniques like visual observations, mass loss measurements, changes in mechanical properties, and changes in molecular weight can provide insights into the degradation process of a polymer, they do not necessarily demonstrate biodegradation [128]. Therefore, respirometric methods, in which the consumption of oxygen and/or the evolution of carbon dioxide (CO_2) is measured, have become the preferred technique for such assessment [79].

10.5 END-OF-LIFE OPTIONS AND ENVIRONMENTAL IMPACT OF PLA

In the recent scientific literature, an aspect of bioplastics biodegradability started to be often discussed, namely the timeframe under which a "biodegradable" polymer actually biodegrades. For example, most packaging materials marked as "biodegradable" completely break down only if composted in industrial units, while they will most probably have limited biodegradation when landfilling (LF) [80]. This is also the case of PLA, its biodegradation being a very controversial problem because it requires temperature around 60 °C and high moisture, conditions that are not reached in normal environments. Some researchers support the ideas that PLA is just a degradable bio-based polymer since it is compostable only in industrial environments and shows very little mineralization (not biodegradation) in the majority of environments [129]. It is stated that PLA does not biodegrade very fast or at all in most environments [130]. Actually, this controversy is driven by the lack of a well-established terminology regarding biodegradability and because until now, its definition does not include the timeframe notion. This aspect is complicated to be achieved especially because is problematic to set a reasonable period of time in which the changes in the material properties start to appear [80]. Moreover, the controversy is accentuated also by the fact that the terms "biodegradable" and "compostable" are often used (in fact misused) interchangeably in the domain of recycling. At this time, in the absence of another definition of biodegradability, we consider that it is not wrong to say that PLA is a biodegradable material.

The lifecycle of PLA is based on a closed-loop system—Figure 10.3. Instead of simply reducing negative environmental impacts, PLA looks to create a positive impact. LCA studies of PLA reveal that it conveys clear benefits in the domains of climate protection and conservation of fossil resources in comparison to fossil-based plastics, whereas negative impacts are often estimated for environmental impacts that result from biomass production. These are in particular potential impacts

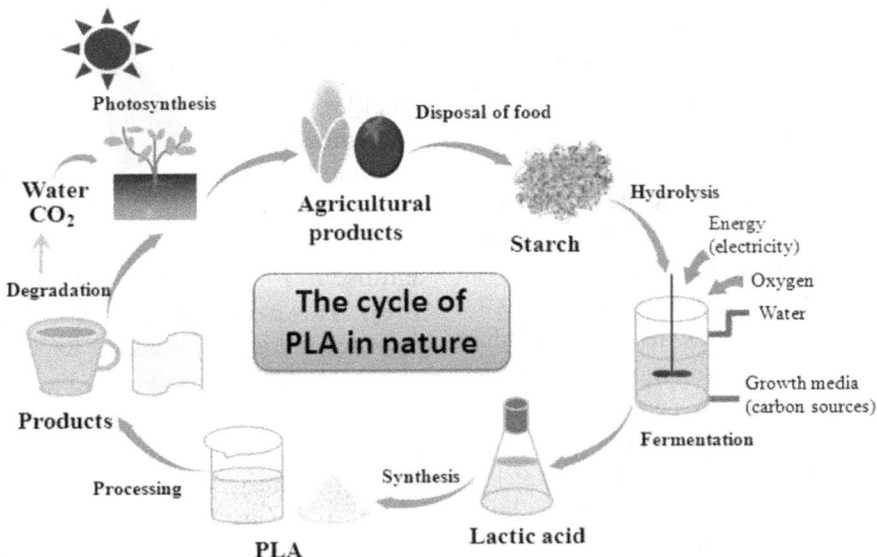

FIGURE 10.3 The cycle of PLA in nature.

Source: Reprinted from Ref. [11]. https://creativecommons.org/licenses/by/3.0/

on eutrophication, acidification, and land use [131–133].

10.5.1 END-OF-LIFE OPTIONS

At present, there are some uncertainties with respect to the end-of-life (EoL) stage since existing waste management systems have to be adapted to PLA and other bio-based plastics [131]. In general, apart from the EoL options suitable for conventional plastics, such as reusing, recycling (reprocessing), and renewable energy recovery (thermal treatment in a waste incineration plant), the EoL options for PLA also include industrial composting/biodegradation, anaerobic digestion, and feedstock recovery (chemical recycling) [134].

From the recycling strategies reported for PLA, mechanical recycling is one of the simplest and most effective energy-saving techniques for reusing PLA waste and is achieved *via* multiple reprocessing steps and employing the same system that produces the primary product. The chemical recycling or feedstock recovery of PLA means breaking biopolymer back into its "building blocks" and reusing those "building blocks." This is generally achieved by hydrolysis, alcoholysis, and thermal depolymerization [135]. A study by Piemonte et al. [136] claims that the production of lactic acid from chemical depolymerization of PLA is preferable, from an energy

point of view to the production of lactic acid by glucose fermentation. Moreover, the authors conclude that the environmental footprint of the analysed process is larger than that of the PLA mechanical recycling. PLA products (mainly long life PLA products) and packaging could be reused by households, assuming the PLA products and packaging maintain the desired properties, functionality, and safety [19].

As early reported in a study by Chien et al. [137], the amounts of polycyclic aromatic hydrocarbons emitted from PLA combustion are significantly lower than those associated with the combustion of other plastics. Moreover, PLA can be combusted with no remaining residue. Thus, the authors conclude that incineration might be a suitable approach for the environmentally friendly disposal of waste PLA. Later was shown that a part of the embedded energy in PLA is recovered, in the cases of thermal treatment in a municipal solid waste incineration plant and treatment, and use as refuse-derived fuel [131, 138]. However, energy recovery does not reduce the demand for raw material used in plastic production [139]. The LCA study conducted by Maga et al. [131] revealed that all investigated PLA recycling technologies (mechanical, chemical, and solvent-based) perform better from an environmental perspective than waste incineration. It is important to be mentioned that this study reflects the German situation and may differ from country to country because the waste collection and sorting system can vary strongly. The previous statement is also supported by the research of Rossi et al. [138] which concludes that food packaging materials heavily contaminated with food may influence the LCA studies.

As reported in different studies until now, industrial composting was evaluated as the worst EoL option for PLA from a climate perspective [138, 140]. Composting neither improves the quality of compost nor allows for energy recovery. However, there are some situations where composting may be a preferable EoL option, such as for packages that have no other desirable path (reuse or recycling) [141]. It would seem that composting would be the optimal EoL option for contaminated PLA. However, there are only a few existing composting facilities that accept biodegradable plastic materials since most are concerned that biodegradable plastics are not easily distinguishable from conventional plastics and that quality control is difficult [19].

The less preferable option to dispose of PLA is LF. The drawback of disposing plastics in landfills lies in the fact that most plastic materials do not degrade in a practical period of time and end up accumulating. Landfills usually do not provide the appropriate environment (conditions

vary considerably by geography) to promote the degradation of PLA in a practical period of time [142].

10.5.2   ENVIRONMENTAL IMPACT

In the study developed by Rossi et al. are considered six EOL options that cover the different levels of the EU waste treatment hierarchy.. Figure 10.4 presents the global warming impacts for PLA (expressed in kg of CO_2 eq per of dry PLA packaging), which was dynamically assessed over a 100-year time horizon. As the study revealed, the industrial composting EoL option has the highest net impact and MR has the lowest net impact. Figure 10.5 shows nonweighted scores of the material production and EoL scenarios for each midpoint category, in which for most impact categories mechanical recycling of PLA was the least-burdening option. Contrariwise, industrial composting and LF were the least favorable options for most impact categories [6].

Similar results were obtained in a more recent study performed by Maga et al. [25] were was revealed that meat trays made from PLA have the highest effect in most of the investigated impact domains and extruded polystyrene (XPS) show the lowest impacts. The study involved the evaluation of LCA for

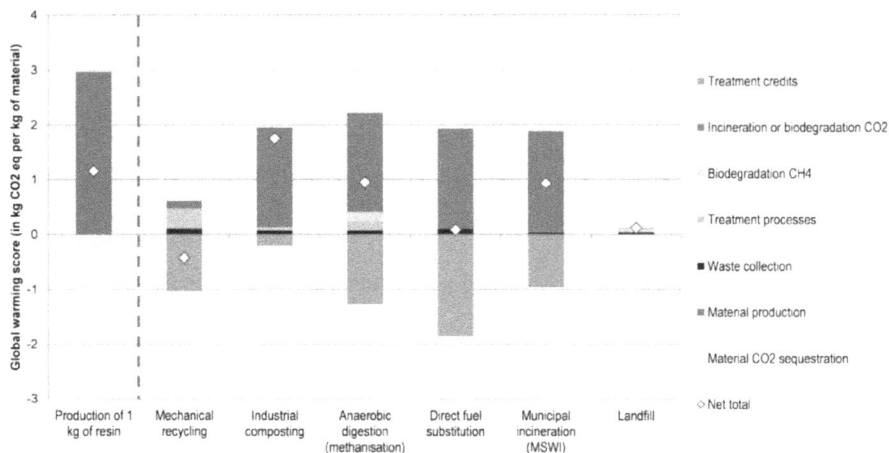

FIGURE 10.4 Comparison of dynamically assessed global warming impacts over 100 years related to the six EoL treatments of PLA (a). The bars on the left side present production impacts of the resin (cradle-to-gate) for comparison purposes.

Source: Reprinted with permission from Ref. [138]. © 2015 Elsevier.

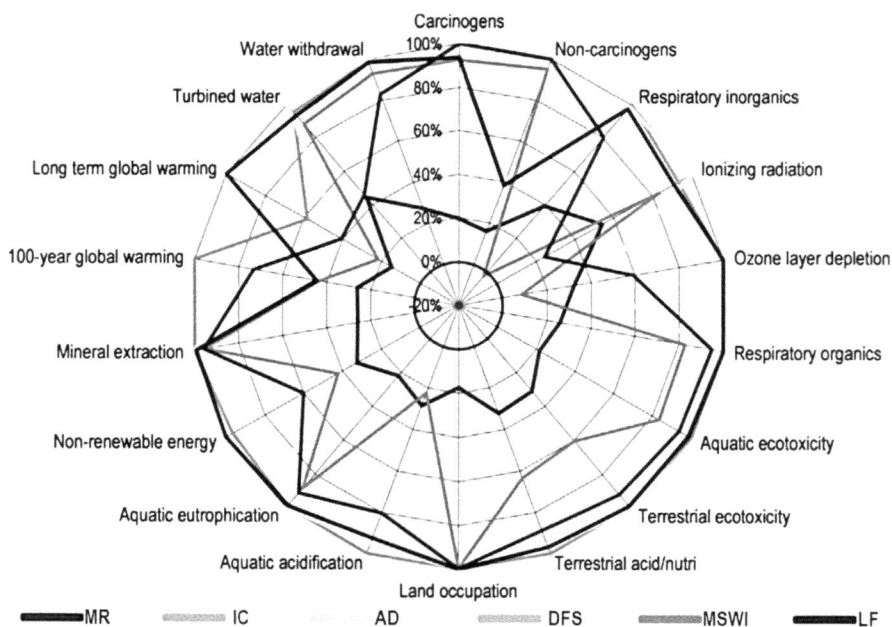

FIGURE 10.5 Comparison of EoL options for PLA for each midpoint category

Source: Adapted with permission from [138] © 2015 Elsevier.

different food packaging, namely, PLA-, XPS-, PP-, and PET-based meat trays. In general, XPS trays allow for reducing the material demand which dominates the life cycle impact assessment results. All of these materials perform worse than XPS trays across all investigated impact categories except for resource depletion. The mentioned study has also limitations since it only focused on the direct environmental impact of the tray itself and not its content (meat), considering that food waste might also be influenced by packaging solutions. Moreover, the choice of tray material and tray design influences the shelf-life of meat and, therefore, could be a relevant aspect to be considered in LCA assessments of food packaging.

When recycling of PLA waste is exclusively compared to thermal treatment the life cycle impact results show environmental benefits (e.g., impact categories—agricultural land occupation, energy-saving, photochemical ozone formation, terrestrial and aquatic eutrophication, and acidification) of all recycling technologies. Environmental advantages are achieved by replacing virgin PLA with PLA recyclates [131].

The results of LCA assessment studies may be different when considering packaging materials based on PLA biocomposites. Beigbeder et al.'s [143] study estimated that the recycling EoL scenario presents the lowest environmental impacts, followed by industrial composting for PLA/flax fibers composite. Energy recovery and landfill are the less interesting EoL scenarios considering the environmental impacts.

10.6 FINAL REMARKS

Currently, the majority of food packaging solutions contain fossil-based plastics, which are produced and consumed in an unsustainable manner. The development of bio-based PLA food packaging is of great interest to both the scientific community and the packaging industry. Even if in the last decade, continuous efforts are being made to increase the CAGR of bio-based polymers, and especially PLA, to achieve sustainable food packaging solutions, the goal is still far to be reached. Certain impediments hamper the large mass production of PLA food packaging, mainly its high production cost, poor policy support for bio-based packaging, inappropriate regulations for the application of materials to food products, and inappropriate or absent composting facilities. Moreover, this delay is also maintained by the controversy that has arisen around PLA biodegradability. Some researchers sustain that PLA is a bio-based compostable polymer and not a biodegradable one. In my opinion, in the absence of an updated definition of biodegradability, it is not wrong to say that PLA is a biodegradable material. Also, more efforts should be made to study the end of life scenarios and their environmental impact on various PLA-based packaging, taking into account that these aspects should be evaluated before they are produced and installed commercially, to ensure a minimal impact of these products all along their life cycle [143].

ACKNOWLEDGMENTS

The author acknowledge the grant of the Romanian Ministry of Education and Research, CNCS—UEFISCDI, project code PN-III-P1-1.1-PD-2019-1101, contract number PD 31/2020, and the co-funding by the European Social Fund, through Operational Programme Human Capital 2014–2020, project number POCU/ 380/6/ 13/123623, project title "PhD Students and Postdoctoral Researchers Prepared for the Labour Market!"

KEYWORDS

- poly(lactic acid)
- food packaging
- bio-based polymer
- biodegradability

REFERENCES

1. Heydari-Majd, M.; Ghanbarzadeha, B.; Shahidi-Noghabi, M.; Najafi, M.A.; Hosseini, M.; A New Active Nano-composite Film Based on PLA/ZnO Nanoparticle/Essential Oils for the Preservation of Refrigerated *Otolithes ruber* Fillets. *Food Packag Shelf Life* **2019**; 19:94–103.
2. Šašek, V.; Vitásek, J.; Chromcová, D.; Prokopová, I.; Brožek, J.; Náhlík, J.; Biodegradation of Synthetic Polymers by Composting and Fungal Treatment. *Folia Microbiol* **2006**; 51(5):425–430.
3. Shimao, M.; Biodegradation of Plastics. *Curr Opin Biotechnol* **2001**; 12: 242–247.
4. Fukushima, K.; Abbate, C.; Tabuani, D.; Gennari, M.; Camino, G.; Biodegradation of Poly (Lactic Acid) and its Nanocomposites. *Polym Degrad Stab* **2009**; 94:1646–1655.
5. Stoleru, E.; Hitruc, E.G.; Vasile, C.; Oprica, L.; Biodegradation of Poly(Lactic Acid)/Chitosan Stratified Composites in Presence of the *Phanerochaete Chrysosporium* Fungus. *Polym Degrad Stab* **2017**; 143:118–129.
6. Castro-Aguirre, E.; Iñiguez-Franco, F.; Samsudin, H.; Fang, X.; Auras, R.; Poly(lactic acid)—Mass production, Processing, Industrial Applications, and End of Life. *Adv Drug Deliv Rev* **2016**; 107:333–366.
7. Datta, R.; Tsai, S.; Bonsignore, P.; Moon, S.; Frank, J.; Technological and Economic Potential of Poly(lactic acid) and Lactic Acid Derivatives. *FEMS Microbiol Rev* **1995**; 16:221–231.
8. Kricheldorf, H.R.; Kreiser-Saunders, I.; Jürgens, C.; Wolter, D.; Polylactides-Synthesis, Characterization and Medical Application. *Macromol Symp* **1996**; 103:85–102.
9. Hartmann, M.H.; High molecular weight polylactic acid polymers.

In: Biopolymers from Renewable Resources; D.L. Kaplan, Ed. Springer: Heidelberg, **1998**, 367–411.

10. Stoleru, E.; Zaharescu, T.; Hitruc, E.G.; Vesel, A.; Ioanid, E.G.; Coroaba, A.; Safrany, A.; Pricope, G.; Lungu, M.; Schick, C.; Vasile, C. Lactoferrin-Immobilized Surfaces onto Functionalized PLA Assisted by the Gamma-Rays and Nitrogen Plasma to Create Materials with Multifunctional Properties. *ACS Appl Mater Interfaces* **2016**; 8 (46):31902–31915.

11. Xiao, L.; Wang, B.; Yang, G.; Gauthier, M.; Poly(Lactic Acid)-based biomaterials: synthesis, modification and applications. In: Biomedical Science, Engineering and Technology; Ghista, D. N., Ed. InTech: Rijeka, Croatia, **2012**. Available from the following: http://www.intechopen.com/books/howtoreference/biomedical-science-engineering-and-technology/poly-lactic-acidbased-biomaterials-synthesis-modification-and-applications.

12. Dorgan, J. R.; Lehermeier, H. J.; Palade, L.-I.; Cicero, J.; Polylactides: Properties and Prospects of an Environmentally Benign Plastic from Renewable Resources. *Macromol Symp* **2001**; 175:55.

13. Gruber, P. R.; O'Brien, M.; Polylactides. NatureWorksTM PLA, in: Biopolymers. Polyesters III. Applications and Commercial Products, 1st ed, Y. Doi, A. Steinbüchel, Eds. Wiley-VCH Verlag GmbH: Weinheim, Germany, **2002**, 235–250.

14. Kawashima, N.; Ogawa, S.; Obuchi, S.; Matsuo, M.; Yagi, T.; Polylactic acid "LACEA". In: Biopolymers. Polyesters III. Applications and Commercial Products, 1st ed, Y. Doi, A. Steinbüchel, Eds. Wiley-VCH Verlag GmbH: Weinheim, Germany, **2002**, 251–274.

15. Msuya, N.; Katima, JHY.; Masanja, E.; Temu, A.K. Poly(lactic-acid) Production from Monomer to Polymer: A

review. *SciFed J Polym Sci* **2017**; 1 (1):1–15.

16. Auras, R.; Harte, B.; Selke, S. An Overview of Polylactides as Packaging Materials. *Macromol Biosci* **2004**; 4:835–864.

17. Sawyer, D. J.; Bioprocessing—No Longer a Field of Dreams. *Macromol Symp* **2003**; 201:271–282.

18. Bogaert, J.-C.; Coszach, P.; Poly(lactic acids): A Potential Solution to Plastic Waste Dilemma. *Macromol Symp* **2000**; 15:287–303.

19. van den Oever, M.; Molenveld, K.; van der Zee, M.; Bos, H. Bio-based and Biodegradable Plastics—Facts and Figures. Wageningen Food & Biobased, The Netherlands, Research number 1722, **2017**, DOI: http://dx.doi.org/10.18174/408350.

20. Commission regulation (EU) (2011). No 10/2011 of 14 January 2011 on Plastic Materials and Articles Intended to Come into Contact with Food.

21. Mistriotis, A.; Briassoulis, D.; Giannoulis, A.; D'Aquino, S.; Design of Biodegradable Bio-Based Equilibrium Modified Atmosphere Packaging (EMAP) for Fresh Fruits and Vegetables by Using Micro-Perforated Poly-Lactic Acid (PLA) films. *Postharvest Biol Technol* **2016**; 111:380–389.

22. Rudeekit, Y.; Numnoi, J.; Tajan, M.; Chaiwutthinan, P.; Leejarkpai, T. Determining Biodegradability of Polylactic Acid Under Different Environments. *J Metals, Mater Miner* **2008**; 18:83–87.

23. Robertson, G.L.; Food Packaging: Principles and Practices. CRC Press: Boca Raton, FL, USA, **2012**.

24. Hottle, T.A.; Bilec, M.M.; Landis, A.E.; Sustainability Assessments of Bio-based Polymers. *Polym Degrad Stab* **2013**; 98:1898–1907.

25. Maga, D.; Hiebel, M.; Aryan, V. A Comparative Life Cycle Assessment of Meat Trays Made of Various

Packaging Materials. *Sustainability* **2019**, 11:5324.

26. Press release: Nova-Institut GmbH (www.nova-institute.eu) Hürth, 27 January 2020: The Global Bio-based Polymer Market 2019—A Revised View on a Turbulent and Growing Market, Available online: http://nova-institute.eu/press/?id=164 (accessed April 26, 2020).

27. https://www.mordorintelligence.com/industry-reports/bio-degradable-polymers-market.

28. Karamanlioglu, M.; Preziosi, R.; Robson, G.D.; Abiotic and Biotic Environmental Degradation of the Bioplastic Polymer Poly(lactic acid): A Review. *Polym Degrad Stabil* **2017**; 137:122–130.

29. https://bioplasticsnews.com/polylactic-acid-or-polylactide-pla/.

30. Hegde, S.; Dell, E.; Lewis, C.; Trabold, T.A.; Diaz, C.A. Anaerobic Biodegradation of Bioplastic Packaging Materials. 21st IAPRI World Conference on Packaging, ISBN: 978-1-60595-046-4.

31. Claro, P.I.C.; Neto, A.R.S.; Bibbo, A.C.C.; Mattoso, L.H.C.; Bastos, M.S.R.; Marconcini, J.M.; Biodegradable Blends with Potential Use in Packaging: A Comparison of PLA/Chitosan and PLA/Cellulose Acetate Films. *J Polym Environ* **2016**; 24:363–371.

32. Kalia, S.; Averous, L.; Biopolymers: Biomedical and Environmental Applications. Wiley: Hoboken, NJ, **2011**.

33. Cong, D.V.; Hoang, T.; Giang, N.V.; Ha, N.T.; Lam, T.D.; Sumita, M. A Novel Enzymatic Biodegradable Route for PLA/EVA Blends under Agricultural Soil of Vietnam. *Mater Sci Eng C* **2012**; 32:558–563.

34. Yuan, M.W.; Qin, Y.Y.; Yang, J.Y.; Wu, Y.; Yuan, M.L.; Li, H.L.; Preparation and Characterization of Poly(L-lactide-co-ε-caprolactone) Copolymer for Food Packaging Application. *Adv Mater Res* **2013**; 779:231–234.

35. Siakeng, R.; Jawaid, M.; Asim, M.; Siengchin, S. Accelerated Weathering and Soil Burial Effect on Biodegradability, Colour and Texture of Coir/Pineapple Leaf Fibres/PLA Biocomposites. *Polymers* **2020**; 12:458.

36. Sharma, S.; Jaiswal, A.K.; Duffy, B.; Jaiswal, S. Ferulic Acid Incorporated Active Films Based on Poly(Lactide) / Poly(Butylene adipate-co-terephthalate) Blend for Food Packaging. *Food Packag Shelf Life* **2020**; 24:1004.

37. Darie, R.N.; Pâslaru, E.; Sdrobis, A.; Pricope, G.M.; Hitruc, G.E.; Poiată, A.; Baklavaridis, A.; Vasile, C.; Effect of Nanoclay Hydrophilicity on the Poly(lactic acid)/Clay Nanocomposites Properties. *Ind Eng Chem Res* **2014**; 53:7877–7890.

38. Darie-Niță, R.N.; Munteanu, B.S.; Tudorachi, N.; Lipșa, R.; Stoleru, E.; Spiridon, I.; Vasile, C. Complex Poly(lactic acid)-based Biomaterial for Urinary Catheters: I. Influence of AgNP on Properties. *Bioinspired, Biomim Nanobiomater* **2016**; 5(4):132–151.

39. Liu, L.-Z.; Ma, H.-J.; Zhu, X.-S.; Preparation and Properties of Polylactide/Nanosilica in Situ Composites. *Pigment Resin Technol* **2010**; 39(1):27–31.

40. Picard, E.; Espuche, E.; Fulchiron, R. Effect of an Organo-modified Montmorillonite on PLA Crystallization and Gas Barrier Properties. *Appl Clay Sci* **2011**; 53(1):58–65.

41. Vasile, C.; Sivertsvik, M.; Mitelut, A.C.; Brebu, M.A.; Stoleru, E.; Rosnes, J.T.; Tanase, E.E.; Khan, W.; Pamfil, D.; Cornea, C.P.; Irimia, A.; Popa, M.E.; Comparative Analysis of the Composition and Active Property Evaluation of Certain Essential Oils to Assess their Potential Applications in Active Food Packaging. *Materials* **2017**; 10:45.

42. Jiang, J.; Gong, L.; Dong, Q.; Kang, Y.; Osako, K.; Li, L.; Characterization of PLA-P3,4HB Active Film Incorporated

with Essential Oil: Application in Peach Preservation. *Food Chem* **2020**; 313:126134.

43. Scaffaro, R.; Maio, A.; Nostro, A.; Poly (lactic acid)/Carvacrol-based Materials: Preparation, Physicochemical Properties, and Antimicrobial Activity. *Appl Microbiol Biotechnol* **2020**; 104(5): 1823–1835.

44. Zeid, A.; Karabagias, I.K.; Nassif, M.; Kontominas, M.G. Preparation and Evaluation of Antioxidant Packaging Films Made of Polylactic Acid Containing Thyme, Rosemary, and Oregano Essential Oils. *J Food Process Preserv*, **2019**; 43(10):e14102.

45. Tawakkal, I.S.M.A.; Cran, M.J.; Miltz, J.; Bigger, S.W.; A Review of Poly(Lactic Acid)-Based Materials for Antimicrobial Packaging. *J Food Sci* **2014**; 79(8):R1477–R1490.

46. Cooper, T.A.; Developments in Bioplastic Materials for Packaging Food, Beverages and Other Fast-moving Consumer Goods. In: Trends in Packaging of Food, Beverages and Other Fast-Moving Consumer Goods (FMCG), T.A. Cooper, Ed. Woodhead Publishing Series in Food Science, Technology and Nutrition: Cambridge, **2013**, 108–152.

47. Dorgan, J.R.; Lehermeier, H.; Mang, M.; Thermal and Rheological Properties of Commercial-Grade Poly(Lactic Acid)s. *J Polym Environ* **2000**; 8:1–9.

48. Bastioli, C.; Global Status of the Production of Biobased Packaging Materials. *Starch–Stärke* **2001**; 53(8):351–355.

49. https://www.greenerpackage.com/ bioplastics/danone_first_switch_pla_ yogurt_cup_germany.

50. Frederiksen, C.S.; Haugaard, V.K.; Poll, L.; et al. Light-induced Quality Changes in Plain Yoghurt Packed in Polylactate and Polystyrene. *Eur Food Res Technol* **2003**; 217:61–69.

51. Jessen, B.; Sustainability and Emerging Topics in Food Research and Education. *Danish Dairy Food Ind* **2007**; 17:22–3.

52. Mattsson, B.; Sonesson, U.; Environmentally-Friendly Food Processing. Woodhead Publishing, Cambridge, **2003**, 195.

53. Mutsuga, M.; Kawamura, Y.; Tanamoto, K.; Migration of Lactic Acid, Lactide and Oligomers from Polylactide Food-Contact Materials. *Food Add Contam* **2008**; 25(10):1283–90.

54. Ahmed, J.; Varshney, S.K.; Zhang, J.X.; Ramaswamy, H.S. Effect of High Pressure Treatment on Thermal Properties of Polylactides. *J Food Eng* **2009**; 93(3):308–12.

55. Halász, K.; Hosakun, Y.; Csóka, L.; Reducing Water Vapor Permeability of Poly(lactic acid) Film and Bottle Through Layer-by-Layer Deposition of Green-Processed Cellulose Nanocrystals and Chitosan. *Int J Polym Sci* **2015**; 2015:954290.

56. NatureWorks, 2016. High Barrier Biobased Flexible Structures. http:// www.natureworksllc.com/News-and-Events/Press-Releases/2016/03-17-16-Metalvuoto-Ingeo-PLA-barrier-film (visited April 24 2020).

57. Klages, A.; KHS Presentation at the BPM-feasible Project Meeting, Utrech, **2013**.

58. Catalá, R.; López-Carballo, G.; Hernández-Muñoz, P.; Gavara, R.; Chapter 10: PLA and Active Packaging. In: Poly(lactic acid) Science and Technology: Processing, Properties, Additives and Applications, vol. 12, Royal Society of Chemistry, Cambridge, UK; Jimenez, A.; Peltzer, M.; Ruseckaite, R. Eds. *RSC Polymer Chemistry Series*, **2015**; 243–265, DOI: 10.1039/9781782624806-00243.

59. Vilela, C.; Kurek, M.; Hayouka, Z.; Röcker, B.; Yildirim, S.; Antunes, M.D.C.; Nilsen-Nygaard, J.; Kvalvåg Pettersen, M.; Freire, C.S.R.; A Concise Guide to Active Agents for Active Food Packaging. *Trends Food Sci Technol* **2018**; 80:212–222.

60. Vasile, C. Polymeric Nanocomposites and Nanocoatings for Food Packaging: A Review. *Materials* **2018**; 11:1834.

61. Papadopoulou, E.L.; Paul, U.C.; Tran, T.N.; Suarato, G.; Ceseracciu, L.; Marras, S.; d'Arcy, R.; Athanassiou, A.; Sustainable Active Food Packaging from Poly(lactic acid) and Cocoa Bean Shells. *ACS Appl Mater Interfaces* **2019**; 11:31317–31327.

62. Sebastien, F.; Stephane, G.; Copinet, A.; Coma, V. Novel Biodegradable Films Made from Chitosan and Poly(Lactic Acid) with Antifungal Properties Against Mycotoxinogen Strains. *Carbohydr Polym* **2006**; 65: 185–193.

63. Del Nobile, M.A.; Conte, A.; Buonocore, G.G.; Incoronato, A.L.; Massaro, A.; Panza, O. Active Packaging By Extrusion Processing of Recyclable and Biodegradable Polymers. *J Food Eng* **2009**; 93(1):1–6.

64. Stoleru, E.; Dumitriu, R.P.; Munteanu, B.S.; Zaharescu, T.; Tanase, E.E.; Mitelut, A.; Ailiesei, G.-L.; Vasile, C. Novel Procedure to Enhance PLA Surface Properties by Chitosan Irreversible Immobilization. *Appl Surf Sci* **2016**; 367:407–417.

65. Safaei, M.; Azad, R.R.; Preparation and Characterization of Poly-Lactic Acid Based Films Containing Propolis Ethanolic Extract to be Used in Dry Meat Sausage Packaging. *J Food Sci Technol* **2020**; 57(4):1242–1250.

66. Miao, L.; Walton, W.C.; Wang, L.; Li, L.; Wang, Y. Characterization of Polylactic Acids-Polyhydroxybutyrate Based Packaging Film with Fennel Oil, and its Application on Oysters. *Food Packag Shelf Life* **2019**; 22:100388.

67. Nampoothiri, K.M.; Nair, N.R.; John, R.P.; An Overview of the Recent Developments in Polylactide (PLA) Research. *Bioresour Technol* **2010**; 101: 8493–8501.

68. Kimura, T.; Ishida, Y.; Ihara, N.; Saito, Y.; High Speed Degradation of Biodegradable Plastics by Composting of Biological Wastes. *Biosci Ind* **2000**; 57:35–36.

69. Bubpachat, T.; Sombatsompop, N.; Prapagdee, B.; Isolation and Role of Polylactic Acid-Degrading Bacteria on Degrading Enzymes Productions and PLA Biodegradability at Mesophilic Conditions. *Polym Degrad Stab* **2018**; 152:75–85.

70. Tokiwa, Y.; Calabia, B.P.; Biodegradability and Biodegradation of Poly (lactide). *Appl Microbiol Biotechnol* **2006**; 72:244–251.

71. Stoleru, E.; Vasile, C.; Oprică, L.; Yilmaz, O. Influence of the Chitosan and Rosemary Extract on Fungal Biodegradation of Some Plasticized PLA-based Materials. *Polymers* **2020**; 12:469.

72. https://all3dp.com/2/is-pla-biodegradable-what-you-really-need-to-know/ (accessed April 21, 2020).

73. Qi, X.; Ren, Y.; Wang, X.; New Advances in the Biodegradation of Poly(lactic) acid. *Int Biodeter Biodegr* **2017**; 117:215–223.

74. Emadian, S.M.; Onay, T.T.; Demirel, B; Biodegradation of Bioplastics in Natural Environments. *Waste Manag* **2017**; 59:526–536.

75. Decorosi, F.; Exana, M. L.; Pini, F.; Adessi, A.; Messini, A.; Giovannetti, L.; Viti, C. The Degradative Capabilities of New Amycolatopsis Isolates on Polylactic Acid. *Microorganisms* **2019**; 7:590.

76. Lim, H.-A.; Raku, T.; Tokiwa, Y. Hydrolysis of Polyesters by Serine Proteases. *Biotechnol Lett* **2005**; 27: 459–464.

77. Thanasak, L.; Srisuda, H.; Rangrong, Y.; Vichien, K.; Co-production of Poly(llactide)-Degrading Enzyme and Raw Starch-degrading Enzyme by Laceyella sacchari LP175 Using Agricultural Products as Substrate, and Their Efficiency on Biodegradation of

Poly(l-lactide)/Thermoplastic Starch Blend Film. *Int Biodeterior Biodegrad* **2015**; 104:401–410.

78. Kawai, F.; Nakadai, K.; Nishioka, E.; Nakajima, H.; Ohara, H.; Masaki, K.; Iefuji, H.; Different Enantioselectivity of Two Types of Poly(lactic acid) Depolymerases Toward Poly(l-lactic acid) and Poly(d-lactic acid). *Polym Degrad Stab* **2011**; 96:1342–1348.

79. Castro-Aguirre, E.; Auras, R.; Selke, S.; Rubino, M.; Marsh, T.; Insights on the Aerobic Biodegradation of Polymers by Analysis of Evolved Carbon Dioxide in Simulated Composting Conditions. *Polym Degrad Stab* **2017**; 137:251–271.

80. Brebu, M.; Environmental Degradation of Plastic Composites with Natural Fillers—A Review. *Polymers* **2020**; 12:166.

81. Elsawy, M.A.; Kim, K-H.; Park, J.-W.; Deep, A.; Hydrolytic Degradation of Polylactic acid (PLA) and its Composites. *Renew Sust Energ Rev* **2017**; 79:1346–1352.

82. Zarzycki, R.; Modrzejewska, Z.; Nawrotek, K. Drug Release from Hydrogel Matrices. *Ecol Chem Eng S* **2010**; 17(2):117–136.

83. Husarova, L.; Pekarova, S.; Stloukal, P.; Kucharzcyk, P.; Verney, V.; Commereuc, S.; Ramone, A.; Koutny, M. Identification of Important Abiotic and Biotic Factors in the Biodegradation of Poly(l-lactic acid). *Int J Biol Macromol* **2014**; 71:155–162.

84. Tokiwa, Y.; Calabia, B.P.; Ugwu, C.U.; Aib, S. Biodegradability of Plastics. *Int J Mol Sci* **2009**; 10:3722–3742.

85. Lipsa, R.; Tudorachi, N.; Darie-Nita, R.N.; Oprica, L.; Vasile, C. et al. Biodegradation of Poly(lactic acid) and Some of its Based Systems with *Trichoderma viride*. *Int J Biol Macromol* **2016**; 88:515–526.

86. William, D.F.; Enzymic Hydrolysis of Polylactic Acid. *Eng Med* **1981**; 10:5.

87. Nakamura, K.; Tomita, T.; Abe, N.; Kamio, Y.; Purification and Characterization of an Extracellular Poly(L-lactic acid) Depolymerase from a Soil Isolate, Amycolatopsis sp strain K104-1. *Appl Environ Microbiol* **2001**; 67:345–353.

88. Apinya, T.; Sombatsompop, N.; Prapagdee, B. Selection of a Pseudonocardia sp. RM423 that Accelerates the Biodegradation of Poly (lactic) acid in Submerged Cultures and in Soil Microcosms. *Int Biodeterior Biodegrad* **2015**; 99:23–30.

89. Kawai, F.; Chapter 27: Polylactic Acid (PLA)-degrading microorganisms and PLA Depolymerases. In: Green Polymer Chemistry: Biocatalysis and Biomaterials, 1043. ACS Symposium Series; American Chemical Society, Washington DC, **2010**; 405–414.

90. Kawai, F.; Polylactic Acid (PLA)-degrading Microorganisms and PLA Depolymerases. *ACS Symp* **2010**; 1043; 405–414.

91. Tomita, K.; Kuroki, Y.; Nagai, K.; Isolation of Thermophiles Degrading Poly(L-lactic acid). *J Biosci Bioeng* **1999**; 87:752–755.

92. Tomita, K.; Nakajima, T.; Kikuchi, Y.; Miwa, N.; Degradation of Poly(L-lactic acid) by a Newly Isolated Thermophile. *Polym Degrad Stab* **2004**; 84:433–438.

93. Hajighasemi, M.; Nocek, B.P.; Tchigvintsev, A.; Brown, G.; Flick, R.; Xu, X.; Cui, H.; Hai, T.; Joachimiak, A.; Golyshin, P.N.; Biochemical and Structural Insights into Enzymatic Depolymerization of Polylactic Acid and Other Polyesters by Microbial Carboxylesterases. *Biomacromolecules* **2016**; 17:2027–2039.

94. Liang, T.W.; Jen, S.N.; Nguyen, A.D.; Wang, S.L.; Application of Chitinous Materials in Production and Purification of a Poly(L-lactic acid)

Depolymerase from *Pseudomonas tamsuii* TKU015. *Polymers* **2016**; 8:98.

95. Jarerat, A.; Tokiwa, Y.; Degradation of Poly(L-lactide) by a Fungus. *Macromol Biosci* **2001**; 1:136–140.

96. Masaki, K.; Kamini, N.R.; Ikeda, H.; Iefuji, H.; Cutinase-like Enzyme from the Yeast Cryptococcus sp Strain S-2 Hydrolyzes Polylactic Acid and Other Biodegradable Plastics. *Appl Environ Microbiol* **2005**; 71:7548–7550.

97. Maeda, H.; Yamagata, Y.; Abe, K.; Hasegawa, F.; Machida, M.; Ishioka, R.; Gomi, K.; Nakajima, T.; Purification and Characterization of a Biodegradable Plastic Degrading Enzyme from *Aspergillus oryzae*. *Appl Microbiol Biotechnol* **2005**; 67:778–788.

98. Pranamuda, H., Tsuchii, A., Tokiwa, Y.; Poly (L-lactide)-degrading Enzyme Produced by Amycolatopsis sp. *Macromol Biosci* **2001**; 1:25–29.

99. Jarerat, A., Tokiwa, Y., Tanaka, H., Production of Poly(L-lactide)-degrading Enzyme by Amycolatopsis Orientalis for Biological Recycling of Poly(L-lactide). *Appl Microbiol Biotechnol* **2006**; 72:726–731.

100. Chomchoei, A., Pathom-aree, W., Yokota, A., Kanongnuch, C., Lumyong, S., Amycolatopsis Thailandensis sp. nov., a Poly(L-lactic acid)-degrading Actinomycete, Isolated from Soil. *Int J Syst Evol Microbiol* **2011**; 61:839–843.

101. Penkhrue, W.; Khanongnuch, C.; Masaki, K.; Pathom-aree, W.; Punyodom, W.; Lumyong, S.; Isolation and Screening of Biopolymer-degrading Microorganisms from Northern Thailand. *World J Microbiol Biotechnol* **2015**; 31:1431–1442.

102. Jarerat, A., Tokiwa, Y., Poly(L-lactide) Degradation by Saccharothrix waywayandensis. *Biotechnol Lett* **2003**; 25: 401–404.

103. Jarerat, A., Tokiwa, Y., Tanaka, H., Poly(L-lactide) Degradation

104. Sukkhum, S., Tokuyama, S., Kitpreechavanich, V.; Development of Fermentation Process for PLA-degrading Enzyme Production by a New Thermophilic Actinomadura sp T16-1. *Biotechnol Bioproc E* **2009**; 14:302–306.

105. Hanphakphoom, S., Maneewong, N., Sukkhum, S., Tokuyama, S., Kitpreechavanich, V., Characterization of Poly(L-lactide)-degrading Enzyme Produced by Thermophilic Filamentous Bacteria Laceyella Sacchari LP175. *J General Appl Microbiol* **2014**; 60:13–22.

106. Konkit, M., Jarerat, A., Khanongnuch, C., Lumyong, S., Pathom-aree, W., Poly(lactide) Degradation by Pseudonocardia alni AS4.1531(T). *Chiang Mai J Sci* **2012**; 39:128–132.

107. Tomita, K., Kuroki, Y., Nagai, K.; Isolation of Thermophiles Degrading Poly(L-lactic acid). *J Biosci Bioeng* **1999**; 87:752–755.

108. Tomita, K., Tsuji, H., Nakajima, T., Kikuchi, Y., Ikarashi, K., Ikeda, N., Degradation of Poly(D-lactic acid) by a Thermophile. *Polym Degrad Stab* **2003**; 81:167–171.

109. Sakai, K., Kawano, H., Iwami, A., Nakamura, M., Moriguchi, M.; Isolation of a Thermophilic poly-L-lactide Degrading Bacterium from Compost and its Enzymatic Characterization. *J Biosci Bioeng* **2001**; 92:298–300.

110. Arena, M., Abbate, C., Fukushima, K., Gennari, M.; Degradation of Poly(lactic acid) and Nanocomposites by Bacillus licheniformis. *Environ Sci Pollut Res* **2011**; 18:865–870.

111. Akutsu-Shigeno, Y.; Teeraphatpornchai, T.; Teamtisong, K.; Nomura, N.; Uchiyama, H.; Nakahara, T.; Nakajima-Kambe, T.; Cloning and Sequencing of a Poly(DL-lactic acid) Depolymerase Gene from Paenibacillus Amylolyticus Strain TB-13 and

its Functional Expression in *Escherichia coli*. *Appl Environ Microbiol* **2003**; 69:2498–2504.

112. Hoshino, A.; Isono, Y.; Degradation of aliphatic polyester films by commercially available lipases with special reference to rapid and complete degradation of poly(L-lactide) film by lipase PL derived from Alcaligenes sp. *Biodegradation* **2002**; 13:141–147.

113. Wang, Z.Y.; Wang, Y.; Guo, Z.Q., Li, F., Chen, S.; Purification and Characterization of Poly(L-lactic acid) Depolymerase from Pseudomonas sp Strain DS04-T. *Polym Eng Sci* **2011**; 51:454–459.

114. Boonluksiri, Y.; Prapagdee, B.; Sombatsompop, N.; An Accelerated Biodegradation of Poly(lactic acid) by Inoculation of Pseudomonas Geniculate WS3 Combined with Nutrient Addition. IOP Conference Series: Materials Science and Engineering, Volume 773, The International Conference on Materials Research and Innovation (ICMARI) December 16–18 2019.

115. Jeon, H.J.; Kim, M.N.; Biodegradation of Poly(L-lactide) (PLA) Exposed to UV Irradiation by a Mesophilic Bacterium. *Int Biodeterior Biodegrad* **2013**; 85:289–293.

116. Petinakis, E.; Liu, X.; Yu, L.; Way, C.; Sangwan, P.; Dean, K.; Bateman, S.; Edward, G. Biodegradation and Thermal Decomposition of Poly(lactic acid)-based Materials Reinforced by Hydrophilic Fillers. *Polym Degrad Stab* **2010**; 95(9):1704–1707.

117. Vasile, C.; Pamfil, D.; Râpă, M.; Darie-Niţă, R. N.; Mitelut, A.C.; Popa, E.E.; Popescu, P.A.; Draghici, M.C.; Popa, M. E. Study of the Soil Burial Degradation of Some PLA/CS Biocomposites. *Compos B Eng* **2018**; 142:251–262.

118. Mosnáčková, K.; Šišková, A.; Janigová, I.; Kollár, J.; Šlosár, M.; Chmela, Š.; Alexy, P.; Chodák, I.; Bočkaj, J.; Mosnáček, J. Ageing of Plasticized Poly(lactic acid)/Poly(β-hydroxybutyrate) Blend Films Under Artificial UV Irradiation and Under Real Agricultural Conditions During Their Application as Mulches. *Chem Pap* **2016**; 70:1268–1278.

119. Mosnáčková, K.; Danko, M.; Šišková, A.; Falco, L.M.; Janigová, I.; Chmela, Š.; Vanovčanová, Z.; Omaníková, L.; Chodák, I.; Mosnáček, J. Complex Study of the Physical Properties of a Poly(lactic acid)/Poly(3-hydroxybutyrate) Blend and its Carbon Black Composite During Various Outdoor and Laboratory Ageing Conditions. *RSC Adv* **2017**; 7:47132–47142.

120. Di Lorenzo, M.L.; Androsch, R. Synthesis, Structure and Properties of Poly(lactic acid). Springer: Berlin, **2017**, 139.

121. Gu, J.D. Microbiological Deterioration and Degradation Of Synthetic Polymeric Materials: Recent Research Advances. *Int Biodeterior Biodegrad* **2003**; 52(2):69–91.

122. Pathak, V.M.; Navneet; Review on the Current Status of Polymer Degradation: a Microbial Approach. *Bioresour Bioprocess* **2017**; 4:15. https://doi.org/10.1186/s40643-017-0145-9.

123. Reineke, W. Aerobic and anaerobic biodegradation potentials of microorganisms. In: Biodegradation and Persistence. The Handbook of Environmental Chemistry, 2, Series: Reactions and Processes, Beek B., ed. Springer: Berlin, **2001**.

124. Iwańczuk, A.; Kozłowski, M.; Łukaszewicz, M.; Jabłoński, S.; Anaerobic Biodegradation of Polymer Composites Filled with Natural Fibers. *J Polym Environ* **2015**; 23:277–282.

125. Yagi, H.; Ninomiya, F.; Funabashi, M.; Kunioka, M.; Anaerobic Biodegradation Tests of Poly(lactic acid) under Mesophilic and Thermophilic Conditions Using a New Evaluation

System for Methane Fermentation in Anaerobic Sludge. *Int J Mol Sci* **2009**; 10:3824–3835.

126. Šmejkalová, P.; Kužníková, V.; Merna, J.; Hermanová, S.; Anaerobic Digestion of Aliphatic Polyesters. *Water Sci Technol* **2016**; 73:2386–2393.

127. Yagi, H.; Ninomiya, F.; Funabashi, M.; Kunioka, M.; Anaerobic Biodegradation of Poly(lactic acid) Film in Anaerobic Sludge. *J Polym Environ* **2012**, 20 (3):673–680.

128. Shah, A.A.; Hasan, F.; Hameed, A.; Ahmed, S.; Biological Degradation of Plastics: a Comprehensive Review. *Biotechnol Adv* **2008**; 26:246–265.

129. https://www.biosphereplastic.com/biodegradableplastic/uncategorized/is-pla-compostable/.

130. Al Hosni, A.S.; Pittman, J.K.; Robson, G.D. Microbial Degradation of Four Biodegradable Polymers in Soil and Compost Demonstrating Polycaprolactone as an Ideal Compostable Plastic. *Waste Manage*, **2019**; 97:105–114.

131. Maga, D.; Hiebel, M.; Thonemann, N. Life Cycle Assessment of Recycling Options for Polylactic Acid. *Resour. Conserv. Recycl.* **2019**; 149:86–96.

132. Weiss, M.; Haufe, J.; Carus, M.; Brandão, M.; Bringezu, S.; Hermann, B.; et al. A Review of the Environmental Impacts of Biobased Materials. *J Ind Ecol* **2012**; 16 (S1):S169–S181.

133. Yates, M. R.; Barlow, C. Y.; Life Cycle Assessments of Biodegradable, Commercial Biopolymers: A Critical Review. *Resour Conserv Recycl* **2013**; 78:54–66.

134. https://www.total-corbion.com/about-pla/pla-lifecycle/.

135. Katiyar, V.; Bio-based Plastics for Food Packaging Applications, ebook, Publisher, Smithers Rapra: Akron, OH, **2017**.

136. Piemonte, V.; Sabatini, S.; Gironi, F.; Chemical Recycling of PLA: A Great Opportunity Towards the Sustainable Development? *J Polym Environ* **2013**; 21:640–647.

137. Chien, Y-C.; Liang, C.; Liu, S.-H.; Yang, S.-H.; Combustion Kinetics and Emission Characteristics of Polycyclic Aromatic Hydrocarbons from Polylactic Acid Combustion. *J Air Waste Management Assoc* **2010**; 60(7):849–855, DOI: 10.3155/1047-3289.60.7.849.

138. Rossi, V.; Cleeve-Edwards, N.; Lundquist, L.; Schenker, U., Dubois, C., Humbert, S., Jolliet, O.; Life Cycle Assessment of End-of-Life Options for Two Biodegradable Packaging Materials: Sound Application of the European Waste Hierarchy. *J Clean Prod* **2015**; 86:132–145.

139. Gabrys, J.; Hawkins, G.; Michael, M.; Accumulation: The Material Politics of Plastic. Routledge and Taylor & Francis Group: Oxon, **2017**, 161.

140. de Andrade, M.F.C.; Souza, P.M.S.; Cavalett, O.; Morales, A.R.; Life Cycle Assessment of Poly(lactic acid) (PLA): Comparison Between Chemical Recycling, Mechanical Recycling and Composting. *J Polym Environ* **2016**; 24(4):372–384.

141. Baldwin, C.J. The 10 Principles of Food Industry Sustainability. John Wiley & Sons: Hoboken, NJ, **2015**, 109.

142. Kiruthika, A.V.; Properties and End-of-Life of Polymers from Renewable Resources. Reference Module in Materials Science and Materials Engineering, **2019**, Doi: 10.1016/b978-0-12-803581-8.10603-4.

143. Beigbeder, J.; Soccalingame, L.; Perrin, D.; Bénézet, J.-C.; Bergeret, A. How to Manage Biocomposites Wastes End of Life? A Life Cycle Assessment Approach (LCA) Focused on Polypropylene (PP)/Wood Flour and Polylactic Acid (PLA)/Flax Fibres Biocomposites. *Waste Manage* **2019**; 83:184–193.

CHAPTER 11

BIODEGRADABLE AND BIOCOMPATIBLE NANOCOMPOSITE HYDROGELS AS EMERGING BIOMATERIALS

RABIA KOUSER[1*], S. I. BHAT[1], HALIMA KHATOON[1], MOHD IRFAN[1], and SAJAD AHMAD DAR[2]

[1]Material Research Laboratory, Department of Chemistry, Jamia Millia Islamia, New Delhi 110025, India

[2]Department of Physics, Government Motilal Vigyan Mahavidyalya College, Bhopal 462008, India

*Corresponding author. E-mail: rabiakhanjmi786@gmail.com

ABSTRACT

The ongoing progress in the development of biopolymeric nanocomposite hydrogels (NCHs) has been an area of active research in the field of biomedical applications especially in drug delivery and tissue engineering. New development in nanotechnology has emphasized the diverse use of NCHs and gained much attention in the biomedical and pharmaceutical industry due to their remarkable properties at the nanoscale level. The novelties of the discovery of NCHs comprise new methodologies toward the fabrication and modification of hydrogels by using various nanofillers with plan polymeric hydrogel matrices. This chapter focuses on biodegradable polymer-based NCHs and also gives an insight into the advancement of biodegradable polymeric NCHs for biomedical applications.

11.1 INTRODUCTION

Biopolymeric hydrogels consisting of self-supporting, water-swollen three-dimensional (3D) structures, and possessing viscoelastic network [1, 2]. These materials have the capability of imbibing water or biological fluids up to several thousand percentages and swell readily without dissolving due to the presence of hydrophilic functional groups (carboxyl, amide, amino, and hydroxyl) in their polymeric chains [3]. Hydrogels have existed for more than half a century, providing one of the earliest reports in the literature "by Wichterle and Lim" in 1954–1960, which described a flexible cross-linked structure of hydroxyethyl methacrylate (pHEMA) based hydrogel [4, 5]. Since then, hydrogel technology has diversified at a large scale, exhibiting good scope for their applications in the field of biomedical sciences, especially enhances since 1970 till date [6, 7]. Thus, the term hydrogel is self-explanatory in terms of its structure, properties, and applications. Hydrogels have fascinated materials scientists and biomedical researchers in terms of their formulations and applications [2, 8]. The swelling properties of hydrogels can be attributed to their higher thermodynamical affinity toward solvents [9], which can be adjusted as per the requirement through the introduction of various cross-linkers via covalent bonds or physical interactions through intramolecular and intermolecular attractions [10]. In addition, these hydrogels show a good capillary action, osmotic pressure, and water diffusion [11]. Biopolymer (such as chitosan, starch, guar gum, alginate,

etc.) based hydrogels show desirable biodegradability and biocompatibility [12–14]. For example, Vashist et al. [15] developed a biocompatible interpenetrating network (IPN) using chitosan, methylmethacrylate (MMA), and oleo polyol for the drug delivery system. Later, Kouser et al. [16] also reported chitosan, acrylonitrile, oleo polyol, and multiwalled carbon nanotube (MWCNT) based nanocomposite hydrogel (NCH) films, which exhibits good mechanical strength, biodegradability, and biocompatibility. Noori et al. reported clay dispersed poly(vinyl alcohol)-chitosan-honey-based novel NCHs for wound dressing applications. These hydrogels are biocompatible and show antibacterial activity [17].

The concept of smart hydrogels is associated with their response toward external stimuli including temperature, pH, ionic concentration, light, magnetic fields, electrical fields, and chemicals [18, 19]. These features result in an enormous potential for their various advanced technological applications, for instance, the controlled drug delivery application which has been an area of active research [7, 20–23]. Many novel hydrogel-based delivery systems have been fabricated to fulfill the ever-increasing demands of the pharmaceutical industries (Figure 11.1). The hydrogels are classification based on their constituents, physical, and chemical structures,

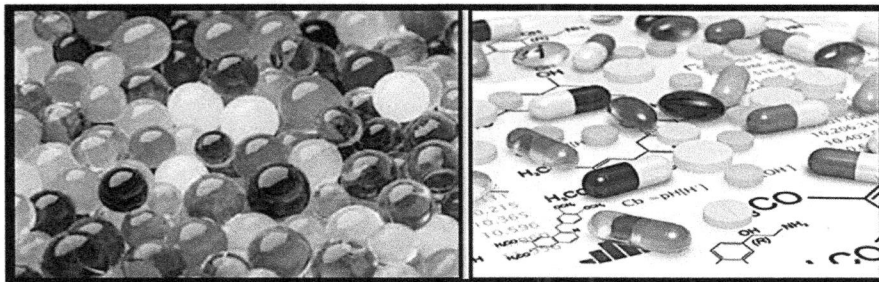

FIGURE 11.1 Hydrogel biomaterials in pharmaceutical applications.

various polymeric forms (like homo and copolymers), blend, IPN, and composites. Their advantages, drawbacks, and applications are further discussed in proceeding units.

11.2 CLASSIFICATIONS OF HYDROGELS

Hydrogels can be classified on the basis of source, configuration, types of cross-link, physical appearance, and electric charges, etc. [10], which has been portrayed in Figure 11.2 and discussed as follows:

11.2.1 CLASSIFICATION ON THE BASIS OF SOURCE

Hydrogels can be classified into natural, synthetic, and semisynthetic [24]. Most of these synthetic hydrogels are synthesized by traditional polymerization of vinyl monomers. Natural hydrogels are obtained through the polymerization of natural monomers like nucleotides, peptides, and polysaccharides. While the semisynthetic hydrogels are somewhat the prepolymer blends of synthetic and natural polymers.

11.2.2 CLASSIFICATION BASED ON POLYMER COMPOSITION

Homopolymeric hydrogels are denoted to polymer network derived from a single species of monomer, which is a basic structural unit comprising of any polymer network. It may have cross-linked network depending upon the nature of monomer and polymerization methods [25].

Copolymeric hydrogels are comprised of two or more different monomer species with at least one hydrophilic component, arranged in a random, block, or alternating configuration along the chain of the polymer network [26].

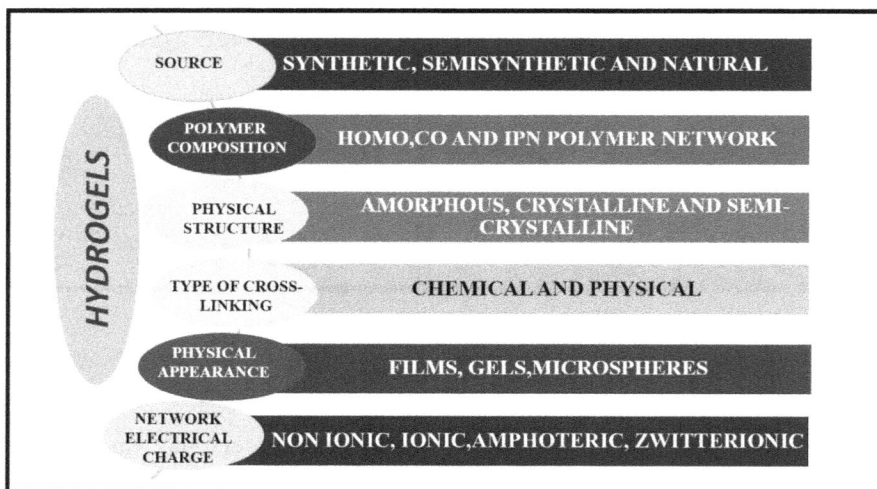

FIGURE 11.2 Classifications of hydrogels.

11.2.3 MULTIPOLYMER INTERPENETRATING POLYMERIC HYDROGEL (IPN)

These are an important class of hydrogels, made of two independent cross-linked synthetic and/or natural polymer components, contained in a network form. IPN can be obtained by grafting two or more polymers forming a network [27].

11.2.4 CLASSIFICATION BASED ON CONFIGURATION

Based on physical, structural, and chemical composition, hydrogels can be classified as amorphous (noncrystalline), semicrystalline (a complex mixture of amorphous and crystalline phases), and crystalline [24].

11.2.5 CLASSIFICATION BASED ON TYPES OF CROSS-LINKING

Hydrogels can be divided into two categories based on the chemical or physical nature of the cross-link junctions. Chemically cross-linked networks have permanent junctions, whereas the physical cross-linked networks have transits junctions that arise from polymer chain entanglement or physical interactions, such as ionic, hydrogen bonding, or hydrophobic interactions [9].

11.2.6 CLASSIFICATION BASED ON PHYSICAL APPEARANCE

Hydrogels can be obtained as a matrix, film, gel, and microsphere, depend on the method of polymerization involved in the preparation process.

11.2.7 CLASSIFICATION BASED ON ELECTRICAL CHARGES

Hydrogels may be divided into four groups based on the presence or absence of electrical charges located on the cross-linked chains [1].

1. Nonionic (neutral).
2. Ionic (including anionic or cationic).
3. Amphoteric electrolyte (ampholytic) containing both acidic and basic groups.
4. Zwitterionic (polybetaines) containing both anionic and cationic groups in each structural repeating units.

11.3 ADVANTAGES OF HYDROGELS

1. Due to their high water-holding capability, they possess a high degree of flexibility very similar to natural tissue [12].
2. Maximum toughness and firmness in the swelling medium and during the storage [28]. Timely release of drugs or nutrients [29].
3. Biocompatible, biodegradable with desirable mechanical strength and can be injected into the human body [16, 30, 31].
4. Colorlessness, odorlessness, and absolute nontoxic nature.
5. The response to external stimuli (i.e., electric signal, pH, temperature, and presence of enzyme or other ionic species) may lead to a change in their physical texture [32].
6. Good transport properties and easy to modify [33].

11.4 DRAWBACKS OF HYDROGELS

Despite the high potential of hydrogels for their use in many cases, these hydrogels have limited biomedical applications. Due to their poor mechanical strength, burst release of the hydrophilic drugs, and improper capability for the loading of the hydrophobic drug are some of the major drawbacks associated with sustained delivery [6, 34–36]. In addition, the limited commercial products of hydrogels are available for their application in tissue engineering and drug delivery [37]. Thus, more progress is needed in these two fields. To overcome these limitations, various modifications in the form of copolymer, interpenetrating polymeric network, and NCHs [29, 38–40] were done. These materials can be utilized in various applications such as electronic applications [41], watering beads for plants [14], in cosmetics industry and plastic surgery [9], environmental applications [42], bacterial culture [43], electrophoresis and proteomic [44], immunotherapy and vaccine [45], scaffolds for bone regeneration [46], cardiac repair [47], ophthalmic applications [48], dental applications, wound healing, and drug delivery applications [49, 50].

Among all aforementioned modifications, NCHs have emerged as an advantageous method to improve the properties of hydrogels within a single unit. Various types of NCHs are discussed in the following sections.

11.4.1   NANOCOMPOSITE HYDROGELS

NCHs can be defined as cross-linked biopolymer networks swollen through the absorption of water in the presence of nanoparticles or nanostructures (10–100 nm). They exhibit superior properties (higher elasticity, mechanical strength) as compare to traditionally made hydrogels [29, 35, 51]. The hard nanoparticles (segment) strengthens the soft organic (segment) polymer matrix, so the final NCHs can exhibits synergistic properties [52]. Literature reveals that the researchers have incorporated various nanoparticles like carbon-based, polymeric, ceramic, and metallic nanofillers, which have tremendously improved the biological properties of these hydrogels [7, 16, 53]. Initially, NCHs

were developed by Harachugi et al. using poly(PNIPAAm) and montmorillonite nanoclay with improved mechanical strength, swelling ability, ocular precision, and stimuli responsiveness [54].

11.4.2   BIODEGRADABLE POLYMER-BASED NCHS

In recent years, researchers have focused on biopolymer-based materials due to their biocompatible nature and tremendous applications in the field of biomedical sciences, especially in drug delivery, tissue engineering, and wound healing [12, 35]. Natural biocompatible polymers like polysaccharides, carbohydrates, modified celluloses, poly(α-amino acids) are being used for the synthesis of hydrogels with reference to their application in the field of biomaterials [55]. The natural polymers are divided into the following major classes based on their chemical structures: (1) polysaccharides (chitosan, guar gum, agar, starch, gum tragacanth, aloe vera gel, hyaluronic acid, alginate, and agarose), (2) polyamides (or polypeptides-collagen, gelatin, fibroin, wheat, soya, and fibrin), and (3) polyesters (poly(3-hydroxyalkanoate))   [47]. Figure 11.3 shows various nanomaterials and biopolymers used in the fabrication of NCHs with improved properties. Many researchers have reported biopolymeric NCHs with

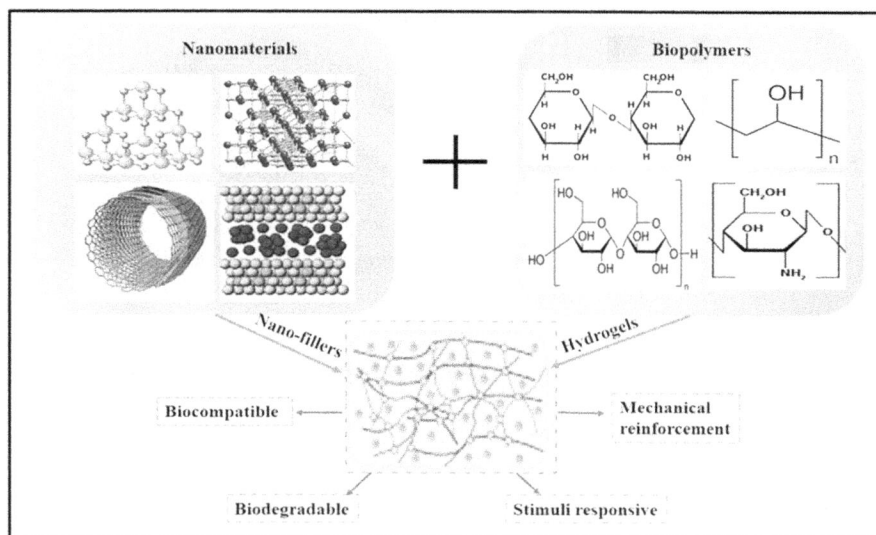

FIGURE 11.3   Various nanomaterials and biopolymer-based NCHs and their properties.

various advantageous properties and applications. For example, Barkhordari et al. [56] prepared hydrogel nanocomposite for controlled delivery of ibuprofen using double-layered hydroxides (LDH)–carboxymethyl cellulose (CMC). Tanpichai et al. [57] prepared NCH films using cellulose nanocrystal (CNC) and poly(vinyl alcohol), with improved mechanical and thermal properties. Peng et al. [58] reported clay dispersed cellulose NCHs exhibit high mechanical strength and superabsorbent properties for methylene blue. Some of the main biopolymer-based NCHs along with their general introduction and recent literature are being discussed in the following section.

11.4.3   CHITOSAN-BASED NCHs

Chitosan is a linear polysaccharide composed of glucosamine and *N*-acetyl glucose amine units linked by b (1–4) glycosides bond (Figure 11.4), created by deacetylation of chitin at acidic condition [27]. The reactive polar groups (amino and hydroxyl) present in the chitosan can be modified to improve their solubility in water or organic solvents. However, it can be soluble in glacial acetic acid due to the protonation of amino groups under an acidic environment [15]. The versatile biological and physicochemical properties of chitosan make it a promising material for various biomedical

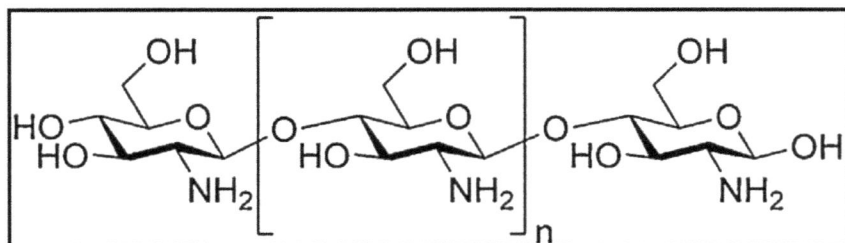

FIGURE 11.4   Chemical structure of chitosan.

applications (Figure 11.5). In spite of these outstanding features, chitosan hydrogels possess inferior mechanical strength [59]. To overcome this drawback, various nanofillers have been dispersed in chitosan that significantly enhances mechanical, thermal, and electrical properties [60–62]. For instance, Tushar et al. reported hybrid materials and nanocomposite ionogels using agarose-chitosan through an ionic liquid via dissolution, regeneration, and sol–gel transition. The uniform dispersion of silver nanoparticles into the chitosan-based polymeric matrix formed nanocomposite having excellent thermal, and mechanical properties, found suitable for the variety of applications such as in actuators,

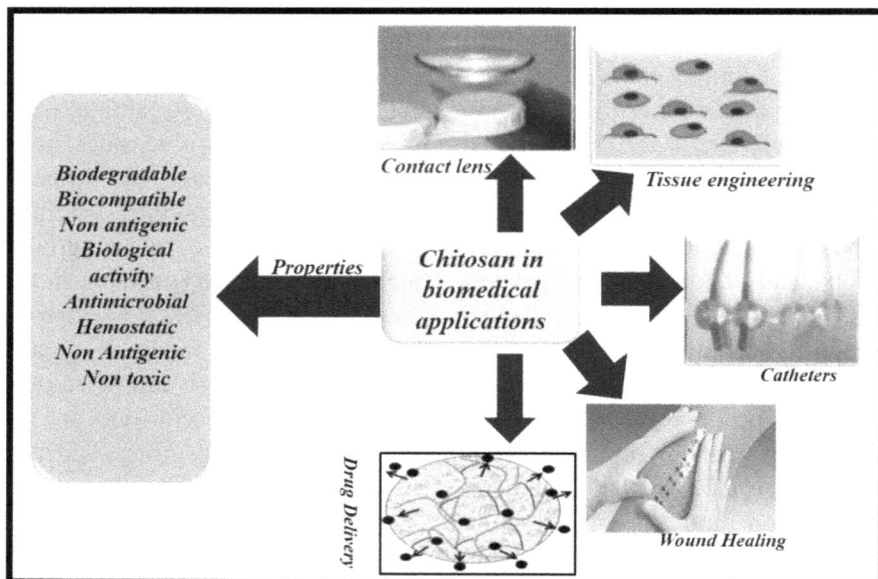

FIGURE 11.5   Properties and biomedical applications of chitosan.

sensors, food packing, wound dressing, and drug delivery system [63]. A list of chitosan-based NCHs using different nanofillers along with their properties and applications are given in Table 11.1.

11.4.4 STARCH-BASED NCHs

Starch is a polysaccharide obtained by alpha glucose consisting of primarily linear amylase and highly branched amylopectin [64]. It is enzymatically fabricated by plants and can be extracted from tubers, seeds, roots, and stems. The structure of starch is shown in Figure 11.6. In recent years, several starch-based hydrogels have been developed for biomedical applications [65]. However, starch exhibits some limitations in the form of poor processability, high brittleness, low water content, poor mechanical strength, and oxygen obstacle properties [65]. In this regard, the incorporation of various nanofillers (hydroxyapatite (HA), MWCNT, nanoclays, metal oxides, etc.) into the starch-based polymer matrix has been carried out to form NCHs [66]. A list of biopolymers (starch and cellulose)-based NCHs with their properties and applications is given in Table 11.2. The biomedical applications, such as tissue engineering, drug delivery, personal care, etc., of starch-based NCHs are given in Figure 11.7. The addition of nanofillers greatly influenced its physicochemical and physicomechanical properties. For example, Spagnol et al. [67] developed superabsorbent hydrogel nanocomposites, using starch-g-poly(sodium acrylate) and nanowhiskers (CNW). The dispersion of CNWs up to 10 wt% in the polymeric matrices would enhance their swelling ability and mechanical strength. Famá et al. prepared NCHs using starch as a matrix reinforced with very small amounts of MWCNTs, which exhibits high elastic modulus (about 100%), while the water vapor permeability value was 43% lower as compared to plain starch hydrogel [68].

11.4.5 CELLULOSE-BASED NCHs

Cellulose is one of the world's most abundant, natural, and renewable biopolymer resources, which is widely present in various forms of biomasses, such as trees, plants, tunicate, and bacteria [69]. Cellulose molecule consists of β-1,4-D-linked glucose chains (Figure 11.8), with molecular formula $(C_6H_{10}O_5)_n$ (n ranging from 10,000 to 15,000) through acetal oxygen covalently bonding C_1 of one glucose ring and C_4 of the adjoining ring [70]. The biocompatible and hydrophilic properties of cellulose-based hydrogels make them the best candidature for food packing, pharmacy, wound healing, etc. [71]. The chemical interactions like H-bonding

TABLE 11.1 List of Chitosan Polymer Used for the Formulation of NCHs with Properties and Applications

S. No	Polymers	Fillers Used	Properties of NCHs	Applications of NCHs	References
1.	Chitosan	MWCNTs	Enhanced optical, electronic, magnetic, chemical and mechanical properties, antimicrobial activity against *Staphylococcus aureus*, *Escherichia coli*, and *C. tropicalis*	Biomedical as well as food science applications	[106]
2.	Chitosan	Silver-loaded nanosilica	Antibacterial activity against *E. coli* and *S. aureus*	Tissue engineering and drug delivery	[107]
3.	Chitosan	ZnO	High thermal and mechanical stability. Antibacterial activities against the growth of *S. aureus*	Biomedical applications (wound dressing)	[108]
4.	Chitosan	MMT/HAP	Biocompatible, biodegradable, high thermal as well as mechanical properties	Tissue engineering applications	[109]
5.	Chitosan	MMT	Biocompatible, high gel strength, optimal and thermal stability, and an increased glass transition temperature (98.2 °C)	Biomedical applications	[110]
6.	Chitosan	Fe_3O_4	Developed a novel tyrosine biosensor with porous network and large surface area	Biosensors to detect catechol	[111]
7.	Chitosan	Zn oxide	Moist environment with good biocompatibility	Wound dressing application	[112]
8.	Chitosan	MMT	High wettability, appropriate pore size, high surface area, biodegradability, and structural reliability	Tissue engineering	[113]
9.	Chitosan	HAP	High biocompatibility, biodegradability, porous structure, suitable for cell ingrowth, osteoconduction, and intrinsic antibacterial nature	Bioactive biomaterials in bone tissue engineering and renowned	[114]
10.	Chitosan	Kaolin clays	Have high swelling capability as well as high biodegradation rate, pH-responsive behavior	Drug delivery applications	[115]
11.	Chitosan		Showed excellent water-solubility. Biocompatible and nontoxic nature. Antimicrobial properties	Pharmaceutical, preservatives and food industry fields	[107]
12.	Chitosan	Carbon-dots	Films are soft but tough biocompatible. High swelling, thermal, and mechanical properties	Biomedical applications	[116]

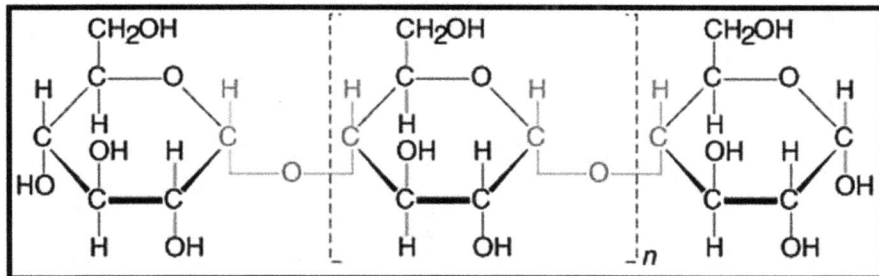

FIGURE 11.6 Chemical structures of starch.

and van der Waals forces within the structure restrict the complete dissolution in the aqueous medium. Therefore, certain chemical modifications have been made to overcome this limitation. The literature revealed that the hydrogel based on cellulose derivatives like methylcellulose, hydroxyl propyl cellulose, hydroxyl propyl methylcellulose, and CMC have been utilized for their chemical modifications [72, 73]. For instance, initially developed cellulose-based nanocomposites hydrogels using bacteria cellulose (BC) and gelatin with high mechanical strength, swelling capacity, and crystallinity [74]. However, the production of BC is very low due to its high cost, which limited its applications. Thus, CNCs were derived from plants having low cost than that of BC. CNCs are highly soluble in water that found a potential scope in the development of NCHs. For instance, researchers developed pH-responsive superabsorbent NCHs using CNC and poly(acrylamide-*co*-acrylate) (PAM-AA) via free-radical

copolymerization technique [70]. Later, Babu et al. [75] fabricated NCHs using CMC, poly(acrylamide-*co*-2-acrylamido-2-ethylpropanesulfonicacid), and silver nanoparticles via free-radical polymerization technique for antibacterial applications [75]. However, Zhang et al. [76] developed graphene oxide (GO) dispersed CMC-polyacrylamide-based NCH. These hydrogels exhibit superior swelling ratios and mechanical strength as compare to that of pristine hydrogel, used for tissue engineering and drug delivery applications.

11.4.6 POLYVINYL-ALCOHOL-BASED NCHs

PVA is a linear synthetic polymer derived from polyvinyl acetate through partial or full hydroxylation. The structure of polyvinyl alcohol (PVA) is given in Figure 11.9. The high degree of hydroxylation and polymerization make PVA insoluble in water. Thus, PVA-based hydrogels

TABLE 11.2   List of Starch and Cellulose Polymers Used for the Formulation of NCHs with Properties and Applications

Sr. No	Polymers	Fillers used	Properties of NCHs	Applications of NCHs	References
1.	Starch	MMT	Enhanced mechanical properties and increased shear moduli of the hydrogels	Wastewater removal application	[117]
2.	Starch	MWCNTs	Improved mechanical strength by wrapping the MWCNTs with starch–iodine complex	Biomedical applications	[118]
3.	Starch	Flax cellulose nanocrystals	These nanocomposite films exhibit improved mechanical properties and water resistance	—	[119]
4.	Starch	Zinc oxide	Show good antibacterial activity	UV-protection antibacterial finishes in medical textiles	[120]
6.	Carboxymethyl starch	MMT	Biodegradable, hydrophilic, significantly enhanced mechanical properties (tensile strength, Young modulus)	Application in agriculture for seed tapes production	[121]
7.	Starch	Fumarate alumoxane	Have the high swelling ability, adsorption capacities for the removal of ammonium ion	Wastewater treatment	[122]
8.	Starch	C, o-doped zinc ferrite nanoparticles	These hydrogel nanocomposite used for drug release with respect to the external magnetic field	Pharmacological application	[123]
9.	Cellulose	Cellulose nanowhisker (CNW)	These materials have enhanced thermal stability, mechanical properties, and a high water vapor transmission rate	Wound dressing	[124]
10.	Methyl cellulose	CNF	Improved the mechanical and structural network properties	As a carrier vehicle for agrochemical controlled release	[125]
11.	CMC	ZnO	pH and salt-sensitive swelling behavior, antibacterial activity against *Escherichia coli* and *Staphylococcus aureus* bacteria	Used effectively for biomedical application	[126]

TABLE 11.2 (Continued)

Sr. No	Polymers	Fillers used	Properties of NCHs	Applications of NCHs	References
12.	CMC	Siler nanoparticles	These composite materials show pH-sensitive high swelling behavior and antibacterial activity	Biological applications	[127]
13.	CMC	Copper oxide	High swelling behavior in various pH values and salt solutions, excellent antibacterial activity against *Escherichia coli* and *Staphylococcus aureus*	Useful for biomedical application	[128]
14.	Cellulose	Graphene	These NCHs significantly enhanced mechanical strength and thermal stability	Applications in the fields of biology and energy	[129]
15	Cellulose	Clays	These NCHs exhibits high mechanical strength, removal of dye efficiency, absorption capacity for methyl blue	Wastewater treatment	[130]
16.	CMC	ZnO	These NCHs have high swelling ratios, show antibacterial activity, and biocompatible nature	Promising wound dressing with sustained release of drug	[131]
17.	CMC	MMT	pH and magnetic sensitive behavior and reduction in swelling ratios	Colon targeted drug delivery	[132]
18.	CMC	Graphene	pH-responsive, high swelling behavior, drug loading efficiency increased	Anticancer drug delivery	[133]
19.	CMC	Reduced graphene oxide	These scaffolds show high thermal, swelling, and porosity and biodegradability	Tissue engineering application	[134]

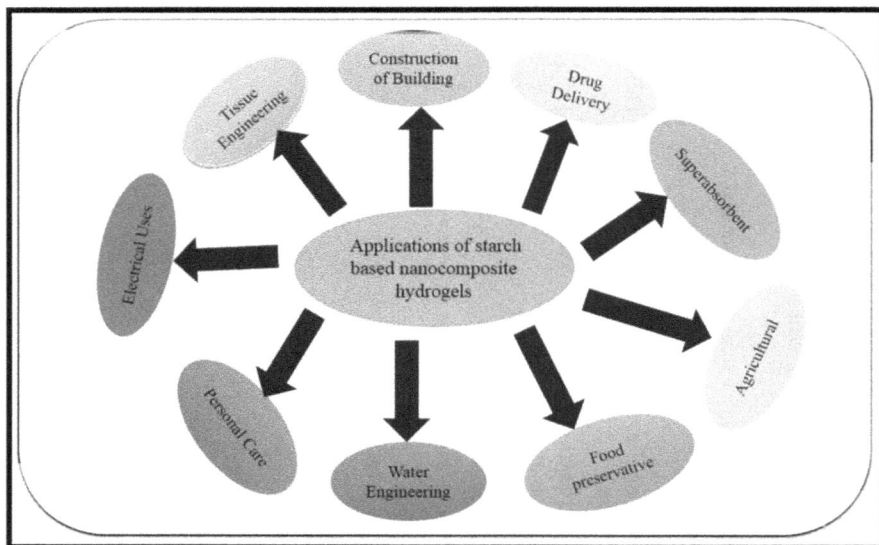

FIGURE 11.7   Application of starch-based NCHs.

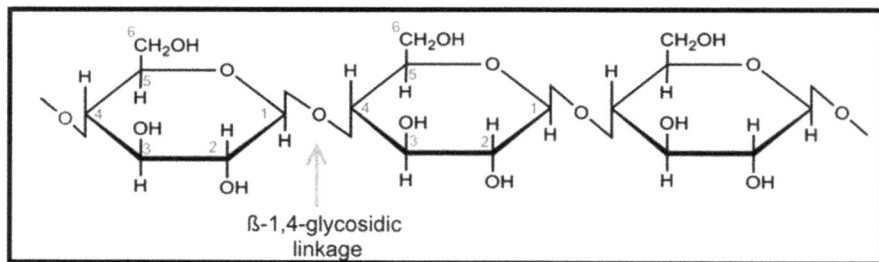

FIGURE 11.8   Structure of cellulose.

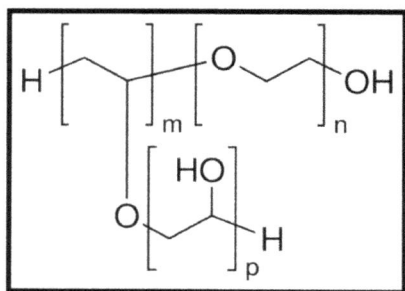

FIGURE 11.9   Structure of polyvinyl alcohol.

provide structural stability in water or any other physiological medium by using it as physical or chemical cross-linkers [77]. They are commonly used in medical devices, paper products, and food industry due to their high hydrophilicity, biocompatibility, and good-film forming properties [78]. However, they exhibit poor mechanical strength, which limits their biomedical applications [79]. The incorporation of nanofillers into these hydrogels to make their nanocomposite is an important strategy used to induce versatile properties [57, 61, 72]. The properties and biomedical applications of these NCHs are given in Figure 11.10.

11.4.7 CHARACTERISTIC PROPERTIES OF VARIOUS NCHs

Literature reveals that the above-cited NCHs have exhibited far superior biocompatibility, biodegradability, mechanical strength, %EWS ratios, drug retention, and external

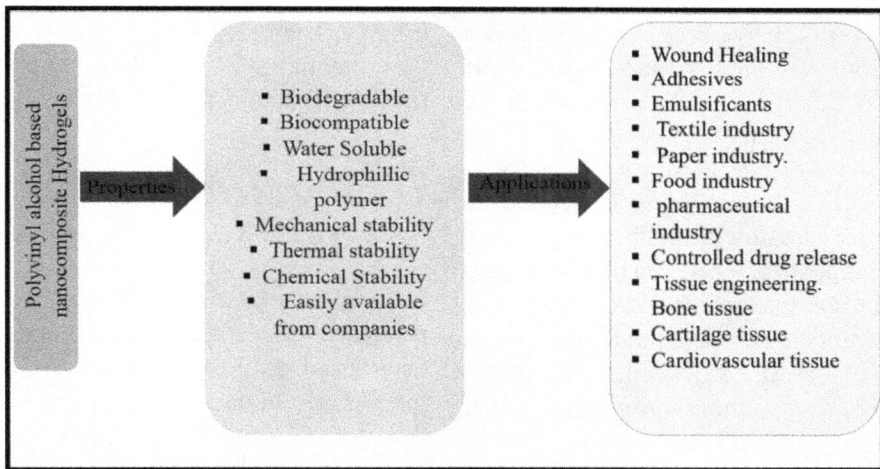

FIGURE 11.10 PVA-based NCHs (properties and applications).

stimuli-responsive behavior [65, 80]. The remarkable improvement in the properties of NCHs has opened a new vista for its various applications at a commercial scale especially in the field of drug delivery, tissue engineering, wound healing, and bone repairing [50]. Wang et al. [81] synthesized NCH using elastin like polypeptides and GO. These NCHs are mechanically flexible to twist, elongate, and coil in different heat intensities. In another study, Yadav et al. [82] synthesized GO, CMC, Alg-based NCH films via solution mixing-evaporation technique. The tensile strength and Young's modulus of these composite films enhanced from 40% to 1128% with the only 1 wt% GO ratios. The same groups further reported another novel NCH films using CMC and GO [83]. The complete dispersion of GO in the CMC system without agglomeration enhances their mechanical as well as thermal stability (i.e., Young's modulus by 67% ± 6% and 148% ± 5%, respectively, from 103 ± 6 to 173 ± 6 MPa and 32.65 ± 5 to 81.12 ± 5 GPa). Pan et al. [84] developed composites using carboxymethyl chitosan (CMC), functionalized GO, fluorescein isothiocyanate, and lactobionic acid (LA) for a targeted drug delivery system. Doxorubicin can be chosen as a model drug to show the pH-responsive drug release behavior from both the LA-functionalized and the LA-free material. The modified GO has exhibits biocompatibility with the liver cancer cells (SMMC-7721) but can induce cell death after treatment with drug-loaded composites. These results show that the composites are strong potential candidates for targeted drug delivery. Zang et al. [85] synthesized NCHs by using sodium alginate (SA), HA, and halloysite nanotubes (HNTs) [85]. These NCHs are loaded with diclofenac sodium via in situ invention of HA nanoparticles during the sol–gel transition of the SA/HNTs suspension. Results showed that the dispersion of HA and HNTs into polymeric matrix enhanced the swelling ratios and the encapsulation efficiency up to 75% as well as drug loading and release rate. Poursamar et al. [86] reported NCHs by using PVA and HA via freeze-drying technique. The combination of bioactive materials with biocompatible and biodegradable polymers is used to developed NCHs with excellent biocompatibility, bioactivity, biodegradability, and mechanical properties. The in vitro results showed that there is an appropriate penetration of the cells into the scaffords pores and also a continuous enhancement in cell aggregation on the scaffords with enhancement in the incubation time, which confirmed the capability of the scaffolds to maintenance cell growth. These types of NCHs have the potential ability for tissue engineering especially bone tissue.

11.4.8 MISCELLANEOUS NANOFILLERS BASED NCHs

Carbon nanotubes, nanosilica, and nanoclay dispersed NCHs have been used as promising materials for various biomedical applications due to their versatile properties like chemical stability, biocompatibility, and controlled release with target drug delivery. For instance, Giri et al. [87] synthesized carboxymethyl guar gum (CMG) chemically tailored MWCNTs-based NCHs for the sustainable release of diclofenac sodium [87]. Significance analysis proved that the interaction between CMG-MWCNTs enhanced rheological property, thermal stability, swelling capacity, and more drug encapsulate and deliberate transdermal release as compared to the CMG without CNTs. Another group reported CNTs-based NCHs with improved mechanical strength, using PVA, polyvinyl pyridine [88]. Rodrigues et al. [89] prepared PVA- and CNTs-based NCHs with an excellent biological reaction as compared to that of neat PVA in osteochondral imperfection renovating. After implantation of these biomaterials, there is an increase in bone growth rate, without affecting any sort of inflammation reaction after 12 days. Jodder et al. [90] also developed MWCNT and alginate-based hydrogels for cell culture, cell therapy, and tissue engineering applications [90]. The resultant MWCNT-alginate scaffords (1–3 mg/mL) have good stability, improved mechanical strength, and enhanced cell proliferation as compare to the alginate alone. Popet et al. [91] reported pH-responsive organic–inorganic systems using positively charged chitosan and mesoporous silica nanoparticles (MSNs) for drug delivery application. In their study, it was found that the drug release (ibuprofen) at pH 7.4 is only 20% and at pH 5.0 it reached 90% after 8 h. Zhu et al. [92] developed composite hydrogels using Cs and MSNs for cartilage regeneration. The introduction of MSNs into chitosan can enhance the gelation rate and increased the elastic modulus (G') from 1000 to 1800 Pa. These hydrogels released gentamicin and bovine serum albumin in a sustained manner. In vitro chondrocyte culture test of

the composite hydrogels showed perform better for chondrocyte growth as compare to neat CS. Later on, Gao et al. [93] prepared tough and stretchable self-healing NCHs using acrylamide, NaMMT via in situ free-radical polymerization technique. The tensile study showed a very high fracture elongation (11,800%) and fracture toughness (10.1 MJ/m^3) due to the strong polymer and clay interactions. Moreover, these NCHs exhibit the capabilities to recover fully after stretching through a drying–reswelling procedure. These properties make them a potential candidature in tissue engineering (i.e., artificial muscle, actuators).

11.5 BIOMEDICAL APPLICATION OF NCHs

The above-mentioned various characteristic properties of NCHs make them more suitable for their applications in the area of biomedical sciences especially bone regeneration, tissue engineering, drug delivery, etc. (Figure 11.11). A brief overview of some of their biomedical applications is being summarized in the proceeding sections.

11.5.1 TISSUE ENGINEERING

In recent years, researchers have received tremendous interest in the

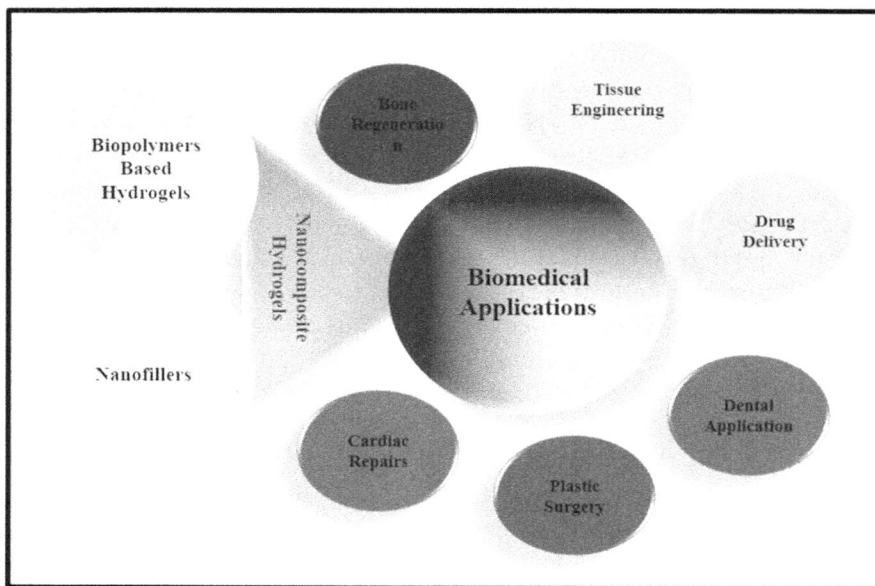

FIGURE 11.11   Biomedical applications of NCHs.

development of NCHs for tissue engineering [94]. The main aim of tissue engineering is to restore diseased or scratched tissue with biomaterials that can be used to balance the normal function [95]. Figure 11.12 shows NCHs for tissue engineering applications. The tailored NCHs provide 3D porous networks, with high hydrophilicity, elasticity, nanostructure surface, and mechanical stiffness. These properties of NCHs provide better cell attachment and cell proliferation [12]. A wide range of natural polymers and nanoparticles has been used to design NCHs with synergistic properties for tissue engineering. For instance, Sinha et al. [96] synthesized NCHs using hydrophilic PVA and hydroxyl apatite (HAP) via freeze-thawing methods for restored and renovate cartilage tissue. Gaharwar et al. [97] fabricated nanoclay-enriched electrospun poly(ε-caprolactone) (PCL)-based scaffolds for differentiation of human mesenchymal stem cells. The enhancement in surface roughness, biodegradability, mechanical strength, thermal properties of scaffolds can be taken place by the addition of nanoclay.

11.5.2   WOUND HEALING

The integration of useful nanofillers into the polymer matrices could produce NCHs with excellent wound dressing properties [5]. The nanocomposite-based material provides the proper level of humidity in wound caring material as compared to that of conventional dressing, which may irritate the dry skin [98]. These materials are highly applicable for wound healing due to their high wettability, biocompatibility, and excellent mechanical properties that are useful in preventing hemorrhage, absorption of exudates, reduction of pain, removal of march, and to protect the wound area from various infections (antibacterial, antifungal) and mechanical injury [9, 99]. In view of this, Kokabi et al. [100] synthesized PVA and clay-based NCHs by using freezing–thawing techniques. The dispersion of nanoclay in the polymer matrices enhanced their swelling and mechanical properties which is applicable for wound dressing. chitosan-based materials are also used in wound healing due to their biodegradability, biocompatibility, and hemostatic properties. Chitra et al. reported silver- and gold-based NCHs [101]. These NCHs have exhibited antibacterial activity and biocompatible nature, which are further used for wound healing applications.

11.5.3 DRUG DELIVERY

The development of a smart NCHs for drug delivery system, comprising micro- or nanoparticles embedded in hydrogels, is an appealing approach

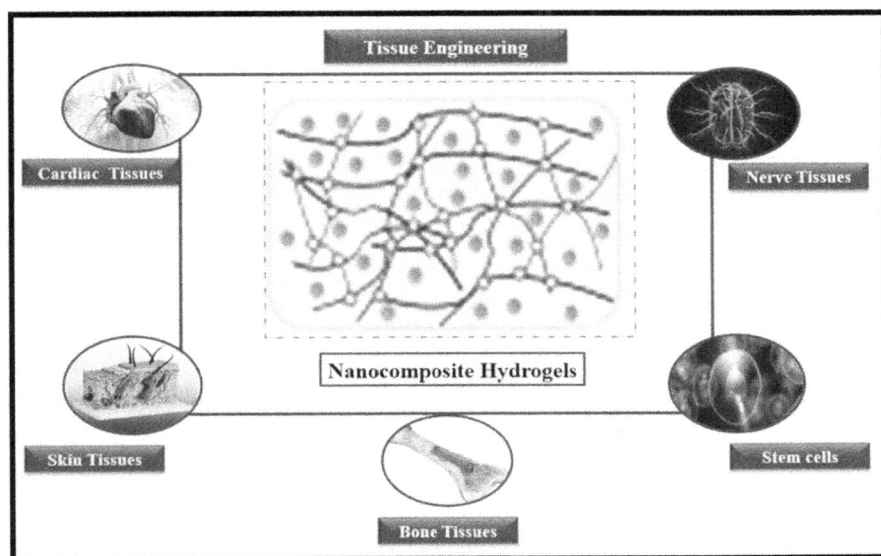

FIGURE 11.12   NCHs for tissue engineering.

[91]. This approach has been used to provide sustained drug delivery of hydrophobic drugs through the incorporation of drugs in particles, followed by the loading of particles in the hydrogel network [53]. The drug releases the rate of NCHs that can depend upon the characteristic properties of nanoparticles, such as swelling behavior, particle size, and dispersion coefficient [87, 102]. The drug can be loaded into a hydrogel and then its release may proceed through several mechanisms: diffusion-controlled, swelling controlled, chemically controlled, and environmentally responsive release [103]. The environment receptive NCHs have a good ability in the drug delivery system in which drug smartly released under peripheral stimuli, such as electric field, magnetic field, temperature, and pH [53]. Recently, researchers reported different biocompatible NCHs for sustained drug delivery without inducing any side effects [28]. For example, Zhu et al. [104] developed an injectable supramolecular hydrogel for sustained delivery of cisplatin in cancer therapy using poly(ethylene glycol)-b-poly(acrylic acid) (PEG-b-PAA) block copolymer nanoparticles. The in vitro cytotoxicity revealed that cisplatin-loaded hydrogels kill the human bladder carcinoma EJ cells. Chien et al. [105] developed a carboxybetain injectable hydrogel. The in vitro cytotoxicity of monomers such as

MMA, HEMA, and PEGMA is done by MTT assay. The % cell viability testing of the synthesized hydrogels demonstrated the potential of the hydrogels as a cell encapsulation scaffold material.

11.6 CONCLUSION

NCH-based drug delivery systems are attractive due to their characteristic properties like water absorption capacity, biodegradability, biocompatibility, and mechanically strong. NCHs are being explored as film material, in wound dressing application, tissue engineering, contact lenses, biosensors, etc. Among the various forms of NCHs based drug delivery systems injectable and orally administered forms are gaining interest. Generally, the use of chemotherapeutic drugs shows limited because of its nonspecific biodistribution and severe side effects. Therefore, it becomes imperative to develop drug delivery systems for the sustained release of these drugs. The objective of this chapter is to develop different NCHs (films and gels) using chitosan, CMC, HEC, and PVA (as a matrix), linseed oil-based polyol (as a cross-linker) and CNT, NaMMT, fumed silica (as a nanofillers) via solution blending and free radical polymerization technique, and to use the prepared NCHs as a potential candidate for their application in the delivery of an anticancer drug (cisplatin). The physicomechanical, spectral, and morphological characterization of these NCHs are discussed in the chapter. The chapter also covers mechanical strength, swelling, biodegradable, and biocompatible properties.

KEYWORDS

- biopolymeric
- nanocomposite hydrogels
- biomedical applications
- pH responsive
- drug delivery
- tissue engineering

REFERENCES

1. Caló, E., & Khutoryanskiy, V. V., Biomedical applications of hydrogels: a review of patents and commercial products, Eur. Polym. J. 65 (2015) 252–267.
2. Hoare, T. R., & Kohane, D. S., Hydrogels in drug delivery: progress and challenges, J. Polym. 49 (2008) 1993–2007.
3. Sornkamnerd, S., Okajima, M. K., & Kaneko, T., Tough and porous hydrogels prepared by simple lyophilization of LC gels, ACS Omega. 2 (2017) 5304–5314.
4. Ahmed, E. M., Hydrogel: preparation, characterization, and applications: a review, J. Adv. Res. 6 (2015) 105–121.
5. Hoffman, A. S., Hydrogels for biomedical applications, Adv. Drug Del. Rev. 43 (2002) 3–12.
6. El-Sherbiny, I. M., & Yacoub, M. H., Hydrogel scaffolds for tissue engineering: progress and challenges, Glob. Cardiol. Sci. Pract. 3 (2013) 316–342.
7. Motealleh, A., & Kehr, N. S., Nanocomposite hydrogels and their applications in tissue engineering, Adv. Healthc. Mater. 6 (2017) 1–19.
8. Cheng, B., Pei, B., Wang, Z., & Hu, Q., Advances in chitosan-based superabsorbent hydrogels, RSC Adv. 7 (2017) 42036–42046.
9. Yahia, L. H., History and applications of hydrogels, J. Biomed. Sci. 4 (2015) 1–23.
10. Hoffman, A. S., Hydrogels for biomedical applications, Adv. Drug Deliv. Rev. 64 (2012) 18–23.
11. Gil, M. S., Thambi, T., Phan, V. H. G., Kim, S. H., & Lee, D. S., Injectable hydrogels-incorporating cancer cell-specific cisplatin releasing nanogels for targeted drug delivery, J. Mater. Chem. B. 5 (2017) 7140–7152.
12. Van Vlierberghe, S., Dubruel, P., & Schacht, E., Biopolymer-based hydrogels as scaffolds for tissue engineering applications: a review, Biomacromolecules. 12 (2011) 1387–1408.
13. Ko, H. F., Sfeir, C., & Kumta, P. N., Novel synthesis strategies for natural polymer and composite biomaterials as potential scaffolds for tissue engineering, Philos. Trans. R. Soc. A Math. Phys. Eng. Sci. 368 (2010) 1981–1997.
14. Cheung, H. Y., Lau, K. T., Lu, T. P., & Hui, D., A critical review on polymer-based bio-engineered materials for scaffold development, Compos. Part B Eng. 38 (2007) 291–300.
15. Vashist, A., Shahabuddin, S., Gupta, Y. K., & Ahmad, S., Polyol induced interpenetrating networks: chitosan–methylmethacrylate based biocompatible and pH responsive hydrogels for drug delivery system, J. Mater. Chem. B. 1 (2013) 168–178.

16. Kouser, R., Vashist, A., Md. Zafaryab, M. Rizvi, A., & Ahmad, S., Biocompatible and mechanically robust nanocomposite hydrogels for potential applications in tissue engineering, Mater. Sci. Eng. C. 84 (2018) 168–179.

17. Noori, S., Kokabi, M., & Hassan, Z. M., Poly(vinyl alcohol)/chitosan/honey/clay responsive nanocomposite hydrogel wound dressing, J. Appl. Polym. Sci. 135 (2018) 1–12.

18. Wang, W., & Wang, A., Nanocomposite of carboxymethyl cellulose and attapulgite as a novel pH-sensitive superabsorbent: synthesis, characterization and properties, Carbohydr. Polym. 82 (2010) 83–91.

19. Zhou, C., & Wu, Q., A novel polyacrylamide nanocomposite hydrogel reinforced with natural chitosan nanofibers, Colloids Surf. B Biointerfaces. 84 (2011) 155–162.

20. Sharma, G., Thakur, B., Naushad, M., Kumar, A., Stadler, F. J., Alfadul, S. M., & Mola, G. T., Applications of nanocomposite hydrogels for biomedical engineering and environmental protection, Environ. Chem. Lett. 16 (2018)113–146.

21. Pandey, M., Mohamad, N., Cairul, M., & Mohd, I., Bacterial cellulose/acrylamide pH-sensitive smart hydrogel: development, characterization, and toxicity studies in ICR Mice Model, Mol. Pharm. 2. 14 (2014) 3596–3608.

22. Zu, Y., Zhang, Y., Zhao, X., Shan, C., Zu, S., Wang, K., Li, Y., & Ge, Y., Preparation and characterization of chitosan-polyvinyl alcohol blend hydrogels for the controlled release of nano-insulin, Int. J. Biol. Macromol. 50 (2012) 82–87.

23. Mulchandani, N., Shah, N., & Mehta, T., Synthesis of chitosan-polyvinyl alcohol copolymers for smart drug delivery application, Polym. Polym. Compos. 25 (2017) 241–246.

24. Garg, A. G., Sweta, Hydrogel : classification, properties, preparation and technical features, Asian J. Biomater. Res. 2016. 2 (2016) 163–170.

25. Varaprasad, K., Malegowd, G., Jayaramudu, T., Mohan, M., & Sadiku, R., A mini review on hydrogels classification and recent developments in miscellaneous applications, Mater. Sci. Eng. C. 79 (2017) 958–971.

26. Takashi Iizawa, S.S., Taketa, H., Maruta, M., Ishido, T., & Gotoh, T., Synthesis of porous poly(N-isopropylacrylamide) gel beads by sedimentation polymerization and their morphology, J. Appl. Polym. Sci. 104 (2007) 842–850.

27. Vashist, A., Gupta, Y. K., & Ahmad, S., Interpenetrating biopolymer network based hydrogels for an effective drug delivery system, Carbohydr. Polym. 87 (2012) 1433–1439.

28. Singh, N. K., & Lee, D. S., In situ gelling pH and temperature-sensitive biodegradable block copolymer hydrogels for drug delivery, J. Control. Release. 193 (2014) 214–227.

29. Giri, A., Bhunia, T., Mishra, S. R., Goswami, L., Panda, A. B., Pal, S., & Bandyopadhyay, A., Acrylic acid grafted guargum-nanosilica membranes for transdermal diclofenac delivery, Carbohydr. Polym. 91 (2013) 492–501.

30. Chang, C., He, M., Zhou, J., Zhang, L., Swelling behaviors of pH- and salt-responsive cellulose-based hydrogels, Macromolecules 44 (2011) 1642–1648.

31. Chrissafis, K., Paraskevopoulos, K. M., Papageorgiou, G. Z., & Bikiaris, D. N., Thermal and dynamic mechanical behavior of bionanocomposites: fumed silica nanoparticles dispersed in poly(vinyl pyrrolidone), chitosan, and poly(vinyl alcohol), J. Appl. Polym. Sci. 110 (2008) 1739–1749.

32. Jung, G., Yun, J., & Kim, H., Temperature and pH-responsive release behavior of PVA/PAAc/PNIPAAm/

MWCNTs nanocomposite hydrogels, Carbon Lett. 13 (2012) 173–177.

33. Caló, E., & Khutoryanskiy, V. V., Biomedical application of nanoparticles: a review of patents and commercial products, Eur. Polym. J. 65 (2018) 252–267.

34. Iliescua, M. H. R. I., Andronescua, E., Ghitulicaa, C. D., Voicua, G., & Ficaia, A., Montmorillonite–alginate nanocomposite as a drug delivery system incorporation and in vitro release of irinotecan, Int. J. Pharm. (2013) 1–10.

35. Wang, X., Du, Y., & Luo, J., Biopolymer/montmorillonite nanocomposite: preparation, drug-controlled release property and cytotoxicity, Nanotechnology 19 (2008) 1–7.

36. Zhang, H., Yang, M., Luan, Q., Tang, H., Huang, F., Xiang, X., Yang, C., & Bao, Y., Cellulose anionic hydrogels based on cellulose nano fibers as natural stimulants for seed germination and seedling growth, J. Agric. Food Chem. 65 (2017) 3785–3791.

37. Ganji, F., & Vasheghani-Farahani, E., Hydrogels in controlled drug delivery systems, Iran Polym. J. 18 (2009) 63–88.

38. Tang, Y., Zhou, D., & Zhang, J., Novel polyvinyl alcohol/styrene butadiene rubber latex/carboxymethyl cellulose nanocomposites reinforced with modified halloysite nanotubes, J. Nanomater. (2013) 1–8.

39. Yu, T., Malugin, A., & Ghandehari, H., Impact of silica nanoparticle design on cellular toxicity and hemolytic activity, ACS Nano. 5 (2011) 5717–5728.

40. Wu, S. H., Mou, C. Y., & Lin, H. P., Synthesis of mesoporous silica nanoparticles, Chem. Soc. Rev. 42 (2013) 3862–3875.

41. Xu, Y., Sheng, K., Li, C., & Shi, G., Self-assembled graphene hydrogel via a one-step hydrothermal process, ACS Nano. 4 (2010) 4324–4330.

42. Deng, G., Li, F., Yu, H., Liu, F., Liu, C., Sun, W., Jiang, H., & Chen, Y., Dynamic hydrogels with an environmental adaptive self-healing ability and dual responsive sol–gel transitions, ACS Macro Lett. 1 (2012) 275–279.

43. Ahearn, D. G., Grace, D. T., Jennings, M. J., Borazjani, R. N., Boles, K. J., Rose, L. J., Simmons, R. B., & Ahanotu, E. N., Effects of hydrogel/silver coatings on in vitro adhesion to catheters of bacteria associated with urinary tract infections, Curr. Microbiol. 41 (2000) 120–125.

44. Görg, A., Weiss, W., & Dunn, M. J., Current two-dimensional electrophoresis technology for proteomics, Proteomics 4 (2004) 3665–3685.

45. Kabanov, A. V., & Vinogradov, S. V., Nanogels as pharmaceutical carriers: finite networks of infinite capabilities, Angew. Chemie, Int. Ed. 48 (2009) 5418–5429.

46. Melek, L. N., Tissue engineering in oral and maxillofacial reconstruction, Tanta Dent. J. 12 (2015) 211–223.

47. Jayakumar, R., Menon, D., Manzoor, K., Nair, S. V., & Tamura, H., Biomedical applications of chitin and chitosan based nanomaterials—a short review, Carbohydr. Polym. 82 (2010) 227–232.

48. Wang, F., Xie, Z., Xing, L., Zhang, H. L. J., Feng, B., Zhang, J. L., Cui, P. F., Qiao, J. B., Shi, K., Cho, C. S., Cho, M. H., Xu, X., & Li, P., Biomaterials biocompatible polymeric nanocomplexes as an intracellular stimuli-sensitive prodrug for type-2 diabetes combination therapy, Biomater. J. 73 (2015) 149–159.

49. Jayaramudu, T., Raghavendra, G. M., Varaprasad, K., Sadiku, R., & Raju, K. M., Development of novel biodegradable Au nanocomposite hydrogels based on wheat: for inactivation of bacteria, Carbohydr. Polym. 92 (2013) 2193–2200.

50. Andronescu, E., Ficai, A., Albu, M. G., Mitran, V., Sonmez, M., Ficai, D., Ion, R., & Cimpean, A., Collagen-hydroxy-apatite/cisplatin drug delivery systems for locoregional treatment of bone cancer, Technol. Cancer Res. Treat. 12 (2013) 275–284.

51. Mahdavinia, G. R., Hasanpour, J., Rahmani, Z., Karami, S., & Etemadi, H., Nanocomposite hydrogel from grafting of acrylamide onto HPMC using sodium montmorillonite nano-clay and removal of crystal violet dye, Cellulose 20 (2013) 2591–2604.

52. Shaabani, Y., Sirousazar, M., & Kheiri, F., Synthetic—natural bion-anocomposite hydrogels on the basis of polyvinyl alcohol and egg white, J. Macromol. Sci. Part B. 55 (2016) 849–865.

53. Gaharwar, A. K., & Peppas, N. A., & Khademhosseini, A., Nanocomposite hydrogels for biomedical applica-tions, Biotechnol. Bioeng. 111 (2014) 441–453.

54. Shibayama, M., Suda, J., Karino, T., Okabe, T., & Takehisa, K. H., Structure and dynamics of poly, Macromolecules 37, (2004) 9606–9612.

55. Aimé, C., & Coradin, T., Nanocompos-ites from biopolymer hydrogels: blue-prints for white biotechnology and green materials chemistry, Polym, J.. Sci. Part B Polym. Phys. 50 (2012) 669–680.

56. Barkhordari, S., Yadollahi, M., & Namazi, H., pH sensitive nano-composite hydrogel beads based on carboxymethyl cellulose/layered double hydroxide as drug delivery systems, J. Polym. Res. 21 (2014) 1–9.

57. Tanpichai, S., & Oksman, K., Cross-linked poly(vinyl alcohol) composite films with cellulose nanocrystals: mechanical and thermal properties, J. Appl. Polym. Sci. (2018) 1–11.

58. Peng, N., Hu, D., Zeng, J., Li, Y., Liang, L., & Chang, C., Superabsorbent cellulose-clay nanocomposite hydro-gels for high efficient removal of dye in water, ACS Sustain. Chem. Eng. 4 (2016) 7217–7224.

59. Liu, H., Wang, C., Li, C., Qin, Y., Wang, Z., Yang, F., Li, Z., & Wang, J., A functional chitosan-based hydrogel as a wound dressing and drug delivery system in the treatment of wound healing, RSC Adv. 8 (2018) 7533–7549.

60. Qiu, X., & Hu, S., "Smart" materials based on cellulose: a review of the prep-arations, properties, and applications, Materials (Basel) 6 (2013) 738–781.

61. Mishra, S. K., Ferreira, J. M. F., & Kannan, S., Mechanically stable antimi-crobial chitosan–PVA–silver nanocom-posite coatings deposited on titanium implants, Carbohydr. Polym. J. 121 (2015) 37–48.

62. Vimala, K., Mohan, Y. M., Varaprasad, K., & Narayana, N., Fabrication of curcumin encapsulated chitosan-PVA silver nanocomposite films for improved antimicrobial activity, J. Biomater. Nanobiotechnol. 2 (2011) 55–64.

63. Trivedi, T. J., Rao, K. S., & Kumar, A., Facile preparation of agarose–chitosan hybrid materials and nanocomposite ionogels using an ionic liquid via disso-lution, regeneration and sol–gel transi-tion, Green Chem. 16 (2014) 320–330.

64. Ali, A. E., & Alarifi, A., Characteriza-tion and in vitro evaluation of starch based hydrogels as carriers for colon specific drug delivery systems, Carbo-hydr. Polym. J. 78 (2009) 725–730.

65. Tang, X., & Alavi, S., Recent advances in starch, polyvinyl alcohol based polymer blends, nanocomposites and their biodegradability, Carbohydr. Polym. 85 (2011) 7–16.

66. Wu, J., Lin, J., Zhou, M., & Wei, C., Synthesis and properties of starch-graft-polyacrylamide/clay superabsorbent composite, Macromol. Rapid Commun. 21 (2000) 1032–1034.

67. Spagnol, C., Rodrigues, F. H. A., Pereira, A. G. B., Fajardo, A. R., Rubira, A. F., & Muniz, E. C., Superabsorbent hydrogel nanocomposites based on starch-g-poly(sodium acrylate) matrix filled with cellulose nanowhiskers, Cellulose 19 (2012) 1225–1237.

68. Famá, L. M., Pettarin, V., Goyanes, S. N., & Bernal, C. R., Starch/multi-walled carbon nanotubes composites with improved mechanical properties, Carbohydr. Polym. 83 (2011) 1226–1231.

69. Siró, I., & Plackett, D., Microfibrillated cellulose and new nanocomposite materials: a review, Cellulose. 17 (2010) 459–494.

70. Zhou, C., & Wu, Q., Recent development in applications of cellulose nanocrystals for advanced polymer-based nanocomposites by novel fabrication strategies, In: Nanocrystals: Synthesis, Characterization and Applications, 2012: 103–120.

71. Chang, C., Zhang, L., Zhou, J., Zhang, L., & Kennedy, J. F., Structure and properties of hydrogels prepared from cellulose in NaOH/urea aqueous solutions, Carbohydr. Polym. 82 (2010) 122–127.

72. Aduba, D., & Yang, H., Polysaccharide fabrication platforms and biocompatibility assessment as candidate wound dressing materials, Bioengineering 4 (2017) 1–16.

73. Chang, C., & Zhang, L., Cellulose-based hydrogels: present status and application prospects, Carbohydr. Polym. 84 (2011) 40–53.

74. Nakayama, B. A., Kakugo, A., Gong, J. P., Osada, Y., Takai, M., & Erata, T., High mechanical strength double-network hydrogel with bacterial cellulose, Adv. Funct. Mater. 14, (2004) 1124–1128.

75. Babu, A. C., Prabhakar, M. N., Babu, A. S., Mallikarjuna, B., Subha, M. C. S., & Rao, K. C., Development and characterization of semi-IPN silver nanocomposite hydrogels for antibacterial applications, Int. J. Carbohydr. Chem. (2013) 1–8.

76. Zhang, H., Zhai, D., & He, Y., Graphene oxide/polyacrylamide/carboxymethyl cellulose sodium nanocomposite hydrogel with enhanced mechanical strength: preparation, characterization and the swelling behavior, RSC Adv. 4 (2014) 44600–44609.

77. Hamidabadi, H. G., Rezvani, Z., Bojnordi, M. N., Shirinzadeh, H., Seifalian, A. M., Joghataei, M. T., Razaghpour, M., Alibakhshi, A., Yazdanpanah, A., Salimi, M., Mozafari, M., Urbanska, A. M., Reis, R. L., & Kundu, S. C., Gholipourmalekabadi, M., Chitosan-intercalated montmorillonite/poly(vinyl alcohol) nanofibers as a platform to guide neuron like differentiation of human dental pulp stem cells, ACS Appl. Mater. Interfaces. 9 (2017) 11392–11404.

78. Hamidi, M., Azadi, A., & Rafiei, P., Hydrogel nanoparticles in drug delivery., Adv. Drug Deliv. Rev. 60 (2008) 1638–49.

79. Lue, S. J., Chen, J. Y., & Yang, J. M., Crystallinity and stability of poly(vinyl alcohol)-fumed silica mixed matrix membranes, J. Macromol. Sci. Part B Phys. 47 (2008) 39–51.

80. Agnihotri, S. A., Mallikarjuna, N. N., & Aminabhavi, T. M., Recent advances on chitosan-based micro- and nanoparticles in drug delivery, J. Control. Release. 100 (2004) 5–28.

81. Wang, E., Desai, M. S., & Lee, S., Light-controlled graphene-elastin composite hydrogel actuators, Nano Lett. 13 (2013) 2826–2830.

82. Yadav, M., Rhee, K. Y., & Park, S. J., Synthesis and characterization of graphene oxide/carboxymethylcellulose/alginate nanocomposites blends, Carbohydr. Polym. 110 (2014) 18–25.

83. Yadav, M., & Ahmad, S., Montmorillonite/graphene oxide/chitosan composite: synthesis, characterization and properties, Int. J. Biol. Macromol. (2015) 1–11.

84. Pan, Q., Lv, Y., Williams, G. R., Tao, L., Yang, H., Li, H., & Zhu, L., Lactobionic acid and carboxymethyl chitosan functionalized graphene oxide nanocomposites as targeted anticancer drug delivery systems, Carbohydr. Polym. J. 151 (2016) 812–820.

85. Zhang, J., Fan, L., & Wang, A., In situ generation of sodium alginate/hydroxyapatite/halloysite nanotubes nanocomposite hydrogel beads as drug-controlled release matrices, J. Mater. Chem. B Pap. 1 (2013) 6261–62.

86. Hossan, M. Y., Molla, M. A. I., Islam, M. S., Rana, A. A., Gafur, M. A., & Karim, M. M., Fabrication and characterization of polyvinyl alcohol-hydroxyapatite biomimetic scaffold by freeze thawing in situ synthesized hybrid suspension for bone tissue engineering, Int. J. Emerg. Technol. Adv. Eng. 2 (2012) 696–701.

87. Giri, A., Bhowmick, M., Pal, S., & Bandyopadhyay, A., Polymer hydrogel from carboxymethyl guar gum and carbon nanotube for sustained transdermal release of diclofenac sodium, Int. J. Biol. Macromol. 49 (2011) 885–893.

88. Ozkahraman, B., Irmak E.T., Carbon nanotube based polyvinylalcohol-polyvinylpyrolidone nanocomposite hydrogels for controlled drug delivery applications, J. Sci. Technol. A Appl. Sci. Eng. 18 (2017) 543–553.

89. Rodrigues, A. A., Batista, N. A., Bavaresco, V. P., Baranauskas, V., Ceragioli, H. J., Peterlevitz, A. C., Santos, A. R., & Belangero, W. D., Polyvinyl alcohol associated with carbon nanotube scaffolds for osteogenic differentiation of rat bone mesenchymal stem cells, Carbon. 50 (2012) 450–459.

90. Joddar, B., Garcia, E., Casas, A., & Stewart, C. M., Development of functionalized based alginate hydrogels for enabling biomimetic technologies, Nature. Com. (2016) 1–12.

91. Popat, A., Liu, J., Lu, G. Q., & Qiao, S. Z., A pH-responsive drug delivery system based on chitosan coated mesoporous silica nanoparticles, J. Mater. Chem. 22 (2012) 11173–11178.

92. Zhu, M., Zhu, Y., Zhang, L., & Shi, J., Preparation of chitosan/mesoporous silica nanoparticle composite hydrogels for sustained co-delivery of biomacromolecules and small chemical drugs, Sci. Technol. Adv. Mater. 14 (2013) 1–9.

93. Gao, G., Du, G., Sun, Y., & Fu, J., Self-healable, tough, and ultrastretchable nanocomposite hydrogels based on reversible polyacrylamide/montmorillonite adsorption, ACS Appl. Mater. Interfaces. 7 (2015) 5029–5037.

94. Zhang, J., & Peng, C.-A., Poly(N-isopropylacrylamide) modified polydopamine as a temperature-responsive surface for cultivation and harvest of mesenchymal stem cells, Biomater. Sci. 5 (2017) 2310–2318.

95. Cha, C., Shin, S. R., Annabi, N., Dokmeci, M. R.., & Khademhosseini, A., Carbon-based nanomaterials: multifunctional materials for biomedical engineering, ACS Nano. 7 (2013) 2891–2897.

96. Sinha, A., Das, G., Sharma, B. K., Roy, R. P., Pramanick, A. K., & Nayar, S., Poly(vinyl alcohol)-hydroxyapatite biomimetic scaffold for tissue regeneration, Mater. Sci. Eng. C. 27 (2007) 70–74.

97. Gaharwar, A. K., Mukundan, S., Karaca, E., Dolatshahi-Pirouz, A., Patel, A., Rangarajan, K., Mihaila, S. M., Iviglia, G., Zhang, H., & Khademhosseini, A., Nanoclay-enriched poly(ε-caprolactone) electrospun scaffolds

for osteogenic differentiation of human mesenchymal stem cells, Tissue Eng. Part A. 19 (2014) 1–15.

98. Hamidi, M., Azadi, A., & Rafiei, P., Hydrogel nanoparticles in drug delivery, Adv. Drug Deliv. Rev. 60 (2008) 1638–1649.

99. Guibal, E., Heterogeneous catalysis on chitosan-based materials: a review, Prog. Polym. Sci. 30 (2005) 71–109.

100. Kokabi, M., Sirousazar, M., & Hassan, Z. M., PVA-clay nanocomposite hydrogels for wound dressing, Eur. Polym. J. 43 (2007) 773–781.

101. Chitra, S. G. G., Franklin, D. S., Sudarsan, S., & Sakthivel, M., Noncytotoxic silver and gold nanocomposite hydrogels with enhanced antibacterial and wound healing applications, Polym. Eng. Sci. (2018) 1–10.

102. Zhao, F., Yao, D., Guo, R., Deng, L., Dong, A., & Zhang, J., Composites of polymer hydrogels and nanoparticulate systems for biomedical and pharmaceutical applications, Nanomaterials 5 (2015) 2054–2130.

103. Chang, C., Duan, B., Cai, J., & Zhang, L., Superabsorbent hydrogels based on cellulose for smart swelling and controllable delivery, Eur. Polym. J. 46 (2010) 92–100.

104. Zhu, W., Li, Y., Liu, L., Chen, Y., Wang, C., & Xi, F., Supramolecular hydrogels from cisplatin-loaded block copolymer nanoparticles and alpha-cyclodextrins with a stepwise delivery property, Biomacromolecules 11 (2010) 3086–3092.

105. Chien, H. W., Tsai, W. B., & Jiang, S., Direct cell encapsulation in biodegradable and functionalizable carboxybetaine hydrogels, Biomaterials 33 (2012) 5706–5712.

106. Venkatesan, J., Jayakumar, R., Mohandas, A., Bhatnagar, I., & Kim, S. K., Antimicrobial activity of chitosan-carbon nanotube hydrogels, Materials 7 (2014) 3946–3955.

107. Mei, N., Xuguang, L., Jinming, D., Husheng, J., Liqiao, W., & Bingshe, X., Antibacterial activity of chitosan coated Ag-loaded nano-SiO$_2$ composites, Carbohydr. Polym. J. 78 (2009) 54–59.

108. Vicentini, D. S., Smania, A., & Laranjeira, M. C. M., Chitosan/poly(vinyl alcohol) films contiaining ZnO nanoparticles and plasticizers, Mater. Sci. Eng. C. 30 (2010) 503–508.

109. Katti, K. S., Katti, D. R., & Dash, R., Synthesis and characterization of a novel chitosan/montmorillonite/hydroxyapatite nanocomposite for bone tissue engineering, Biomed. Mater. 3 (2008) 1–12.

110. Kabiri, K., Mirzadeh, H., Zohuriaan-Mehr, M. J. & Daliri, M. Chitosan-modified nanoclay–poly(AMPS) nanocomposite hydrogels with improved gel strength, Polym. Int. 58 (2009) 1252–1259.

111. Wang, S., Tan, Y., Zhao, D., & Liu, G., Amperometric tyrosinase biosensor based on Fe$_3$O$_4$ nanoparticles–chitosan nanocomposite, Biosens. Bioelectron. 23 (2008) 1781–1787.

112. Kumar, P. T. S., Raj, N. M., Praveen, G., Chennazhi, K. P., Nair, S. V., & Jayakumar, R., In vitro and in vivo evaluation of microporous chitosan hydrogel/nanofibrin composite bandage for skin tissue regeneration, Tissue Eng. Part A. 19 (2013) 380–392.

113. Zhang, J., Wang, L., & Wang, A., Preparation and properties of chitosan-g-poly(acrylic acid)/montmorillonite superabsorbent nanocomposite via in situ intercalative polymerization, Ind. Eng. Chem. Res. 46, (2007) 2497–2502.

114. Venkatesan, J., & Kim, S., Chitosan composites for bone tissue engineering—an overview, Mar. Drugs. 8, (2010) 2252–2266.

115. Kumar, A., Kumar, P., & Kumar, P., Biodegradability and swelling capacity

of Kaolin based chitosan-g-PHEMA nanocomposite hydrogel, Int. J. Biol. Macromol. 74 (2015) 620–626.

116. Konwara, A., Gogoia, N., Majumdarb, G. & Chowdhury, D., Green chitosan-carbon dots nanocomposite hydrogel film with superior properties, Carbohydr. Polym. (2014) 1–27.

117. Guclu, G. Ebru-Al, Emik, S., Iyim, T. B., Ozgumus, S., & Ozyurek, M., Removal of Cu^{2+} and Pb^{2+} ions from aqueous solutions by starch-graft-acrylic acid/montmorillonite superabsorbent nanocomposite hydrogels, Polym. Bull. 65 (2010) 333–346.

118. Famá, L. M., Pettarin, V., Goyanes, S. N., & Bernal, C. R., Starch/multi-walled carbon nanotubes composites with improved mechanical properties, Carbohydr. Polym. J. 83 (2011) 1226–1231.

119. Cao, X., Chen, Y., Chang, P. R., Muir, A. D., & Falk, G., Starch-based nanocomposites reinforced with flax cellulose nanocrystals, eXPRESS Polym. Lett. 2 (2008) 3144.

120. Vigneshwaran, N., Kumar, S., Kathe, A. A., Varadarajan, P. V., & Prasad, V., Functional finishing of cotton fabrics using zinc oxide–soluble starch nanocomposites, Nanotechnology 17 (2006) 5087–5095.

121. Wilpiszewska, K., Antosik, A.K., & Spychaj, T. Novel hydrophilic carboxymethyl starch/montmorillonite nanocomposite films, Carbohydr. Polym. (2015) 1–32.

122. Shahrooie, B., Rajabi, L., Dereakhshan, A. A., & Keyhani, M., Fabrication of a new starch-based hydrogel nanocomposite for the removal of ammonia from wastewater, Proc. 5th Int. Congr. Nanosci. Nanotechnol. (2014) 2–24.

123. Lima-tenório, M. K., Tenório-neto, E. T., Garcia, F. P., Nakamura, C. V., Guilherme, M. R., Muniz, E. C., Pineda, E. A. G., & Rubira, A. F., Hydrogel nanocomposite based on starch and co-doped zinc ferrite nanoparticles that shows magnetic field-responsive drug release changes, J. Mol. Liq. (2014) 1–6.

124. Gonzalez, J. S., Luduena, L. N, Ponce, A., & Alvarez, V. A., Poly(vinyl alcohol)/cellulose nanowhiskers nanocomposite hydrogels for potential wound dressings, Mater. Sci. Eng. C. 34 (2014) 54–61.

125. Aouada, F. A., De Moura, R., Orts, W. J., & Mattoso, L. H. C., Preparation and characterization of novel micro- and nanocomposite hydrogels containing cellulosic fibrils, J. Agric. Food Chem. 59, (2011) 9433–9442.

126. Aghazadeh, M., Gholamali, I., Namazi, H., & Aghazadeh, M., Synthesis and characterization of antibacterial carboxymethyl cellulose/ZnO, Int. J. Biol Macromol. (2014) 1–27.

127. Yadollahi, M., Namazi, H., & Aghazadeh, M., Antibacterial carboxymethyl cellulose/Ag nanocomposite hydrogels cross-linked with layered double hydroxides, Int. J. Biol. Macromol. (2015) 1–31.

128. Aghazadeh, M., Gholamali, I., Namazi, H., & Aghazadeh, M., Synthesis and characterization of antibacterial carboxymethylcellulose/CuO bio-nanocomposite hydrogels, Int. J. Biol. Macromol. (2014) 1–28.

129. Xu, M., Huang, Q., Wang, X., & Sun, R., Highly tough cellulose/graphene composite hydrogels prepared from ionic liquids, Ind. Crop. Prod. J. 70 (2015) 56–63.

130. Peng, N., Hu, D., Zeng, J., Li, Y., Liang, L., & Chang, C., Superabsorbent cellulose-clay nanocomposite hydrogels for highly efficient removal of dye in water, ACS Sustain. Chem. Eng. 4 (2016) 7217–7224.

131. Rakhshaei, R., & Namazi, H., A potential bioactive wound dressing based

on carboxymethyl cellulose/ZnO impregnated MCM-41 nanocomposite hydrogel, Mater. Sci. Eng. C. 73 (2017) 456–464.

132. Mahdavinia, G., Afzali, A., Etemadi, H., & Hosseinzadeh, H., Magnetic/pH-sensitive nanocomposite hydrogel based carboxymethyl cellulose-g-polyacrylamide/montmorillonite for colon targeted drug deliver, Nanomedicine Res. J. 2 (2017) 111–122.

133. Rasoulzadeh, M., & Namazi, H., Carboxymethyl cellulose/graphene oxide bio-nanocomposite hydrogel beads as anticancer drug carrier agent, Carbohydr. Polym. (2017) 1–19.

134. Chakraborty, S., Ponrasu, T., Chandel, S., Dixit, M., & Muthuvijayan, V., Reduced graphene oxide-loaded nanocomposite scaffolds for enhancing angiogenesis in tissue engineering applications, R. Soc. Open Sci. 5 (2018) 1–14.

CHAPTER 12

NEW PERSPECTIVES ON DEVELOPMENT OF NANOCOMPOSITES BASED ON BIODEGRADABLE POLYMERS AND THEIR TISSUE ENGINEERING APPLICATIONS

NICA SIMONA LUMINITA[1]*, DELIA MIHAELA RATA[2], and CRISTIAN LOGIGAN[1]

1Chemical Company SA, 14 Chimiei Bvd., 707252 Iasi, Romania

2Faculty of Medical Dentistry, "Apollonia" University of Iasi, Iasi 700399, Romania

Corresponding author. E-mail: nica.simona@icmpp.ro

ABSTRACT

Recently developed nanocomposite technology permits the development of new materials with improved mechanical, thermal, electrical, and various other properties, in comparison with those of a virgin polymer. The nanocomposites based on biodegradable and biocompatible polymers have been used in various biomedical applications such as drug delivery, antimicrobial properties, tissue engineering (TE), wound dressings, stem cell therapy, cancer therapy, cardiac prosthesis, artificial blood vessels, biosensors, and enzyme immobilization. This chapter describes the current developments in the field of TE, particularly regenerative medicine reviews green polymers composites. After a short description of the commonly used biopolymers and their physicochemical characteristics, this chapter presents the preparation, characterization, properties, and applications of nanocomposites based on biodegradable polymers. Efficient strategies to design new materials with enhanced properties as compared with classical biomaterials are presented. This offers the possibility to overcome the limitations imposed by synthetic polymers and solves important medical problems. The shape, properties, and functionalization routes of the nanofillers allowed modulation of pursued properties of great importance in regenerative medicine.

12.1 INTRODUCTION

Different health problems are caused by damaged tissues or organs. Various factors such as traumatic injury, disease, and wounds caused by surgery interventions affect damaged tissue [1, 2]. Restoration/replacement of the affected hard or soft tissue implies rapid identification and treatment of the infection in a valid medical clinic.

Tissue engineering (TE) deal with the growth of new tissues and organs, by combining cells from the patient itself, a three-dimensional (3D) scaffold, growth factors, and signals [3]. The role of growth factors is to direct cell behavior with the final scope of developing fully functional organs or tissues [4, 5]. Stem cells are very attractive for TE applications. This is due to their unique capacity to self-renew and differentiate into different specialized cells [6]. The behavior of a stem cell is influenced by its microenvironment [7]. The cell microenvironment represents the extracellular matrix (ECM) [8]. The ECM provides structural support and biochemical factors for surrounding cells [9]. The natural ECM is a complex network [10] composed of proteins, polysaccharide, and water. Researchers hypothesized that scaffolds should have the structural and functional integrities of their natural counterparts [11]. However, a broad debate concerning the importance given by spatial relationship between components of ECM (collagen, elastins, glycosaminoglycans, etc.) [12] and fully composition of ECM [13] can be found in the cited resource [9]. Collagen represents the first choice as a scaffold material since it constitutes a major part of ECM in the human body. Collagen is a protein containing around 28 different types [14]. Generally, collagen is found in the tissue of cartilage, bone, and tendon [15]. It was reported that collagen has low antigenic properties [16] making highly biocompatible natural polymer. Cellulose represents a suitable alternative to mimic the ECM in TE [16]. Nontoxicity, fibrous character, fibril stiffness, surface chemistry, gel formation, and capability to form porous structures are the features required for this process [14].

Depending on the origin of tissue and organs subjected to the repairing process, the biomaterial chosen as scaffold should accomplish specific requirements. Strength, biodegradability, fabrication technique, bioactivity, and cytocompatability are envisaged [8]. Ceramics, natural and synthetic polymers, metals/metal oxide nanoparticles (silver, gold, and iron oxide) are considered important biomaterials to be used in biomedical applications [17]. Biomaterials can be degradable or nondegradable depending on their specific use. Additionally, hydrogels received great attention from researchers for the fabrication of scaffolds. Hydrogels are a 3D network of either natural or synthetic polymer insoluble in water [18]. In summary, biopolymers are used to construct 3D scaffolds for regenerating damaged tissue such as skin tissue [1], bone tissue [2]—particularly, constituting the main focus in this chapter, nerve tissue [3], cartilage tissue [19]. However, applications of biomaterials in drug delivery systems [20, 21], dental [22], corneal, and retina TE [23] will not be discussed here. Even if naturally derived polymers used alone support very well cells attachments [24], their main drawbacks referring to immunogenicity and complex structural composition [25] limits their in vivo applications [26]. As a consequence, natural polymers could be recognized as foreign materials by the immune system of the patient [25]. Contrariwise, synthetic polymers [27] are characterized by a more controllable structure [28] which can be easily adapted to permit desired physicochemical characteristics to habilitate them for TE. But their biocompatibility (the lack of cell recognition signals) still remains a difficult task to achieve [29]. Low thermal stability and poor mechanical properties in addition to inadequate barrier properties represent also challenges of using only biopolymers alone in tissue scaffold constructs [30].

The strategy of incorporating materials at nanoscale dimensions in different matrices types—metallic-, ceramic-, and polymer-based—seems to be a proper method for improvement of physicochemical properties of final constructs in TE applications. Advantages [31] are provided in Scheme 12.1.

In this chapter, we focus on the selection of materials for bone TE, different preparation techniques for scaffold construct, and some important proprieties of the final products.

12.2 CANDIDATE MATERIALS FOR TE SCAFFOLDS

12.2.1 BIOPOLYMER COMPOSITE FOR SCAFFOLD TE

The structures of the scaffold constitute the most important aspect in

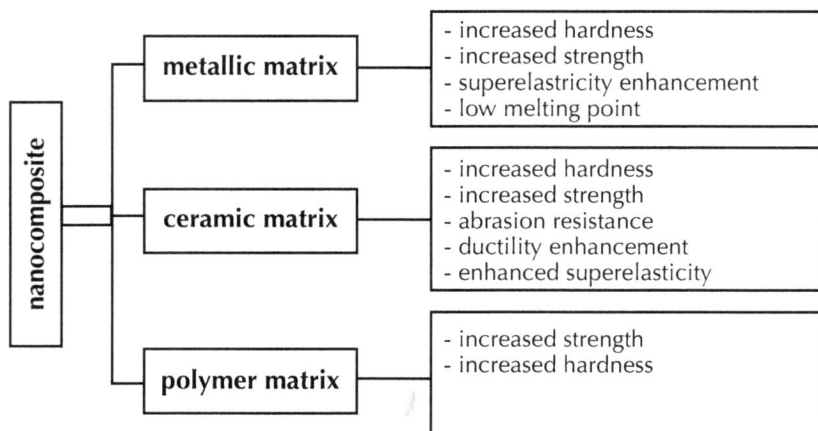

SCHEME 12.1   Classification of nanocomposite matrices and advantages.

bone TE. Its morphology influences cell migration [32]. However, the material composition of the scaffold is not the only one that must be taken into consideration to induce bone formation. The two important parameters such as pore size and interconnectivity of the porous structure also affect the bone regeneration process [33]. In bone TE, the porosity required for scaffold construction is >90% [34] being optimal for regeneration of the new bone tissue [35]. It is believed that a minimum pore size diameter at around 100 μm [36] assures cell migration and nutrient transport [37]. However, a porous structure with pore size under 100 μm or small porosity is possible not to have enough space for bone ingrowth [38].

The most utilizable biopolymers in bone TE [39] are shown in Scheme 12.2.

A nanocomposite formed with different components enhances different properties of biomaterials. Natural polymers—proteins or polysaccharides—have a major role in cell adhesion and function [40]. One of the main disadvantages of these materials refers to the lack of control of mechanical properties. Many studies concentrated on different aspects of natural cellulose-based materials [41] as well as nanocellulose (NC)-based materials [42]. The utilization of cellulose as a biomaterial receives validation from its unique chemical structure. The empirical formula of cellulose is $(C_6H_{10}O_5)_n$. It is an unbranched homopolysaccharide of linear glucan chain consisting of thousands of $\propto$-D-glucapyranoseunits linked via $\beta(1 \rightarrow 4)$-glycosidic bonds [43]. A large description of its chemical structure can be found in the cited resource [44].

NC refers to the cellulose with one dimension in the nanometer range [45]. Depending on the source of cellulose—plants or bacteria—and its processing conditions [42]. NC can be classified into three major categories: nanocrystalline cellulose, nanofibrillated cellulose, and bacterial cellulose (BC). Despite their similarity in composition [46], important differences can be found in morphology [47], particle size [48], and crystallinity [49] leading to different material properties. BC rweceived great attention from the researchers due to its high crystallinity (74%–96%) [50], high mechanical strength (tensile strength 200–300 MPa) [51], excellent stability, and low cytotoxicity [52]. Additionally, high porosity [53], great biological cell affinity, higher flexibility, capacity to hold water, and high resistant degradation recommend this natural polymer to form composites. The first scientific rapport regarding the production of BC belongs to Brown [54]. Since then numerous researchers have tried to produce BC using different type of bacteria such as *Gluconacetobacter*, *Pseudomonas*, *Salmonella*, and *Sarcina ventriculi* [55].

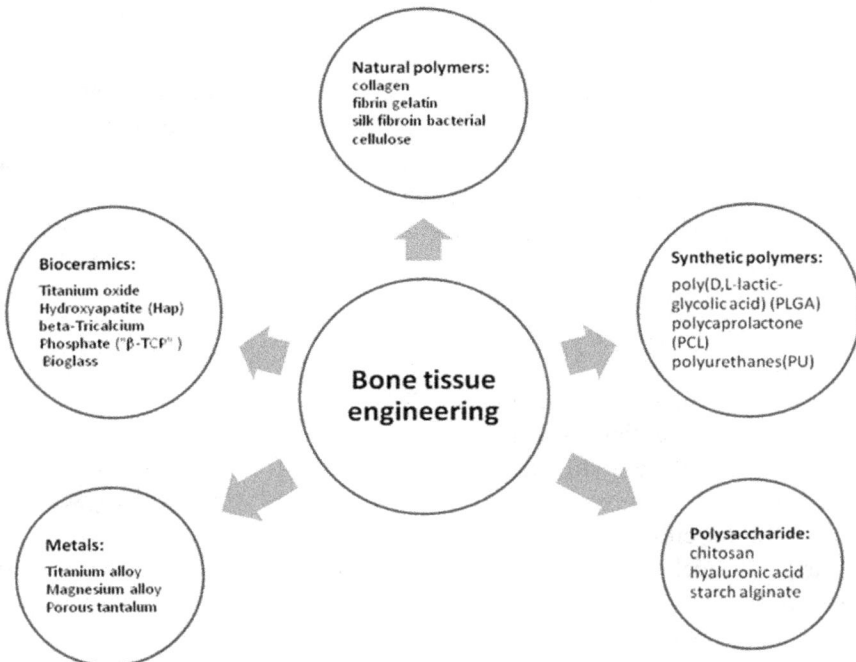

SCHEME 12.2   The most utilized polymers in bone TE.

Some researchers explored the concept of fully bio-derived nano-composite using either natural polymers or synthetic polymers. Most methods are centered to form collagen/BC composites based on physical bonds (aggregation, adsorption, coating, etc.). Few examples of forming BC-based composite scaffolds are taken into account.

BC could be considered both matrix and filler reinforcement. Noh et al. [56] reported the preparation of composite scaffolds composed of a different ratio of bacterial cellulose and collagen (1:1, 3:1, 5:1). They reported that composite scaffold with higher BC contents (5:1 content) is more resistant to the contraction in wet conditions than scaffolds made only by collagen. Also, final scaffolds could provide an optimum microenvironment for the osteogenic differentiation potential of umbilical cord blood-derived mesenchymal stem cells (UCB-MSCs) viability and also proliferation in vitro. As for in vivo study, they concluded that neovascularization occurs in the composite.

Another group of researchers has prepared porous collagen–carboxy-methyl cellulose/hydroxyapatite (ColCMC/HA) composite [57]. The technique chosen for scaffolds preparation such as bimolecular coassembly route favored deposition of nanosized hyaluronic acid. They reported improved mechanical properties for final composite (7.06 MPa)

scaffolds and porosity between 71% and 85%. This parameter favors cell attachment, growth, and nutrient delivery for tissue regeneration. In vitro, experiments showed that the ColCMC/HA composite is highly biocompatible.

12.2.2   MAGNETIC POLYMERS COMPOSITE FOR SCAFFOLD TE

The development of magnetic polymer composite as scaffolds for TE gained a lot of attention in the literature data. Generally, magnetic particles have the tendency to agglomerate in the polymeric matrix. Numerous preparation techniques have been utilized to achieve a homogenous dispersion in the biopolymer matrix, such as copre-cipitation, microemulsion, thermal decomposition, ultrasonic irradiation, sol–gel [58]. The characteristics of the host matrix—its morphology, porosity, surface area—influence the filler features, such as size and shape.

The inclusion of magnetic nanoparticles, particularly magnetite or ferro-/ferri-magnetic material, in BC matrix could be considered a promising composite scaffold for bone TE. This effect could be possible due to the susceptibility of guiding the filler to the site of interest. Additionally, biocompatibility, nontoxicity in the human body affects the osteoinduction property of the scaffold in the presence or in the absence of the external magnetic field [59].

Research performed by Torgo [60] proposed development of the nanocomposite scaffold for bone TE using BC as a matrix and a complex of magnetite and HA nanoparticles as fillers. The selected method of preparation was ultrasonic irradiation. Researchers revealed a uniform distribution of magnetic nanoparticles in the matrix. For the same research, other properties are mentioned below:

1. the decrease of crystallinity index of BC from 82.5% to 62% in the composite,
2. decrease of saturation magnetization of the nanocomposite from 15.84 to 3.94 emu/g,
3. significant decrease of swelling ability as the nanoparticles were incorporated in the polymer matrix, and
4. the degree of porosity was maintained at a higher degree at about 80% in the nanocomposite scaffold.

Research conducted by Bigham et al. [61] developed multifunctional magnetic 3D scaffolds nanocomposite for use in bone TE. Firstly, they successfully synthesized magnetic Mg_2SiO_4–$CoFe_2O_4$ nanoparticles with a core–shell structure. In order to form a nanocomposite material as a 3D scaffold polymer sponge technique was implied. Bioactivity, biodegradability,

mechanical properties, hyperthermia capability, controlled release potential, antibacterial activity, cell compatibility, and attachment were evaluated. It was revealed that the scaffold has a high porous structure along with interconnected porosity (84%) and optimum mechanical properties very similar to cancellous bone.

The results obtained by researchers recommend Mg_2SiO_4–$CoFe_2O_4$ nanocomposite scaffold applied in the bone TE.

12.2.3 CERAMIC POLYMERS COMPOSITE FOR SCAFFOLD TE

Researchers tried to develop ceramic–polymer nanocomposite as scaffolds in bone TE. Specific ceramics such as bioactive glasses can maintain cell growth [62]. The main drawbacks of bioactive glass mention their low mechanical properties and the capacity to be very brittle [63]. However, the advantage of controlling chemical composition made them attractive as scaffold materials [64].

Dharmalingam et al. [65] prepared cellulose/zinc-sulfate–calcium-phosphate nanocomposites. The preparation method of the material was via microwave-assisted synthesis method. The nanocomposite showed no influence on the proliferation of osteosarcoma (MG63 cells).

Another group of researchers reported the fabrication of bioactive and bioresonable porous scaffolds, based on poly(3-hydroxybutyrate) for bone TE [66]. A thermally induced phase-separation technique was implied to produce porous structure. Incorporation of HA into the matrix enhanced osteoinductivity and osteoconductivity of the final construct. Results showed improvement of porosity at about 92%.

Draghici et al. [67] fabricated composite scaffolds based on calcium phosphate and barium titanate using BC as polymeric support. Results have showed a high porosity.

12.3 TECHNIQUES FOR FABRICATION NANOCOMPOSITE SCAFFOLDS

Coupling 3D scaffold fabrication technique with the selection of materials are key requirements for achieving optimum scaffold properties in targeted hard or soft tissues. Conventional techniques and more recently, rapid prototyping (RP) [68] methods are considered for the fabrication of final scaffold constructs. The RP approaches will not be discussed. Depending on the applied method, different scaffolds with different features—various conformation of the polymer chain (if the scaffold material is considered a polymer) or distribution of nanosized inorganic materials—will be obtained [3]. However, conventional techniques for scaffold construction imply the development of porous polymer structures such as support for cell adhesion.

However, complex structures are very difficult to produce using conventional techniques. Generally, conventional techniques refer to:

1. solvent casting/particulate leaching,
2. electrospinning,
3. freeze-drying, and
4. phase-separation method.

However, solvent casting/particulate leaching and electrospinning methods will be further discussed.

12.3.1 SOLVENT CASTING/ PARTICULATE LEACHING

Solvent casting/particulate leaching is a facial method to produce materials with enhanced properties. These approaches involve casting a mixture of the polymer solution and porogen particles (which are insoluble in water or nontoxic solvents) and inorganic granules to construct 3D networks [69]. Porogen particles utilized are sugar, sodium chloride, and saccharide. The main advantage of the particulate leaching methods is that the size of the porogen particles and polymer/porogen ratio control the internal pore size and porosity of the final scaffold, respectively. After evaporation of the solvent, the dried scaffold material is fractionated in a proper solvent in order to remove the particulate. However, some limitations could be retained: (1) incomplete solvent removal by evaporation process; (2) presence of an open-pore structure and absence of interconnectivity. It is reported that scaffolds obtained through solvent casting/particulate leaching method have a pore size of 30–300 μm and porosity about 20%–50% [70].

Sola et al. [71] have developed innovative 3D porous scaffolds in order to mimic the bone marrow niche in vitro. Materials subjected to the research were two polymers with different backbone flexibilities: a rigid one—polymethyl methacrylate (PMMA) and polyurethane (PU)—a flexible polymer. Porogen used by researchers was NaCl, in the form of a common salt table. Polymer-to-salt-ratio (1:4 for both PMMA and PU) revealed a higher interconnected porosity (82.1 vol% and 91.3 vol%). Additionally, it was reported improved mechanical properties having elastic modulus between 29 and 1283 kPa.

Lin et al.'s study [72] reported the obtaining of polyvinyl alcohol active layers by the above-mentioned method. In this study, porogen was considered a salt solution.

Research provided by Mao et al. [73] focuses on the fabrication of porous PLA-based composite scaffold using a combination of solvent casting/particle leaching and sol–gel method. They report the fabrication of poly(lactic acid)-based porous materials characterized by high porosity with hyaluronic acid fillers. It observed an improvement in mechanical properties. Additionally,

the resultant scaffolds showed well-defined interconnected porous structure with a pore size of 200–400 μm and rough pore wall. According to literature data, a material having such a size pore is recommended as support for cell adhesion and ingrowth.

12.3.2 ELECTROSPINNING

The electrospinning method is utilized for producing nanofibrous scaffold materials for TE. The structures of the fibers are affected by factors such as:

1) properties of the solution— concentration, viscosity, conductivity, and surface tension.
2) process factors—applied potential, collection distance, and emitting electrode polarity.
3) environmental parameters—temperature and relative humidity [74].

Fibers produced by this method have high porosity. This technique was used to fabricate scaffolds for tissue regeneration applications.

The disadvantage of this technique is that it involves the use of organic solvents.

Shao et al. [75] constructed a porous scaffold for bone TE consisting of multilayer nanofiber fabrics by weaving nanofiber yarns of polylactic acid (PLA) and Tussah silk fibroin (TSF). Electrospinning was involved as a method for the fabrication of the final product.

Researchers found that the mechanical properties of the resulting scaffold were determined by the ratio between PLA and TSF (with Young's modulus at 417.65 MPa and tensile strength at 180.36 MPa).

Additionally, scaffolds also support the adhesion and proliferation of mesenchymal stem cells.

12.3.3   FREEZE-DRYING METHOD

The freeze-drying technique is utilized to remove residual solvent from material in order to produce a dry powder as to be loaded into cells. The material is dissolved in the solvent and then is subjected to the frozen process in a dry-ice bath. After the sublimation of the solvent, it can be removed by vacuum. A dry powder is obtained. During freeze-drying, the temperature is maintained sufficiently low that any remixing of the phase-separated polymer solution is prevented. Some studies reported scaffold fabrication for tissue regeneration.

Govindan et al. [76] reported the fabrication of 3D porous gelatin (G) and phosphate glass-reinforced gelatin composite scaffolds. The method chosen for material preparation was the free-drying technique for its simplicity and cost effective.

Researchers reported the size pores (between 100 and 500 μm), a high porosity (~73%) along with compression modulus of 4.89 MPa. According to biological tests, developed composite scaffolds by the researchers can mimic the ECM for cell adhesion and proliferation.

The main disadvantage of the freeze-drying technique is referring to the incapacity to maintain a uniform porosity. The main advantage of the freeze-drying technique refers to high porosity and proper interconnectivity of the material.

12.3.4   PHASE-SEPARATION TECHNIQUE

In this method, it is necessary to dissolve an organic solvent in a solution made by polymer with/or without ceramic particles. In order to induce phase-separation, it is necessary that the solution must be rapidly cooled in the first step, inducing the solidification of the solvent. After the evaporation of the solvent after the sublimation process, a porous scaffold could be obtained.

Advantages of the phase-separation technique mention a more controllable structure of a scaffold material along with controllable pore size. Qu [77] has reviewed the fabrication of different porous scaffolds for their use in bone TE. They were fabricated through the phase-separation technique.

12.4 CONCLUSION

Research in the field of TE nano-composite and their applications in TE have been focused mainly on the challenges of creating biopolymers-based composite materials containing either natural or synthetic polymers. Different materials for the scaffold construct offer enhanced mechanical properties. Integration of inorganic nanoparticles provides integration bioactivity. However, the discover of new products to be used as scaffolds in bone TE still represent a challenge.

Development of natural materials, like BC nanofibers, for TE, nevertheless, it seems likely that the high aspect ratio of inorganic nanoparticles will be applied to engineering structurally oriented tissues, such as skeletal muscle, tendons, ligaments, and nerves, in order to produce 3D scaffolds.

ACKNOWLEDGMENT

The financial support of European Social Fund for Regional Development, Competitiveness Operational Programme Axis 1—Project "Petru Poni Institute of Macromolecular Chemistry—Interdisciplinary Pol for Smart Specialization through Research and Innovation and Technology Transfer in Bio(nano)polymeric Materials and (Eco) Technology," InoMatPol (ID P_36_570, Contract 142/10.10.2016, cod MySMIS: 107464) is gratefully acknowledged.

KEYWORDS

- **nanocomposites**
- **biodegradable polymers**
- **tissue engineering**

REFERENCES

1. Eming, S.A.; Martin, P.; Tomic-Canic, M.; Wound repair and regeneration: mechanisms, signaling, and translation. *Sci. Transl. Med.* **2014**, 6, 265sr6.
2. Sumbria, D.; Berber, E.; Rouse, B.T.; Factors affecting the tissue damaging consequences of viral infections. *Front. Microb.* **2019**, 10, 2314–2327.
3. Eltom, A.; Zhong, G.; Muhammad, A.; Scaffold techniques and designs in tissue engineering functions and purposes: a review. *Adv. Mater. Sci. Eng.* **2019**, DOI: 10.1155/2019/3429527.
4. Caddeo, S.; Boffito, M.; Sartori, S.; Tissue engineering approaches in the design of healthy and pathological in vitro tissue models, *Front. Bioeng. Biotechnol.* **2017**, 5, 40–62.
5. Rață, D.M.; Cadinoiu, A.N.; Atanase, L.I.; Bacaita, S.E.; Mihalache, C.; Daraba, O.-M.; Gherghel, D.; Popa, M.; In vitro behaviour of aptamer-functionalized polymeric nanocapsules loaded with 5-fluorouracil for targeted therapy. *Mater. Sci. Eng. C.* **2019**, 103, 109828–109838.

6. Zakrzewski, W.; Dobrzyński, M.; Szymonowicz, M.; Rybak, Z.; Stem cells: past, present, and future. *Stem. Cell. Res. Ther.* **2019**, 10, 2–22.

7. Brown, P.T.; Handorf, A.M.; Jeon, W.B.; Li, W.-J.; Stem cell-based tissue engineering approaches for musculoskeletal regeneration. *Curr. Pharm. Des.* **2013**, 19, 3429–3445.

8. Barthes, J.; Özçelik, H.; Hindié, M.; Ndreu-Halili, A.; Hasan, A.; Vrana, N.E.; Cell microenvironment engineering and monitoring for tissue engineering and regenerative medicine: the recent advances. *Biomed. Res. Int.* **2014**, 2014, 921905–92124.

9. Kusindarta, D.L.; Wihadmadyatami, H.; The role of extracellular matrix in tissue regeneration. In: Tissue Regeneration. Kaoud, H.A.E., Ed.; IntechOpen: London, 2018, 65–73

10. Theocharis, A. D.; Skandalis, S. S.; Gialeli, C.; Karamanos, N.K.; Extracellular matrix structure. *Adv. Drug. Deliv. Rev.* **2016**, 1, 4–27.

11. De Witte, T.-M.; Fratila-Apachitei, L.E.; Zadpoor, A.A.; Peppas, N.A.; Bone tissue engineering via growth factor delivery: from scaffolds to complex matrices. *Regen. Biomat.* **2018**, 5, 1–15.

12. Ramanathan, G.; Singaravelu, S.; Muthukumar, T.; Thyagarajan, S.; Perumal, P.T.; Sivagnanam, U.T.; Design and characterization of 3D hybrid collagen matrixes as a dermal substitute in skin tissue engineering. *Mater. Sci. Eng. C.* **2017**, 72, 359–370.

13. Yue, B.; Biology of the extracellular matrix: an overview. *J. Glaucoma.* **2014**, 23, S20–S23.

14. Mäkitie, R.E.; Costantini, A.; Kämpe, A.; Alm, J.J.; Mäkitie, O.; New insights into monogenic causes of osteoporosis. *Front. Endocrinol. (Lausanne).* **2019**, 10, 70–127.

15. Franchi, M.; Trirè, A.; Quaranta, M.; Orsini, E.; Ottani, V.; Collagen structure of tendon relates to function. *Sci. World J.* **2007**, 7, 404–420.

16. Ferreira, A.M.; Gentile, P.; Chiono, V.; Ciardelli, G.; Collagen for bone tissue regeneration. *Acta Biomat.* **2012**, 8, 3191–3200.

17. Bagde, A.D.; Kuthe, A.M.; Quazi, S.; Gupta, V.; Jaiswal, S.; Jyothilal, S.; Lande, N.; Nagdeve, S.; State of the art technology for bone tissue engineering and drug delivery. *IRBM.* **2019**, 40, 133–144.

18. Zaman, M.; Hydrogels, their applications and polymers used for hydrogels: a review. *Int. J. Biol., Pharm., Allied Sci.* **2015**, 4, 6581–6603.

19. Armiento, A.R.; Stoddart, M.J.; Alini, M.; Eglin, D.; Biomaterials for articular cartilage tissue engineering: learning from biology. *Acta Biomater.* **2018**, 65, 1–20.

20. Chen, J.-C.; Li, L.-M.; Gao, J.-Q.; Biomaterials for local drug delivery in central nervous system. *Int. J. Pharm.* **2019**, 560, 92–100.

21. Yang, C.; Blum, N.T.; Lin, J.; Qu, J.; Huang, P.; Biomaterial scaffold-based local drug delivery systems for cancer immunotherapy. *Sci. Bull.* **2020**, doi. org/10.1016/j.scib.2020.04.012.

22. Raţă, D.M.; Cadinoiu, A.N.; Darabă, O.; Mihalache, C.; Mihalache, G.; Burlui, V.; Metronidazoleloaded chitosan/poly(maleic anhydride-alt-vinyl acetate) hydrogels for dental treatments. *Int. J. Med. Dent.* **2016**, 6, 92–97.

23. Couture, C.; Desjardins, P.; Zaniolo, K.; Germain, L.; L.Guérin, S.; Enhanced wound healing of tissue-engineered human corneas through altered phosphorylation of the CREB and AKT signal transduction pathways. *Acta Biomater.* **2018**, 73, 312–325.

24. Hickey, R.J.; Pelling, A.E.; Cellulose biomaterials for tissue engineering. *Front. Bioeng. Biotechnol.* **2019**, 7, 45–60.

25. Andorko, J.I., Jewell, C.M.; Designing biomaterials with immunomodulatory properties for tissue engineering and regenerative medicine. *Bioeng. Transl. Med.* **2017**, 2, 139–155.

26. Altomare, L.; Bonetti, L.; Campiglio, C.E.; De Nardo, L.; Draghi, L.; Tana, F.; Farè, S.; Biopolymer-based strategies in the design of smart medical devices and artificial organs. *Int. J. Artif. Organs.* **2018**, 41, 337–359.

27. Siracusa, V.; Microbial degradation of synthetic biopolymers waste. *Polymers (Basel).* **2019**, 11, 1066.

28. Asadi, N.; Rahmani, A.; Del Bakhshayesh, Davaran, S.; Akbarzadeh, A.; Common biocompatible polymeric materials for tissue engineering and regenerative medicine. *Mater. Chem. Phys.* **2019**, 7, 122528.

29. Williams, D.F.; Challenges with the development of biomaterials for sustainable tissue engineering. *Front. Bioeng. Biotechnol.* **2019**, 7, 127–136.

30. Andorko, J.I.; Jewell, C.M.; Designing biomaterials with immunomodulatory properties for tissue engineering and regenerative medicine. *Bioeng. Transl. Med.* **2017**, 2, 139–155.

31. Bramhill, J.; Ross, S.; Ross, G.; Bioactive nanocomposites for tissue repair and regeneration: a review. *Int. J. Environ. Res. Public. Health.* **2017**, 14, 66–87.

32. Hakkinen, K.M.; Harunaga, J.S.; Doyle, A.D.; Yamada, K.M.; Direct comparisons of the morphology, migration, cell adhesions, and actin cytoskeleton of fibroblasts in four different three-dimensional extracellular matrices. *Tissue. Eng. Part A.* **2011**, 17, 713–724.

33. Abbasi, N.; Hamlet, S.; Love, R.M.; Nguyen, N.-T.; Porous scaffolds for bone regeneration. *J. Sci.: Adv. Mater. Dev.* **2020**, 5, 1–9.

34. Maleki, H.; Shahbazi, M.-A.; Montes, S.; Hosseini, S.H.; Eskandari, M.R.; Zaunschirm, S.; Verwanger, T.; Mathur, S.; Milow, B.; Krammer, B.; Hüsing, N.; Mechanically strong silica-silk fibroin bioaerogel: a hybrid scaffold with ordered honeycomb micromorphology and multiscale porosity for bone regeneration. *ACS Appl. Mater. Interfaces.* **2019**, 11, 17256–17269.

35. Wu, C.; Ramaswamy, Y.; Zreiqat, H.; Porous diopside (CaMgSi2O6) scaffold: a promising bioactive material for bone tissue engineering. *Acta Biomater.* **2010**, 6, 2237–2245.

36. Murphy, C.M.; O'Brien, F.J.; Understanding the effect of mean pore size on cell activity in collagen-glycosaminoglycan scaffolds. *Cell. Adh. Migr.* **2010**, 4, 377–338.

37. Karageorgiou, V.; Kaplan, D.; Porosity of 3D biomaterial scaffolds and osteogenesis. *Biomaterials* **2005**, 26, 5474–5491.

38. Iviglia, G.; Kargozar, S.; Baino, F.; Biomaterials, current strategies, and novel nano-technological approaches for periodontal regeneration. *J. Funct. Biomater.* **2019**, 10, 1–36.

39. Chocholata, P.; Kulda, V.; Babuska, V.; Fabrication of scaffolds for bone-tissue regeneration. *Materials (Basel)* **2019**, 12, 568– 593.

40. Harjunpää, H.; Asens, M.L.; Guenther, C.; Fagerholm, S.C.; Cell adhesion molecules and their roles and regulation in the immune and tumor microenvironment. *Front. Immunol.* **2019**, 10, 1078–1102.

41. Hickey, R.J.; Pelling, A.E.; Cellulose biomaterials for tissue engineering.

Front. Bioeng. Biotechnol. **2019**, 7, 45–60.

42. Luo, H.; Cha, R.; Li, J.; Hao, W.; Zhang, Y.; Zhou, F.; Advances in tissue engineering of nanocellulose–based scaffolds: a review. *Carbohy. Polym.* **2019**, 224, 115144–115159.

43. Samir, A.; Alloin, F.; Dufresne, A.; Review of recent research into cellulosic whiskers, their properties and their application in nanocomposite field. *Biomacromolecules* **2005**, 6, 612–626.

44. Tayeb, A.H.; Amini, E.; Ghasemi, S.; Tajvidi, M.; Cellulose nanomaterials—binding properties and applications: a review. *Molecules* **2018**, 23, 2684–2708.

45. Klemm, D.; Kramer, F.; Moritz, S.; Lindström, T.; Ankerfors, M.; Gray, D.; Dorris, A.; Nanocelluloses: a new family of nature-based materials. *Angew. Chem. Int. Ed. Engl.* **2011**, 50, 5438–5466.

46. Phanthong, P.; Reubroycharoen, P.; Hao, X.; Xu, G.; Abudula, A.; Guan, G.; Nanocellulose: extraction and application. *Carbon Res. Convers.* **2018**, 1, 32–43.

47. Yang, X.; Han, F.; Xu, C.; Jiang, S.; Huang, L.; Liu, L.; Xia, Z.; Effects of preparation methods on the morphology and properties of nanocellulose (NC) extracted from corn husk. *Ind. Crops Prod.* **2017**, 109, 241–247.

48. Maiti, S.; Jayaramudu, J.; Das, K.; Reddy, S.M.; Sadiku, R.; Ray, S.S.; Liu, D.; Preparation and characterization of nano-cellulose with new shape from different precursor. *Carbohydr. Polym.* **2013**, 98, 562–567.

49. Daicho, K.; Saito, T.; Fujisawa, S.; Isogai, A.; The crystallinity of nanocellulose: dispersion-induced disordering of the grain boundary in biologically structured cellulose. *ACS Appl. Nano Mater.* **2018**, 1, 5774–5785.

50. Park, S; Baker, J.O.; Himmel, M.E.; Parilla, P.A.; Johnson, D.K.; Cellulose crystallinity index: measurement techniques and their impact on interpreting cellulase performance. *Biotechnol. Biofuels.* **2010**, 3, 1–10.

51. Feng, X.; Ullah, N.; Wang, X.; Sun, X.; Li, C.; Bai, Y.; Chen, L.; Li, Z.; Characterization of bacterial cellulose by gluconacetobacter hansenii CGMCC 3917. *J. Food Sci.* **2015**, 80, E2217.

52. Chen, Y.M.; Xi, T.; Zheng, Y.; Guo, T.; Hou, J.; Wan, Y.; Gao, C.; In vitro cytotoxicity of bacterial cellulose scaffolds used for tissue-engineered Bone. *J. Bioact. Compat. Polym.* **2009**, 24, 137–145.

53. Tang, W.; Jia, S.; Jia, Y.; Yang, H.; The influence of fermentation conditions and post-treatment methods on porosity of bacterial cellulose membrane. *World J. Microbiol. Biotech.* **2010**, 26, 125–131.

54. Brown, A.-J; On an acetic ferment which forms cellulose. *J. Chem. Soc., Trans.* **1886**, 49, 432–439.

55. Klemm, D.; Kramer, F.; Moritz, S.; Lindström, T.; Ankerfors, M.; Gray, D.; Dorris, A.; Nanocelluloses: a new family of nature-based materials. *Angewandte Chemie.* **2011**, 50, 5438–5466.

56. Noh, Y.K.; Dos Santos Da Costa, A.; Park, Y.S.; Du, P.; Kim, I.-H.; Park, K.; Fabrication of bacterial cellulose-collagen composite scaffolds and their osteogenic effect on human mesenchymal stem cells. *Carbohydr. Polym.* **2019**, 219, 210–218.

57. He, X.; Tang, K.; Li, X.; Wang, F.; Liu, J.; Zou, F.; Yang, M.; Li, M.; A porous collagen-carboxymethyl cellulose/ hydroxyapatite composite for bone tissue engineering by bi-molecular template method. *Int. J. Biol. Macromol.* **2019**, 137, 45–53.

58. Ito, A.; Kamihira, M.; Tissue engineering using magnetite nanoparticles.

Prog. Molec. Biol. Transl. Sci. **2011**, 104, 355–395.

59. Majewski, P.; Thierry, B.; Functional-ized magnetite-nanoprticles synthesis, properties, and bio-application, *Crit. Rev. Sold State Mater. Sci.* **2007**, 32, 203–215.

60. Torgbo, S.; Sukyai, P.; Fabrication of microporous bacterial cellulose embedded with magnetite and hydroxy-apatite nanocomposite scaffold for bone tissue engineering. *Mater. Chem. Phys.* **2019**, 237, 121868–121878.

61. Bighama, A.; Aghajanian, A.H.; Behzadzadeh, S.; Sokhani, Z.; Shojaei, S.; Kaviania, Y.; Hassanzadeh-Tabrizi, S.A.; Nanostructured magnetic Mg$_2$SiO$_4$-CoFe$_2$O$_4$ composite scaffold with multiple capabilities for bone tissue regeneration. *Mater. Sci. Eng. C.* **2019**, 99, 83–95.

62. Wheeler, D.L.; Stokes, K.E.; Park, H.M.; Hollinger, J.O.; Evaluation of particulate bioglass in a rabbit radius ostectomy model. *J. Biomed. Mater. Res.* **1997**, 35, 249–254.

63. Hench, L.L.; Jones, J.R.; Bioactive glasses: frontiers and challenges. *Front. Bioeng. Biotechnol.* **2015**, 3, 194–206.

64. Gerhardt, L.-C.; Boccaccini, A.R.; Bioactive glass and glass-ceramic scaf-folds for bone tissue engineering. *Materials (Basel)* **2010**, 3, 3867–3910.

65. Dharmalingam, K.; Padmavathi, G.; Kunnumakkarab, A.B.; Anandalak-shmia, R., Microwave-assisted synthesis of cellulose/zinc-sulfate-calcium-phos-phate (ZSCAP) nanocomposites for biomedical applications. *Mater. Sci. Eng. C.* **2019**, 100, 535–543.

66. Espostia, D.; Chiellinic, M.; Bondiolid, F.; Morsellia, F.; Fabbria, D.; Highly porous PHB-based bioactive scaffolds for bone tissue engineering by in situ synthesis of hydroxyapatite. *Mater. Sci. Eng. C.* **2019**, 100, 286–296.

67. Draghici, A.-D.; Busuioc, C.; Mocanu; A. Nicoara, A.-I.; Iordache, F.; Jinga; S.-I.; Composite scaffolds based on calcium phosphates and barium titanate obtained through bacterial cellulose templated synthesis. *Mater. Sci. Eng. C.* **2020**, 110, 110704–110733.

68. Melchels, F.P.W.; Domingos, M.A.N.; Klein, T.J.; Malda, J.; Bartolo, P.J.; Hutmacher, D.W.; Additive manu-facturing of tissues and organs. *Prog. Polym. Sci.* **2012**, 37, 1079–1104.

69. Hayati, A.N.; Rezaie, H.R.; Hosse-inalipour, S.M.; Preparation of poly (3-hydroxybutyrate)/nano-hydroxy-apatite composite scaffolds for bone tissue engineering. *Mater. Lett.* **2011**, 65, 736–739.

70. Hutmacher, D.W.; Scaffolds in tissue engineering bone and cartilage. *Bioma-terials* **2000**, 21, 2529–2543.

71. Sola, A.; Bertacchini, J.; D'Avella, D.; Anselmi, L.; Maraldi, T.; Marmi-roli, S.; Messori, M.; Development of solvent-casting particulate leaching (SCPL) polymer scaffolds as improved three-dimensional supports to mimic the bone marrow niche. *Mater. Sci. Eng. C.* **2019**, 96, 153–165.

72. Lin, W.; Li, Q.; Zhu, T.; Study of solvent casting/particulate leaching technique membranes in pervaporation for dehy-dration of caprolactam. *J. Ind. Eng. Chem.* **2012**, 18, 941–947.

73. Mao, D.; Li, Q.; Li, D.; Chen, Y.; Chen, X.; Xu, X.; Fabrication of 3D porous poly(lactic acid)-based composite scaf-folds with tunable biodegradation for bone tissue engineering. *Mater. Des.* **2018**, 142, 1–10.

74. Shirobokov, K.P.; Sentyakov, B.A.; Svyatskii, V.M.; Factors affecting fiber formation. *Russ. Eng. Res.* **2010**, 30, 1210–1212.

75. Shao, W.; He, J.; Han, Q.; Sang, F.; Wang, Q.; Chen, L.; Ding, B.; A biomi-metic multilayer nanofiber fabric

fabricated by electrospinning and textile technology from polylactic acid and Tussah silk fibroin as a scaffold for bone tissue engineering. *Mater. Sci. Eng. C.* **2016,** 67, 599–610.

76. Govindan, R.; Gu, F.L.; Karthi, S.; Girija, E.K.; Effect of phosphate glass reinforcement on the mechanical

and biological properties of freeze-dried gelatin composite scaffolds for bone tissue engineering applications. *Mater. Today Commun.* **2019,** 22, 100765.

77. Qu, H.; Additive manufacturing for bone tissue engineering scaffolds. *Mater. Today Commun.* **2020**, 24, 101024.

CHAPTER 13

BIONANOCOMPOSITES FOR BIOMEDICAL APPLICATIONS

SHRIKAANT KULKARNI

*Adjunct Professor, Vishwakarma University, Pune, India;
E-mail: shrikaant.kulkarni@vit.edu.*

ABSTRACT

Bionanocomposites are biphasic materials and, therefore, combine the properties of both the matrix and the filler which together constitute it. In fact, in isolation, the matrix (base material) and filler (reinforcing material) have their own but limited properties. However, to broaden the range of properties in them they are coupled in the requisite proportions. Bionanocomposites are comprised of biomaterials necessarily and, therefore, biodegradable once their shelf life gets over as well as biocompatible apart from a host of other meaningful properties by virtue of which they find widespread applications in different walks of life. This chapter takes a review of the applicability of bionanocomposites in the frontier area of biomedicine and discusses the desired property profile of the biomaterials used in the making of such bionanocomposites for biomedical applications in particular.

13.1 INTRODUCTION

Composite materials have been integrated into our lives. Composites have numerous applications, right from sports equipment to transportation systems. Composites by virtue of their unique properties to use them in making tennis rackets, kayaks, boats, and a host of other recreational sports equipment because they are lightweight and strong. They find use in aircraft parts, such as the seat frames to wings [1, 2]. Composites are used in producing the rotor blades of helicopters due to their far better resistance to fatigue stresses as compared to metals. Automobile tires consist of rubber embedded with threads of steel to extend durability and to enhance performance efficiency at higher speeds. Historic and more contemporary examples of composite materials because of their characteristic mechanical properties find many applications since historic times and contemporarily too. However, nonmechanical composites have also come up and have been gaining momentum over time given the demand for electrical, thermal, magnetic, and optical properties in our lives [3]. Such materials may not be visible normally as they are encased by protective barriers, but still, they exhibit key functions in various systems, like personal computers.

There has been scaling down of electronic devices to the nanoregime. The electrical conductivity possessed by carbon nanotubes (CNTs) and their use in making insulators is also the area of significant interest. The use of polymer-enabled nanocomposites in such application areas has diversified holding a lot much of potential in the applications such as photovoltaic cells, supercapacitors, sensors, printable conductors, light-emitting diodes, field-effect transistors electromagnetic interference shielding, transparent conductive coatings, electrostatic dissipation, electromechanical actuators, and numerous electrode applications [4–6].

13.2 APPLICATIONS IN DENTAL TREATMENT

According to the National Institutes of Health published report on the dental materials more than 10 crore people do not have teeth and there is a great demand for implants in dentistry. Over the last 5 decades, researchers' focus was on the synthesis of biomaterials that can interface with biological systems to assess, treat, enable, or even substitute any tissue, organ, or function of the body-related dental implants in order to treat crores of patients over the years. Teeth are primarily the hardest materials present in our body consisting of hydroxyapatite $(Ca_5(PO_4)_3(OH))$. National Bureau of Standard (US) has laid down standards for appropriately selecting and grading implants of dental amalgam. From the host of materials, polymer-based materials have dragged a lot much of attention because of their versatility in property profile and processing, keeping in view the conditions prevailing in the body, susceptibility of polymers under the given conditions, such as elevated temperature, electromagnetic radiation, and atmospheric oxygen, and therefore used on a wider scale in dentistry in the form of composites (resin–ceramic) as materials for the restoration of body-based dental implants, implants, dental cements, as well as dental bases and teeth [7, 8]. Such resin-based composites have an average particle size far >1 μm, and microfillers are used for a dental restoration that is adhesive so as to prepare the minimum cavity and esthetically sound, materials for restoration, cavity liners, sealants for pit and fissure plugging, cores and buildups, crowns, temporary restorations, cementing materials for single or multiteeth prostheses and orthodontic devices, endodontic sealing agents, post root canal, etc. [9], due to their easy fabricability and tenability in congruence with specific requirements as restorative, cementing, provisional materials, sealers, etc. The monomer base material used on priority in making dental composites for commercial purposes has been bis-GMA, because of its high viscosity and is blended with other dimethacrylates namely triethylene glycol dimethacrylate, diurethane dimethacrylate (UDMA), etc. [10]. Macrofill materials, however, have limited applications because of difficulty in polishing them and the presence of the rough surface. For obtaining durable esthetics, nanofills are preferred as they are normally weak because of their comparatively lower filler content, and therefore a balance has to be struck to derive enough mechanical strength by enhancing the filler contents by embedding more densely the prepolymerized resin fillers within the body of matrix which is added upon by some microfill particles [11].

These days, organically modified ceramics composites that can flow contain adhesive monomers, namely Vertise Flow (Kerr), Fusio Liquid Dentin (Pentron Clinical); and epoxy-based silicone fluids have been synthesized for getting more sound dental implants [12]. Such materials are characterized by improved mechanical properties and offer resistance to shrinkage as against dimethacrylate-based resins used typically, attributed to the curing reaction of epoxide which brings about the opening of oxirane ring and is presently used in preference as liners and smaller restorations, as well as the precursors for obtaining universal self-adhesive composites [13, 14].

13.3 APPLICATIONS IN ORTHOPEDICS

Bone is a naturally occurring extracellular matrix (ECM) living tissue consisting of organic–inorganic complex with composition as 60% mineral (nanohydroxyapatite (nHA), with formula $Ca_{10}(PO_4)_6(OH_2)$, and 30% matrix in the form of collagen, which is a primary structural protein of connective tissue), and remaining 10% water by weight. It is characterized by excellent strength in addition to special elastic properties thereby making it a right candidate for developing structural frameworks and is providing the protective barrier to internal organs and bone marrow, regulating blood pH and maintaining the calcium and phosphate levels for smooth metabolic processes to take place, as well as it supplies reservoirs of ions essential for bringing about normal bodily activities and is a part and parcel of endoskeleton in vertebrates [15–18]. Injuries to bones cause hospitalization at times because of fractures and grafting exercises have been considered as a global phenomenon.

To address such issues intervention at right time is a must, various materials have been in use to either repair or replace bone and should have requisite mechanical properties apart from regulated degradation rates and biocompatibility while interacting with the living tissues [19]. Hydroxyapatite-based $(Ca_{10}(PO_4)_6(OH)_2)$ polymeric nanocomposites have been synthesized for the purpose of bone repair and developing implants. Hydroxyapatite, a main ingredient of hard tissue, shows certain unwanted mechanical properties on employing directly, and hence, polymer-based composites are in demand. Biodegradation of the matrix is a must to provide for infiltration of fresh bone growth at the site of the repair. Naturally occurring polymers such as polysaccharides, polypeptides, collagen, chitosan, etc., or synthetically derived biodegradable polymers are

frequently used as the matrix in such investigations. Collagen-derived gelatin and poly-2-hydroxyethyl-methacrylate/poly(3-caprolactone) nanocomposites depending upon the use of hydroxyapatite are the right candidates for bone repair purpose.

Many synthetic materials, like polymers, and composites have been used due to their biofunctionality and biocompatibility in tandem with the requirements and the kind of function that is intended to be replaced. Most often hydroxyapatite-based bioabsorbable nonallogenic bone substitutes (ceramics) are available in the market for clinical purposes due to their biocompatibility and further, their strength can be exacerbated by employing natural and synthetic polymers using coprecipitation technique [20–23]. Polymer composites containing plaster of Paris (PP) $(CaSO_4 \cdot 1/2H_2O)$ possess a porous structure that is appropriate for osteoblast proliferation and as a bone substitute material for clinical use. When gypsum $(CaSO_4 \cdot 2H_2O)$ is heated to 150 °C, loss of its water of hydration takes place to form calcium sulfate hemihydrate $(CaSO4 \cdot \frac{1}{2}H_2O)$ as per the following equation, and on grinding it into powder form, it is called PP [24].

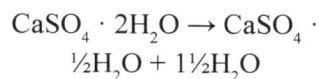

$$CaSO_4 \cdot 2H_2O \rightarrow CaSO_4 \cdot \frac{1}{2}H_2O + 1\frac{1}{2}H_2O$$

Various experimental works showed that the potential of the biomaterial says dentin, which is predominantly derived from hydroxyapatite and PP, enhances the stability of material used in grafting and promoting bone healing, preserving soft tissue, and membrane function [25].

The nanoscale modeling of hybrids produced by synthetic methods and use of composites is still in the infancy stage, mimicking naturally occurring microstructures although synthetic molecules may come up as new generation materials that are compatible in toughness characteristics will with their natural counterparts. The optimization of their morphology and bioactivity in such innovative hybrid composites is a major challenge [26]. Incumbent orthopedic implants typically are specifically comprised of bioinert materials like metals, ceramics, or polymers and offer coupling of two or more constituents [27]. The design of bonelike composites with marked mechanical properties and exacerbated biocompatibility demands a biomimetic approach with natural bone as an ideal material [28].

13.4 APPLICATIONS IN TISSUE ENGINEERING

The engineered bionanocomposites have a semblance with the engineered nanocomposites as they also are amalgamations of hard and soft components similar to the natural tissues [29–33]. The materials so developed possess outstanding properties and exceptionally high mechanical strength leading bionanocomposites to acquire a pivotal importance in tissue regeneration [34–36]. The bionanocomposites are developed in order to meet the specific needs of the frontier technologies to form matrices, tissue regeneration, and drug delivery systems and will play a key role in furthering the development of innovative therapeutics. The huge challenges posed in the development of bionanocomposites in particular for tissue regeneration are acquiring sound mechanical strength in addition to high biocompatibility and biological activity [37, 38].

For such applications, numerous biopolymers are coupled with nanofillers in a host of materials as matrices such as silicate-based ones like clays and SiO_2 nanoparticles, ceramics-based ones like nHA, calcium triphosphate $(Ca_3(PO_4)_2)$ nanoparticles, synthetically derived layered double hydroxides (LDHs), graphene and CNTs such as carbon-based nanomaterials, [39], and metals like Au, Ag, etc., and metal oxides like iron oxide (Fe_2O_3) [40]. Moreover, to enhance the mechanical strength of the biopolymers, all of these nanofillers also provide added functionality to the ultimate composite derived. For example, CNTs impart electrical conductivity, inorganic fillers, or silicate-based fillers are primarily minerals, which are responsible for maintaining various bodily functions [41], for example, Ca is instrumental in bone formation, silicon in ceramics for skeletal development. Silicon makes human stem cells to facilitate the synthesis of collagen type I [42]. Therefore, embedding such kinds of functional nanofillers within the body of biopolymers impart them bioactivity essential for both tissue regeneration and their maintenance.

The outstanding mechanical property profile of bionanocomposites makes it possible in mimicking the bone structure as they provide integration of materials at various levels, structural, and biological properties possessed by the coupling of polymer and nanofiller. Indeed, numerous bionanocomposites acquire a structure resembling bones by the coupling of natural bone mineral/ceramic (nHA) and collagen. A few other natural biopolymers are also amalgamated with HAs, such as silk fibroin, bovine serum albumin, alginate, and chitosan.

The synthetic polymers consist of both categories such as degradable and nondegradable polymers, namely, polylactic glycolic acid (PLGA), polylactic acid (PLA), poly-l-lactic acid (PLLA), poly-caprolactone (PCL), polyurethane, polyethylene, poly(etherketone), and polyvinyl alcohol (PVA). Incorporating nanofillers like HAs enhance surface topography and

alters porosity, close to natural bone, and promotes the spread of osteoblast and regrow bone. Other inorganic ceramic-based nanofillers that resemble in structure with the natural bone are tricalcium phosphate, bioactive glass, glass–ceramic (apatite–wollastonite) may exhibit potentially high mechanical strength and mineralization enabling scaffold formation for bone tissue engineering purpose. Moreover, by the amalgamation of natural and synthetic polymers with nHA, mimicking mechanical property profile and hierarchical structure resembling natural bone is possible to obtain. For example, an n-HA/collagen/ PLA composite has a compressive modulus, 1 MPa and an elastic modulus, 47.3 MPa by varying the PLA ratio in the composite, which lies within the lower compression range of cancellous bone and within the elastic range of trabecular bone, respectively. It was found that on implanting it in a rabbit with segmental defect, it could integrate well with the defect and to a certain extent replaced with the new bone after about 12 weeks [43, 44].

Future research initiatives should be aimed at the substitution of HA in biopolymer-based implants by inorganic, or organic–inorganic, substrates as alternatives. Apart from nHA or nHA-based ceramic nanofillers, SiO_2-based ones [45] such as $Mg_2 SiO_4$ (forsterite) are also used for the preparation of bionanocomposites to be employed in bone tissue engineering. The silanol groups from silicon-based ceramics react with Ca^{+2} and PO_4^{-3} ions depositing an amorphous calcium phosphate $(Ca_3(PO_4)_2)$ akin to the natural bone. The bionanocomposites made out of silica (or bioglass)-based polymers exhibit excellent bioactivity; however, where do they stand against nHA-based bionanocomposites are yet to be unraveled. The bionanocomposites of forsterite–polymer exhibit excellent mechanical properties over either nHA–polymer or bioglass–polymer bionanocomposites [46].

Layered or fibrous clay-based silicate nanoparticles are coupled with both natural or synthetic polymers are used to produce bionanocomposites used in the field of tissue engineering. Clays intercalate within natural polymers, like proteins namely, gelatin, collagen, or zein (corn protein), and polysaccharides, namely, chitosan, cellulose, starch, or alginate, better the mechanical strength and lowers the rate of degradation as well. Typically, bionanocomposites of chitosan–clays have been developed in various forms to exacerbate their properties so as to make them amenable to use in the field of tissue engineering. The clay part many times has been constituted by montmorillonite (MMT) in the nanocomposites.

The natural polymers are preferred in use because they constitute the naturally occurring ECM ingredients like collagen or resemble structurally with natural ECM ingredients like gelatin, which is nothing but denatured collagen, or chitosan, which resembles the proteoglycans typically using it as a bone and tissue matrix. Indeed, a matrix obtained from merely the natural polymers is normally embedded as a weak inclusion of a nanofiller that enhances their mechanical stability. Further, such biopolymers possess polar functional groups, which interact through hydrogen bond or van der Waals forces with silanol groups present in clay or other nanofillers [47, 48].

Moreover, little amounts of surface-modified clay nanoparticles (natural MMT layered silicate) if added to brittle polymers such as PLGA can enhance its properties like toughness and elongation when tested for tensile tests ranging from 7% for the pure polymer to 210% for the polymer bionanocomposite [49]. The rise in properties such as reinforcement and toughness has been ascribed to the cross-linking network formed between polymer chain molecules and MMT silicate nanoparticles by means of repeated crazing and shear yielding. For example, more increase in properties when MMT is combined with PLLA [50, 51] and PLA [52].

However, in some cases, it has been observed that if the clay is added it decreases the crystallinity of polymers and exacerbates the degradation rate of the process. Finally, the coupling of fibrous clays like sepiolite (magnesium silicate), with biopolymers like collagen produce hybrid materials that possess a great degree of orderliness [53, 54]. The better chemical affinity between collagen and sepiolite fibers results in a higher degree of alignment with the collagen chain molecules. Further, in the cross-linked polymers, the degradation rate is low but they improve mechanical properties resulting in a better service line on implanting them in the body.

Apart from strengthening the properties of materials in the bulk, hydrogel properties can also be enhanced by using the nanofillers. The addition of optimal amounts of nanofillers to hydrogels leads to broadening the horizons of their biomedical applications. Bionanocomposites can be derived by mixing with physical and mechanical means the hydrogel precursors and the already synthesized nanoparticles and bringing about gelation thereafter. Nanoparticles can act as cross-linkers in some situations and thereby on blending with the hydrogel precursors can bring about gelation. In both cases, the nanoparticles are encapsulated by the hydrogel precursor components. However, the nanofillers here demand a change in their surface chemistry by functionalization in the form of attachment of the polymeric chain molecules to better the interaction between the hydrophilic end groups in the network and the nanoparticles. The hydrogel bionanocomposites present a unique tenability in the hydrogel properties with the help of the appropriate nanoparticles used in the bionanocomposites. For example, metal colloids of gold, silver, or platinum when used are instrumental in changing the optical behavior of hydrogels.

If iron oxide particles are incorporated, they change magnetic behavior while CNTs alter conductive behavior. By possessing such specific properties make them quite meaningful in the application areas, such as biosensing, nerve tissue conduits, and stimuli sensitive system., for example, coated CNTs with methacrylated gelatin (GelMA) and introduced into GelMA having a porous structure. In total, 0.5% CNT addition to the GelMA leads to an increase in the tensile strength and enhanced three times sudden rise in beating frequency of the seeded cardiomyocytes. Moreover, the cells get largely aligned and can be excited at the expense of lower electrical potential, which is due to the introduction of CNTs which are highly conductive. Further, an engineered cardiac tissue has shown contraction cyclically on exposing to the fluid environment and releasing from the substrate. The surface chemistry CNTs' are modified with amyloid fibers to better the dispersion as well as interaction with hydrogels that possess fibrous structure. Using such approaches, hydrogels with tailored properties can be designed to meet a host of biomedical applications. Still, they need to be explored for their cytocompatibility-based applications.

Carbon-based nanofillers namely graphene have been used in the tissue engineering application. Graphene sheets conjugated with networks of hydrogel like polypeptides or other synthetic counterparts of hydrogels have been used to develop nanocomposites that respond to external stimuli as well as highly resilient. Such nanocomposites possessing higher tensile strength can be used to prepare scaffolds of elastomers in tissues subjected to mechanical stress and electrical impulses like a heart.

Silicate-based nanoclays have been in use to develop bionanocomposite hydrogels possessing properties such as high mechanical strength and elasticity. Nanoclays add ultra-elongation and super-stiffness to the hydrogels because of better interaction with biopolymer ascribed to their anisotropic, plate-like morphological features along with higher aspect ratio. Both natural (MMT,

and halloysite) or synthetic polymers (laponite) are used frequently. A synthetic polymer such as laponite is better placed than natural MMT clay and is attributed to their monolayer nanoparticles dispersions, ultra-purity, and gelation tendencies. MMT has a plate-like layered structure, while halloysite has a hollow nanotube, and given their structure that they are useful to the advantage for developing controlled drug delivery systems. Synthetically derived silicate nanoparticles or platelets can be used to design robust and tissue adhesive systems on blending with hydrogels [55–57]. Such systems are also shear thinning and can be employed for invasive therapies.

Designing magnetic scaffolds is gaining a lot of interest in tissue engineering. Recently, hydrogel–gold–iron oxide scaffolds were found to have high biocompatibility and provide an opportunity to study the effect of magnetic fields on cells. Moreover, cells decorated on the collagen-iron oxide scaffold were found to acquire magnetism making them vulnerable to magnetic cell levitation and examination of cells in 3D natural environments [58].

A key factor other than the composition of the bionanocomposite is its architecture at large. A major factor involved in preparing a tissue engineering scaffold is to create a porous bed having an array of pores highly interconnected. Thus, methods employed in making traditional tissue engineering scaffolds also have been made use of to produce bionanocomposites, such methods include fiber bonding, phase separation, solvent casting (SC), particle leaching (PL), gas foaming (GF), and emulsification/freeze-drying and have been employed to prepare bionanocomposites that are akin to foams having a requisite porosity and interconnected network of pores. The foam-like bionanocomposites is prepared to regulate the pore size and orientation. Pore sizes vary from nano to micron regime are essential for smooth transport processes to take place and to attach cells better. In contrast, unidirectional pores give layered/hierarchical structure resembling natural bionanocomposites like nacre. One of the approaches adopted to get unidirectional pores is to allow the composite-generating mixture to freeze in one direction, which hastens the process of aligned pores formation.

The right choice of method to design the bionanocomposite framework affects pore structure, porosity as well as the bioactivity of the bionanocomposite at large. For example, bionanocomposites synthesized by coupling of SC/PL possess the potential risk of being laced with the residual organic solvent, which can influence cell activity. On distinguishing the properties of bionanocomposites, designed by either SC/PL or GF/PL presented scaffold made with the latter to be mechanically and biologically stronger. The compressive and tensile strength of the GF/PL scaffolds showed an increase of 99% and 1331%, respectively, as against the SC/PL scaffolds [59]. In some other experiments, fibers have been formed by using fiber-spinning to develop bionanocomposites.

Fiber-based bionanocomposites are found to possess better structural stability and are amenable to develop bone and cartilage tissue ligaments. This is by virtue of the porosity and high mechanical properties possessed by the fiber scaffolds which facilitate cellular infiltration. Some other parameters governing the designing and which further influence the performance efficiency of bionanocomposite in the form of scaffolds or tissue rejuvenation abilities are as follows:

1. Nanocomposites-to-polymer ratio,
2. aspect ratio of nanofillers, and
3. structure and organizational order in the inorganic nanofiller material, for example, both size and shape of the nHA have a major impact upon mechanical strength and cell response in particular osteoblast.

It has been reported through earlier studies that nanocomposite coating of nHA and PCL have grown over calcium phosphate scaffold (basic) was largely instrumental in integrating and enhancing mechanical strength and cell response in osteoblast when nanosized needle-shaped HA particles were preferably used as against micron-sized rod or spherical shaped particles [60]. Finally, highly performing tissue rejuvenating bionanocomposites that are multifunctional can be developed by the addition and release of drugs simultaneously, growth parameters, and cytokines.

Such a kind of strategy is very handy wherein the same material offers both the support matrix and push for differentiation as well as proliferation. These strategies are followed in nHA–polymer composites for the regeneration of bone tissue wherein the bone morphogenic protein on immobilization in nHA helped in regulating the osteogenic process [61, 62]. Adsorption of iron on the clay as another example was brought about to develop a magnetic bionanocomposite, so as to exacerbate the osteoblast proliferation and aligning them with the electrospun fibers [63]. Further examples involve a morphogenetic protein to facilitate tissue regeneration in an HA–alginate–collagen system added upon by a vitamin in a Ca-deficient HA–chitosan nanocomposite. Such customized composites having multifunctionality are very vital

tools for developing the systems for tendon or bone repair or tissue engineering at large.

13.5 APPLICATIONS IN DRUG DELIVERY AND RELEASE

Drug delivery systems technology development is one of the frontier areas of science, which involves drawing knowledge from different disciplines, to further the cause of human health care. The treatment of acute or chronic ailments has been made possible by delivering the drugs at the proper site to the patients over the years. The conventional techniques and pathways of drug administration led to a drug uptake efficiency that is governed by the properties of the drugs such as solubility, charge, molecular size, etc., as well as the properties of the administration site like pH, surface area, enzymes present, transport mechanisms, etc. The introduction of polymer nanotechnology and nanocomposites to biomedical/biotechnological application areas is emerging fast. The properties like biocompatibility and size in the nanoscale regime of the particles are used to the advantage possessed by bionanocomposites that can be used for drug delivery applications. The major advantage of polymer-based bionanocomposites in drug delivery applications is it follows a cumbersome diffusion path for the encapsulated drug to be taken to the site against a barrier but delivers it in a sustained manner with abysmally low or no release by bursting [64]. Polymer-based nanocomposites have been preferably recommended for the said purpose. The nanoparticles added to allow the drug to be released slowly in a better-controlled manner, with decreased swelling and enhanced mechanical properties of hydrogel-based nanocomposites.

Numerous examples of drug delivery systems developed with the help of bionanocomposites are in existence. Organic/inorganic nanofillers like clays and ceramic-based nanofillers are commonly used. Moreover, silicate [65] as well as carbon-based nanofillers such as graphene and CNTs are also have been good candidates in the controlled drug delivery systems development. CNTs are gaining momentum because of their capabilities to diffuse through the barrier such as blood–brain barrier; comparatively more surface area, the hollowness of the core, and can be readily functionalized as well as can encapsulate multifunctional moieties and thereby become versatile and tools which hold lot much of promise and potential in drug delivery applications. Normally, the nanofiller plays an important role in acting as a drug carrier. Both the loading and release of drug by bionanocomposite relies upon the chemistry of

nanofillers such as ionic bonding, hydrogen bonding, intercalation, concentration of nanofiller, extent of dispersion, and aspect ratio of the particles [66].

Clays have been in use for since long as drug excipients, which facilitate the formation of stable suspension, emulsion, and keeping the stability intact for a long. Based on the kind of clay/nanofiller and type of drug, the clay may adsorb the drug at its surface with the help of mechanisms such as ion-exchange or intercalation within the lamellae. The drug loading on the clay surface can be improved and regulated by modifying the surface chemistry of the clays using various biopolymers. Biopolymer nanoclay composites present a host of advantages like the ability to form film, bioadhesion, and cell uptake efficiency. Fibrous clays, namely, sepiolite, palygor-skite, and layered clays, namely, MMT and laponite have been in use as reinforcing materials or fillers, which prohibit water from entering (thereby control drug release) into biopolymers or drug containers made up of bionanocomposites. Clay bionanocomposites formed with chitosan, alginate, and carboxy-methyl cellulose (CMC) in the form of microspheres as containers have been in use for the delivery of beta-blockers and numerous nonsteroidal antiinflammatory drugs in a controlled manner, for example, chitosan–palygorskite, or alginate–chitosan–clay, microspheres were also used as carriers for the diclofenac delivery. The particles showed resistance to degradation in the presence of gastric fluids although underwent gradual degradation followed by controlled release of diclofenac at the proper site, that is, the intestine [67].

Silica-based bionanocomposites obtained using methods like spray drying form nanospheres also have been used as drug delivery carriers. The silica–alginate hybrid type of nanoparticles shows biocompatibility and endocytosed by the fibroblast can be used as targeted drug release carriers for the future. Similarly, carrageenan–silica aerogels were prepared by drying using supercritical CO_2 as a solvent. Indeed, targeted drug delivery can also be done by use of the magnetic nanoparticles in the bionanocomposites possessing multifunctionality and biodegradability. The drug then gets accumulated and delivered at the targeted site precisely by taking advantage of their paramagnetic properties and capacity to release the carried drug [68].

Injectable drug delivery can also be achieved for smart polymers that respond to the external stimuli with nanoparticle composites, for example, Pluronic, a polymer sensitive to temperature, is not a preferred candidate used for long-term drug delivery because of its rapid dissolution rate.

However, coupling Pluronic with laponite (synthetically derived layered clay) nanoparticles brought about a major shift in the Pluronic phase transition temperature and thereby the hydrogel show a marked increase in resistance to dissolution, which significantly delayed the release of drug-loaded [69]. Moreover, the bionanocomposites also find a use for gene and vaccine delivery. Choy and his teammates have shown that DNA-loaded LDH can be used efficiently as a nonviral vector in gene therapy. They later observed that DNA can be intercalated within the interlayer space provided by a magnesium–aluminum LDH following an ion-exchange mechanism, thus DNA double-helical chains get aligned parallel to the nodal plane of the LDH. The bionanocomposites promote diffusion into the cells by shield the negative charge carried by DNA, thus the low pH of lysozyme brings about the dissolution of the LDH followed by transfer of DNA to the cell nucleus [70, 71]. DNA binding to nanofillers can also be obtained using hydrogen bonding, say in clays in some other types of bionanocomposites.

Inorganic–organic bionanocomposites are used recently for vaccine delivery and this field is yet to be explored fully. The nanofillers which may be useful in such applications are iron oxide, calcium phosphate, and LDH [72]. Sepiolite clay surface chemistry was changed and with xanthan was used in one of the cases for delivering influenza vaccine. The cue was taken from the fact that the influenza microbes get adsorbed on a surface resembling with nasal mucosa, where influenza get stabilized as its natural site [73]. The bionanocomposites tend to form stable dispersions and thereby are helpful in making the nasal delivery of vaccine [74]. The bionanocomposites possess a high degree seroprotection and, therefore, marked efficacy of the vaccine may be the fallout. Advances over time have resulted in bionanocomposite-based vaccine delivery systems that can offer better thermal stability of the vaccines, which is very vital from the point of view of storing vaccines during prepandemic periods [75].

13.6 APPLICATIONS IN WOUND DRESSING AND HEALING

Bionanocomposite exhibiting excellent drug release properties holds a lot much of promise and potential in wound dressing applications also, attributed to their more uptake efficiency of water and noncytotoxicity along with mucoadhesion property. Additionally, a wound dressing material ought to have robustness, higher flexibility and elasticity, wear-resistance, and self-healing tendency. Many bionanocomposites meet these requirements thereby making them ideal materials in wound dressing applications.

A host of metallic-, ceramic-, and polymeric-based nanoparticles are being studied for their applicability in expeditious wound healing [76–78]. The interest in the preparation of bionanocomposites in wound dressing and healing applications has been developed due to the unique properties possessed by the nanofillers. Latest advances in effective and enhanced antimicrobial wound dressing and healing incorporated silver nanoparticles [79–82]. Moreover, ZnO nanoparticles and chitosan also possess better antimicrobial properties; silicate clays, namely, MMT exhibits hemostatic characteristics. Combining the healing properties of such nanoparticles with a suitable elastic polymer results into the right kind of material for wound dressing and healing [83–85].

Additionally, because of multifunctionality of bionanocomposites, they find use as tissue adhesives, hemostats, and antimicrobial agents. Silver nanoparticles (AgNPs) decorated on the bacterial cellulose surface were found to be markedly effective in making antimicrobial wound dressing materials [86]. Further, AgNPs were grown on mussel-inspired poly(dopamine methacrylamide-co-methyl methacrylate, MADO) nanofibers also by electrospinning process [87]. The process ensured homogeneous growth of the AgNPs on the surface of the polymer, while they are prevented from oxidation by using catechol. This provided the wound dressing a sustained antimicrobial activity and release. The ECM-type matrix in combination with antimicrobial properties was found to be effective enough in healing wounds expeditiously. The other polymers such as PVA, chitosan [88], and electrospun CMC fibers can also be put together for the said purpose.

Wound dressing by AgNPs is used for the clinical purpose and is nontoxic too; however, reducing the requisite amount AgNPs and enhancing efficacy with the help of bionanocomposites has a therapeutic advantage [89]. Ceramic nanoparticles due to their semicrystallinity as well as easy ionic species release on dissolving have exhibited rapidity in wound healing and angiogenesis. Some such materials of importance finding use as wound dressing materials are silica, silicate clays, bioglass, and ZnO NPs. Silica nanoparticles functionalized with amino groups carry N-diazeniumdiolate groups which have been used for the release of nitric oxide (NO) to promote wound healing [90]. Introduction of the functional groups in a PEG and chitosan matrix furthered the control on the release of NO. Wounds treated by the composite which release NO exhibited wound healing within 12 days, while the untreated wounds take too much time in healing [91]. Additionally, silica nanoparticles can be encapsulated by collagen for wound dressing applications to get over the problem of burst release of drugs if the dressings are done using the only collagen. Moreover, silica nanoparticles enhanced the life of collagen dressing by reducing the degradation rate of enzymes of collagen by means of metalloprotease [92].

The blood clotting capability of composites formed by combining aluminosilicate clay, namely MMT, MMT, and laponite nanoparticles with biopolymer has been used to advantage for its hemostatic activity. Further, being charged particles, these firmly bond with growth factors like vascular endothelial growth factor (VEGF) delivery and have been in use for exacerbating angiogenesis and expeditious wound healing. The VEGF-functionalized clay if added to collagen imparts stability to the hydrogel network which is attributed to the involvement of strong physical forces. Introduction of VEGF within interlayers of laponite–collagen gel increases angiogenesis in vivo and can be worthwhile in facilitating rapid healing of wounds [93]. Further, nanosilicate-based hydrogels are employed as injectable systems for faster hemostasis by about 78% in traumatic injury situation may be because of nanosilicates pulling the plasma to the wound site. Additionally, gelatin–MMT was prepared for the controlled release antimicrobial wound dressing by intercalating ciprofloxacin with the help of electrostatic attraction within interlayers of silicate MMT nanocomposites [94]. Similar to collagen–clay nanocomposites, degradation of gelatin was inhibited due to MMT.

Finally, bacterial cellulose and MMT when combined led to the formation of robust bionanocomposites that are vulnerable to highly swelling and can be used as wound dressing materials. Moreover, ions such as sodium, calcium, and copper can also be loaded using the ion-exchange behavior of the MMT. Bionanocomposite loaded with Cu-MMT possesses maximum antimicrobial activity in particular against Gram-negative (*Escherichia coli*) and Gram-positive (*Staphylococcus aureus*) bacterial species [95]. AgNPs and ZnO NPs both have antimicrobial properties, which are by virtue of the size and capability of nanoparticles to release Zn^{2+} ions from the ZnO sol [96]. ZnO NPs introduced into chitosan/chitin wound dressing bandages showed blood clotting and antibacterial properties [97, 98]. Further investigations demonstrated that PVA macroporous cryogel synthesized by using repeated freeze-thaw method introducing ZnO NPs in the networks of cryogel exhibited antimicrobial activity against both gram-positive and

gram-negative bacteria; species quite effectively [99]. ZnO is approved by the FDA; however, further research on the safety aspects of ZnO needs to be undertaken when used in the form of nanoparticles [100].

13.7 APPLICATIONS IN BIOSENSING

Conjugated polymers containing numerous nanofillers have been studied for their suitability in sensor applications such as gas sensors, biosensors, and chemical sensors. The nanofillers used are nanowires of metal oxide, CNTs, nanogold, silver, nickel, copper, platinum, and palladium particles [101]. CNTs bring about significant change in the electrical resistance on exposure to typical gases namely NO_2 and NH_3 [102]. A nanocomposite made out of single-walled CNT (SWCNT)/polypyrrole combination produced a gas sensor showing sensitivity akin to SWCNT in isolation [103]. The sensing ability of such nanocomposites can be dependent upon variations in the conductivity on exposure to gas or chemicals of either the nanofiller or the conjugated polymer, variations in pH, electrochromic or electrooptical, catalytic activity, chemiluminescence, or biological recognition, etc., for example, sensors for dopamine detection consists of poly (anilineboronic acid)/CNT [104] and polyaniline/gold composite hollow microsphere systems [105].

13.8 CONCLUSION

Bionanocomposites by virtue of choosing the right combinations of matrix and a reinforcing materials and thereby the properties of these materials can also be combined so as to cater to the needs of a host of applications in different frontier areas. Bionanocomposites have been used in dental treatment, tissue engineering, drug delivery and release containers or carriers making, developing wound dressings and healing materials, biosensing, etc. A few representative applications of bionanocomposites have been discussed in this chapter. But they can be explored further by coupling the right kind of matrix with a suitable nanofiller so as to make the derived bionanocomposites suitable for innovative applications.

KEYWORDS

- **bionanocomposites**
- **biphasic**
- **matrix**
- **reinforcing material**
- **filler**
- **biomedicine**

REFERENCES

1. Bowen, D. H. A., in: Kelly, A. (Ed.), Concise Encyclopedia of Composite Materials, Elsevier: Oxford, 1994, pp. 7–15.
2. Haresceugh, R. F. A., in: Kelly, A. (Ed.), Concise Encyclopedia of Composite Materials, Elsevier: Oxford, 1994, pp. 1–7.
3. Newnham, R. E., & Giniewicz, J. R., Comprehensive Composite Materials 1 (2000) 431–463.
4. Baibarac, M., & Gomez-Romero, P.J., Nanoscience and Nanotechnology 6 (2006) 1–14.
5. Baughman, R. H., Zakhidov, A. A., & De Heer, W.A., Science 297 (2002) 787–792.
6. Moniruzzaman, M., & Winey, K. I., Macromolecules 39 (2006) 5194–5205.
7. Williams, D. F., Biodegradation of Medical Polymers, in: Williams, D. F. (Ed.), Concise Encyclopedia of Medical and Dental Materials, Pergamon Press: Oxford, 1990, pp. 69–74.
8. Donachie, M., in: Davis, J. R. (Ed.), Biomaterials, Metals Handbook Desk Edition, 2nd ed., ASM International: Cleveland, OH, USA, 1998, pp. 702–709.
9. Sun-Hong, M., Ferracane, J. L., & In-Bog, L., Dental Materials 26 (2010) 1024–1033.
10. Peutzfeldt, A., European Journal of Oral Sciences 105 (1997) 97–116.
11. Bayne, S. C., Heymann, H. O., & Swift, E. J. Jr., Journal of the American Dental Association 125 (1994) 687–701.
12. Wolter, H., Glaubitt, W., & Rose, K., Materials Research Society Symposia Proceedings 271 (1992) 719–724.
13. Weinmann, W., Thalacker, C., & Guggenberger, R., Dental Materials: Official Publication of the Academy of Dental Materials 21 (2005) 68–74.
14. Ilie, N., & Hickel, R., Dental Materials Journal 25 (2006) 445–454.

15. Bundela, H., & Bajpai, A. K., eXPRESS Polymer Letters 2 (3) (2008) 201–213.
16. Vuong, J., & Hellmich, C., Journal of Theoretical Biology 287 (2011) 115–130.
17. Fritsch, A., Hellmich, C., & Dormieux, L., Journal of Theoretical Biology 260 (2009) 230–252.
18. Sowjanya, J. A., Singh, J., Mohita, T., Sarvanan, S., Moorthi, A., Srinivasan, N., & Selvamurugan, N., Colloids and Surfaces B: Biointerfaces 109 (2013) 294–300.
19. Song, J., Malathong, V., Carolyn, R., & Bertozzi, J., Journal of the American Chemical Society 127 (2005) 3366–3372.
20. Bauer, T. W., Geesink, R. C., Zimmerman, R., & McMahon, J. T., Journal of Bone and Joint Surgery 73-A (1991) 1439–1452.
21. Bucholz, R. W., Clinical Orthopaedics and Related Research 395 (2002) 44–52.
22. Frame, J. W., Journal of Dentistry 3 (1975) 177–187.
23. Kanellakopolou, K., & Giamarellos–Bour boulis, E. J., Drugs 59 (2000) 1223–1232.
24. Merkx, M. A. W., & Maltha, J. C., Oral and Maxillofacial Surgery 32 (2006) 1–6.
25. Chandara, C., Azizli, K. A. M., Ahmad, Z. A., & Sakai, E., Waste Management 29 (2009) 1675–1679.
26. Mendes, S. C., Reis, R. L., Bovell, Y. P., Chunna, A. M., Blitterswijk, C. A., & Bruijn, J. D., Biomaterials 22 (2001) 2057–2064.
27. Ma, X. D., Qiam, X. F., Yin, J., & Zhu, Z. K., Journal of Materials Chemistry 12 (2002) 663–666.
28. Bajpai, A. K., Bajpai, J., & Shukla, S., Reactive and Functional Polymers 50 (2001) 9–21.
29. Gao, H. J., Ji, B. H., Jager, I. L., Arzt, E., & Fratzl, P., Proceedings of the National Academy of Sciences of the United States of America 100 (2003) 5597–5600.
30. Wu, C. J., Gaharwar, A. K., Schexnailder, P. J., & Schmidt, G., Materials 3 (2010) 2986–3005.
31. Sahoo, N. G., Pan, Y. Z., Li, L., & He, C. B., Nanomedicine 8 (2013) 639–653.
32. Fratzl, P., & Weinkamer, R., Progress in Materials Science 52 (2007) 1263–1334.
33. Tang, Z. Y., Kotov, N. A., Magonov, S., & Ozturk, B., Nature Materials 2 (2003) 413–U418.
34. Sheldon, B. W., & Curtin, W. A., Nature Materials 3 (2004) 505–506.
35. Vaia, R., & Baur, J., Science 319 (2008) 420–421.
36. Vaia, R. A., & Wagner, H. D., Materials Today 7 (2004) 32–37.
37. Stodolak-Zych, E., Fraczek-Szczypta, A., Wiechec, A., & Blazewicz, M., Acta Physica Polonica 121 (2012) 518–521.
38. Darder, M., Aranda, P., & Ruiz-Hitzky, E., Advanced Materials 19 (2007) 1309–1319.
39. Mendes, R. G., Bachmatiuk, A., Buchner, B., Cuniberti, G., & Rummeli, M. H., Journal of Materials Chemistry B 1 (2013) 401–428.
40. Aime, C., & Coradin, T., Journal of Polymer Science Part B: Polymer Physics 50 (2012) 669–680.
41. Hoppe, A., Guldal, N. S., & Boccaccini, A. R., Biomaterials 32 (2011) 2757–2774.
42. Gaharwar, A. K., & Peppas, N. A., Khademhosseini, A., Biotechnology and Bioengineering 111 (2014) 441–453.
43. Liao, S. S., & Cui, F. Z., Tissue Engineering 10 (2004) 73–80.
44. Liao, S. S., Cui, F. Z., Zhang, W., & Feng, Q. L., Journal of Biomedical Materials Research—Part B 69b (2004) 158–165.
45. Calandrelli, L., Annunziata, M., Della Ragione, F., Laurienzo, P., Malinconico, M., & Oliva, A., Journal of Materials

Science: Materials in Medicine 21 (2010) 2923–2936.

46. Diba, M., Kharaziha, M., Fathi, M. H., Gholipourmalekabadi, M., & Samadi-kuchaksaraei, A., Composites Science and Technology 72 (2012) 716–723.

47. Dawson, J. I., & Oreffo, R. O. C., Advanced Materials 25 (2013) 4069–4086.

48. Ruiz-Hitzky, E., Darder, M., Fernandes, F. M., Wicklein, B., Alcantara, A. C. S., & Aranda, P., Progress in Polymer Science 38 (2013) 1392–1414.

49. Xu, W., Raychowdhury, S., Jiang, D. D., Retsos, H., & Giannelis, E. P., Small 4 (2008) 662–669.

50. Lee, J. H., Park, T. G., Park, H. S., Lee, D. S., Lee, Y. K., Yoon, S. C., & Nam, J. D., Biomaterials 24 (2003) 2773–2778.

51. Ray, S. S., Yamada, K., Okamoto, M., & Ueda, K., Nano Letters 2 (2002) 1093–1096.

52. Ozkoc, G., Kemaloglu, S., & Quaedflieg, M., Polymer Composites 31 (2010) 674–683.

53. Olmo, N., Lizarbe, M. A., & Gavilanes, J. G., Biomaterials 8 (1987) 67–69.

54. Olmo, N., del Pozo, A. M., Lizarbe, M. A., & Gavilanes, J. G., Collagen and Related Research 5 (1985) 9–16.

55. Gaharwar, A. K., Schexnailder, P., Kaul, V., Akkus, O., Zakharov, D., Seifert, S., & Schmidt, G., Advanced Functional Materials 20 (2010) 429–436.

56. Gaharwar, A. K., Kishore, V., Rivera, C., Bullock, W., Wu, C. J., Akkus, O., & Schmidt, G., Macromolecular Bioscience 12 (2012) 779–793.

57. Schexnailder, P. J., Gaharwar, A. K., Bartlett, R. L., Seal, B. L., & Schmidt, G., Macromolecular Bioscience 10 (2010) 1416–1423.

58. Souza, G. R., Molina, J. R., Raphael, R. M., Ozawa, M. G., Stark, D. J., Levin, C. S., Bronk, L. F., Ananta, J. S., Mandelin, J., Georgescu, M. M., Bankson, J. A., Gelovani, J. G., Killian, T. C., Arap, W., & Pasqualini,

R., Nature Nanotechnology 5 (2010) 291–296.

59. Kim, S. S., Park, M. S., Jeon, O., Choi, C. Y., & Kim, B. S., Biomaterials 27 (2006) 1399–1409.

60. Roohani-Esfahani, S. I., Nouri-Khorasani, S., Lu, Z. F., Appleyard, R., Zreiqat, H., Biomaterials 31 (2010) 5498–5509.

61. Jeon, B. J., Jeong, S. Y., Koo, A. N., Kim, B. C., Hwang, Y. S., & Lee, S. C., Macromolecular Research 20 (2012) 715–724.

62. Liu, Y., Lu, Y., Tian, X. Z., Cui, G., Zhao, Y. M., Yang, Q., Yu, S. L., Xing, G. S., & Zhang, B. X., Biomaterials 30 (2009) 6276–6285.

63. Lewkowitz-Shpuntoff, H. M., Wen, M. C., Singh, A., Brenner, N., Gambino, R., Pernodet, N., Isseroff, R., Rafailovich, M., & Sokolov, J., Biomaterials 30 (2009) 8–18.

64. Cypes, S. H., Saltzman, W. M., & Giannelis, E. P., Journal of Controlled Release 90 (2003) 163–169.

65. Slowing, I. I., Trewyn, B. G., Giri, S., & Lin, V. S. Y., Advanced Functional Materials 17 (2007) 1225–1236.

66. Aguzzi, C., Cerezo, P., Viseras, C., & Caramella, C., Applied Clay Science 36 (2007) 22–36.

67. Wang, Q., Zhang, J. P., & Wang, A. Q., Carbohydrate Polymers 78 (2009) 731–737.

68. Tan, S. T., Wendorff, J. H., Pietzonka, C., Jia, Z. H., & Wang, G. Q., ChemPhysChem 6 (2005) 1461–1465.

69. Wu, C. J., & Schmidt, G., Macromolecular Rapid Communications 30 (2009) 1492–1497.

70. Choy, J. H., Kwak, S. Y., Park, J. S., Jeong, Y. J., & Portier, J., Journal of the American Chemical Society 121 (1999) 1399–1400.

71. Choy, J. H., Kwak, S. Y., Jeong, Y. J., & Park, J. S., Angewandte Chemie International Edition 39 (2000) 4042–4045.

72. Xu, Z. P., Zeng, Q. H., Lu, G. Q., & Yu, A. B., Chemical Engineering Science 61 (2006) 1027–1040.

73. Ruiz-Hitzky, E., Darder, M., Aranda, P., del Burgo, M. A. M., del Real, G., Advanced Materials 21 (2009) 4167–4171.

74. Amorij, J. P., Hinrichs, W. L. J., Frijlink, H. W., Wilschut, J. C., & Huckriede, A., The Lancet Infectious Diseases 10 (2010) 699–711.

75. Brandau, D. T., Jones, L. S., Wiethoff, C. M., Rexroad, J., & Middaugh, C. R., Journal of Pharmaceutical Sciences 92 (2003) 218–231.

76. Rieger, K. A., Birch, N. P., & Schiffman, J. D., Journal of Materials Chemistry B 1 (2013) 4531–4541.

77. Balkawade, R. S., & Mills, D. K., The FASEB Journal (2012) 26.

78. Kalashnikova, I., Das, S., & Seal, S., Nanomedicine-UK 10 (2015) 2593–2612.

79. Jain, J., Arora, S., Rajwade, J. M., Omray, P., Khandelwal, S., & Paknikar, K. M., Molecular Pharmaceutics 6 (2009) 1388–1401.

80. Maneerung, T., Tokura, S., & Rujiravanit, R., Carbohydrate Polymers 72 (2008) 43–51.

81. Chaloupka, K., Malam, Y., & Seifalian, A. M., Trends in Biotechnology 28 (2010) 580–588.

82. Song, J. L., Birbach, N. L., & Hinestroza, J. P., Cellulose 19 (2012) 411–424.

83. Vaiana, C. A., Leonard, M. K., Drummy, L. F., Singh, K. M., Bubulya, A., Vaia, R. A., Naik, R. R., & Kadakia, M. P., Biomacromolecules 12 (2011) 3139–3146.

84. Li, X. Q., Wang, H. F., Rong, H. L., Li, W. H., Luo, Y., Tian, K., Quan, D. Q., Wang, Y. G., & Jiang, L., Journal of Colloid and Interface Science 445 (2015) 312–319.

85. Chu, C. Y., Peng, F. C., Chiu, Y. F., Lee, H. C., Chen, C. W., Wei, J. C., & Lin, J. J., PLoS One (2012) 7.

86. Wu, J., Zheng, Y. D., Song, W. H., Luan, J. B., Wen, X. X., Wu, Z. G., Chen, X. H., Wang, Q., & Guo, S. L., Carbohydrate Polymers 102 (2014) 762–771.

87. GhayamiNejad, A., Unnithan, A. R., Ramachandra, A., Sasikala, K., Samarikhalaj, M., Thomas, R. G., Jeong, Y. Y., Nasseri, S., Murugesan, P., Wu, D. M., Park, C. H., Kim, C. S., ACS Applied Materials & Interfaces 7 (2015) 12176–12183.

88. Li, C. W., Fu, R. Q., Yu, C. P., Li, Z. H., Guan, H. Y., Hu, D. Q., Zhao, D. H., & Lu, L. C., International Journal of Nanomedicine 8 (2013) 4131–4145.

89. Walker, M., & Parsons, D., International Wound Journal 11 (2014) 496–504.

90. Shin, J. H., Metzger, S. K., & Schoenfisch, M. H., Journal of the American Chemical Society 129 (2007) 4612–4619.

91. Friedman, A. J., Han, G., Navati, M. S., Chacko, M., Gunther, L., Alfieri, A., & Friedman, J. M., Nitric Oxide-Biology and Chemistry 19 (2008) 12–20.

92. Alvarez, G. S., Helary, C., Mebert, A. M., Wang, X. L., Coradin, T., Desimone, M. F., Journal of Materials Chemistry B 2 (2014) 4660–4670.

93. Dawson, J. I., Kanczler, J. M., Yang, X. B. B., Attard, G. S., & Oreffo, R. O. C., Advanced Materials 23 (2011) 3304.

94. Kevadiya, B. D., Rajkumar, S., Bajaj, H. C., Chettiar, S. S., Gosai, K., Brahmbhatt, H., Bhatt, A. S., Barvaliya, Y. K., Dave, G. S., & Kothari, R. K., Colloids and Surfaces B: Biointerfaces 122 (2014) 175–183.

95. Ul-Islam, M., Khan, T., Khattak, W. A., & Park, J. K., Cellulose 20 (2013) 589–596.

96. Raghupathi, K. R., Koodali, R. T., & Manna, A. C., Langmuir 27 (2011) 4020–4028.

97. Kumar, P. T. S., Lakshmanan, V. K., Anilkumar, T. V., Ramya, C., Reshmi, P., Unnikrishnan, A. G., Nair, S. V., &

Jayakumar, R., ACS Applied Materials & Interfaces 4 (2012) 2618–2629.

98. Kumar, P. T. S., Lakshmanan, V. K., Raj, M., Biswas, R., Hiroshi, T., Nair, S. V., & Jayakumar, R., Pharmaceutical Research-Disorder 30 (2013) 523–537.

99. Chaturvedi, A., Bajpai, A. K., Bajpai, J., & Singh, S. K., Materials Science and Engineering C: Materials for Biological Applications 65 (2016) 408–418.

100. Soenen, S. J., Parak, W. J., Rejman, J., & Manshian, B., Chemical Reviews 115 (2015) 2109–2135.

101. Hatchett, D. W., & Josowicz, M., Chemical Reviews 108 (2) (2008) 746–769.

102. Kong, J., Franklin, N. R., Zhou, C., Chapline, M. G., Peng, S., Cho, K., Dai, H., et al., Science 287 (2000) 622–625.

103. An, K. H., Jeong, S. Y., Hwang, H. R., & Lee, Y. H., Advanced Materials 16 (2004) 1005–1009.

104. Ali, S. R., Ma, Y., Parajuli, R. R., Balogun, Y., Lai, W. Y. C., & He, H., Analytical Chemistry 79 (2007) 2583–2587.

105. Feng, X., Mao, C., Yang, G., Hou, W., & Zhu, J. J., Langmuir 22 (2006) 4384–4389.

PART III
Environmental Opportunities and Risks

CHAPTER 14

CYCLODEXTRIN-BASED POLYMERS FOR POLLUTANT REMOVAL

TÂNIA F. COVA*, DINA MURTINHO, ROBERTO GARCIA, ALBERTO A. C. C. PAIS, and ARTUR J. M. VALENTE*

Department of Chemistry, University of Coimbra, CQC, 3004-535 Coimbra, Portugal

Corresponding author. E-mail: tfirmino@qui.uc.pt; avalente@ci.uc.pt.

ABSTRACT

Cyclodextrins (CDs) and CD-containing materials, including polymers, hydrogels, and nanoparticles, possess unique properties that have paved the way for the development of innovative and greener strategies for environmental remediation. These intrinsic structural and chemical properties, biocompatible and bioabsorbent character, and the ability to interact selectively with a wide range of molecules, forming inclusion complexes, and higher order supramolecular structures, make these materials highly attractive to remove persistent and emerging pollutants. The individual or simultaneous effective removal of inorganic, organic, and pharmaceutical active compounds derived from anthropogenic activities is still struggling with different techno-logical, economic, and social issues. These include compound specificity, limited availability of environmental data and poor resource management, and lack of regulatory stringency for establishing the diversity of contaminants. This chapter provides a critical and timely compilation of the key contributions and advances in environmental remediation technologies based on natural/modified polymers and CD, with selected examples of adsorbents and target matrices and adsorbed compounds. Also described are the (1) benefits of CD-based polymers in outperforming conventional systems, (2) structural variations, (3) preparation strategies and conditions, (4) interaction mechanisms, and (5) removal performances.

14.1 INTRODUCTION

As Europe Commission prepares to unveil an ambitious Green Deal, a strategy that includes 50 specific policy measures for tackling and adapting to climate change and that envisions climate neutrality by 2050 (i.e., by reducing greenhouse emissions, ensuring clean air, clean water and healthy food, creating high-paying jobs, and eradicating all types and forms of oppression), it is crucial that scientists working in different areas of chemistry, materials science, and engineering seek to improve and even rethink the most exciting, interesting, or challenging aspects related to the development of greener, safer, and cost-effective strategies for environmental remediation. The latter has relied mainly on employing several technologies, such as filtration, adsorption, absorption, chemical reactions and photocatalysis, to remove contaminants, such as particulate matter, metal ions, heavy metals, pesticides, toxic gases, industrial and household effluents, and a plethora of other inorganic and organic compounds from different environmental media including the atmosphere, water, and soil [136, 174].

The development of systems directed at removing different contaminants in complex mixtures, covering metals, dyes, hydrocarbons and their derivatives, nanomaterials, pharmaceuticals, and other micropollutants, as well as the proof-of-concept to scale-up such systems for field testing, considering an assessment of the energy efficiency, and the respective environmental benefits and risks are still in its infancy [17, 145, 245, 255, 279, 288, 298, 337, 384].

Resolving the legacy of persistent and emerging contaminants worldwide has become urgent and the international community, including the European Union has called for prompt and further action, in order to implement strategies for adaptation and for preventing the unprecedented and irreversible damage caused by pollution and climate change.

Several approaches involving different types of materials, such as those of polymeric nature, have been explored for this purpose at different functional scales [9, 210, 231, 232, 335, 355, 375]. The most recent contributions have focused on the use of nanobiomaterials, as the detection, capture, and degradation of contaminants has been challenging [137, 199, 281]. This is due to various factors, such as the low reactivity, high volatility, concentration, toxicity, complexity and coexistence of multiple compounds, and specificity of the environment.

Recent literature focused on the adsorption of toxic compounds by modified polysaccharides and cyclodextrins (CDs) suggests that there is an increasing interest in the synthesis of new low-cost adsorbents to be employed in different environmental matrices (e.g., natural water and wastewater) [83, 335].

This chapter presents a historical perspective of the key advances in the design of CD-based materials as a versatile tool for removing pollutants from different environmental matrices, as the CD supramolecular chemistry has offered attractive strategies for preparing CD derivatives, including poly(CD) and grafted or coated polymeric materials.

An exhaustive account of the concepts and applications published in the early years of natural CD in environmental applications falls beyond the scope of this chapter. The reader is referred to [47, 83, 85, 93, 174] for a full description of these efforts.

The increasing number of studies published each year shows recent and continuing interest in applications of CDs in environmental remediation, corresponding to an annual growth rate of ca. 4.32% since 1994. Also, the large-scale commercialization of CDs and CD derivatives has led to a cost reduction, which is favorable to the increment of environmental applications. In the context of water remediation, a successful example is CycloPure, Inc., a leading innovator in water purification technologies and the developer of the eco-friendly DEXSORB™ filtration/adsorbent products (US9624314

patent) directed at removing micropollutants, such as perfluorintated compounds (PFOA and PFOS), residues of pesticides, pharmaceuticals, and other toxic compounds from drinking water.

The first report on a pilot-plant scale demonstration of CD-based sorbents corresponds to the CD-bead-polymer sorption-technology, a tertiary treatment tested in a biomachine-type municipal wastewater treatment plant. This pilot-plant allowed removing estradiol, ethinyl estradiol, estriol, diclofenac, ibuprofen, bisphenol A (BPA), and cholesterol with more than 80% efficiency from 300 L effluent, reducing the respective concentration from ~5 µg L^{-1} to lower than 0.001–1 µg L^{-1}), and, in general, within a short time [120]. A correlation between sorption performances and binding constants of CD-micropollutant inclusion complexes was found, demonstrating that inclusion complex formation was the main driving force in the sorption mechanism.

Until 10 years ago, only a few hundred studies on the use of CD in environmental remediation were published, resulting from the contributions made over three decades. In 2019, a cumulative number of ca. 560 articles in the Web of Science database included these keywords, corresponding to an increase in ca. 80% for just one decade. In this chapter, there is room to mention only a small fraction of these applications.

Despite the increasing number of studies covering relevant progress (preparation procedures, structural variations, supramolecular properties, complexation behavior, adsorption mechanisms, removal performances, and benefits) on these CD-based macromolecular systems, comprehensive, and systematic rationales of the involved interaction patterns from multidisciplinary studies, aiming at improving the design of functional CD-cleaning materials, are still a challenging and fertile proving-ground, from both academic and commercial perspectives.

14.1.1 CD CONTRIBUTIONS CO-OCCURRING IN ENVIRONMENTAL REMEDIATION

Scientific literature covering CD-based polymers, grafted and coated materials for dealing with the detection, capture, and degradation of pollutants has increased exponentially in recent years. However, the formulation and understanding of a complete, integrated picture of the major target environmental needs and research contributions, as well as the spotting of cutting-edge synthetic, analytical, and scaling-up trends in the design of multifunctional CD-materials for leading with (1) the coexistence of persistence and emerging pollutants and with (2) the specificity of the contaminated environmental matrices (atmosphere, water, and soil) are critical tasks. The challenge relies on how the analysis of these efforts with hundreds of published studies, reveals the most prominent applications supported by CD and CD-polymeric materials [87, 93, 174, 229, 232, 240, 265].

Figure 14.1 presents a holistic picture of the insights generated in the 1900–2019 period (panel a) and in the last 5 years (panel b), ranked in the research domain of "Science Technology" of the Web of Science database.

By assessing the different facets of CD and CD-based materials in environmental remediation considering specific target contaminants (e.g., aromatic compounds, dyes, pesticides, pharmaceutical compounds, metal ions, or heavy metals) and target matrices (atmosphere, water, and soil) it is possible (1) to spotting the major environmental scenarios that have benefited most from the application and removal performances of CD-adsorbents, and those that still lack of such an improvement, as evidenced by the type of material and outcome, (2) to identify the most relevant types of pollutants and CD-adsorbents in each environmental matrix, and (3) to assess the dynamics of persistent and emerging contaminants and the respective CD-cleaning strategies employed over the last century, and how these

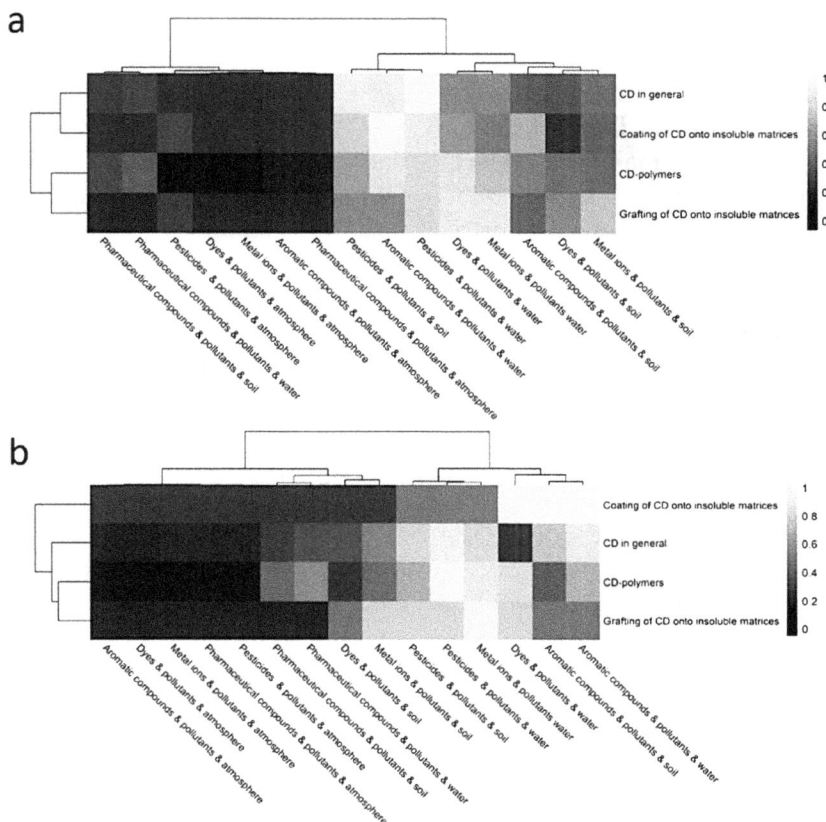

FIGURE 14.1 An overview of the major scientific contributions comprising CD and CD-based materials (polymers, CD grafted, or coated onto insoluble matrices) for removing inorganic and organic pollutants from the atmosphere, water, and soil. The clustering heatmap displays the relative counts of studies dedicated to the development and use of CD-based adsorbents for target matrices and adsorbed compounds and published in the 1900–2019 (10 December) period. Data are expressed as fractions of the highest number of publications, including articles, reviews, books, editorials, letters, and patents containing specific co-occurring keywords (see the text for details), and following a standard normalization. Hierarchical clustering using Euclidean distances and Ward linkage was conducted on both CD-materials and type of application related to a specific contaminant/matrix relation. A dark blue-to-yellow color scheme is used to represent co-occurrences (1—yellow and 0—dark blue reflect the highest and lowest relative contributions, respectively).

are related, giving rise to important research directions.

An extensive literature search on the application of CD and CD-materials for removing five selected types of pollutants (aromatic compounds, dyes, pesticides, pharmaceutical compounds, and metals) in three different matrices (atmosphere, water/wastewater, and soil) is thus carried out, using a global set of 75 co-occurring keywords, each composed of four main terms, type of CD material, type and class of pollutants, and the environmental matrix in which pollutants occur (e.g., first co-occurrence: CD-polymer AND Pesticides AND Pollutants AND Water, second co-occurrence: CD (in general) AND Metal (excluding organic compounds) AND Pollutant AND Soil.

A total of 560 contributions on CD for pollutant removal, reported in 184 distinct sources (including articles, reviews, books, editorials, letters, and patents), collecting 15,118 citations (average citations per documents of ca. 28.95), involving more than 2100 distinct authors and displaying a collaboration index of 3.82, were published between 1904 and 10 December 2019 and can be found in the worldwide Web of Science database (2019 is the most productive year on the topic with ca. 81 contributions).

The most relevant sources possessing at least 10 articles are *Journal of Hazardous Materials* (45), *Chemosphere* (35), *Chemical Engineering Journal* (31), *Carbohydrate Polymers* (21), *Environmental Science & Technology* (21), *Environmental Science and Pollution Research* (17), *Environmental Pollution* (13), *Science of the Total Environment* (12), *Journal of Colloid and Interface Science* (11), and *Journal Inclusion Phenomena and Macrocyclic Chemistry* (10).

Considering the collected data and the selected pollutants and matrices, four different types of materials including natural and modified CD, CD-polymers, immobilized matrices with grafted and coating CD that embrace the most frequent chemical and environmental challenges are defined.

The heatmap represented in Figure 14.1 reflects the impact of each type of outcome attributed to each type of CD material on the development of strategies for decontamination of atmosphere, water, and soil.

The analysis of co-occurring keywords is performed aiming at finding the number of publications reported simultaneously in the selected environmental matrix. This relation is established with a greater or lesser impact depending on the frequency of each set of keywords in the last century (panel a) and in the last 5 years (panel b).

The natural groups established based on the most relevant co-occurring relationships are also identified (see Figure 14.1). The dendrograms corresponding to the scientific production on different target pollutants and the respective environmental matrices, covering the last century (panel a) and to the last 5 years (panel b) show that pollutants and matrices are organized in two main groups (panel a), which discriminates, in general, pollutants present in the atmosphere, and pharmaceutical compounds captured in water and soil, from those studies focused on metals (metal ions and heavy metals), pesticides, dyes, and aromatic compounds present in water and soil. This structure indicates similar outcomes within the group displaying air matrices (Group 1) irrespective of the type of CD-material. Group 2 composed only of water and soil contaminants has benefitted from a significant production of CD-polymers, and also grafted and coating technologies.

The most representative outcomes in Group 2 are associated with pesticides and aromatic compounds in aqueous environments.

The removal of pesticides has been prompted, in general, by the different CD-materials. A more pronounced effort related to the development and application of CD-polymers and CD-coating technologies for removing aromatic compounds from water is also identified. CD-grafting strategies have had a significant impact on the clean-up of water matrices containing dyes and metals. A similar trend is observed when the purpose is to remove dyes and metals from soil matrices. Within the latter, coating strategies have been mainly employed for dealing with pesticides and aromatic compounds.

When the time period is narrowed to the last 5 years, an examination of the similarity between the type of CD outcomes reveals that increasing attention has been given to emergent pollutants including pharmaceutical compounds present in both water and soil matrices.

Researchers are still improving (1) CD-coating materials for removing aromatic compounds from water (panels a and b), (2) CD-polymers for removing pesticides from water and soil, and metal ions from water, and also optimizing (3) CD-grafting approaches for capturing dyes and metal ions from water environments.

Historically, synthetic dyes (e.g., cationic dye methylene blue (MB)) and metals (e.g., Pb(II), Cu(II), and Cd(II)) are two classes of hazardous contaminants that often coexist in industrial wastewaters and drinking water, posing critical concerns to the environment and public health, and leading to the development of bifunctional adsorbents (e.g., EDTA-cross-linked β-CD [386]) capable of capturing metals and cationic dyes simultaneously [199].

The range of applications engaging CD-based polymers and the complexity of environmental matrices and pollutants are now quite extended as a result of a deep understanding and characterization of the interaction patterns intrinsic to these types of materials.

Despite some initial debate and low competing removal performances, the use of CD in environmental remediation has been accelerated and maturated in recent years essentially due to their suitability to new applications and industry needs.

Recently, it was shown that CD (natural β-, γ-CDs and methyl- and hydroxypropyl-CD derivatives) can also entrap mycotoxins (toxic secondary metabolites of filamentous fungi fostered by climate change) including the masked mycotoxin zearalenone-14-glucoside (Z14G, reaching $K_{bind} \approx 10^3$ L mol^{-1} in γ-CD-Z14G complex) in aqueous solutions, including beverages. Z14G is one of the most challenging masked mycotoxin produced from zearalenone, as it possesses a large hydrophilic glucose moiety. CD was able to encapsulate Z14G and the insoluble β-CD bead polymer decreased the Z14G content in aqueous solutions [113].

Previous studies [17, 34, 65, 254, 296] have also reported the significant affinity of CDs with zearalenone and zearalenols [17, 34, 65, 254, 296] allowing, for example, the removal of zearalenone by an insoluble β-CD bead polymer from aqueous buffers and spiked corn beer samples.

Based on these observations, CD technology is a promising tool to develop mycotoxin adsorbents that can remove both parent mycotoxins and some of their metabolites (e.g., masked mycotoxins) from aqueous solutions (including beverages).

There are also several lab- and field-testing strategies for soil remediation employing CDs as solubilizers of organic contaminants [135]. Soil flushing with hydroxypropyl-β-CD (HP-β-CD) [41, 45, 46] has been recognized as a potential technology for removing crude oil containing common alkanes and polyaromatic hydrocarbons (PAHs) from porous media [124].

Carboxymethyl β-CD has been able to washing soils containing mixed organic and inorganic pollutants [218, 301, 360]. The remediation mechanism relies on salt formation with metals (Cd, Cu, Mn, and Pb), and on the formation of inclusion complexes with organic pollutants (PCBs with ca. 100% of removal capacity) [111].

The first studies applying CD derivatives to recognize and remove pollutants from the environment, including aromatic or phenolic compounds of varying structural complexity have appeared at the end of the 1990s [93, 229]. These contributions are described in several publications [22, 26, 86, 159, 175, 231, 307, 314, 339, 363] and patents [18, 32, 49, 60, 259, 287, 315, 316, 322] emphasizing the critical role of the CD cavities and suggesting chemisorption, promoted by the formation of an inclusion host–guest complex, as the main force in sorption mechanisms.

More recently, it has been demonstrated that sorption patterns cannot be explained solely by the inclusion phenomenon, since other mechanisms external to the host cavity and involving the polymer network have been proposed. It has been thus suggested that adsorbent-pollutant association is promoted by the cooperative effect of the grid of the polymer network, surface sorption, hydrophobic interactions, van der Waals forces, hydrogen bonding, electrostatic interactions, and among others [232].

14.2 THE CRITICAL ROLE OF CD-BASED POLYMERS IN POLLUTANT REMOVAL

During the last four decades, polysaccharides, including CD, have received special attention in environmental applications due to their high availability, biodegradability, and biocompatibility, nontoxicity, excellent chemical reactivity and facile synthesis, adsorption capabilities, potential for large scale production, and harmless to the environment [83, 85, 93].

The chain of polysaccharides confers the polymer a high hydrophilicity. The respective flexible structure and the presence of functional and reactive groups (acetamido and primary amino and hydroxyl groups) facilitate the adsorption behavior of polysaccharides.

The inclusion process based on host–guest interactions has played a key role in the modification of the characteristics of the contaminants, since CD present steric relationships with pollutant molecules (e.g., organic compounds), which are governed by van der Waals forces, enthalpy, and entropy contributions [82–85]. These macrocyclic hosts can also remove heavy metals from aqueous media by forming CD-metal complexes as a result of the covalent binding between the deprotonated hydroxyl groups on the outer surface of CD and the metal ions.

More recently, various polymeric networks have been prepared from polymerization or copolymerization reactions involving CD, crosslinking agents, and other grafting agents, for capturing different classes of pollutants [16, 38, 86, 155, 161, 164, 178, 228, 326, 336, 355, 385]. The alkylation and carboxymethylation of natural CDs have allowed generating CD derivatives displaying unique properties for host–guest binding, stabilizing, separating, or absorbing contaminants from wastewater, soil, and air.

Crosslinking and immobilizing CD molecules have also been employed for producing water-insoluble CD-based materials useful for separation and decontamination purposes [298].

Recently, Valente et al. [85] have provided a summary of the key progress in developing CD-based materials, including polymers, hydrogels, and nanocomposites, used as sorbents for removing toxic compounds from wastewater. The preparation procedures of these macromolecular systems, as well as structural variations, supramolecular properties, adsorption and complexation behaviors, and also the removal performances for different pollutants, were duly discussed.

Different CD molecules have been modified for producing derivatives with a high affinity toward contaminants in aqueous solutions. Modification relies for example on (1) the direct modification of β-CD by crosslinking using a coupling agent or on (2) the covalent binding of β-CD molecules to a preexisting insoluble matrix, including highly crosslinked synthetic polymers, chitosan (CS), alginates, nanofibers, textiles, silica beads, and zeolite [85]. In the first strategy, the molecules are concatenated using a coupling agent for producing an insoluble 3D network, displaying the desired characteristics, such as porosity, density, surface area, adsorption and chemical behavior, and considering the type of CD and crosslinking spacer, and molar ratio. In the second strategy, β-CDs are immobilized on insoluble materials, acting as filters. Another alternative is based on the crosslinking reaction using epichlorohydrin (EPI), which possesses two reactive groups able to bind to CD, generating hydrophilic

polymeric mixtures (hydrogels) with multiple CD molecules linked by repeating glyceryl linkers [38, 104, 164, 166, 194, 249, 252, 253, 290, 297, 298, 336].

Crosslinked CD materials exhibit high adsorption capability, are efficient in the removal of water contaminants present at trace levels, the kinetics are relatively fast, and are suitable for dealing with slow pollutant uptake and poor removal of hydrophilic pollutants [148, 155, 178, 203, 228, 270, 316, 326, 336, 337, 355, 356, 386].

Despite CD-polymers have been easily regenerated (e.g., by multiple extraction cycles with solvents) without a significant loss of the complexation and adsorption capacity, the adsorption mechanism is still poorly understood, as it involves various cooperative interactions with different contributions.

Adsorption of pollutants benefits, in general, from the formation of the CD-pollutant complex and from other interactions external to the cavity, which are influenced by the contact time, the type of sorbent, and the chemical structure of the pollutant. The formation of the host–guest complex, due to the presence of CD, the association due to the cooperative effect of the grid of the polymer network, hydrogen bonding, van der Waals forces, hydrophobic and electrostatic interactions, and also the physical adsorption in the polymer network, has been raised as the mainstay of the adsorption mechanism [232].

14.2.1 HISTORICAL BACK-GROUND

Although CDs were discovered in 1891 by Villiers, it was only at the end of the 1980s that CD started to be applied in areas such as chromatography, food industries, and pharmaceuticals. Since then, CDs have been used in numerous applications in areas as diverse as agriculture, pharmacy, food, cosmetics, medicine, chromatography, catalysis, biotechnology and textile industries, and among others. A recent review about the most relevant fields of application of CD was published by Crini et al., including selected comprehensive reviews and book chapters [93].

Adsorption methods are the most commonly used in pollutant removal due to the lower cost, comparing with other methods, and due to the possibility of adsorbate regeneration. The most used adsorbate in remediation is probably activated carbon because it has a high surface area and absorption capacity for some pollutants. However, its use is limited because regeneration is difficult and expensive [231]. The search for alternative materials has been constant, and CD, due to its ability to form inclusion complexes

with many organic and inorganic molecules, has become attractive for this purpose. Furthermore, the inclusion process is reversible which allows CD recycling. It is therefore not surprising that at the end of the last century CD began to be used in environmental remediation [298].

Nevertheless, native CDs and some of their derivatives are soluble in water, which limits their application, especially in wastewater treatment. To circumvent this problem and to facilitate the removal and recycling process, insoluble CD-based materials have been developed.

The first studies on the use of CD-based polymers for pollutant removal appeared in the literature at the end of the 1990s [94, 95, 232].

The synthesis of CD-based polymers was first described in 1965 by Solms and Egli. The crosslinked copolymers were prepared by reaction of CD with EPI, in basic medium. Later on, Wiedenhof et al. prepared microparticles of the same crosslinked copolymers that are used for chromatography [303].

Since these pioneering works, several other authors reported the synthesis of CD-EPI crosslinked polymers that can be obtained as water-soluble or insoluble, with high or low molecular weights and with different morphologies (hydrogels, particles, foams, sponges, etc.) [230, 232].

Insoluble CD-EPI polymers can swell in water and, due to the presence of CDs, can form inclusion complexes with a variety of compounds, including pollutants. Due to these characteristics, CD-EPI polymers were the first to be used for pollutant removal, as described in patents by Tabushi and Friedman [163].

Other crosslinking agents, such as diisocyanates, diacids, and their derivatives were posteriorly used to prepare CD-based polymers that were tested for their capacity to adsorb several types of contaminants in the air, water, and soil [174, 298]

Besides these crosslinked CD-based materials, several others have appeared in the literature and were used in remediation. For example, the immobilization of CDs in natural and artificial organic polymers and the anchorage in inorganic polymers such as silica have been reported in numerous publications. More recently, the use of some innovative materials, including nanomaterials (nanosponges, magnetic nanoparticles (MNPs), carbon nanotubes (CNTs), graphene, among others), to anchor CDs and also to allow an easy regeneration and recycling of the materials has also been referred [122, 171, 174, 231].

14.2.2 PREPARATION AND SUPRAMOLECULAR PROPERTIES

CDs can be considered polyfunctional monomers since the macrocyclic host possess several reactive hydroxyl groups at the 2, 3, and 6 positions of the glucose unit, susceptible to experience substitution reactions and/or chemical modification. The primary hydroxyl groups at the C6 positions are usually more reactive than the secondary ones (C2 and C3 positions), although reaction conditions such as temperature and alkalinity can reverse this reactivity [230].

Due to these intrinsic characteristics, CDs can be directly copolymerized with other monomers or attached to a myriad of materials. Moreover, CDs can also be modified in order to obtain derivatives with other functionalities that can be used as monomers or linked to several substrates. Different types of architectures can thus be obtained, including linear, branched, and crosslinked networks.

Two main types of CD-based polymers can be prepared: the first involves the use of CD or CD derivatives as monomers and their reaction with a coupling agent in order obtain linear polymers or crosslinked structures; in the second type, CDs are attached to a matrix (organic or inorganic) via covalent linkage or physical entrapment [95, 167, 223].

Depending on the application, water-soluble or insoluble CD-based polymers can be synthesized. For the synthesis of water-soluble polymers, dilute solutions and low ratios of the CDs/other reactants are used, and usually, almost linear polymers or polymers with pendent groups are obtained. Water-insoluble polymers can be prepared using a higher concentration of CDs or decreasing the ratio of CDs/crosslinking agents [167].

Most of the literature concerning the preparation of CD-based materials for remediation refers to the use of three major types of materials: CD-based polymers and grafting or coating of the CD onto insoluble matrices.

In this section, CD-based materials will be divided into two main groups: CD-polymers and CD linkage to polymeric matrices.

14.2.2.1 CD POLYMERS

The majority of CD transformations, including polymer formation, is based on the hydroxyl group reactivity. Hydroxyl groups can act as nucleophiles or electrophiles and react with a variety of other functional groups, forming ethers, esters, halides, etc. CD hydroxyl groups can be deprotonated using strong bases, thus originating alkoxide ions, strong nucleophiles that can undergo SN2 reaction. CD ethers can be prepared by protonation of the hydroxyl groups by acids, converting the poor leaving group OH− to H_2O, a better leaving group, and thus acting as electrophiles in the reaction with other alcohols. CD esters can be prepared by

reaction with acids, acid chlorides, or anhydrides. The reaction of CDs with isocyanates or isothiocyanates results in the synthesis of urethanes or thiourethanes, respectively. Due to the broad reactivity of the hydroxyl groups, innumerous derivatives have been prepared. CDs are also polyhydroxylated compounds able to form polymers by reaction with other monomers.

CD-crosslinking polymers are particularly important in the context of remediation due to the cooperative effect between the CD cavity and the polymeric network that can be observed in the sorption of pollutants. Depending on the type of CD, the crosslinking agent, and the reaction conditions, materials with different characteristics can be prepared [174].

As mentioned above, CD-EPI polymers were one of the first to be used for pollutant removal. These polymers are prepared by reaction of CDs with EPI, under heating, catalyzed by a base, usually NaOH or KOH, in a one-pot reaction (Scheme 14.1).

The mechanism of this reaction is well-known and depending on the reaction conditions, polymers with different degrees of crosslinking can be obtained. Thus, soluble or insoluble polymers can be prepared, and in the latter case, it is possible to obtain the polymer as gels, particles, or nanoparticles (NPs), as already mentioned [91, 129, 274, 275].

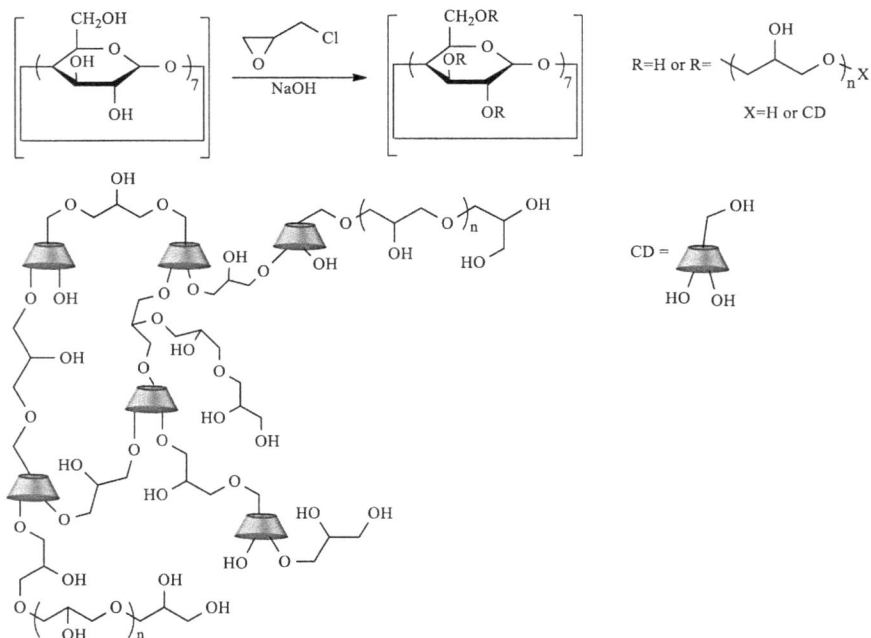

SCHEME 14.1   CD-epichlorohydrin polymers.

In addition to the patents cited in the previous section, Kiji et al. [163] published one of the first papers reporting the use of β-CD-EPI polymer for the sorption of 4-methylbiphenyl, sodium dodecylbenzenesulfonate, and phenol.

In 1996, Shao et al. [294] prepared β-CD-EPI and HP-β-CD-EPI crosslinked gels and used them to sorb acid, direct, mordant and reactive textile dyes from aqueous solutions. The effect of pH, salt, and anionic surfactants on the gel sorption capacity was evaluated. The authors suggested that the sorption of the dyes in CD-EPI gels resulted not only from host–guest

inclusion complex formation but also from dye adsorption to the polymer network [293].

Later on, Crini et al. [90] used similar β-CD-EPI polymers, with different ratios of β-CD/EPI, for the sorption of several aromatic pollutants. The good sorption capacities of these materials were explained considering inclusion complex formation, physical adsorption to the polymer network, and hydrophobic guest–guest interactions.

At the same time, Murai et al. [239] reported the use of a β-CD-EPI polymer for the recovery of ionic and nonionic surfactants. Good adsorption percentages were obtained and the adsorbents could easily be recycled. The formation of inclusion complexes and interactions with the polymer network was pointed as responsible for the sorption mechanism.

Since these first papers reporting the use of CDs-EPI in the removal of dyes and aromatic pollutants, several other publications appeared in the literature using these polymers for the removal of contaminants, namely metals, pesticides, surfactants, aromatic pollutants (including PHAs and PCBs), pharmaceuticals, and among others [74, 90, 92, 194, 255, 267, 269, 372, 377].

CD-EPI polymers have been widely explored as absorbents due to the ease of preparation and efficiency in the removal of a broad range of pollutants. The sorption mechanisms for CD-EPI polymers have been intensively discussed and are well documented in a recent review by Crini et al. [232]. The authors refer that most of the sorption results obtained for these polymers can be explained essentially by the presence of the CD and the degree of crosslinking. The prevailing sorption mechanism seems to be chemisorption, namely the formation of pollutant inclusion complexes with the CDs. However, the degree of crosslinking, which affects the swelling properties, also contributes to the sorption mechanism, and therefore physisorption acts in synergy with chemisorption.

The sorption properties of CD-EPI polymers can be modulated using CD derivatives or by performing the polymerization reaction in the presence of other compounds or polymers, thus extending the scope of application of these materials [85].

Several references are also made in the literature to the use of several CD derivatives that were crosslinked with EPI, namely hydroxypropyl, carboxymethyl, aminoethyl, methyl, and sulfonyl, for pollutant removal [205, 266, 377, 382].

The use of CD-EPI polymers for pollutant removal is extensively documented in the literature in several review articles on this topic [129, 230, 232].

CD can also be copolymerized using other monomers besides EPI. Diisocyanates can react with CDs to form polyurethanes (Scheme 14.2) and depending on the crosslinking agent and its relative molar ratio, materials with different surface areas, pore size distribution, mechanical properties, and sorption ability can be obtained. Various diisocyanates can be used as crosslinking agents, namely 1,6-hexamethylene diisocyanate (HDI), 2,4-toluene diisocyanate (TDI), 1,4-phenylene diisocyanate (PDI), 4,4'-dicyclohexylmethane diisocyanate (CDI), 4,4'-diphenyl-methane diisocyanate (MDI) and 1,5-naphtalene diisocyanate (NDI), see Scheme 14.2 [225, 342].

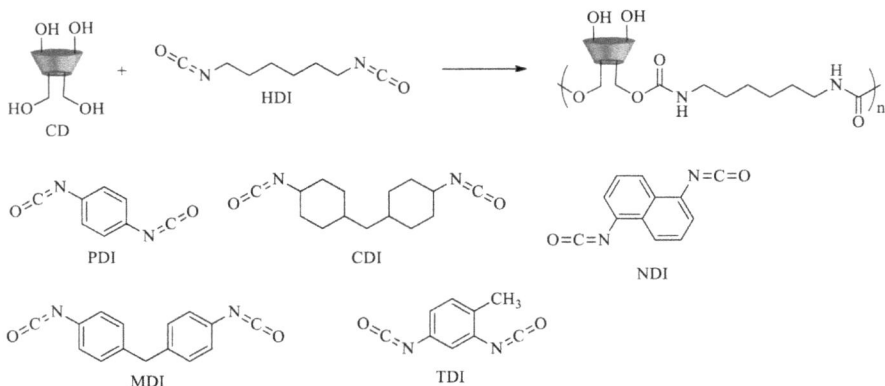

SCHEME 14.2   CD-polyurethane polymers.

Usually, polar aprotic solvents such as dimethylsulfoxide or dimethylformamide, and temperatures from 10 °C to reflux are used to prepare these polyurethanes. If an excess of the diisocyanate is used, highly cross-linking polymers can be obtained. Recently, CD polyurethanes were also synthesized using microwave irradiation and when compared with the conventional method, higher yields were obtained and lower reaction times (3–10 min) were needed [40].

These polymers, also called nanosponge CD polyurethanes, proved to be effective in the removal of several pollutants. In the early 1980s, Shono et al. studied the interaction of β-CD-1,3-bis(isocyanatomethyl)cyclohexane and β-CD-1,3-bis(isocyanatomethyl) benzene resins with several organic compounds [222, 361].

Later on, in 1999 and 2000, Li and Ma described the synthesis of nanoporous CD-HDI polymers for the removal of organic contaminants from water and these materials were superior to activated carbon or molecular sieves in the removal of trace levels of the contaminants [179, 180].

After these initial works, several research groups have used CD polyurethanes to remove various types of contaminants [177].

Yilzmaz et al. used β-CD-MDI, β-CD-HDI, and β-CD-MDI as sorbents for the removal of several dyes and aromatic amines from aqueous solutions. It was proposed that host–guest interactions, physical adsorption, and hydrogen bonding were responsible for the sorption mechanism [257, 258].

Mamba and Krause et al. prepared several polyurethanes by the reaction of CDs or CDs derivatives with TDI and HDI that were used for the sorption of water contaminants (organic matter, p-nitrophenol, pentachlorophenol, 2-methylisoborneol, etc.) [206, 248].

Wilson et al. and Mohamed et al. reported the synthesis of various CD-based polyurethanes (β-CD-PDI, β-CD-HDI, β-CD-MDI, β-CD-CDI, and β-CD-NDI) that were used as sorbents for naphthenic acids, methyl chloride, *p*-nitrophenol, and chlorinated aromatic compounds. In general, a direct correlation between the sorption capacity of these polyurethanes and their surface area was observed [226].

Besides the research groups mentioned above, several other authors used CD polyurethane-based materials for the removal of pollutants, including aromatic and perfluorinated compounds (PFCs), dyes, pesticides, etc. [177, 327]

CD polyesters constitute another important class of CD-polymers that can be obtained by reaction of CDs or CD derivatives with diacids, diacid chlorides, or dianhydrides.

One of the first examples of CD-derived polyester synthesis was reported by Zemel and Koch. These authors prepared crosslinking polymers by the reaction of CD with succinyl chloride, glutaryl chloride, and adipoyl chloride [367].

In 2005, Flores et al. [121] prepared β-CD-polyesters using different ratios of adipoyl chloride as a crosslinking agent (Scheme 14.3), which were used for the inclusion complexation of phenolic compounds.

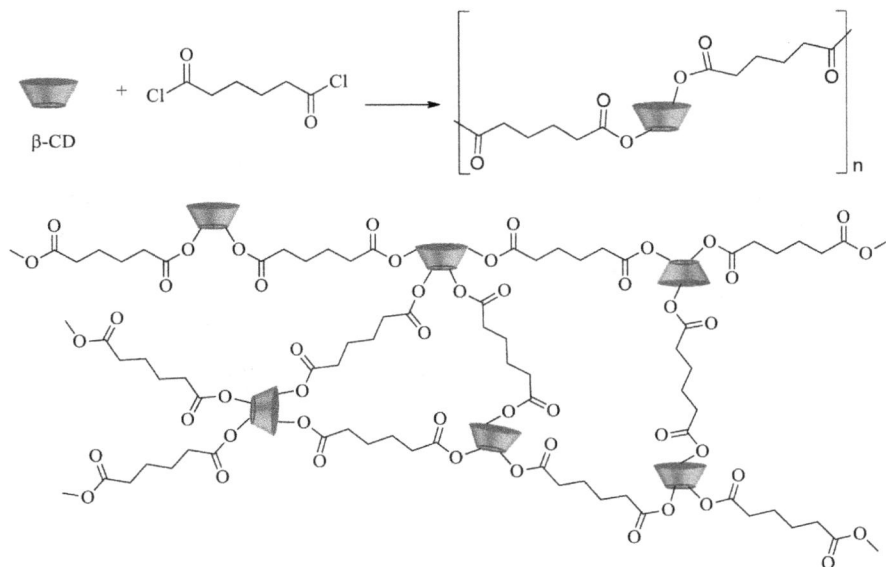

SCHEME 14.3 β-CD-polyester with adipoyl chloride as a crosslinking agent.

In 2002, Martel et al. [341] patented the synthesis of CD polyesters using several polycarboxylic acids as crosslinking agents [341].

In 2005, the same authors described the synthesis of water-soluble polymers and gels obtained by reaction of CDs with several polycarboxylic acids: citric acid (CA), 1,2,3,4-butanetetracarboxylic acid, and poly(acrylic acid). Using CA as comonomer and NaH_2PO_4 as the catalyst, soluble and insoluble polyesters were obtained. Soluble polymers were obtained using milder reaction conditions (temperatures below 150 °C) and insoluble polymers were prepared when more drastic reaction conditions were used (temperatures above 150 °C) [212].

Also in 2005, Girek et al. prepared β-CD crosslinked with succinic anhydride that was used for the removal of Cu(II), Zn(II), and Cd(II) from aqueous solutions using a flotation method. The polymers were synthesized using NaH to deprotonate β-CD, thus forming an oxoanion or an epoxide group that can react with the anhydride, forming an ester linkage [131].

In a previous study, the same authors prepared β-CD crosslinked with maleic anhydride (MA) and it was observed that higher molecular weights were obtained with the increase of the reaction temperature (130 °C, 4 h, the molar ratio of 1:7:7 for β-CD:NaH:MA) and that these polymers were water-insoluble. At temperatures below 100 °C, water-soluble polymers were obtained [132].

After these preliminary studies, several other CD-based polyesters have been reported using acid chlorides, acids, or anhydrides as crosslinking agents.

CD crosslinked with acid chlorides such as succinyl chloride, sebacoyl chloride, and terephthaloyl chloride was used in the removal of small organic compounds and dyes [126].

CD-polycarboxylic polymers using CA, 1,2,3,4-butanetetracarboxylic acid, and EDTA (Scheme 14.4) as crosslinkers have been used in the sorption of several classes of contaminants, namely, metals, small organic pollutants, dyes, and herbicides [148, 380].

Anhydrides such as pyromellitic anhydride or 1,8,4,5-naphthalic dianhydride were used in the synthesis of CD polyesters for the retention of heavy metals and dyes [35].

More recently, linkers like 4,4-difluorodiphenylsulfone, tetrafluoroterephthalonitrile (TFP), 4,4'- bipyridine, decafluorobiphenyl, bis (4-hydroxyphenyl) sulfone, and 4,4'-bis (chloromethyl)biphenyl have also been reported [185, 337, 347].

14.2.2.2 CD LINKAGE TO POLYMERIC MATRICES

CD can be attached to organic or inorganic polymers by covalent linkage or

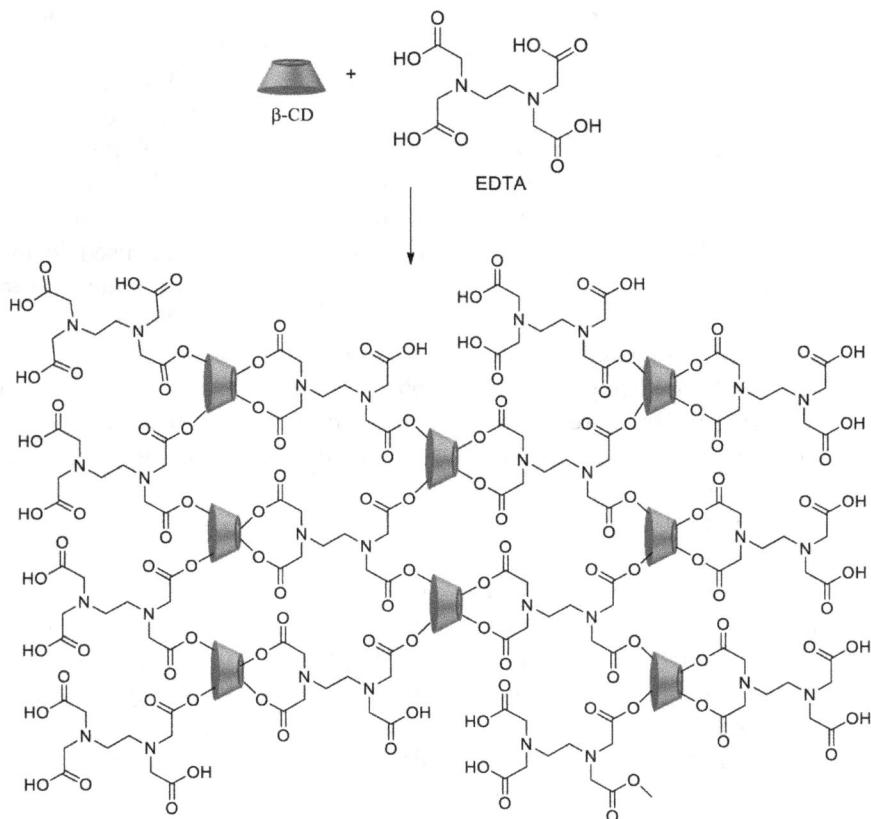

SCHEME 14.4   CD-polycarboxylic polymer using EDTA as crosslinker.

physical entrapment. The formation of covalent bonds between CD and the support is the common method used to prepare materials for remediation. Usually, water-insoluble supports are used, the most common being low-cost organic polymers or inorganic materials [95].

The sorption mechanisms of these materials are complex and are dependent not only on guest–host interactions of the sorbates with the CD but also on the chemical properties and network structure of the support. Parameters like porosity and surface area, as well as the type of functional group present in the matrix structure, have special relevance.

One of the strategies used to prepare organic polymers containing CD units as pendent groups consists of the preparation of a CD deriva-tive that can be copolymerized with another monomer or grafted onto a matrix. Artificial and natural polymers can be used as matrices. Although CDs have been immobilized onto several artificial polymers for diverse purposes, with particular emphasis on chromatography, it was only in the 1990s that several authors mentioned the use of these materials for remediation.

Sreenivasan prepared a β-CD modified with hydroxyethylmethacylate (HEMA), by a reaction of the hydroxyl groups of HEMA with toluene-2,4-diisocyanate, followed by a reaction of the other isocyanate group with one hydroxyl group of the β-CD. This monomer was then grafted onto a polyurethane using AIBN as an initiator. The resulting polymer was used for the sorption of several steroids [305].

β-CD, monotosylated in the 6th position of the glucose unit, has also been reacted with poly(allylamine), Scheme 14.5, originating linear polymers with pendent CD groups. The complexation of pyrene with the polymer, in aqueous solutions, was studied and it was observed that in polymers with low CD content (DS lower than 5%), a 1:1 complex was formed and with higher CD content, a 2:1 complex was observed [146].

Crini et al. [100] also used monotosylated β-CD to obtain macroporous beads of polyamine functionalized with CDs. Polymers with various degrees of substitution were prepared and used in the removal of organic pollutants from aqueous solutions. High sorption capacities were obtained for chlorophenols and it was pointed out that in the sorption mechanism, acid–base interaction between the polyamine network and the pollutants were involved, as well as the formation of inclusion complexes with the CD and guest–guest hydrophobic interactions.

The homopolymerization of a CD monomer and its copolymerization with a comonomer (HEMA) was described by Janus et al. [153], for the sorption of 4-*tert*-butyl benzoic acid. 2-hydroxy-3-methacryloyl-oxy-propyl-β-CD was homopolymerized or copolymerized in different proportions with HEMA, using potassium persulfate as initiator [153].

Crosslinking structures can also be prepared using artificial polymers. For example, PVA crosslinked

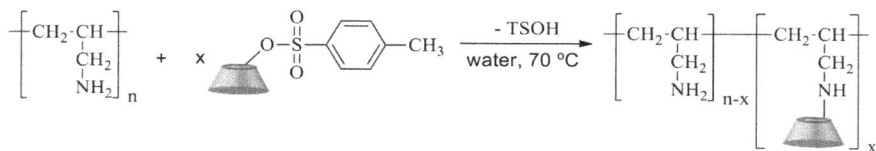

SCHEME 14.5   β-CD grafted onto poly(allylamine).

with β-CD, using CA as linker, was reported for the adsorption of aniline, 1-naphthylamine, and MB (Scheme 14.6) [381].

In addition to the above-mentioned polymers, CDs were also anchored to other artificial polymers, namely, polyether polyol, polyester, polyethylene terephthalate (PET), poly(styrene-alt-maleic anhydride), chloromethylated polystyrene-*co*-divinylbenzene, among others, that were used in the adsorption of several contaminants [107].

The immobilization of CD onto natural polymers has been equally described for remediation purposes. Usually, low-cost natural polymers like cellulose, starch, or CS are the choice. Several synthetic

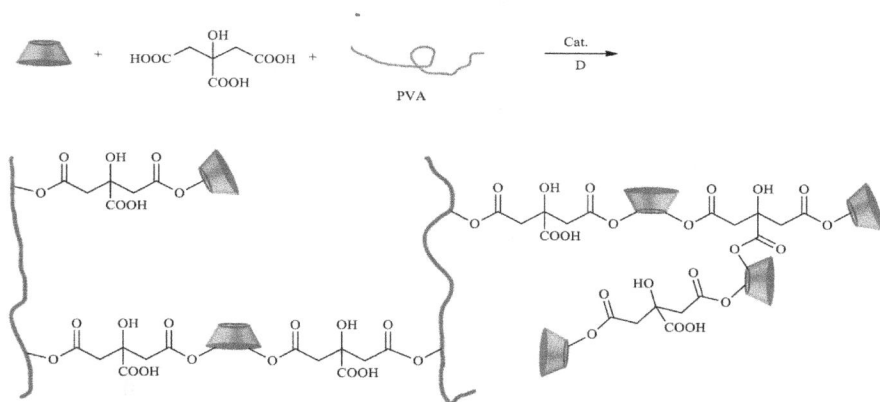

SCHEME 14.6   PVA-β-CD crosslinked with citric acid.

procedures can be used with this objective, including the synthesis of CD derivatives with appropriate reactive groups or the use of linkers. CD modified with various functional groups such as amino, tosyl, acid, and aldehyde have been prepared with this aim. Linkers like diacids, diisocyanates, and EPI are the most frequently used to attach CDs to natural polymers. Depending on the degree of CD substitution or the linker molar ratio, crosslinked structures can be obtained [357].

CS, a cationic biopolymer, can be easily obtained through the deacetylation of chitin, a natural polymer found in crustaceans. The presence of primary amine groups in this compound, which exhibits extensive reactivity, is a determining factor for its wide range of applications, including remediation. Therefore, it is not surprising that several authors, in the end of the last century and beginning of this one, reported the use of CS modified with CD for the removal of pollutants. For example, Tojima et al. prepared water-insoluble CS beads using hexamethylene diisocyanate as crosslinker (Scheme 14.7). To this polymer, α-CD was anchored through reaction with *O*-formylmethyl-α-CD, in the presence of sodium cyanoborohydride. The formation of inclusion complexes between this material and *p*-nitrophenol was studied [312].

The same polymer was also used in the removal of phatalate esters from an aqueous solution [70].

Crosslinked CS-CD-polymers were reported by Martel et al. β-CD functionalized with monochlorotriazinyl, with a degree of substitution of 2.8, was reacted with CS, via a substitution reaction. The polymers were used for the decontamination of aqueous solutions containing textile dyes [211].

Aoki et al. [21] prepared crosslinked CS modified with β-CD by the reaction of CS with succinic anhydride, followed by reaction with *mono*-6-amino-mono-6-deoxy-β-CD and a carbodiimide. The adsorption of two endocrine disruptors,

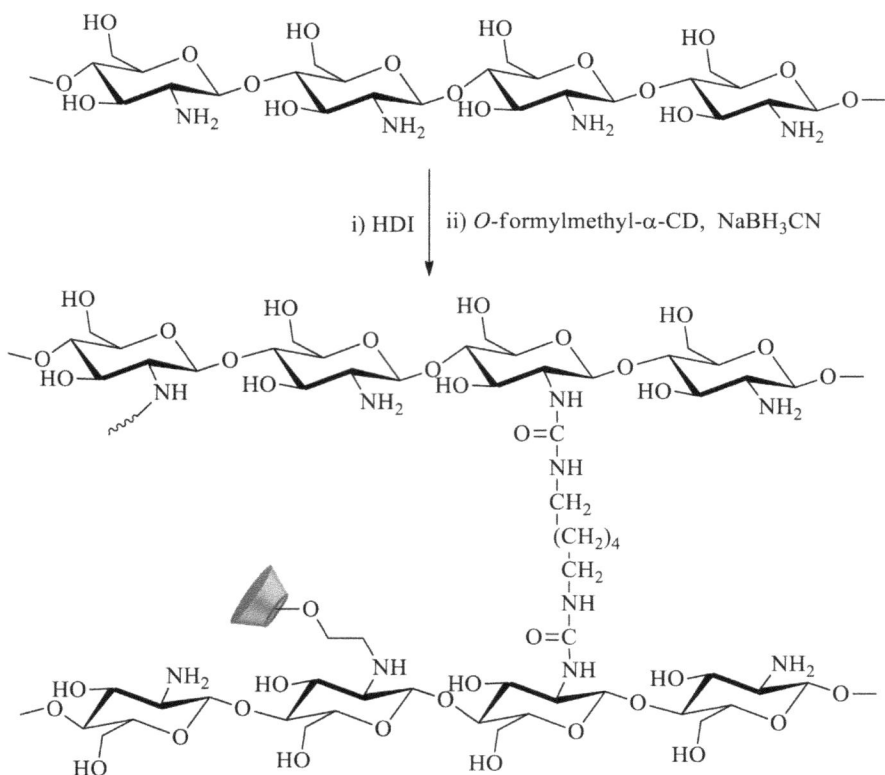

SCHEME 14.7 Chitosan crosslinked with HDI and modified with α-CD.

p-nonylphenol and BPA, was studied [21].

CS beads obtained by crosslinking with glutaraldehyde and treatment with 1,6-hexamethylene diisocyanate and β-CD were also reported and used to absorb hydroquinone [369].

More recently, several other materials obtained by the immobilization of CDs onto CS have been reported for the adsorption of metals, dyes, and small organic molecules [369].

Other crosslinked networks have been prepared, using CDs as monomers and natural or modified natural polymers as support. Carboxymethylcellulose (CMC), a modified cellulose polymer, in conjugation with EPI as the crosslinking agent, has been widely used as an absorbent in remediation [346].

These polymers are prepared through the reaction of CD and CMC with EPI in a basic medium. A crosslinked polymer is thus obtained. In 2002, Crini et al. used CMC-CD-EPI polymers for the sorption of β-naphthol with promising results. Similar absorption yields were obtained with CD-EPI and CMC-CD-EPI polymers (70%). The authors concluded that several types of interactions, namely, hydrogen bonds, interactions with the EPI network, physical adsorption, and inclusion complexes were responsible for the sorption mechanism [97].

Later on, CMC-CD-EPI polymers were used by the same group for the removal of cationic dyes (C: I. basic Blue 9, Malachite Green, C.I. Basic Blue 3, Basic Violet 3, and Basic Violet 10) with good results. The sorption capacity of these polymers is pH-dependent, which reveals that electrostatic interactions between the carboxylate groups of the polymer network and the cationic dyes play an important role in the sorption mechanism [88].

Although CS and CMC are among the most cited natural polymers and their derivatives, reference is also made in the literature to the use of other natural materials, such as starch, wood flour, sawdust, cotton, and cellulose, for the removal of various pollutants [283].

CDs have also been anchored to inorganic materials and used as adsorbents for environmental applications. Materials like silica and, more recently, magnetic particles have been reported in the literature.

Silica-based materials are particularly attractive because these materials have low-cost, high surface areas, and porosity and good physical, chemical, and mechanical properties.

CD can be grafted or coated onto silica and. CDs are located on the surface of the material which increases the accessibility of adsorbates. However, this method has the disadvantage that the amount of bound CD is low and not uniformly distributed. Porous silica can instead be used, which greatly increases the amount of CD that can be attached.

Mesoporous silica, which is characterized by having ordered pore structures, can be obtained in multiple forms, spheres, rods, discs, powders, etc. These materials can be prepared by hydrolysis and condensation of alkoxysiloxanes, using a catalyst (normally a base or an acid) and a template, usually a surfactant [243].

The characteristics of the porous structure and the macroscopic morphology are determined by the silica source and the synthetic conditions (type of surfactant, ionic strength, pH, temperature, composition of the reaction mixture, and reaction time). These siloxane-based polymeric networks have a large number of silanol groups that be used to functionalize these materials [130].

Silica gel materials modified with CDs have been prepared and these organic–inorganic hybrid materials have been used for the separation of organic compounds, including chiral discrimination. Some of these materials are nowadays used in HPLC and GC columns as stationary phases [95].

In the context of remediation, mesoporous silica-based materials are the most often used. The first report of the incorporation of CDs in mesoporous silica was made by Mercier et al. The mesoporous materials were prepared in a one-step procedure: monochlorotriazinyl-β-CD

was reacted with 3-aminopropyl-triethoxysilane and then TEOS, docecylamine, and 1,3,5-trimethyl-benzene (TMB, as pore-expanding agent) were added (Scheme 14.8). The resulting mesoporous material (CD-HMS) was used for the absorption of small organic compounds, pesticides, and dyes from aqueous solutions [150].

Silicon-based polymers, prepared by the sol–gel method, were described by Lambert et al. First, β-CD monomers were prepared by reaction with allyl bromide (substitution occur on positions 2 and 6 of each glucose units), followed by hydrosilylation of the allyl groups with trialkoxysilanes or methylation of the hydroxyl group of the 3-position and then hydrosilylation. These monomers and calixarene monomers or mixtures of all of these were polymerized by the sol–gel method. The materials containing CD were used for the removal of 4-nitro-phenol from water. The obtained polymers had a random network and low surface areas but were better absorbents than charcoal [173].

Using the same monomers, the authors also prepared mesoporous materials by cocondensation with tetraethoxysilane (TEOS) in the presence of a surfactant. These materials, with high surface areas and mesoporous with diameters ca. 3 nm, presented better sorption results [192].

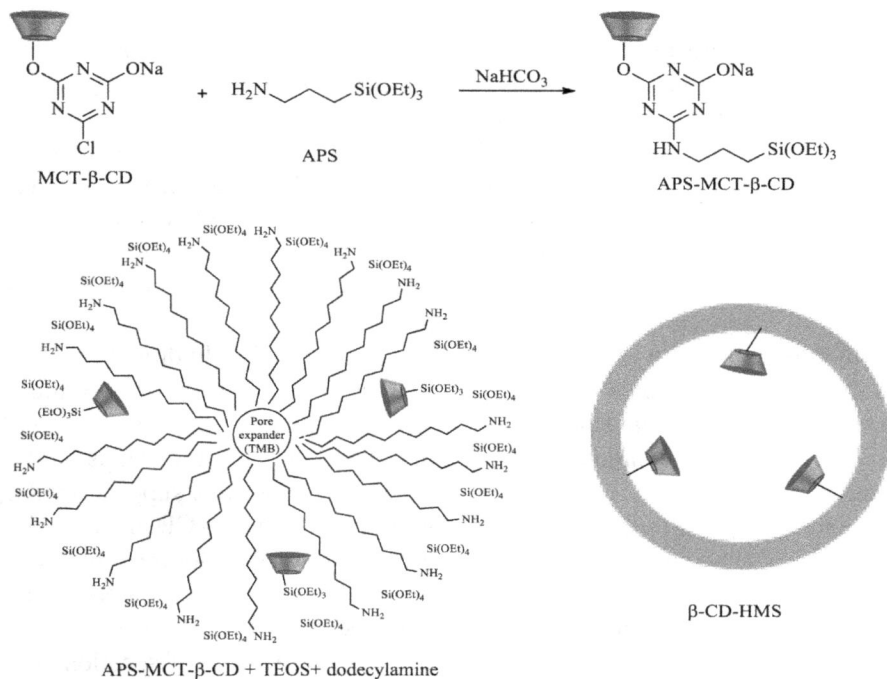

SCHEME 14.8   Schematic synthesis of β-CD-HMS.

Liu et al. used β-CD-mesoporous silica for the adsorption of humic acids. A β-CD monomer was prepared through reaction with 3-isocyanatopropyltriethoxysilane, followed by treatment with TEOS in basic medium, using cetyltrimethylammonium bromide (CTAB) as a template. A hybrid material with 4% of CD, with a high BET surface area, was efficient on the removal of two different types of humic acids from water [193].

It was observed that these CD-mesoporous materials were superior to unmodified mesosilicas with respect to their capacity to absorb organic compounds. However, the absorption capacity was dependent on the CD loading (usually 2%–8%) and CD contents higher than 8% seem to decreased the absorption capacity of the materials. This can be due to the destruction of the mesoporous structures, with high CD content, leading to the formation of amorphous materials [128].

More recently, other mesoporous silica modified with CD have also been prepared by reacting tetraethyl orthosilicate in the presence of a surfactant (CTAB is the most used), using a basic medium (usually NH_3). The products are then treated with APTS and then the amine groups of the modified silica can be reacted with CD derivatives (monotosyl-CD) or linkers (1,1'-carbonyldiimidazole), Scheme 14.9. These hybrid materials were used for the removal of dyes [73].

APTS can be substituted for 3-chloropropyltriethoxysilane and, in this case, a CD modified with amino groups (e.g., mono-6-ethylenediamine-β-CD or mono-6-ethylamine-β-CD) is used [295].

The use of highly porous silica-based aerogels containing β-CD has also been reported by Durães et al. for the sorption of phenol-derived compounds [215].

Commercial or natural silica materials have also been used for pollutant removal. *Silochrome C-120 silica* has been modified with aminopropyl groups and then mono-6-tosyl-CD or bromoacetyl or thiosemicarbazidoacetyl derivatives of tosyl-CD were linked to this silica. This material with nanosized pores was applied in the removal of toxic ions [33].

Attapulgite, a hydrated octahedral layered magnesium aluminum silicate mineral with siloxane groups in the bulk and silanol groups on the surface, was used as support for the immobilization of β-CD and used for the absorption of chlorophenols from water. A mesoporous material was obtained with a high surface area, and the mechanism of sorption was

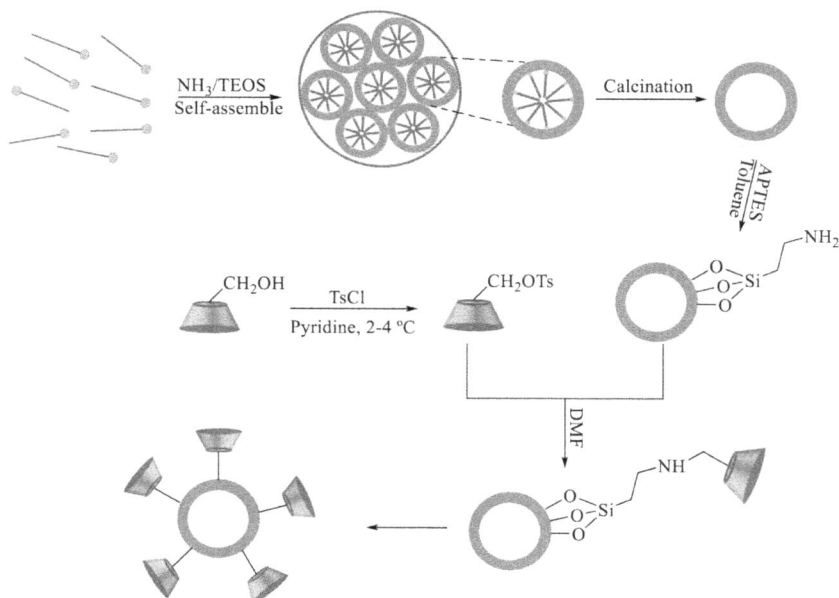

SCHEME 14.9   Schematic synthesis of mesoporous silica modified with CD.

pointed to be mainly by hydrogen bonding–electrostatic interaction and hydrophobic interaction [260].

Crosslinking agents such as CA, HDI, and EPI have also been reported to link CD to silica. These materials, with a high content of CD, are easy to prepare, low-cost, and have interesting adsorption properties [61].

In recent years, MNPs, especially those containing Fe_3O_4, have been employed in several fields, including biotechnology, biomedical, material science, engineering, and environmental areas. These materials, with high surface area, small diffusion resistance, and fast response under an applied external magnetic field, can be easily separated and recovered from aqueous solutions, which makes them very attractive for application in environmental remediation. The modification of MNPs with CD allows the formation of inclusion complexes with the adsorbates and thus facilitates the removal of pollutants [261].

There are several methods for the preparation of Fe_3O_4, although the most used is the coprecipitation method. Typically, $FeCl_2 \cdot 4H_2O$ and $FeCl_3 \cdot 6H_2O$ are dissolved in water, the solution is heated and stirred, in an inert atmosphere, and then NH_4OH is added. The heating and stirring are maintained during a period of time to complete the growth of the NP crystals.

In order to prepare the modified CD-MNPs, two principal approaches have been used: a two-step approach, in which the MNPs are first prepared and then the CDs are attached to them by using a CD derivative or a linker; or a one-step coprecipitation method where the CD derivative is mixed with the iron compounds, and then the NH_4OH is added.

Badruddoza et al. [29] reported the synthesis of carboxymethyl-β-CD-MNPs employing the one-step and two-step coprecipitation method (Scheme 14.10) described above for the removal of MB. Carbodiimide was used to activate the carboxyl group of the β-CD that was then reacted with the hydroxyl groups of the MNPs.

The attachment of carboxymethyl-β-CD onto the magnetic surface was confirmed by FTIR, TGA, and zeta potential analysis. The carboxymethyl-β-CD-MNPs prepared by the one-step method exhibited higher adsorption capacity than those prepared by the two-step method, which was explained by the higher content of CD present in the former one, thus having more active sites for adsorption [29].

The same authors also reported the use of carboxymethyl-β-CD-MNPs prepared by the two-step method or the one-step method for the removal of Cu ions or Pb, Cd, and Ni, respectively. The authors concluded that the adsorption process involved the formation of surface complexes between the copper ions and oxygen atoms of the CD [31].

Two-step method:
(*carbodiimide method*)

CM-β-CD

One-step method:
(*precipitation method*)

$Fe^{2+} + 2Fe^{3+} + CM-\beta-CD \xrightarrow[90\ °C,\ 1h]{NH_4OH,\ N_2}$

CM-β-CD-MNPs

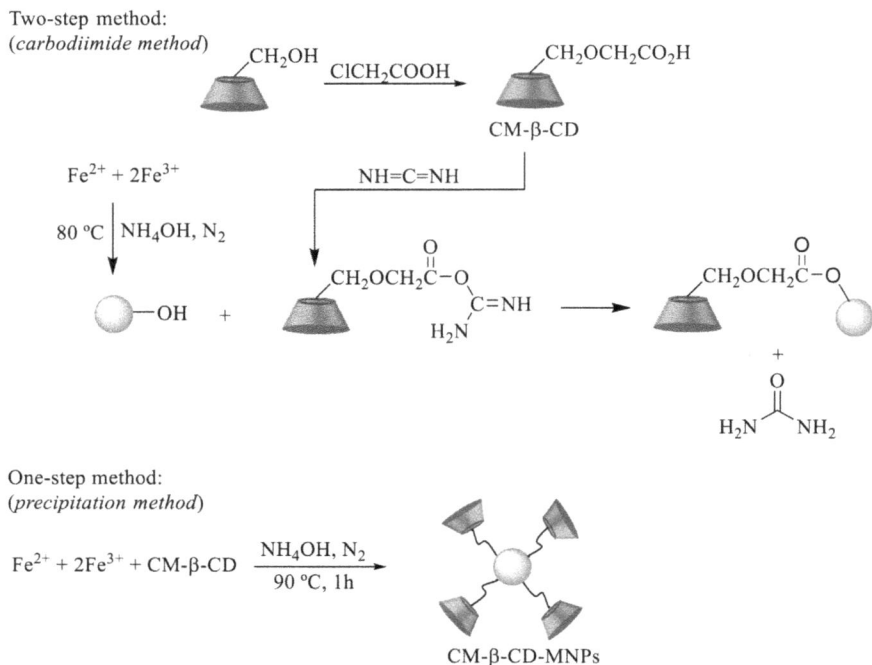

SCHEME 14.10   Synthesis of carboxymethyl-β-CD-MNPs using one-step or two-step method.

In another report, carboxymethyl-β-CD-MNPs were prepared by a different method. MNPs were synthesized by thermal decomposition of FeOOH in 1-octadecene, using oleic acid as an end-capping agent. These MNPs were then reacted with carboxymethyl-β-CD and used in the simultaneous removal of organic pollutants and arsenic from aqueous solutions. Organic pollutants can interact with the CD cavity to form host–guest inclusion complexes and As can be adsorbed on the surface of the iron, and thus be simultaneously removed [66].

The use β-CD-CS-MNPs, for the absorption of methyl blue and hydroquinol, has been reported by Fan et al. β-CD-CS was prepared by reaction of maleoyl-β-CD with CS via activation with EDC/DMAP. Then, β-CD-CS was linked to the MNPs, obtained by the coprecipitation method, using glutaraldehyde (Scheme 14.11). A spherical polymer with a high surface area was obtained [116].

The one-step coprecipitation method was used by Zhang et al. where nonderivatized CD was used to prepare β-CD-MNPs, which simplifies the synthetic procedure and

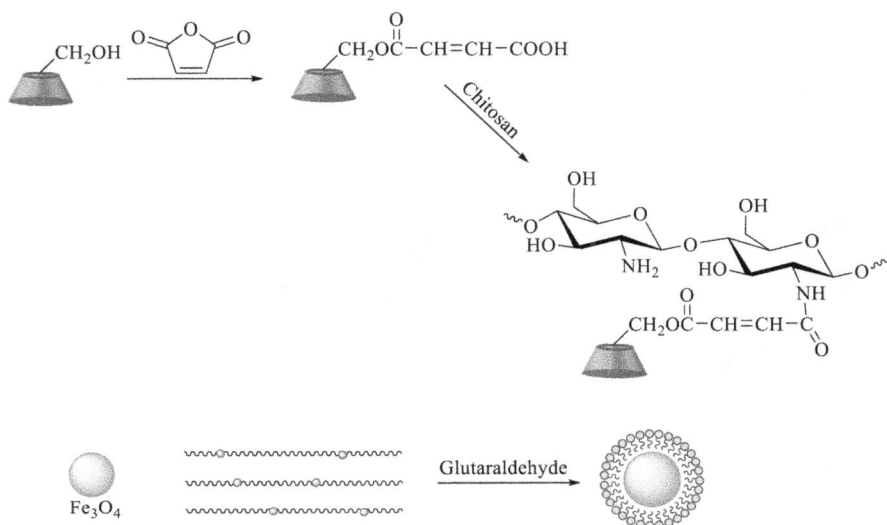

SCHEME 14.11   Synthesis of β-CD-chitosan-MNPs.

reduces costs. β-CD is linked to MNPs by electrostatic interactions and this material was used to remove Co (II) and 1-naphthol [376].

In another study, β-CD functionalized with [3-(2,3-epoxypropoxy) propyl]trimethoxysilane was linked to MNPs using tetraethyl orthosilicate (Scheme 14.12). These modified MNPs were used for the adsorption of azo dyes [23].

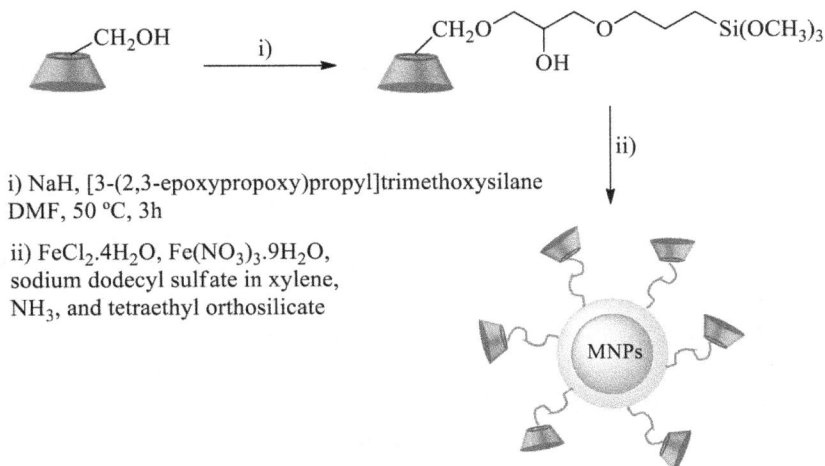

i) NaH, [3-(2,3-epoxypropoxy)propyl]trimethoxysilane
DMF, 50 °C, 3h

ii) FeCl₂.4H₂O, Fe(NO₃)₃.9H₂O,
sodium dodecyl sulfate in xylene,
NH₃, and tetraethyl orthosilicate

SCHEME 14.12   Synthesis of β-CD-immobilized MNPs.

MNPs modified with hexamethylenediamine and reacted with carboxymethyl-β-CD via EDC/NHS activation and have also been used for PCBs removal in soil [398].

In addition to the above-mentioned materials, several other CD derivatives (triazinyl-β-CD, hydroxypropyl-β-CD, succinyl-β-CD) have been linked to MNPs and employed in the removal of heavy metals and dyes from aqueous solution [4].

Crosslinked CD-based polymers were also immobilized on MNPs using anhydrides, EPI, diphenylcarbonate, or diisocyanate as crosslinking agents, for the removal of heavy metals, dyes, pesticides, PFCs, and BPA [56].

Zero-valent iron powders (Fe^0) were also used to prepare magnetic particles (NZVI). Mixing CS-β-CD gel (obtained by reaction of carboxymethyl-β-CD sodium salt with CS, using a carbodiimide as an activator) with NZVI that was syringe dropped into a NaOH solution, under stirring, allowed the obtention of the composite. The NZVI particles were entrapped into CS-β-CD pores and these materials were efficient in the removal of copper and arsenic ions [56].

Superparamagnetic iron-oxide nanoparticles (SPIONs) modified with β-CD were obtained *in situ* by reaction of $FeCl_3$ and $FeCl_2$ (2:1 stoichiometric ratio) in an aqueous ammonia solution followed by treatment with a surfactant (tetramethylammonium hydroxide) and β-CD. This SPION-β-CD was used to remove BPA and Malachite Green dye and oil spill remediation.

Nanofibers, like other nanomaterials, have extremely small dimensions, which gives them unique physical and chemical properties and makes them candidates for applications in areas such as electronics, catalysis, energy storage, biomedical applications, water treatment, and environmental remediation. Nanofibers are one-dimensional (1D) materials with high surface area, porosity, and tensile strength, and can be prepared using several techniques, electrospinning being the most used.

The electrospinning apparatus usually consists of a syringe with a nozzle, an electric field source, and a pump. This technique is based on electrostatic interactions, more specifically electrostatic repulsion forces that are used for the nanofiber preparation. A peristaltic pump is used to push the solution from the needle, which is subject to a high voltage power supply. The solution droplet ejected at the tip of the needle will be collected in a counter electrode, the collector. Evaporation of the solvent leads to nanofiber formation. The physical properties of the nanofibers are dependent on the solution properties and technical and environmental variables. Several materials have been used to produce nanofibers, including polymers, ceramics, small molecules, and their composites [188].

Several papers, in the last decade, refer to the use of nanofibers containing CD for pollutant removal, with good adsorption capacities. Teng et al. reported the preparation of silane-based nanofiber membranes to be used as an adsorbent for indigo carmine (q_{max} of 495 mg g^{-1}). Monochlorotriazinyl-β-CD reacted with 3-aminopropyltriethoxysilane in the presence of CTAB as a template, then tetraethyl orthosilicate was added, followed by HCl, as a catalyst. The reaction mixture was stirred for 2 h, PVA was added and the obtained solution was subjected to electrospinning (Scheme 14.13). Mesoporous nanofiber membranes were obtained

(β-CD content of 0.19 mmol g^{-1}), after drying and extraction of the template with acetone. Electrostatic interactions between the dye and the protonated amino groups of the composite, hydrogen bonding, π–π interactions, and formation of inclusion complexes are responsible for the high absorption capacity [310].

PVA was used in the preparation of sericin-β-CD-PVA nanofibers for the adsorption of MB from an aqueous solution (q_{max}=261.10 mg g^{-1}). Sericin, β-CD, PVA, and CA, as the crosslinking agent, were mixed and stirred for 2 h at 85 °C. Upon cooling, the mixture was subject to electrospinning, followed by thermal crosslinking (Scheme 14.14). The nanofibers could be easily separated and are recyclable [389].

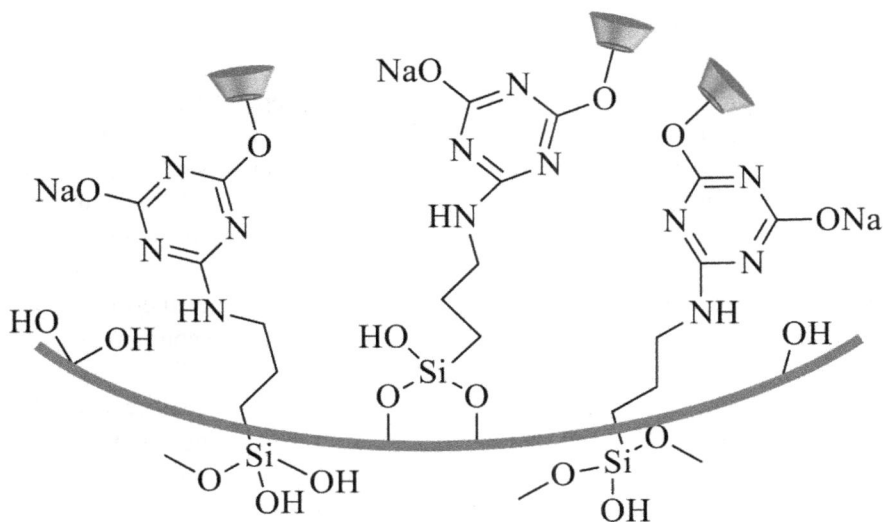

SCHEME 14.13 Schematic representation of the pore of silane-based nanofibers.

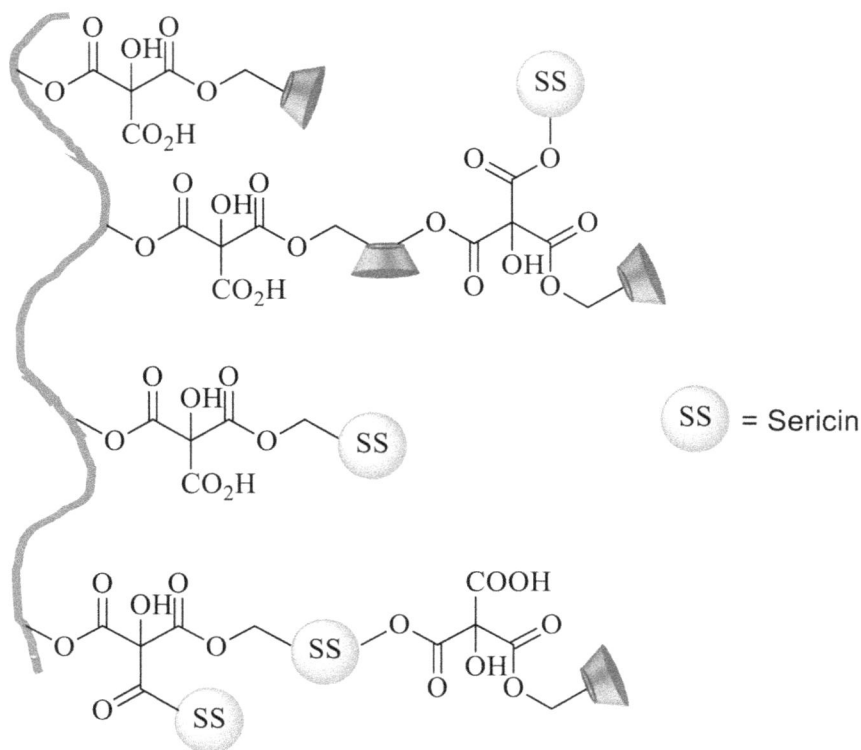

SCHEME 14.14   Crosslinked sericin-β-CD-PVA nanofibers.

Poly(ethylene terephthalate) nanofibers modified with α-CD, β-CD, and γ-CD have equally been described. Mixtures of the polymer with the CD in TFA/dichloromethane (1:1) were subjected to electrospinning and the corresponding nanofibers were prepared. No CD aggregates or inclusion complexes between CD and PET were observed in the nanofibers. The entrapment of aniline from the vapor phase was studied and better results were obtained with γ-CD [162].

The preparation of β-CD-poly(acrylic acid) nanofibers by electrospinning was reported for the efficient removal of MB from water solutions (q_{max}=826.45 mg g^{-1}).

Nanofibers were obtained after electrospinning and thermal crosslinking of a solution containing β-CD, PAA, and CA in DMF. The materials were characterized before and after thermal crosslinking and it was observed that the surface area and the pore diameter decreased after the thermal crosslinking, while resistance and tensile strength increased. These nanofibers were also capable of separating the anionic dye MO from the cationic dye MB due to the presence of carboxylic groups [391].

Several other CD nanofibers were posteriorly prepared for MB removal using the electrospinning method but lower adsorption capacities were obtained [63].

Solutions with different ratios of α-CD, β-CD, and γ-CD and nylon 6,6 in formic acid were used for the synthesis of nanofibers using the electrospinning technique. Materials with CDs content of 17%–19% and 33%–35% were obtained when 25% or 50% of the CDs were used with respect to nylon 6,6. Better results were obtained with 50% β-CD-nylon 6,6 for toluene vapor entrapment and it was observed that the efficiency was dependent on the type of CD

The electrospinning of a solution of a β-CD polymer (10, 15, and 50 wt% with respect to the polyacrylonitrile) and polyacrylonitrile in DMF has also been reported for the removal of reactive dye. The β-CD polymer was prepared by a crosslinking CD with CA and PEG as the modifier. Nanofibers with an increased diameter and higher content of β-CD were obtained [43].

Nanofibers were also obtained from a mixture of polyethersulfone and β-CD in DMF by electrospinning. Fibers with 518 ± 133 nm diameter were prepared and used for estradiol and chlorpyrifos removal. Leaching of CD for the aqueous solution was observed [286].

The preparation of regenerated cellulose nanofibers by two different methods, the physical mixing method and the grafting method, was reported by Guohao et al. The first method consists in the electrospinning of mixtures of cellulose acetate with α-CD, β-CD, and γ-CD in DMF/acetone, followed by drying and immersion in a NaOH solution. In the second method, cellulose acetate was subjected to electrospinning, treated with NaOH, and CDs were grafted onto the fibers using CA as a crosslinking agent and sodium hypophosphite as the catalyst. The nanofibers containing γ-CD and prepared by the grafting method proved to be more efficient in the removal of toluene from aqueous solutions [365].

Carbon nanofibers (CNFs) modified with β-CD were prepared and used for the adsorption of tetracycline (TC) and MB.

First, pyromellitic dianhydride and 4,4'-oxydianiline were polymerized in DMF, followed by electrospinning and thermal treatment, to obtain the polyimide nanofibers. Then, β-CD was mixed with the nanofibers in water, transferred to an autoclave and heated (180 °C, 12 h). This material was then carbonized by heat treatment, under nitrogen. The composite presented a mesoporous structure and high surface area, which contributed to the high adsorption capacities obtained for TC and MB (q_{max}=543.48 and 746.27 mg g^{-1}, respectively) [184].

The use of nanofibrous membranes modified with CDs for remediation has also been reported. Polyvinylpyrrolidone and β-CD were dissolved in DMF and then glutaraldehyde (to crosslink CD) and ammonium persulfate (to crosslink PVP) were added. After electrospinning, the obtained membrane was cured. When compared to PVP nanofibrous membranes, those modified with crosslinked CD are thickener and less porous but presents higher adsorption capacity for methyl orange (MO) [348].

Nanofibrous membranes of poly (L-latic acid) (PLA) modified with polydopamine (PDA) and β-CD were equally reported and used for the removal of small organic molecules and dyes. PLA nanofibrous membranes were prepared by electrospinning of a polymer solution in trifluoroethanol. The obtained membranes were first coated with PDA and then reacted with mono-6-deoxy-6-ethylenediamine-β-CD. It was observed that PDA particles form polydisperse aggregates that uniformly coat the PLA nanofibers. PDA and β-CD impart added hydrophilicity to the membrane due to the presence of polar groups. The membranes exhibit high hydrophilicity and oleophobicity when immersed, resulting in a material suitable for separating oil/water mixtures. Good separation efficiencies were obtained for toluene/water mixtures and selective adsorption for cationic dyes in mixtures of cationic/anionic dyes [160].

As already mentioned, activated carbon is the most commonly used commercial adsorbent due to its good adsorption capacity for organic contaminants. CNTs and graphene oxide (GO), other carbon-based materials, have also been explored as adsorbents for the removal of pollutants.

CNTs, discovered in 1991 by Iijima, have a hollow and layered structure and large specific surface area, which make them attractive for remediation. CNT can be classified into two types: single-walled (SWCNT) or multi-walled (MWCNT). CNT can be functionalized, mainly by oxidation, thus introducing hydroxyl and carbonyl groups to the sidewalls of the CNT. These groups can allow the attachment of other compounds to the CNT, including CD. The combination of these two materials provides composites that have high stability, ease of recycling and recovery, properties that are conferred by CNT, coupled with a good ability to adsorb various organic compounds (due to the presence of CDs), make these composites very useful for the removal of pollutants [152].

MWCNT oxidized with a mixture of sulfuric and nitric acids

in order to introduce carboxylic groups were modified with β-CD by reaction with hexamethylene diisocyanate (Scheme 14.15) and used by Salipira et al. as adsorbents for the removal of *p*-nitrophenol and trichloroethylene.

High removal efficiency, 99%, was obtained for a 10 mg L^{-1} solution of *p*-nitrophenol, which is better than those obtained with activated carbon and CD-polymers. The porous surface of the material and the formation of π–π interactions between the *p*-nitrophenol and the CNT may be responsible for the removal efficiency. These MWCNT-β-CD polymers, with high thermal stability, were also able to adsorb trichloroethylene from water in very low concentrations (ppm) and were also used for the removal of cobalt and lead from water solutions [282].

In another report, the same group prepared phosphorylated MWCNT modified with β-CD for the absorption of cobalt and 4-chlorophenol. Oxidized MWCNTs were first reacted with oxalyl chloride and then with 4-aminophenyl methylphosphonate and 6-amino-hexanol (to provide possible sites for the polymerization reaction with CD). This material was then polymerized with β-CD using HDI as a linker (70 °C, 24 h). The polymer, with a spongy network, seems to be more adequate to remove low concentrations of pollutants [207].

MWCNTs grafted with β-CD were also prepared by treatment of MWCNT under N$_2$ plasma conditions (N$_2$ plasma of 10 Pa, power of 70 W, voltage of 650 V, and electrical current of 60 mA). Then the grafting reactor was heated (80 °C), β-CD was added, and the reaction was continued for 24 h, under stirring. MWCNT-g-β-CD, a more environmentally friendly material, was obtained and used in the removal of PCBs from aqueous solutions [291].

β-CD was also immobilized on MWCNTs modified with glycine. The hydroxyl groups of oxidized MWCNTs were reacted with glycine and then β-CD was linked, at room temperature, to the amino groups of glycine using EPI as the crosslinker. The adsorption of several dyes from water and wastewater was tested with good results (maximum adsorption

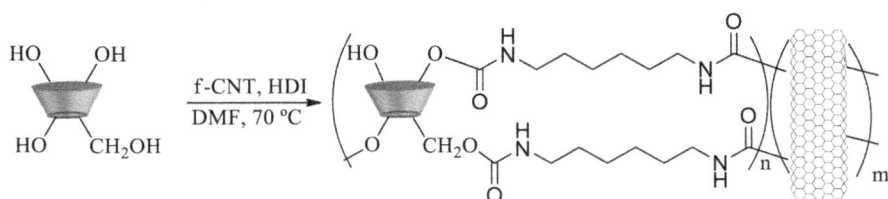

SCHEME 14.15 Synthesis of MWCNT-β-CD, using HDI as the linker.

capacity of 500 mg g^{-1} was obtained for Disperse Red 1) [227].

The use of MWCNT modified with magnetic particles and β-CD in the removal of several contaminants were also described. Oxidized MWCNT, dissolved in ethylene glycol, were mixed with FeCl$_3$ in the presence of trisodium citrate/sodium acetate and PEG 20000. The reaction mixture was transferred to an autoclave and maintained at 200 °C, for 8 h. MWCNTs modified with magnetic particles were then reacted with β-CD, at 85 °C, using HDI as crosslinker (Scheme 14.16). This composite was used in the adsorption of *p*-nitrophenol with the advantage of easy separation and recovery of the adsorbent by magnetic separation [195].

Oxidized MWCNTs were reduced with hydrazine in the presence of β-CD and then treated with FeCl$_3$,

FeSO$_4$, and ammonium hydroxide in an inert atmosphere. The obtained magnetic composite containing 17 wt% of CD and 23.5 wt% of iron oxide particles was used in the adsorption of MB (q_{max} = 196.5 mg g^{-1}) [76].

Zero-valent iron NPs were also anchored to nitrogen-doped MWCNTs and then reacted with β-CD using TDI as the crosslinker. This nanocomposite was used in the removal of BPA from aqueous solutions [238].

Nanoscale bimetallic particles of nickel on iron were also supported on CNTs by mixing ferric nitrate, nickel nitrate, and oxidized CNTs, followed by heating under argon flux at 450 °C and reduction of the metal oxides with 10% H$_2$ Ar^{-1}. The modified CNTs were then reacted with β-CD and HDI (70 °C, inert atmosphere, 24 h). The NPs, with an average diameter of 16 nm, presenting

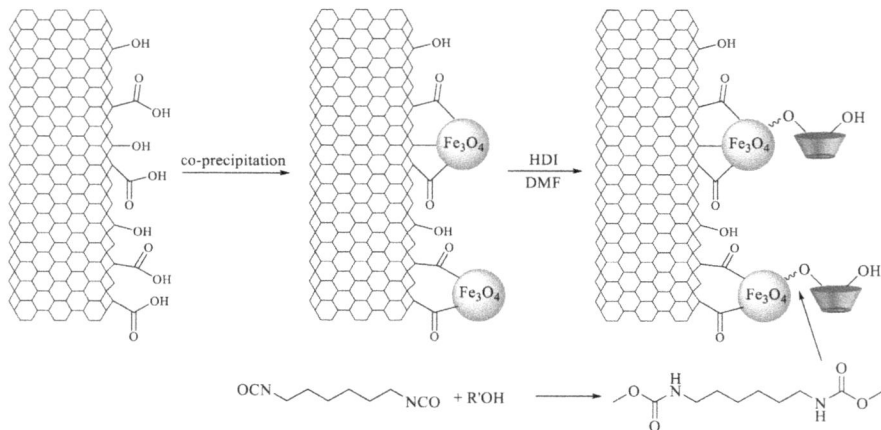

SCHEME 14.16   Schematic representation of the synthetic route for MWCNTs-Fe$_3$O$_4$-CD.

meso- and micro-pores were used for the degradation and removal of trichloroethylene in water [168].

Multiwalled inorganic nanotubes, namely halloysite nanotubes (HNTs), modified with CDs were also used in the removal of contaminants from water. First, HNTs were reacted with 3-chloropropyl-trimethoxysilane (70 °C) and then with potassium xanthogenate (12 h, 80 °C). Subsequent reaction with glycidyl methacrylate (GMA) and 2,2-azobisisobutyronitrile as initiator allowed the synthesis of HNTs-poly(GMA) via RAFT polymerization. Finally, HNTs-poly(GMA) was treated with mono-6-deoxy-6-hexanediamine-β-CD, and the HNTs modified with β-CD were obtained (Scheme 14.17).

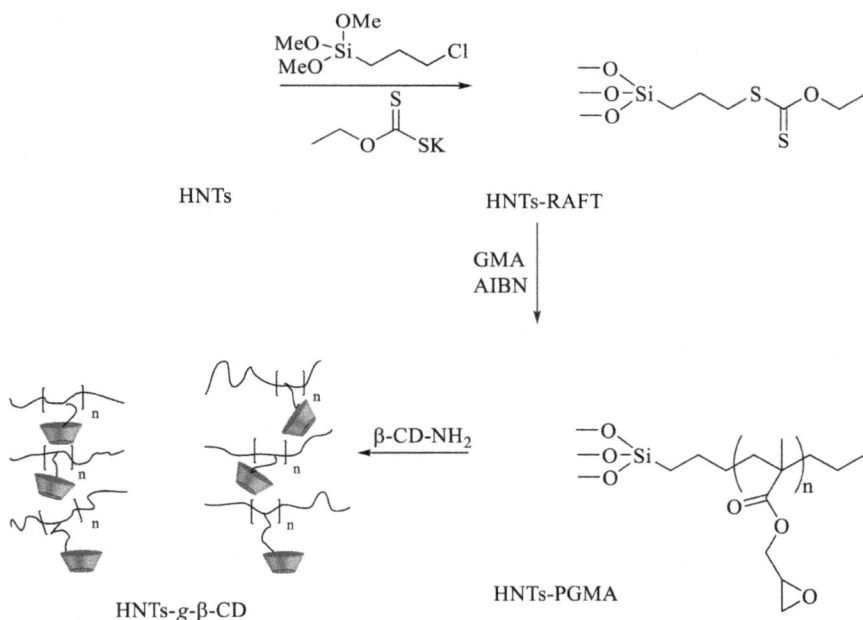

SCHEME 14.17 Schematic route for HNTs-β-CD composites.

The amount of β-CD grafted onto the HNTs-poly(GMA) was 80.9 mg g^{-1} and HNTs were covered with an organic layer 25 nm thick. The composite was used in the removal of MB from aqueous solutions [58].

β-CD was also grafted on HNTs using a different strategy. β-CD was first modified with (3-glycidyloxypropyl)trimethoxysilane in a basic medium. After acidification, the β-CD derivative was refluxed in water with the HNTs to obtained β-CD-HNTs hybrid material. Different contents of β-CD-HNTs were incorporated on polyvinylidene fluoride membranes, using polyvinylpyrrolidone as pore former and dimethylacetamide as the solvent. The membranes presented a uniform distribution of the β-CD-HNTs, an asymmetric porous structure and higher porosity were obtained for membranes with higher contents of β-CD-HNTs. A better removal ratio of Cu(II) was obtained for membranes with higher contents of β-CD-HNTs. The Cu(II) removal by the membranes was attributed to chelation reactions and electrostatic interactions between the β-CD-HNTs and the contaminant [201].

GO is synthesized by chemical oxidation of graphite with strong oxidants, usually, potassium permanganate in sulfuric acid, followed by exfoliation in a particular solvent using sonication. GO, like graphene, is a single-atomic layered material, although much cheaper, with oxygen-containing functional groups on the carbon surface (epoxide, hydroxyl, and carboxyl). The exact structure and composition of GO are not completely known, since it is dependent on the synthetic conditions. Nonetheless, this material, due to its layered structure, negatively charged surface, and large surface area, is quite suitable to be used as an adsorbent. The modification of GO with other compounds, namely CDs, can improve the properties of this material [171].

Fan et al. [115] prepared a composite of magnetic CD-CS-GO

to be used in the removal of MB from water. Maleoy-β-CD was first reacted with CS and subsequently with magnetic particles, at pH 8.0–9.0, and glutaraldehyde (55 °C for 1.5 h). The carboxylic groups of GO were then activated with EDC/NHS prior to the addition of magnetic β-CD-CS and more glutaraldehyde and the reaction proceeded for 2 h, at 65 °C (Scheme 14.18). This material exhibits an adsorption capacity of 84.32 mg g^{-1}. Better results were obtained at higher pH that was explained by the occurrence of electrostatic interactions between the surface of the composite, which is negatively charged, and the cationic dye.

This adsorbent was also used by the same research group, for the removal of hydroquinone and Cr (VI) [183].

In another report, the same research group described a magnetic β-CD-GO composite that was prepared by reacting magnetic particles, NH_2-β-CD and glutaraldehyde (at 50 °C for 0.5 h), and then GO dispersion was added. The obtained product, used in MB removal, could be easily recycled and reused [181].

Another magnetic β-CD-GO material was synthesized by Liu et al. using a different methodology. Magnetic GO was prepared using the chemical precipitation method, GO was mixed with $FeCl_2$ and $FeCl_3$ in water and NH_4OH solution was

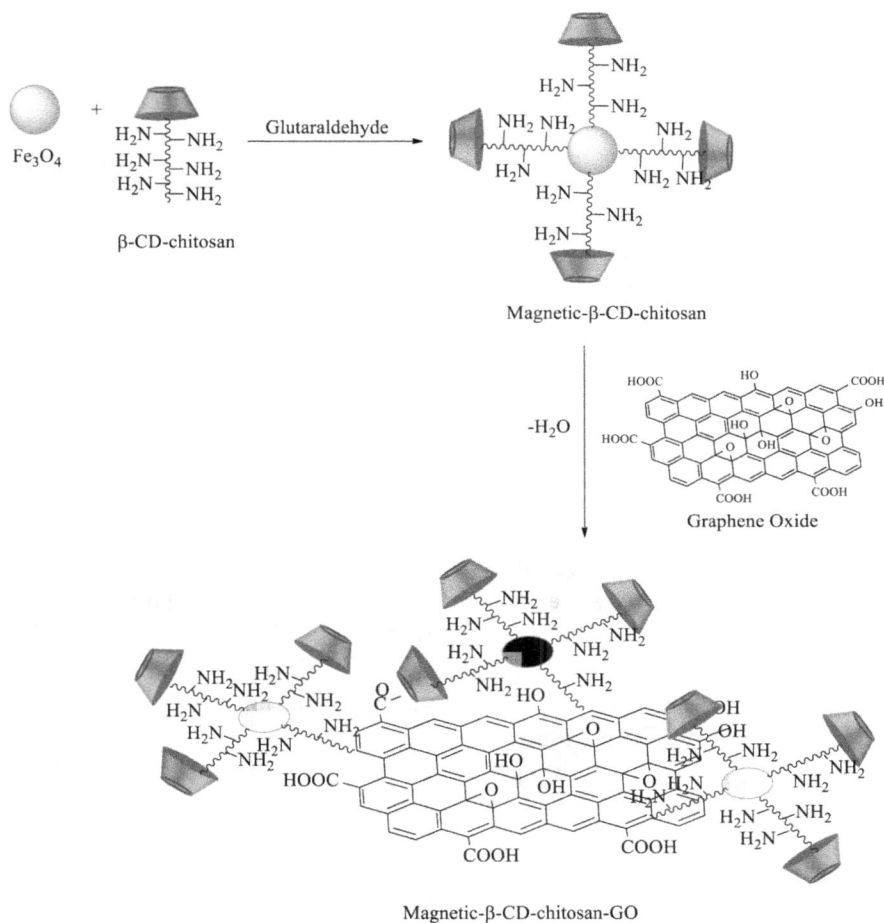

SCHEME 14.18 Schematic synthesis of magnetic-β-CD-chitosan-GO.

added dropwise. β-CD was then immobilized onto the magnetic GO by mixing both in water and heating at 60 °C, under argon. The removal of Acid Fuchsin and Rhodamine 6G from aqueous solutions was investigated [196].

In another paper, GO was first reacted with β-CD in a basic medium (pH 9–10) and then β-CD-GO was further polymerized with pyrrole in the presence of $FeCl_3$. The obtained material exhibited a maximum adsorption capacity of 666.67 mg g^{-1} (45 °C) for Cr(VI). This high value was attributed to the large surface area and the substantial number of cavities and OH groups present in the material, besides the presence of polypyrrole [68].

Magnetic β-CD-GO was also prepared using Fe_3O_4 modified with aminopropyltriethoxysilane that was first reacted with CM-β-CD using EDC to activate the carboxyl groups of the CD and then this modified material was reacted with the GO carboxylic groups activated with EDC/NHS. The nanocomposites were used as adsorbents for *p*-phenylenediamine and malachite green with good results (Q_{max} = 892.85 mg g^{-1} and 990.10 mg g^{-1}, respectively) [323].

Magnetic NPs modified with 3-chloropropyl-trimethoxysilane were reacted with potassium xanthogenate and GMA, as described above for HNTs. These particles were then reacted with GO and β-CD to obtain a nanoadsorbent for MB removal from water solutions [57].

The use of β-CD-GO materials for remediation, without the presence of magnetic particles has also been reported. Song et al. prepared β-CD-GO by mixing GO, β-CD at pH 11, and then hydrazine (to reduce GO) was added (60 °C, 4 h). The β-CD-GO material obtained was used for the removal of Co(II) and BPA. The presence of oxygen-containing functional groups on the surface of the material was pointed out as responsible for it sorption ability [304].

GO was also modified with EPI by a substitution reaction and then the epoxide groups of EPI were reacted with β-CD (Scheme 14.19). The β-CD-GO composite thus obtained was used in the removal of several dyes [332].

In another paper, GO-EPI-β-CD was prepared in a similar way but then was reacted with hydrazine to reduce the GO. This material presented good adsorption capacities for methyl blue, MO, and basic fuchsin (580.4, 328.2, and 425.8 mg g^{-1}, respectively) [309].

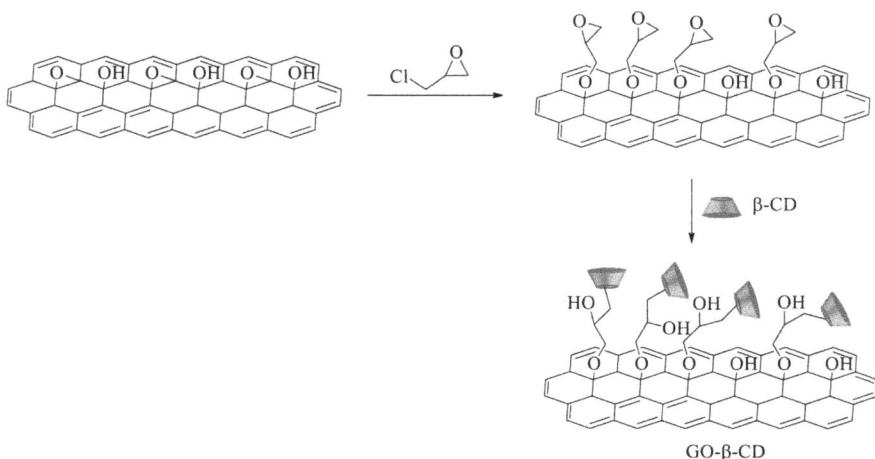

GO-β-CD

SCHEME 14.19   β-CD-GO composite synthesis.

A composite of β-CD-GO-isophorone diisocyanate was also reported and used to remove MO (adsorption capacity of 83.40 mg g^{-1}) from a wastewater solution. In a first step, GO was reacted with isophorone diisocyanate, using a catalytic amount of dibutyltin laurate, T-12, (80 °C, 3 h, under nitrogen). Then, the obtained material was reacted with β-CD in the presence of T-12, at 70 °C for 3.5 h [354].

β-CD-GO-isophorone diisocyanate composite, synthesized as described above, was also linked to diatomaceous earth (activated using NaOH 2.5 M, 2 h, 100 °C) through reaction with the carboxylic groups of GO, in the presence of CTAB. This composite was used in the removal of MB (Q_{max} = 110.50 mg g^{-1}) [345].

The same authors reported the use of a β-CD-GO-CS material prepared using maleoyl-β-CD that was activated with EDC/DMAP and then reacted with CS. Glutaraldehyde was then added to the β-CD-CS material (60 °C, then 85 °C, 2.5 h). GO, which was first activated with EDC/NHS, was then reacted with the β-CD-CS material and more glutaraldehyde. The removal of Mn(II) from the water was evaluated. Ion exchange and complexation mechanisms were responsible for the good adsorption capacity of the material.

Another β-CD-GO-CS material was synthesized using a different approach: GO was mixed with β-CD in water and then CS and acetic acid were added. After stirring, genipin (crosslinker) and sodium ascorbate (reducing agent) was added. The mixture was then heated (90 °C, 12 h) in an autoclave. A very high adsorption capacity (1134 mg g^{-1}) was obtained with this hydrogel for MB. The adsorption is a result of film diffusion and intraparticle diffusion and is pH and ionic strength dependent, suggesting that electrostatic interaction occurs between the hydrogel and the dye [198].

The use of a slide ring hydrogel (SRHG) for MB removal from wastewater has also been reported. SRHG was prepared by reaction of GO functionalized with isocyanate (obtained by reaction of GO with toluene-2,4-diisocyanate) with a polypseudorotaxane of CD (obtained using α-CD and poly(ethylene glycol) 6000), followed by reaction of this hydrogel with acrylamide. Moderate adsorption capacity (92.3 mg g^{-1}) was obtained for this material [302].

Recently, a GO-PDA-β-CD ultrafiltration membrane with a dual function, adsorption of Pb (II), and rejection of MB was prepared. GO and dopamine were mixed (pH 11) and this solution was spread onto nonwoven fabrics, using the drop-coating method, and heated at 80 °C, 4 h. This material was then reacted with CD and glutaraldehyde. A rejection coefficient of 99.2% was obtained for MB and a maximum adsorption capacity of 101.6 mg g^{-1} for Pb (II) [330].

In addition to the principal CD-containing materials referred above, many others like dendrimers, molecular imprinted polymers, titanium dioxide NPs, zeolites, etc., have been used for CD immobilization and applied in remediation [15].

14.3 REMOVAL OF TARGET POLLUTANTS FROM DIFFERENT MATRICES

In what follows, major contributions of CD-polymer for removing contaminants from soil and water environmental matrices will be summarized.

14.3.1 SOIL

Soil pollution is a major environmental issue once may have a significant impact on the economy and human health [141]. Soil pollution arises, in general, from anthropogenic activities, being mining [318, 319] and agriculture [62] among the major sources of agrochemical and industrial pollution [221], respectively. Due to the environmental impact of this phenomenon, soil remediation became an important issue for researchers all over the world. However, the soil is a highly complex reaction media once is characterized by a variety of factors and properties. Soils may differ greatly in permeability, porosity, structure, composition, and density. When choosing a

clean-up technology to treat a contaminated substrate, all those properties, plus the nature (inorganic, phenolic, aliphatic, and among others) and state (dissolved, adsorbed, aggregated, etc.) of the pollutants to be removed, have to be carefully taken into account. Hence, diverse in situ and ex situ strategies of soil remediation have arisen. For instance, in situ treatments have included soil flushing [27], electrokinetic separation [397], phytoremediation [8], and biodegradation, generally with bacteria [306]. Ex situ methods have involved transporting the polluted soils to an external location, as in the case of washing [351], thermal desorption [379], chemical oxidation and/or reduction [118], and photodegradation [396]. Soil vapor extraction has been carried out both in situ and ex situ [10].

14.3.1.1 SOLUBILIZATION

The first studies reporting the potential applications of CD in soil remediation addressed them as pollutant-solubilizing additives for flushing, in lieu of conventional surfactants, acids, or organic cosolvents whose remains could be harmful. This remediation method is schematized in Figure 14.2. Brusseau et al. [50], from the University of Arizona, clearly enhanced the elution of anthracene, pyrene, and 2,4,4'-trichlorobiphenyl (TCB), by using hydroxypropyl-β-CD (HPβCD). Remarkably, the retardation factor of

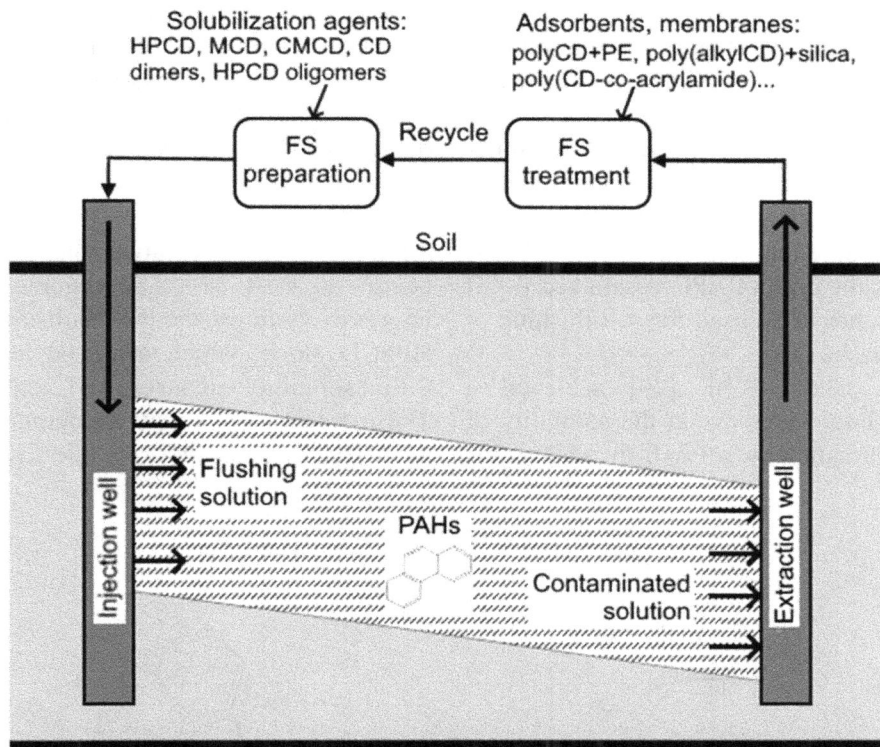

FIGURE 14.2 Diagram of a soil flushing process, indicating the potential usage of CD derivatives.

TCB through sandy soil decreased from 828 to 1.6. Brusseau's group extended the successful application of CD derivatives to phenanthrene and cadmium [51] and, later, to trichloroethene and tetrachloroethene [48].

Since a high aqueous solubility is appreciated for usage in soil flushing and soil washing, CD-functionalized polymeric matrices are disregarded in favor of isolated CD as solubilizing agents. Likewise, hydrophilic derivatives such as HPβCD, carboxymethyl-β-CD (CMβCD [67]), and glycine-β-CD [324] usually outperform the less soluble β-CD. Methyl-β-CD (MβCD) is another widely studied derivative for soil remediation because of its interaction with the soil, increasing the adsorption energy of water [156]. α-CD and γ-CD have been scarcely studied, but the latter has achieved better results than the former for the removal of norflurazon [321], diazinon [80], and cycloalkyl halides [244].

Currently, the ability of CD derivatives to form inclusion complexes is frequently conducted to aid advanced oxidation techniques, electrokinetic separation, and biological treatments [71, 72].

Interestingly enough, a dimer formed by two β-CD units linked by a hydrophilic agent may be more soluble than unmodified β-CD. This is the case of EPI crosslinked β-CD dimer to be used for soil flushing or washing.

Choi et al. [79] achieved a 97-fold increase in the solubility of fluoranthene, a PAH, by adding this product. The authors estimated an association constant of 33×10^3 M^{-1}. Later, the same group synthesized very soluble oligomers consisting of HP-β-CD units crosslinked by EPI (Figure 14.3), which were able to increase 178-fold the solubility of fluoranthene, the association constant being 160×10^3 M^{-1}. The oligomer impressively improved the results of HPβCD alone, which achieved an 8-fold solubility enhancement [262]. These recent results may encourage other researchers to test soluble CD

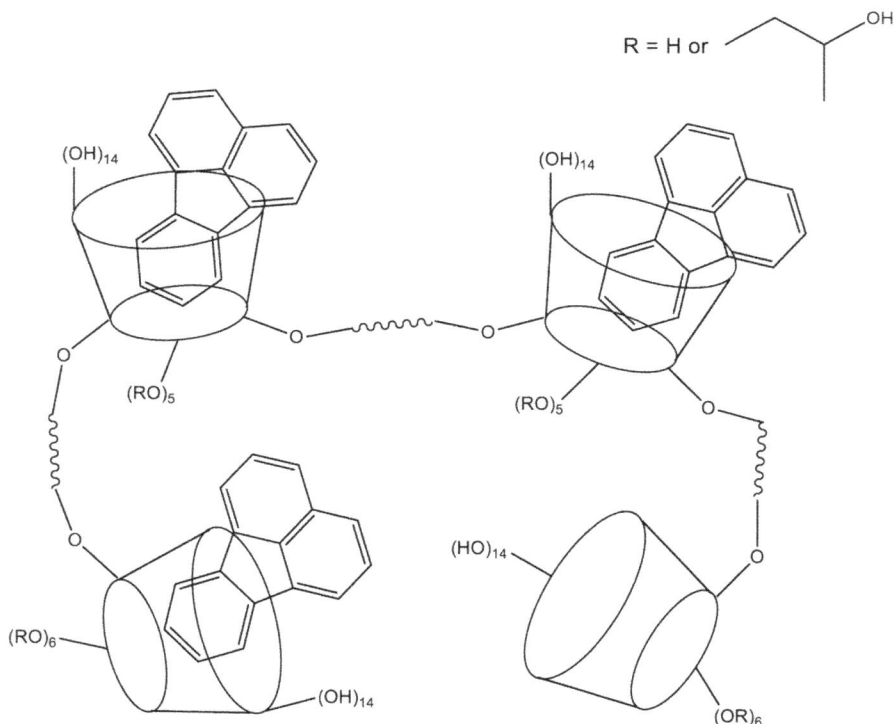

FIGURE 14.3   One of the possible structures (minimum crosslinking) for a hydroxypropyl-β-CD tetramer [262], with fruoranthene entrapped into some CD cavities. Wavy bonds represent the crosslinking moiety (e.g., EPI). Secondary –OH groups are shown on the larger base of the truncated cones.

oligomers as solubilization aids for soil remediation.

Highly soluble CD derivatives can also be combined with highly soluble polymers. An aqueous solution of both CM-β-CD and carboxymethyl CS achieved, in soil washing, was found to attain higher removal efficiencies than a solution of either of them [359]. Removal of PAHs went from 17% to 53% with the single solutions to 69%–84% upon mixing, removal of fluorinated compounds went from 27% to 53% to 72%–83%, removal of Ni went from 6% to 22% to 35%–44% and removal of Pb went from 18% to 43% to 58%–69%, achieving similar enhancements for other heavy metals. By a two-way ANOVA analysis, the authors proved that the interaction between the polymer and the CD was significant, but did not venture into the nature of this interaction. Even though testing a cationic CS covalently functionalized with CM-β-CD could be tempting, it would likely not be soluble enough for soil washing.

14.3.1.2 WASHWATER TREATMENT

In any case, the contaminated water from soil flushing, washing, and other common remediation techniques has to be treated (Figure 14.3), and insoluble CD-containing polymers can play a crucial role as effective adsorbents for persistent, troublesome pollutants. It has been mentioned here that CD derivatives achieved extraordinary removal efficiencies of PAHs, which are particularly ecotoxic and stable in soils and groundwater, since CDs are insoluble and not exposed to sunlight. For instance, polyethylene membranes that have been coated with poly-α-CD, poly-β-CD, or poly-γ-CD were shown to decrease phenanthrene concentration from 1.8 to roughly 0.3 ppm, while the unmodified polyethylene membrane decreased it to 0.7 ppm [161]. The crosslinker was CA in all cases and, once again, γ-CD resulted in the best performance. With the addition of poly-CD, size exclusion, and adsorption onto the plastic substrate were complemented with a molecular entrapment mechanism.

In a similar way that CD improved the removal efficiency of PAHs by plastic membranes, these molecules can improve the performance of ceramic adsorbents, and the retention into hydrophilic gels is enhanced. As an example of the former, β-CD and pentynyl-β-CD have been polymerized with EPI in presence of silica microparticles, giving out flower-like structure particles with silica noncovalently trapped in them [78]. The removal efficiency of silica alone was 78%, which was increased to 96% by blending with the pentynyl-β-CD. The authors alleged that the hydrophobicity conferred by pentynyl groups enhanced the entrapment of phenanthrene and eased the recovery of the adsorbent from water (Figure 14.4). Regarding gels, Topuz and Uyar [313] produced a poly-CD cryogel in which CD units were crosslinked by poly(ethylene glycol) diglycidyl ether. It was aimed at the removal of PAHs and achieved q_e values ranging from 0.105 (anthracene) to 1.25 mg g^{-1} (fluoranthene).

A CD-polymer structure can work as the support of a photodegradation catalyst, instead of using ceramic supports, which are known to enhance the activity of TiO$_2$ alone [241]. Provided the catalyst is either fully ceramic or biodegradable, degradation by light is one of the most popular advanced oxidation techniques proposed for the removal of pollutants from surface soils and groundwater [328]. As an example, β-CD (triethoxysilylpropyl) urethanes can be polymerized

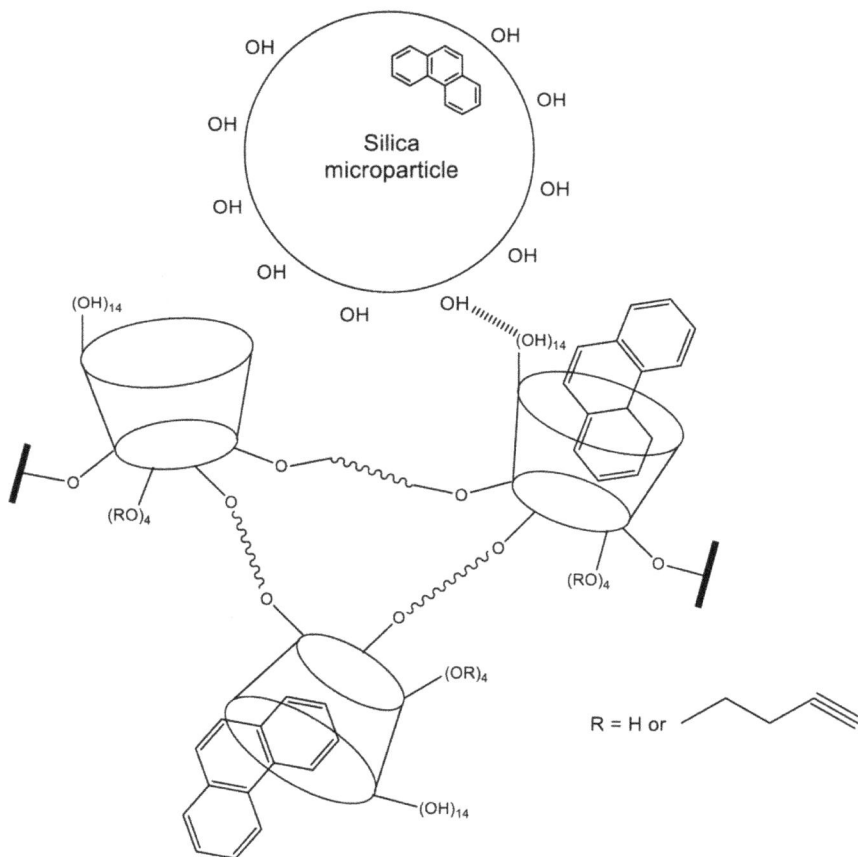

FIGURE 14.4   Simplified scheme of poly(pentynyl-β-CD) with silica microparticles, plus some phenanthrene molecules either entrapped into the CD cavities or diffusing through silica.

in acidic media and grafted to the anatase surface, by reacting with the –OH groups surrounding TiO$_2$ [273]. This catalyst was tested with MO. The adsorption capacity reached 0.9 mg g^{-1} while that of nonsupported TiO$_2$ NPs was 0.05 mg g^{-1}, which resulted in a faster photodegradation. A simpler nanocomposite comprising β-CD, polycaprolactone, and TiO$_2$ was able to attain a removal efficiency of Pb(II) up to 98%, at pH 9.7, for an adsorbent dosage of 0.05 g and 10 ppm of heavy metal [289]. Adsorption data were shown to fit a Langmuir isotherm and kinetics was well described by a pseudo-second-order rate equation.

14.3.1.3 PROTECTION OF SOILS

It is well-known that superabsorbent hydrogels can improve soil properties, particularly for agriculture and forestry. These hydrogels can protect seedlings and crops from drought [151], confer controlled release of pesticides [277], and reduce the loss of nutrients by leaching [2]. Generally, hydrogels are built by crosslinking hydrophilic polymer chains in a 3D network, being able to hold as much water as possible, but not dissolving in the aqueous medium. Although the primary reason for choosing CD is their polyhydroxy, 3D structure, it has been suggested that hydrated β-CD can entrap up to 10 water molecules without distortion [268]. The product resulting from the copolymerization of β-CD, sodium alginate, and acrylic acid, using N,N'-methylenebis(acrylamide) as the crosslinking agent, achieved water absorbency values over 200 g g^{-1} at pH 6 [149]. Moreover, CD units clearly improved the biodegradability of the hydrogel.

Besides creating the structure by covalent bonds, hydrogels can be made, for instance, by "threading" a hydrophilic polymer chain through the cavities of CD. These pseudo-polyrotaxanes and other supramolecular CD-containing gels have been extensively covered by Zhao et al. [388].

14.3.1.4 MOLECULAR IMPRINTING

In other matters, CDs have also been applied in the synthesis of molecularly imprinted polymeric adsorbents, taking advantage of their ability to form inclusion complexes with a target pollutant. Among other targets, this has been exploited for benzylparaben in wastewater [25], forchlorfenuron in fruits [77], and prometryne in a rice field and surface soil [138]. For the latter, the authors functionalized silica gel with β-CD and 8-hydroxy-quinoline moieties, obtaining this way the supporting matrix. Then, the pollutant molecule allegedly formed an inclusion complex with the β-CD moieties, and small and reactive monomers, namely acrylamide or methacrylic acid, reacted to form a tailored polymer over the supporting template. Finally, the host pollutant was washed away. The removal efficiency of prometryne, previously extracted with acetone from the surface soil sample, was 87.3% with acrylamide as the monomer, attaining an adsorption capacity at the equilibrium of 45 $\mu mol\ g^{-1}$, and 94.9% with methacrylic acid as the monomer, for which q_e reached nearly 50 $\mu mol\ g^{-1}$.

Recently, a molecularly imprinted electrochemical detector of BPA, consisting of polyacrylate, β-CD and reduced GO, and covalently bonded, has been suggested [11]. The resulting product was used to coat a glassy carbon cylinder, proposed as the working electrode in a three-electrode system in which the auxiliary one was a platinum wire and the reference one was Ag/AgCl. The limit of detection was reported to be 8nM. Another CD-functionalized sensor had been designed by Zengin et al. [368] for the detection of specific PAHs. The authors prepared a polymer brush by reversible addition-fragmentation chain transfer polymerization of GMA over thin silicon substrates, then attached β-CD moieties and formed an inclusion complex with either anthracene or pyrene. Measurements needed a Raman microscope and the limits of detection were as low as 0.8 nM for pyrene and 2.4 nM for anthracene, respectively.

All of the described above is summarized in Table 14.1.

Taking into account, the complexity of soils, their diversity, and the variety of strategies developed to treat them, it is not surprising that CDs have been used to obtain completely different products, from nonpersistent solubilizing agents to highly efficient solid adsorbents. Generally, those very different products exploit the formation of inclusion complexes, be it to solubilize nonpolar organic compounds, to entrap and/or detect a certain pollutant, or to build the template for a tailored polymer structure. The ease to create highly porous materials by crosslinking is undoubtedly useful for membranes, catalyst supports, and hydrogels. Finally, similarly to other saccharides, CD promotes the growth of microbes, thus supporting the biodegradation of the whole material.

14.3.2 WATER

The appearance or persistence of contaminants in natural water, and in wastewater, and the respective adverse environmental effects, have boosted the number of studies focused on adsorption of water pollutants by modified CDs [85].

Pollutants found in water and wastewater comprise a plethora of chemical species and compounds, including residues of heavy

TABLE 14.1 Usage of Polymeric CD Derivatives for Soils

Task	Properties Desired	Structure	References
Protection of soils by hydrogels	High water holding capacity, not becoming soluble	3D structure, crosslinks, CD copolymerized with hydrophilic monomers	[149]
Soil flushing or washing	Solubility, ability to form soluble inclusion complexes	CD oligomers with hydrophilic functional groups	[262]
Photocatalytic degradation	Mechanical endurance, fast adsorption, high adsorption capacity	CD-containing polymer as support and TiO$_2$ as an active site	[272]
Groundwater or washwater treatment: membrane filtration	Adequate pore size, water resistance	CD or poly-CD attached to a hydrophobic substrate	[161]
Groundwater or washwater treatment: adsorption	Fast adsorption, high adsorption capacity, water resistance	Poly-CD (crosslinked) or CD-containing polymer, insoluble	[78] [313]
Detection of a certain soil pollutant	Low limit of detection (and quantification), user-friendliness, specificity to target pollutant	Molecularly imprinted polymer, using a CD-pollutant inclusion complex as template	[11] [368]

metals (e.g., chromium, cobalt, lead), organic compounds (e.g., aromatic and polychlorinated compounds), pharmaceuticals (e.g., analgesics, anti-inflammatory, antibiotics, and psy-chiatric drugs), active compounds present in personal care products (e.g., fragrances), steroids and hormones, pesticides and herbicides, surfactants, dyes and plasticizers, and fungal metabolites (mycotoxins). These are in general rarely monitored and controlled, toxic or carcinogenic, and nonbiodegradable [85].

The major classical concern on the effect of contaminants in the environment and human health has been associated with heavy metals and arsenic ions, active ingredients in pesticides, and other persistent organic pollutants. Those detected in water and wastewater have included (1) inorganic pollutants, including heavy metals and arsenic ions, and (2) organic pollutants composed of polycyclic aromatic compounds, benzene and dibenzofuran derivatives, phenol derivatives (e.g., naphthol derivatives), polychlorobiphenyls, surfactants, pharmaceutical residues, endocrine disruptors, household, and industrial chemicals. Within the latter, organic dyes have been discharged into the environment from the paper, textile, and cosmetics industries.

14.3.2.1   METAL IONS

Heavy metal pollution is an anthropogenic unavoidable issue that humanity has struggled to deal with. Heavy metal is a poorly defined term that refers to a group of elements with a density greater than four, including metals and metalloids like arsenic [109]. Though naturally occurring and some being biologically essential (e.g., copper and chromium are micronutrients [42], pollution emissions raise the concentration of these elements in natural environments to dangerous concentrations [5, 6]).

Recently a comprehensive review on the remediation of heavy-metal-containing residues has been published highlighting the priority of this topic in terms of the sources of pollution and different available technologies for their treatment [319]. Due to its relevance, the application of CD on heavy metal treatment was also challenging for many researchers. However, as we will note later, the effect of CD on the retention and treatment of heavy metals is not remarkable. This comes directly out from the molecular structure of CD—the hydroxyl groups, which occurred at both rims of CD, have a low ability to coordinate with predominant cations [54, 55, 276].

A pioneer work on this issue was done by Buvári and Barcza [54] who have studied the complexation and solubility of β-CD in the presence of aqueous solutions of inorganic salts based on alkaline cations (Li$^+$, Na$^+$, and K$^+$) [54]. However, it has also been concluded that although the interactions CD-cations cannot be neglected, anions are the main counterparts on the complexation. The interaction between other metal ions and β-CD has been reported two decades later. The most investigated metal ion is the Cu(II), probably due to its economic value [44, 216, 217, 224]. However, the interaction of other metal ions, such as Mn(III), Cr(III), Ni(II), and Co(II), with CD have also been studied [219, 242].

Nevertheless, some of these complexes were synthesized in the solid-state, preventing any assessment on the association constant (K_a) or the contribution of the anion/cation on the interaction with the CD. In 2002, Norkus et al. [250] reported the interaction between Pb(II) and β-CD, showing that it is highly improved ($\log K_a=15.5$) at basic pH (10–11.5).

The development of materials based on CD-based or CD-functionalized polymers for the remediation of water containing metal ions is thus a logical step [85, 87].

CD can be polymerized by, for example, copolymerization and crosslinking [96, 98, 352]. For example, CA—a polycarboxylic acid—can be used as a crosslinker for CD-polymerization (CA-β-CD).

This reaction can occur, via condensation, at relatively low temperature (ca. 140 °C) without organic solvents or additives [393]. The adsorption of Cu(II) onto CA-β-CD leads to a maximum adsorption capacity of 31.4 mg g^{-1}, for an initial concentration of 100 mg L^{-1} [148]. However, by testing the effect of different aromatic compounds (BPA and MB) on the simultaneous removal with Cu(II), it was found that the maximum adsorption capacity of Cu(II) remains similar in the presence of BPA (31.9 mg g^{-1}) but decreases in the presence of MB (24.8 mg g^{-1}). This suggests that the complexation mainly occurs in the free carboxylic groups, occurring a competitive process between the MB and the Cu(II).

However, natural polymers show some disadvantages that limit their use in practical wastewater treatment applications, such as their low surface area and difficult separation from the liquid phase. Magnetic sorbents on the other hand have a relatively high surface area and are easy to separate and manipulate in complex multiphase systems with an external magnetic field [30]. Based on this principle, carboxymethyl-β-CD-polymer, obtained by copolymerization with EPI, grafted on Fe_3O_4 NPs were prepared and their performance on metal ions (Pb(II), Cd(II) and Ni(II)) adsorption was evaluated [30]. The maximum adsorption capacity (q_m) of the adsorbents for those metal ions (64.5, 27.7, and 13.2 mg g^{-1}, respectively), at pH 5.5, is improved (ca. two-fold higher) by coating the magnetic NPs with the CD-based polymer. A Langmuir sorption isotherm shows that CDs may have a significant role in the adsorption of metal ions. By performing a multicomponent adsorption experiment, it can be concluded that the affinity of the adsorbent toward the metal ions decreases as follows: Pb(II), Cd(II), (Ni(II). This behavior was justified on the basis of the acidity behavior of ions (hard of soft ions [247, 264] and their ability to form covalent or ionic bonding [329].

Although CDs can be used as a polymer and, consequently, as an adsorbent, most of the applications of CD occurs via functionalization of polymers. The most common one is the CS. CS (Scheme 14.20) is a polycationic (pH higher than 6.5) polysaccharide found naturally in the shellfish and crustaceous exoskeleton [271]. CS is used as a bioadsorbent for a wide range of environmental contaminants. This comes out as a consequence of the presence of amino and hydroxyl groups in its structure, which can react directly with some pollutants or, alternatively, be modified in order to improve its versatility [170, 214]. Heavy metals are among the pollutants well adsorbed by CS or CS blends/composites [204, 373].

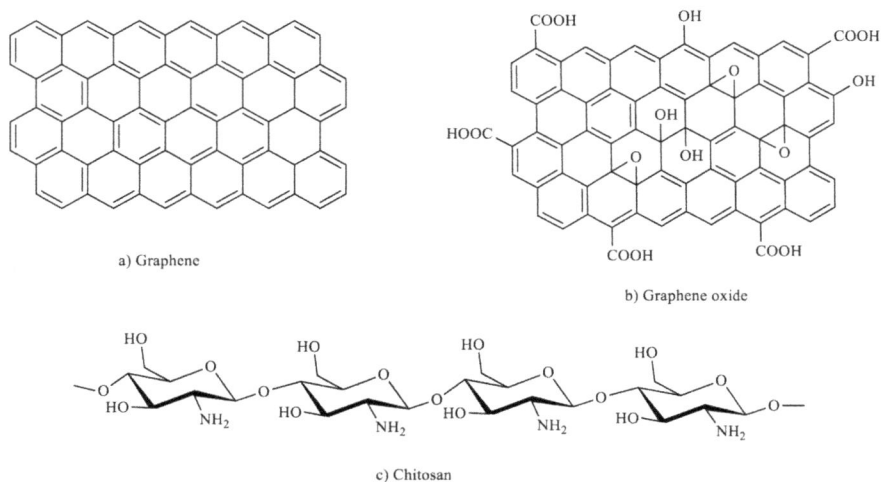

a) Graphene

b) Graphene oxide

c) Chitosan

SCHEME 14.20   Structures of (a) graphene, (b) GO, and (c) chitosan.

Magnetic β-CD-CS NPs were synthesized by Fan et al. [114]. Initially, by using maleic acid, in the presence of 4-dimethylamino-pyridine, the β-CD is functionalized to form maleoyl-β-CD [39]. The latter compound is then used to functionalize the CS and to form the β-CD-CS adsorbent. Following that Fe_3O_4 NPs are coated by β-CD-CS. An original approach was described by Elanchezhiyan et al. [112] who have synthesized CS-β-CD, by using glutaraldehyde as crosslinker [256], in the presence of metal ions: lanthanum, zirconium, and cerium. The obtained polymer was success-fully applied for the adsorption and recovery of commercial cutting oil.

The use of CS shows some draw-backs as low compressive strength and the difficulty to produce mono-liths. Such limitations, important for practical applications, can be over-come by the incorporation of GO on the polymer structure [172, 374]. GO is a hydrophilic derivative of graphene (see Scheme 14.20b and c) that has a lower content of carbon in favor of a higher amount of oxygen in the form of hydroxyl, carboxyl, and epoxy groups. Such modifica-tion of the graphene structure makes the GO highly hydrophilic and, consequently, highly dispersible in aqueous media. Furthermore, GO retains much of the properties of the highly valued nanomaterial pristine graphene, being easier to prepare and process and cheaper [101] and showing good performance as an adsorbent of different types of pollut-ants [110, 172]. It should be stressed that the grafting of CD to GO sheets can also be used as an adsorbent for metal ions adsorption [102, 147].

In order to improve the stability of CS, as described above, β-CD-CS/GO materials have been prepared by Li et al. [182]. The procedure consists of reacting the amine group of CS with the GO's carboxyl group, forming an amide. The obtained material shows good sorption ability to Cr(VI). Of course that the sorption process of this metal ion is slightly different from others once Cr(VI) is present in solution as anions, that is, CrO_4^{2-}, $Cr_2O_7^{2-}$, or $HCrO_4^-$ [338]. At Ph higher than 6.8 only the former is stable, whilst at pH lower than 6.8 the predominant species is the hydrogen chromate anion [395]. Consequently, in general, the most reliable Cr(VI) quantification procedure must include all these three species. The maximum adsorption capacity of this material is 67.66 mg g^{-1} and the mechanism of adsorption is complexed involving initially the electrostatic interaction between the protonated amine groups of CS and the chromium-containing anions; following that, the Cr(VI) is reduced to Cr(III), being able to interact directly with either carboxylic groups of GO or the β-CD [395].

Nowadays, there is an industry demand to develop new product applications for finding new markets. Li et al. [186] have modified filter paper with β-CD, using CA as the crosslinker. The functionalization of paper is based on the presence of hydroxyl groups of cellulose in the paper allowing the esterification with an acid. The obtained modified filter paper shows a significant adsorption capacity for Cu(II) (39.1 mg g^{-1}) when compared with the unmodified filter paper (0.62 mg g^{-1}); however, it seems that the CA-β-CD has the main role in the sorption process, once studies carried out using CA-β-CD has adsorbent showed that the Cu(II) reaches maximum adsorption equal to 39.4 mg g^{-1}. Thus, the advantage of using a modified filter paper might be the possibility of using a complementary filtration process [186].

The synthesis of bi- or multifunctional adsorbents (i.e., adsorbents able to adsorb different types of pollutants) has also generated the interest of researchers. In this field, different approaches were taken.

For example, Qin et al. [270] have shown that a copolymer of a modified β-CD and 1-vinylimidazole (VI) were able to simultaneously adsorb Cd(II), congo red (CR), and rhodamine B (RB). It is known that the adsorption mechanism between dyes (ionic or neutral) and metal ions with CD-containing adsorbents may follow one of the following driving forces: electrostatic interaction electrostatic interactions, co-ordination, π–π interactions, and hydrophobic interactions. Thus, to improve the ability of CD to complex ions, CDs should be functionalized with chelate groups [383, 385]. Therefore, β-CD was initially modified to per-6-(*tert*-butyldimethylsilyl)-β-CD [133] followed by subsequent synthesis of 2,3-di-*O*-methacrylated-6-*tert*-butyldimethylsilyl)-β-CD (MCD) [280, 292]. MCD was then polymerized by using different routes: by reacting with 2,2'-azobis(2-methylpropionitrile) to form the homopolymer (PMCD) or with 1-vinylimidazole to form a copolymer (P(MCD-VI). Comparing the maximum removal capacities of both polymers toward the above-mentioned polymers, it has been found that the use of the copolymer always leads to higher values of q_{max}: 175, 711.7, and 110 mg g^{-1}, compared with 40, 20, and 10 mg g^{-1} obtained for PMCD, for RB, CR, ad Cd(II), respectively (the q_{max} values are approximated). It is interesting to find out that these authors clear suggest that the CD has no effect on the Cd(II) adsorption, once the adsorption of this metal ion occurs by coordination with 1-vinylimidazole.

The effect of the dye on the simultaneous adsorption of Cd(II) and dye was also evaluated. In the presence of RB (a cationic dye), there is a synergetic effect on the adsorption process, that is, the adsorption of both dye and Cd(II) is higher than when isolated; however, in the presence of an anionic dye (CR) the adsorption of Cd(II) is highly enhanced, which may be ascribed to the possible interaction of the metal with the anionic dye, thus increasing the interaction sites (both VI and CD, via interaction with the dye).

14.3.2.2 DYES

More than 100 tons of organic dyes are produced annually [123, 308]. This demand is due to the need for colorants for a broad number of industries, including textile [358], leather [300], food [353], and solar cells [364]. Furthermore, a significant part of dyes is lost, or wasted, in the dyeing process [343]. For these reasons, dyes are serious contaminants present in water, leading to significant aesthetic and ecological issues. In fact, dyes are, in general, nonbiodegradable, oxidizing agents, and potentially carcinogenic and toxic [208]. Furthermore, dyes are also light and heat resistant, and therefore may act as a barrier for light and air with the consequent interference in the biotic of aquatic resources [202].

Although many resources have been applied for improving the dye efficiency and the development of eco-friendly dyes [106, 263], which may lead to a decrease in toxicity and dye-containing waste-waters, many works must be done to treat properly the actual levels of dye contaminations. There are many physicochemical treatment methods for the removal of dyes from industrial wastewater, such as advanced oxidation processes (AOPs) [24, 28], electrochemical [213], and adsorption [220]. The latter method has been extensively used for the discoloration and removal of dyes from wastewaters due to its low price, easy availability, and efficiency [36].

CDs are among the most used compounds for the removal of dyes and other contaminants from different waste sources [87, 99, 108, 117]. Herein, we review all matrices based on, or containing, CD for the removal of dyes from wastewaters. Schemes 14.21 and 14.22 show the molecular structures of the dyes mentioned in this section.

In the previous section, we have described the application of 1-vinylimidazole for the polymerization of a modified β-CD (2,3-di-O-methacrylated-6-*tert*-butyldimethylsilyl)-β-CD) [270]. The obtained polymer shows an adsorption capacity for CR and RB of 712 and 175 mg g^{-1}. Authors have also shown that the copolymer of a modified β-CD and 1-vinylimidazole (VI) were able to simultaneously adsorb Cd(II), CR, and RB. These results for dye adsorption cannot be easily explained on the basis of the binding constants for the equilibrium between β-CD and CR and RB: 1667 M^{-1} and 4266 M^{-1} (i.e., assuming a 1:1 stoichiometry for both cases), respectively [144, 197]. However,

whilst hydrophobic and π–π interactions are the predominant mechanism for RB adsorption, for the CR electrostatic interactions between the sulfonate groups of the dye and the easily protonated groups of VI play the main role in the adsorption process [270].

The use of polyurethane foams shows advantages as adsorbents due to their network structure and high ratio surface area/volume [140, 299]. Following that, polyurethane-β-CD-polymers have been synthesized. The synthetic route consisted in the reaction of β-CD (β-CD), polyether polyol (NJ-210) and isophorone diisocyanate, in the presence of a foaming agent and a catalyst, n-hexane, and dibutylbis (lauroyloxy) tin, respectively [106]. The adsorption capacity of the PU-derivative was tested by using Eriochrome Black T (EBT). This dye is a persistent azo dye pollutant [1]. The PU-derivative showed the best adsorption efficiency (93%) for a β-CD content of 7% (w/w). The mechanism of interaction between the polymer and the EBT occurs via ion–dipole interaction between the positive nitro groups (R-NO$_2^+$) and the nonbonding electrons of the amide C=O group, respectively. This interaction is in agreement with the occurrence of adsorbent/adsorbate interaction driven by the entropy [62].

Polyethylenimine (PEI) is an aliphatic polyamine characterized by the –(C$_2$H$_5$N)– repeat unities; PEIs

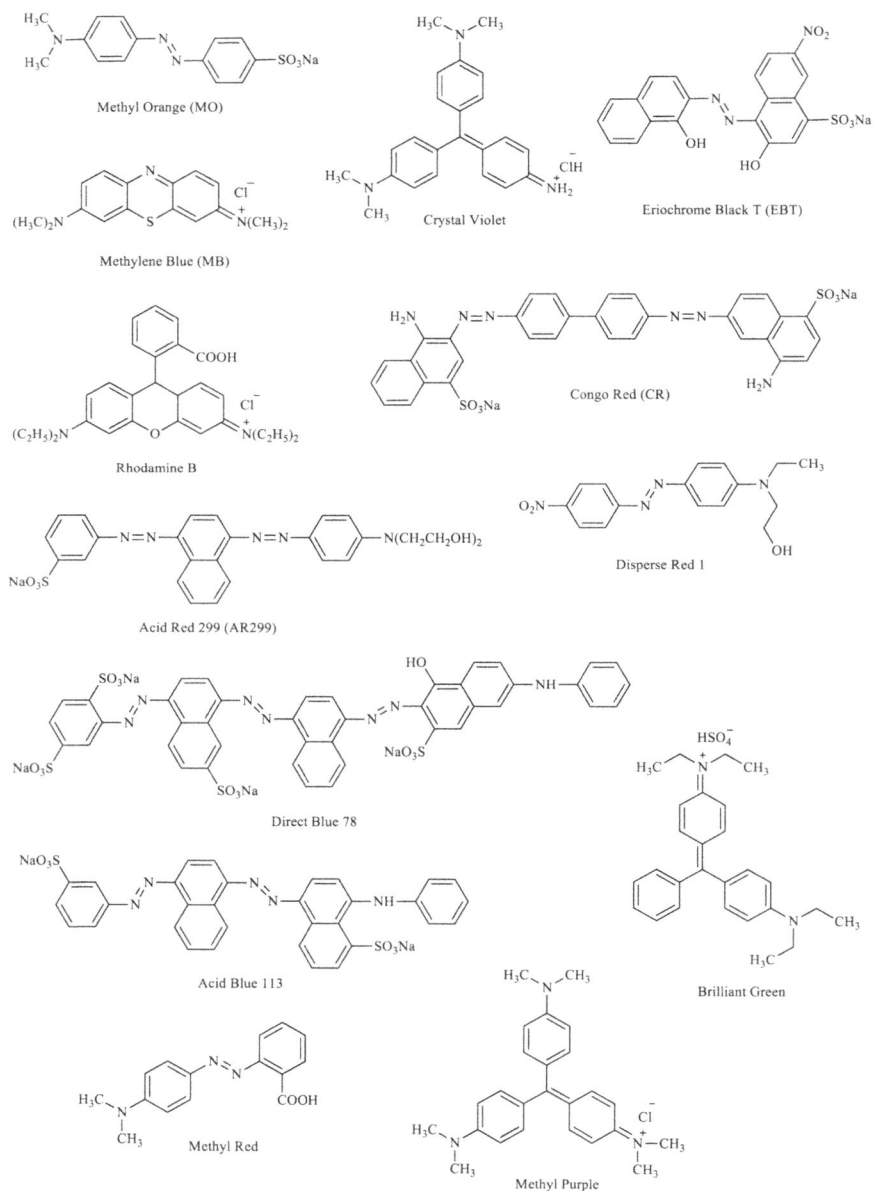

SCHEME 14.21 Structure of dyes mentioned in the section (Part I).

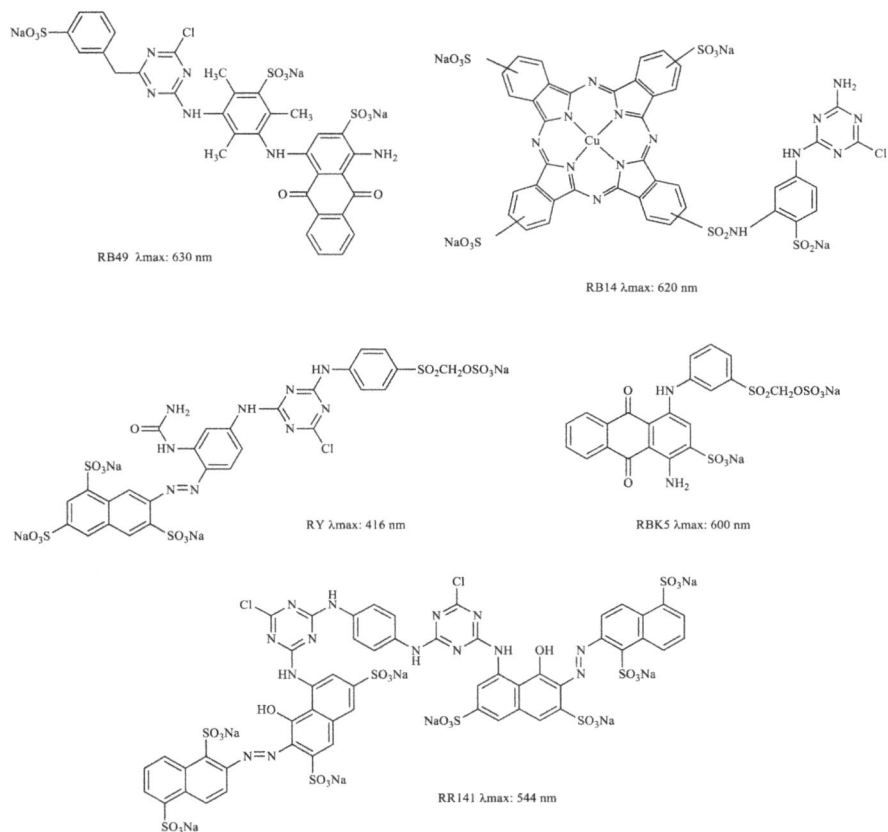

SCHEME 14.22 Structure of dyes mentioned in the section (Part II).

are polycationic electrolytes that can be synthesized in branched (viscous liquid) or linear (crystal solid) state. Branched PEIs have shown a high potential for the adsorption of a large variety of pollutants [3, 103, 176]. Due to the solubility in water of both PEI and β-CD, these materials must be immobilized onto solid material for practical purposes. Chen et al. [69] used Fe_3O_4 magnetic NPs as matrices to support PEI and CDs. The authors started to synthesize PEI

grafted Fe_3O_4 NPs by using glutaraldehyde as anchor [191]; after that, the Fe_3O_4-PEI composite reacts with maleoyl-β-CD [39] to form Fe_3O_4-PEI/CD (Scheme 14.23). The composite shows a significant maximum adsorption capacity for methylene orange (192.2 mg g^{-1}), obtained at low pH: 3; as pH increases the amino groups of PEI becomes less protonated and, consequently, the electrostatic interaction between the negatively charged MO and PEI

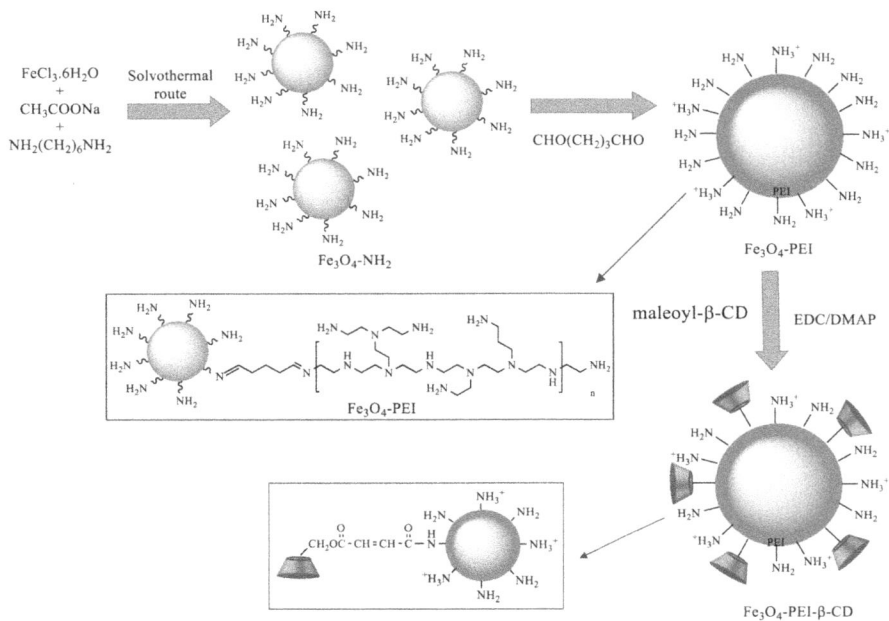

SCHEME 14.23 Fe_3O_4-PEI/CD composite.

becomes weaker, decreasing the amount of adsorbed dye. Despite this, it is expected that MO can interact with CD, in a 1:2 (MO:CD) stoichiometry, with the corresponding association constants K_1 and K_2 equal to 2970 M^{-1} and 100 M^{-1}, respectively [59]. Similar tests have been carried out with Pb(II) and the maximum adsorption capacity was equal to 7.31 mg g^{-1}, at pH 6, that is, when Pb(II) is hydrolyzed [317].

A different approach is the use of graphene-based materials as an adsorbent for pollutant removal [13, 143]. Among these materials graphitic carbonitride (g–C_3N_4) has been shown a promising material

for wastewater treatment, joining its low cost to chemical stability and photocatalysis activity [334, 371]. A g-C_3N_4 functionalized with –CD was synthesized to produce a dual function material to act as a microfilter and as a photocatalyst. Liu et al. [200] have shown, by using MB as a dye model, that an adsorption capacity of 69.7 mg g^{-1} is reached; this value is approximately three times higher than the adsorption capacity obtained for g-C_3N_4 alone. Using the material as a filter of the remaining (supernatant) solution, it can be found that an amount lower than 1% of MB remains in solution (the initial concentration of MB in solution is 6.4 mg L^{-1}). In the end, the photocatalytic performance of β-CD/g-C$_3$N$_4$ has also been assessed. The use of β-CD/g-C$_3$N$_4$ leads to a photodegradation efficiency of 98.5%, after 3 h of exposure to a 664 nm light irradiation, whilst by using bulk g-C$_3$N$_4$, in similar conditions, only 37.9% is obtained [200].

Wu et al. [345] have functionalized GO with β-CD in the presence of diatomaceous (also called diatomite), a natural siliceous (ca. 89% SiO$_2$) material, also containing aluminum and ferric oxides [345]. The highest adsorption efficiency and the maximum adsorption capacity of the adsorbent, using MB as a dye model, were 98% and 110.5 mg g^{-1}, respectively.

The use of CNTs as matrices for the adsorption of dyes has also been attempted and some papers have been published [142, 169]. The functionalization of multiwalled carbon nanotubes (MWCNTs) with β-CD has been reported by Mohammadi and Veisi [227]. The modification of MWCNT consists of the reaction of MWCNT with glycine, for esterification of the hydroxyl groups of the CNTs; subsequently, β-CD is attached to the amino groups of the amino acid, using EPI as crosslinker agent. The adsorption of four different dyes (MB, acid blue 113, MO, and dispersed red 1—see Scheme 14.21) were evaluated. The adsorption of the dyes occurs via a Langmuir mechanism, with maximum adsorption capacities equal to 90.9, 172.4, 96.2, and 500 mg g^{-1}, respectively. The thermodynamic analysis of the adsorption process has shown that the interactions are exothermic and enthalpic driven (i.e., $|\Delta H°|$ higher than $|T\Delta S°|$).

14.3.2.2.1   Poly-CD

A different approach, that takes into consideration all advantages of using CD, in particular, the formation of host–guest complexes with dyes (among other pollutants) consists in the polymerization of CD (as it was referred before). Following that the use of poly(β-CD), PCD, synthesized using, for example, EPI as a crosslinker, as adsorbent of dyes was deeply analyzed in 2013 by Morini-Crini and Crini [229]. However, other crosslinkers have also been used to polymerize β-CD as, for example, 4,4'-bis(chloromethyl) biphenyl [356].

More recently, EPI was used to polymerize β-CD and hydroxypropyl-β-CD (HP-β-CD). The polymers have been evaluated as adsorbents for Direct Red 83:1 and Direct Blue 78 [240; 265]. For an initial concentration of dye equal to 300 mg L^{-1}, the maximum adsorption capacity for β- and HP-β-CD-polymers are 13.4 and 3.3 mg g^{-1}, following a Freundlich mechanism; the higher efficacy of β-CD when compared with HP-β-CD was similar to that found for α-CD and HP-α-CD [266]. The reasons for such significant differences are not, however, well understood. The same group has carried out similar studies using Direct Blue 78 as adsorbate and PCD as an adsorbent; the maximum adsorption capacity found for this system was 23.47 mg g^{-1} and the adsorption mechanism is similar to that reported before. However, the study is complemented by the use of AOPs to remove the remaining dyes. By using the techniques in a sequential way, the removal efficiency of Direct Blue 78 is higher than 99%.

The development of adsorbents for the simultaneous adsorption of dyes (or other pollutants) is always challenging once it depends, in general, on different types of interactions.

Zhou et al. [394] have developed a poly(β-CD) with the ability for simultaneous adsorption of anionic and cationic dyes. The adsorbent was synthesized in two different steps: initially β-CD was polymerized via esterification with CA; the second step consists of the grafting of the 2-dimethylamino ethyl methacrylate. The obtained material contains, besides the hydrophobic cavity of CD, carboxylic groups, and tertiary amino groups providing, in this way, a multitude of sites for interaction of both cationic and anionic dyes, respectively. The adsorption studies of MB, MO, and BPA onto the

PCD grafted with 2-dimethylamino ethyl methacrylate, showed high q_m values: 335.5, 165.8, and 79 mg g^{-1}, respectively. However, these values were obtained at different pH conditions: 11, 4, and 6.5, respectively. This prevents the broad use of this pH-responsive material, once the maximum adsorption capacity is dependent on the pH. For that reason, authors have checked the effect of the simultaneous adsorption of BPA and MB, and BPA and MO, at different pH values: 4 and 11, respectively. The adsorption capacity of both adsorbates did not change significantly when the adsorption test was performed with a single or two components.

14.3.2.2.2 Saccharide-based Polymers

CS is a polycationic polysaccharide found naturally in the exoskeleton of shellfish and crustaceous. CS is the most commonly used cationic biopolymer (positively charged at pH lower than 6.5) [14]. The CS is nontoxic, biocompatible, and is able to form films and, therefore it has found many applications, including as adsorbent for a wide range of environmental contaminants [52], such as dyes [14, 20, 89]. This happens as a consequence of the high content of amino and hydroxyl groups existing in its structure. The performance of

CS reinforced with β-CD previously modified with CA for the adsorption of a set of dyes was evaluated: C.I. Reactive Blue 49 (RB49), Reactive Yellow 176 (RY176), Reactive Blue 14 (RB14), Reactive Black 5 (RBK5), Reactive Red 141 (RR141) (see Scheme 14.22) [387]. All these dyes are significantly removed by the CS-based polymer, with values higher than 90% of removal efficiencies. These removal efficiencies correspond, just as an example, to a maximum adsorption capacity of the polymer toward Reactive Blue 49 of 498 mg g^{-1}. This performance is justified by the occurrence of electrostatic interactions between the sulfonate groups of all dyes and the protonated ammonium groups of CS (pK_a ca. 6.5) [331] along with hydrogen bonding between the amine groups of dyes and the hydroxyl groups of CS.

Dan et al. report a study where CS/β-CD-polymer, as a membrane, is prepared by using vinyltriethoxysilane as a crosslinker. The membranes show high porosity as a consequence of electrostatic repulsion (or steric hindrance) between the CD and the CS [362]. Consequently, the membrane shows a maximum adsorption capacity for Acid Red 299 (strong acid dye, used in textile and leather dyeing) equal to 13.4 mg g^{-1}. The model that better describes the isotherm of the acid dye to the membrane is the Langmuir equation

and the process is endothermic, which might be related to the need to overtake a polymer interface to reach the binding sites.

By using a similar approach γ-CD was grafted to starch (from corn), by using EPI as the crosslinker. The adsorption capacity of the material was tested toward three different dyes: MB, methyl purple, and CR; the first two dyes are cationic and the third one (CR) is anionic. Despite the size of γ-CD cavity [81], host–guest inclusion compounds, with higher association constants, with dyes are formed; for example, the association constants for CR and MB are 3800 M^{-1} [236] and 2.95×10^7 M^{-2} [139], respectively. The removal efficiencies for all three dyes are around 60% [142].

A different method has been followed by preparing a GO-containing CS/β-CD hydrogel [198]. Liu et al. [198] have shown that after the synthesis of GO is possible to prepare a 3D hydrogel, with a sheet-like morphology, by using a so-called "one-step green-assemble strategy." In a summary way, such a strategy consists of mixing of β-CD, CS, and GO in the presence of sodium ascorbate and genipin [203]. Both these compounds are crosslinking agents promoting the linkage between the hydroxyl and carboxyl groups of GO and the hydroxyl groups of both CD and CS. By using MB as a dye model, it has been found that using an initial MB concentration equal to 50 mg L^{-1}, the adsorption capacity of CD-containing gel is ca. 225 mg g^{-1}, 50% higher than the hydrogel with CD, after 4 h adsorption. However, the maximum adsorption capacity observed by the hydrogel was 1134 mg g^{-1}. This significant adsorption capacity can be justified by the occurrence of a mesoporous structure associated with the electrostatic interactions between the dye and the adsorbent. In fact, it has been found a relevant dependence of the adsorption capacity of MB on the pH and ionic strength; for example, it has been reported a decrease in the adsorption capacity of MB with an increase in the ionic strength (i.e., increasing the NaCl concentration), indicating the occurrence of the screening effect on the electrostatic interactions [119].

14.3.2.2.3   Cellulose-based Materials

Cellulose is an abundant, renewable, and eco-friendly polymer produced by plants and trees. Cellulose molecules, although insoluble in water [189], have many active hydroxyl groups, thus it can easily be chemically modified [251] and in such a way be capable to form supramolecular interactions with dyes, in particular organic dyes.

In the previous section, the use of modified filter paper for adsorption of

Cu(II) was described [251]. The use of that type of material is of outmost importance once the papers companies are looking for new innovative products and markets. Besides Cu(II), the performance of paper adsorption for MB, Brilliant Green, and Rhodamine-B has also been tested. The modified filter paper was able to adsorb 124.5, 130.4, and 99.7 mg g^{-1} of MB, RB, and BG, respectively. These values drastically decrease to 41.4 and 29.6 mg g^{-1} (for MB and RB, respectively), when the adsorbent material contains 70% of the pristine (nonmodified) filter paper, clearly indicating the importance of CD on the adsorption process.

Cotton fiber, a natural fiber made from cellulose, is a renewable, degradable, cheap, and nonpolluting material. Furthermore, cotton fiber presents a large specific surface area and hollow, flat-banded structure, which is beneficial for adsorption. Cotton fibers have also been tested as efficient adsorbents for micropollutants. Based on that, Yue et al. [366], grafted an amino-terminated hyperbranched polymer was obtained by melt polycondensation of methyl acrylate and diethylene triamine [370] and β-CD to cotton.

In summary, the chemical modification of cotton occurs after the initial oxidation with sodium iodate to obtain dialdehyde-cotton; subsequently, the modified cotton is reacted with β-CD using EPI; at the end of the process, the hyperbranched polymer is added, and the interaction with the modified cotton occurs via hydrogen bonding between amino groups of the hyperbranched polymer with the carbonyls of amide groups of cotton. The resulting polymer shows a higher removal efficiency and high maximum adsorption capacity for CR (94% and 300.8 mg g^{-1}, respectively); however, for MB, the performance is more modest leading to 42% and 98.7 mg g^{-1}, respectively. Such behavior is explained by the possibility of MB aggregation [366], which hampers adsorption.

In the present cases, the sorption mechanism follows a Freundlich model based on the occurrence of a heterogeneous adsorption surface and showing that the adsorption process is highly favorable.

As described before, CA is commonly used as a crosslinker for polymerization of cellulose-based materials and CD by forming an ester linkage.

Zhou et al. [393] have studied the functionalization of pine sawdust (PS) with β-CD by esterification with CA for aniline remediation (Scheme 14.24). Aniline and derivatives (see Scheme 14.25) are frequently used as an organic intermediate in the process of producing dyes, rubbers, medicine, and paint. To its toxicity and recalcitrant properties, it is highly demanding

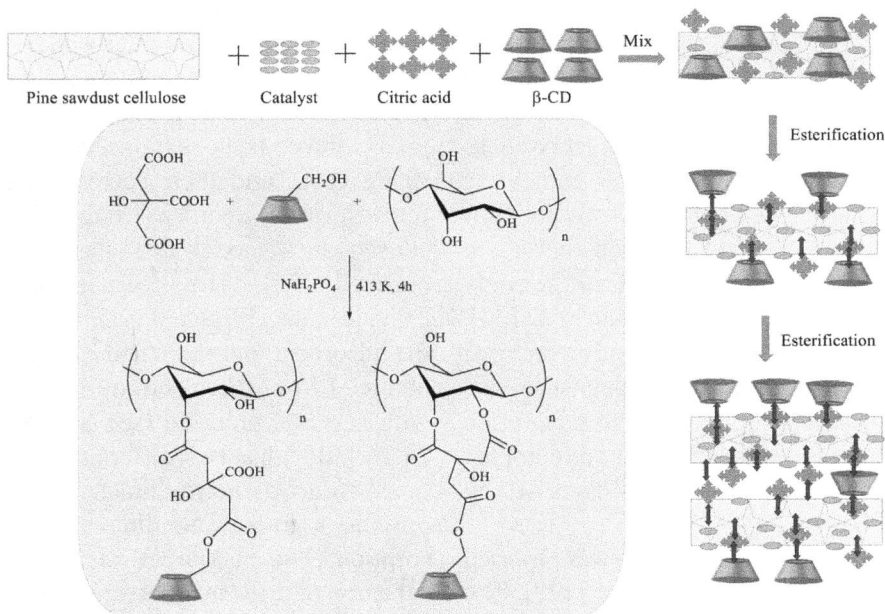

SCHEME 14.24   Pine sawdust modified with β-CD. Adapted with permission from [393]. Copyright 2020 American Chemical Society.

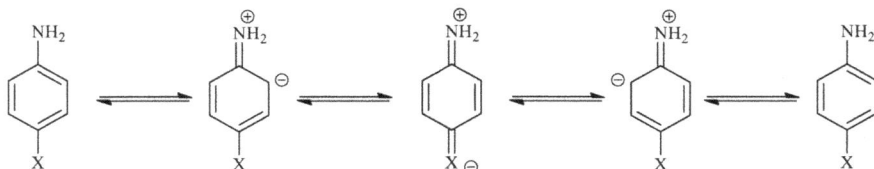

SCHEME 14.25   Resonance structures for aniline derivatives.

to develop strategies for removing aniline from aquatic media [19]. The adsorption capacity of PS-CA-CD toward aniline is 12.3 mg g^{-1}, more than twice the adsorption capacity of PS functionalized with CA, PS-CA, (5.8 mg g^{-1}). Such improvement has been justified by the structure of the sawdust, that is, cell walls of sawdust mainly consist of cellulose and lignin and many hydroxyl groups are present such as in tannins or other phenolic compounds; besides, the presence of CD improve the sorption capacity through host–guest hydrophobic interactions as well as hydrogen-bonding via carboxylic groups, whereas PS-CA only contains carboxyl groups. Additionally, aniline can be protonated, and its pK_a strongly depends on the substituents [134], promoting a further sorption process: the ion-exchange mechanism. This clearly suggests that the adsorption process and mechanism are complex and have been intensely discussed by Crini et al. [99].

A possible approach was reported by Ghemati and Aliouche [127]. The authors modified the β-CD through its reaction with *N*-methylol acrylamide, to form acrylamidomethylated-β-CD.

After that, the modified CD was grafted onto cellulose. The obtained polymer shows a maximum adsorption capacity for MO and MB of around 11 and 14 mg g^{-1}, respectively. The q_m values were obtained for solutions at pH 3 and 11.

The removal of chiral pollutants is challenging since enantiomers present similar physical properties but can differ regarding their toxicity and interaction with other chiral substances [12]. Chiral microspheres based on optically active helical polyacetylene and modified with β-CD have been synthesized by Liang et al. and their performance for adsorption of methyl red (MR), chosen as a model dye, has been checked [87]. The polyacetylene provides the chiral environment to the adsorbent and the CD derivative obtained from the reaction of β-CD with acryloyl chloride (see Scheme 14.26 [105]) has two different functions: to act as a crosslinker and to provide a further bonding site for sorption. The maximum adsorption capacity of MR in microspheres is 52 mg g^{-1}, for an initial concentration of 100 mg L^{-1}. The addition of CD derivative as crosslinker is significant on the adsorption mechanism, once the sorption of MR on microspheres with the CD leads to a $q_m = 17.7$ mg g^{-1}. Approximately 80% of the total sorbed amount of MR is adsorbed in the first hour of the experiment, showing a great affinity of the adsorbent toward the organic dye.

14.3.2.2.4 Fibers

In the last decade, an emphasis has been placed on the development of polymer-based fibers for high dye removal performance. The fibers can be obtained by different techniques, being the most common one electrospinning [37] and are

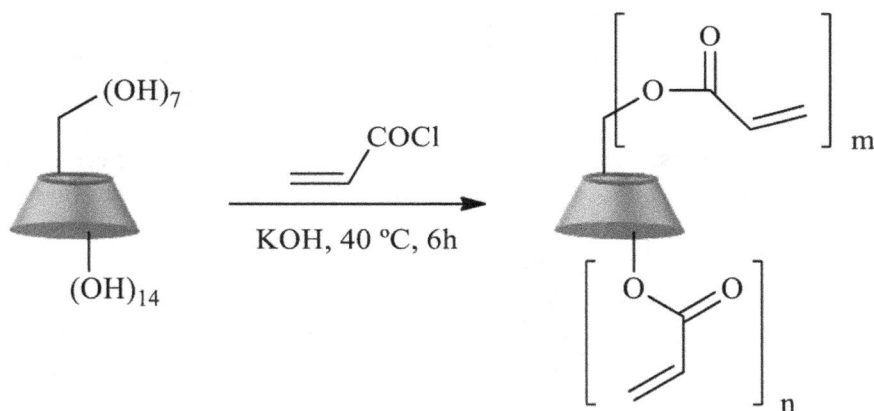

SCHEME 14.26 Synthesis of acryloyl-β-CD.

characterized by their long length and diameters ranging from 50 nm to some micrometers [209]. Additionally, fibers are also characterized by their good mechanical properties, high surface-to-volume or surface-to-weight ratios, and ease functionalization along with their ability to form supramolecular structures [64, 278]. Consequently, fibers have been explored in areas such as electrical and optical applications [320], biomedical [7], and other industrial applications [53]. Among them, the

treatment of wastewater [285], and in particular those containing dyes, will be focused here. In this respect, Zhao has prepared a set of composite nanofibers for the removal of MB and MO [389].

The first work has involved the formation of sericin/β-CD/poly(vinyl alcohol) composite nanofibers [389].

Sericin is a natural macromolecular protein containing a significant number of polar (hydroxyl, carboxyl, and amino) groups and, therefore, able to interact and sorb organic dyes. Being sericin and CD both hydrophilic, these molecules are not able to be used as an adsorbent for dyes. Thus, an aqueous mixture of sericin, β-CD, and PVA has been prepared. The in-situ thermal crosslinking was achieved by adding CA to the solution. The obtained fiber shows a maximum adsorption capacity of MB equal to 187.97, 229.89, and 261.10 mg g^{-1}, at 293, 313, and 333 K, respectively, and at pH 8. The adsorption capacity of this fiber is 1.5- and 2.5-fold higher than the adsorption capacity of PVA-CD and PVA-CA fibers, respectively.

A few months later, an electrospun nanofiber membrane based on a mixture of poly(acrylic acid)—PAA—and β-CD, and using CA as a crosslinker, was prepared [390]. The assessment of fiber properties showed that the crosslinking led, as expected, to a decrease in the surface area and pore area, and volume; on the other hand, the tensile strength and elongation at break increased with crosslinking. The maximum adsorption capacity of the MB onto PAA-CA-β-CD fiber is 826.45 mg g^{-1}, reached at pH 9. The adsorption mechanism for these fibers, as well as for the sericin-containing fibers, follows a Langmuir isotherm, indicating that the adsorption occurs via monolayer formation, and a pseudo-second-order kinetic model, suggesting a chemisorption adsorption process. It should be highlighted

that these fibers show much better performance than those previously described. However, these fibers are able to adsorb 12.41 mg g^{-1} of MO, at the same pH 9. Rationale behind such phenomenon is based on electrostatic repulsion between MO and carboxylic groups which also prevent the formation of host–guest interaction between the dye and the CD.

Looking for the development of environmentally friendly fibers, Chen et al. [75] have used gelatin-β-CD mixtures. Gelatin is a peptide-based polymer resulting from the hydrolysis of collagen [246]. The electrospun fibers were obtained by using glutaraldehyde as crosslinker and trifluoroethanol as the solvent. The adsorption capacity of these fibers was tested by using MB and a modest value of $q_m \approx 15$ mg g^{-1} has been found. As expected the maximum adsorption capacity increases by increasing the pH and by decreasing the temperature. These values suggest that the adsorption process is exothermic in agreement with a Langmuir-type isotherm.

Using also bio-based polymers, Guo et al. have reported the synthesis of fibers of ε-polycaprolactone (PCL) and poly(β-CD) (PCD), where the PCD was synthesized using EPI as crosslinker [165]. The dependence of the amount of PCD, from 10% to 50%, in the fibers composition, on the maximum adsorption capacity of MB was assessed and increase from

4.4 to 10.5 mg g^{-1}. These values are, even so, higher than those found for the adsorption of MB onto PCL (3.8 mg g^{-1}). However, q_m values for MB are 2–3 times lower than q_m values found for 4-aminoazobenzene (14.0–20.2 mg g^{-1} when PCD increases from 10% to 50% in the blend). Based on these values, the authors have hypothesized that the interaction of MB with fibers mainly occurs by adsorption whilst with the 4-aminoazobenzene host–guest interactions, involving the CD, are occurring.

Electrospun CNFs were also modified with β-CD to improve its capacity for removing TC (an antibiotic) and MB. The process is based on using polyimide as a support scaffold and the β-CD was chosen as a carbon precursor; afterward carbon NPs were deposited on the surface of the polyimide followed by its carbonization to get the CNFs [184]. These NFs show a much higher BET specific area (3562 m² g^{-1}) when compared with that for pure polyimide-based CNFs (716 m² g^{-1}). The maximum adsorption capacity of these fibers for MB is 543.5 mg g^{-1} (pH 11) and the sorption isotherm follows the Redlich–Peterson equation [344], suggesting that the adsorption is favorable although occurs via monolayer formation. The maximum adsorption capacity is significantly higher than that found for polyimide-based CNFs (272.5

mg g^{-1}), obtained in similar conditions [378]. However, here the β-CD is only used as a source of carbon and, consequently, does not play any role in the MB adsorption process, which mainly occurs via π–π interactions [184].

More recently, electrospun fibers composed of β-CD and poly(N-isopropylacrylamide-co-methacrylic acid) blends have been prepared. The electrospinning was followed by thermal crosslinking [154]. Fibers with different molar ratios of copolymer and PCD have been prepared and crystal violet (CV) has been used as a model molecule to test the adsorption performance. The obtained fibers show thermal responsive properties once the N-isopropylacrylamide monomers became hydrophobic at 55 °C. This leads to an increase in the maximum adsorption capacity of about 15%, from 1049.9 to 1253.8 mg g^{-1} (for an initial [CV] = 700 mg L^{-1}), when temperature increases from 25 to 55 °C. As it happens for all other systems cases (except where otherwise indicated), the adsorption of CV occurs via monolayer formation and the adsorbent can be assumed as having a homogeneous interface. This sorption mechanism is corroborated by a pseudo-second-order kinetics.

Due to the lightweight, high surface area and interconnected porous structure nanofibers can be used as membranes for environmental remediation [333]. Nanofibrous membranes composed of poly(butylene succinate-*co*-terephthalate) (PBST) and PCD have been synthesized for the adsorption of MB [340]. The PBST has been polymerized according to the procedure described in the ref. [340]. The PBST membrane is then functionalized in situ with PCD, previously synthesized using CA as the crosslinker. The PBST/PCD membrane, independently of the amount of PCD used in the membrane functionalization, shows removal efficiency higher than 90% (with a maximum removal efficiency of 98.9%), and a significant removal capacity is reached in less than 1 h. The sorption mechanism seems to essentially involve the CD once the maximum sorption capacity of MB onto PCD (after 12 h immersion) is 7.61 mg g^{-1} and onto the membrane is 7.91 mg g^{-1} (by using the membrane with a mass ratio PBST/PCD equal to 1.2).

14.3.2.3   PESTICIDES

Pesticides are the main organic pollutants worldwide, possessing, in general, a common characteristic: poor solubility in water, which difficult the respective capture. For this reason, removal strategies based on CD and CD-polymers have a great advantage in solubilizing and removing such contaminants [125, 210].

There are a few studies focused on the design and application of CD-based adsorbents for pesticides.

For instance, Cai et al. [56] [194] have proposed a rational screening of CD-based polymer based on the relationship between adsorption potential and adsorbent adsorbate, with the aim of adsorbing and separating a mixture of 10 pesticides from water. The spherical porous CD-polymers were prepared via one-fold, or composite CD as complex, and EPI as the cross-linker. The adsorption kinetics and isotherms of the polymers toward pesticides suggested that adsorbents with a homogeneous open network structure were able to absorb pesticides through multiple adsorption interactions, including inclusion in CD, loading into swelling water, and also physical adsorption on the network. Individual contributions to adsorption performances of the CD-polymer properties were discriminated by multivariate regression analysis. It was concluded that CD content, swelling capacity, and pore size were the most relevant factors. This relationship was then imposed for screening the multiplex CD-polymer and obtaining higher adsorption potential.

CDs inside of nanoporous carbon have also displayed great efficiency for removing the *p,p'* substituted diphenyl class of pesticides (e.g., DDT, DDD, and DDE). These pesticides showed significant adsorption due to their suitable geometric fit within the cavity of the CD, leading to the formation of stable inclusion complexes. In a similar study, CD-polymers were introduced in ceramic membranes composed of Al_2O_3, SiC, and TiO_2 [15].

Other adsorbents containing CD-functionalized mesoporous silica for hexachlorocyclohexane-based, hexachlorobicycloheptene-based, and p,p' substituted biphenyl-based pesticides (with a mass concentration range of 0.060–0.270 μg mL^{-1}) in solution, have also been investigated by Sawicki and Mercier [284]. The authors have highlighted the importance of controlling the position and orientation of the binding sites within the structure of the adsorbent, together with the maximization of the respective number. It was found that adsorbents containing low to intermediate amounts of CD molecules displayed optimal adsorption affinity toward the pesticides, with higher specificity toward p,p' substituted diphenyl-based pesticides.

In another study, β-CD was cross-linked with rigid aromatic groups, yielding a high-surface-area, mesoporous polymer, displaying adsorption capacity rate constants of 15–200 times greater than nonporous β-CD and activated carbon adsorbent [210]

On the other hand, beads of an EPI-crosslinked β-CD polymer were prepared and characterized (using FTIR) by Orprecio and Evans [255]. The CD-polymer was employed for packing a column directed at trapping molecular models of pesticides (e.g., 2-naphthol with a trapping efficiency of 70%) that are subjected to acid–base equilibria in natural waters. The best trapping efficiencies, determined with fluorescence spectroscopy, were obtained for beads with a nominal CD/EPI ratio of 1:29.

The sorption properties of CD-polymers in the form of nanosponges with pesticides, such as 4-chlorophenoxyacetic acid and 2,3,4,6-tetrachlorophenol have been evaluated and compared with other materials (e.g., granulated activated carbon) [281]. These designed nanosponges have displayed favorable sorption capacities for chlorinated aromatic guests.

The sorption capacity and binding affinity of the sorbents were greater for nanosponges, despite granulated activated carbon possessing a greater surface area. It was also suggested that nanosponges possess additional properties, such as serving as stabilization enhancers of NPs.

A magnetic copper-based metal-organic framework, based on a Fe_4O_3–GO–β–CD nanocomposite (as the magnetic core and support), was prepared and used for the adsorption/removal of neonicotinoid insecticides from aqueous solution, with adsorption capacities for clothianidin, nitenpyram, thiacloprid, thiamethoxam, acetamiprid, imida-cloprid, and dinotefuran (at 100 mg L^{-1}) of 1.77, 2.56, 2.88, 2.88, 2.96, and 3.11 mg g^{-1}, respectively [190].

More recently [157], the coating of anionic CD-polymer on PET textile was carried out by crosslinking between β-CD and CA (CTR) in order to remove paraquat herbicide from aqueous solution. This is a popular hydrosoluble herbicide displaying water solubility of 620 g L^{-1} at 25 °C, lethal dose for human of 35 mg kg^{-1}, and maximum permissible concentration for drinking water and for the surface of 0.1 and 1–3 μg L^{-1}.

Among the various methods employed to remove paraquat from water, adsorption

has demonstrated higher efficiency, straightforward operation, and low-cost.

Junthip et al. [157] have proposed and tested the adsorption efficiency of paraquat in aqueous solutions using textile coated with anionic CD-polymer, where CA was used as a bridging agent. The removal performance was assessed considering different conditions, including the initial concentration of pesticide, pH of the solution, and adsorption temperature. It was found that neutral pH and lower temperatures favored the adsorption of the pesticide on the functionalized textile. Pseudo-second-order model and Langmuir models were selected for describing adsorption kinetics (achieved at 420 min) and isotherm, respectively. The

estimated thermodynamic parameters suggested a spontaneous process ($\Delta G°$ lower than 0), an exothermic process ($\Delta H°$ lower than 0), and an enhanced disorder ($\Delta S°$ higher than 0). Also relevant was the reusability of the functionalized textile in methanol which reached 78.6% after six regeneration cycles [157].

Moniri et al. [311] have recently prepared and employed a grafted β-CD/thermo-sensitive polymer onto modified $Fe_3O_4@SiO_2$ NPs for removing fenitrothion from aqueous solution. Best sorption of fenitrothion (ca. 30 mg g^{-1}) on the grafted CD-polymer was achieved under a temperature of 25 °C and pH of 6, and the best sorption time was 5 min. Adsorption efficiency was higher than 99%.

14.3.2.4 PHARMACEUTICAL COMPOUNDS

Pharmaceuticals, including analgesic, anti-inflammatory, and antibiotics, have been introduced into the environment via several routes, including farming, households, medical facilities, hospitals, and pharmaceutical industries. Specifically, these include those compounds that are frequently used and found in wastewater, including ibuprofen, acetaminophen, amoxicillin (β-lactams), streptomycin (amino-glycosides), ciprofloxacin (fluoroquinolones), azithromycin and clarithromycin (macrolides), penicillin, penicillin/streptomycin combinations, and TCs, just to name a few examples [85].

Among the most consumed pharmaceuticals worldwide, ibuprofen is a nonsteroidal anti-inflammatory drug largely distributed without prescriptions. The respective occurrence in environmental matrices has been markedly observed in the industrial and agricultural waste stream, municipal/hospital wastewater, and surface water.

On the other hand, high concentrations of veterinary drugs such as ionophore antibiotics have been detected in sediments and in water (in lower concentrations). Ibuprofen, diclofenac, and gemfibrozil have also been identified in sewage sludge.

Insoluble CD-polymers have been employed as cost-effective adsorbents in these scenarios due to the attractive removal performances and a higher number of cycles when compared to other convention strategies [355].

Shahgaldian et al. [359] have prepared three water-insoluble CD-based polymers aiming at removing three selected pharmaceutically active ingredients, levofloxacin, aspirin, and acetaminophen, in aqueous solutions. Results suggested that the sorption of the pharmaceuticals on the polymers followed the Freundlich model. The highest sorption capacity was obtained with β-CD polymer that displayed a higher affinity for aspirin. In general,

the interaction kinetics indicated an increase of the adsorbed equilibrium amount of 70% within 10 min.

Moulahcene et al. [233] have proposed insoluble CD-polymers with high adsorption capacities, containing different types of CD molecules crosslinked with CA, toward the removal of ibuprofen in aqueous solutions. Ibuprofen retention in the CD-polymers decreased following the order: α-γ-CD, α-CD, γ-CD, β-CD, α-γ-β-CD. The effect of various operating variables, such as the pH, ionic strength, contact time, drug concentration, and mass of adsorbent, was evaluated. The maximum adsorption capacity was achieved in acidic pH, in which ibuprofen was in its molecular form, favoring the electrostatic interaction, and the formation of the inclusion complex. The ionic strength also contributed to increase the removal performance of the CD-polymers.

In a similar study [235], the removal capacity in aqueous solutions of various CD-polymers containing different CD molecules crosslinked with CA, toward progesterone, was investigated. It was concluded that (1) the CD-polymers, and the β-CD polymer, in particular, exhibited higher adsorption capacities and cycle number than activated carbon, and short adsorption time toward progesterone, (2) no significant effect of initial pH, initial concentration, and adsorbent amount, on the adsorption capacity of β-CD polymer was found, (3) adsorption increased with ionic strength and temperature, (4) kinetic extraction was successfully modeled by a pseudo-second-order model, and (5) adsorption of progesterone on CD-polymers was a spontaneous and endothermic process.

Ibuprofen was also the target contaminant in another study [158] in which β-CD-containing nanofilters possessing different chemical composition and thickness (1.5–3.5 mm) were designed. The aim was to measure the adsorption capacity of β-CD polymer beads and nanofilters containing the former. It was expected that the proposed method was suitable for determining several organic micropollutants displaying well-defined chemical structure and composition and also detecting organic contaminants released from the nanofilters.

Mhlanga et al. [237] have conducted kinetic and thermodynamic studies of the sorption and removal of aspirin and paracetamol from aqueous solutions, employing nitrogen-doped carbon nanotubes (N-CNTs) and Fe anchored N-CNTs (Fe/N-CNTs) copolymerized with β-CDs. The sorption process for the drugs was described by Freundlich and Langmuir isotherms (aspirin and paracetamol, respectively). Maximum sorption capacities of 71.9 and 41.0 mg g^{-1} onto N-CNTs-β-CD, and 101.0 and 75.2 mg g^{-1} onto Fe/N-CNTs-β-CD, for aspirin and paracetamol,

respectively, were obtained at 298 K. Drugs were removed within 30 min of contact time by nanocomposites. The pseudo-second-order and Elovich kinetic models were used for describing the sorption kinetics of both drugs onto the nanocomposites. The sorption process was also spontaneous and exothermic.

The detection and quantification of diclofenac in water and wastewater samples (from 1 M, 0.3 mg L^{-1} to 50 M, 15.9 mg L^{-1}) has also been explored [350] resorting to a CD-polymer (optimal concentration of 50 g mL^{-1}) containing a fluorescent dye (100 Nm).

The nucleophilic substitution of β-CD with TFP has been conducted by Zhou et al. [392] aiming at preparing a β-CD polymer for removing BPA, chloroxylenol (PCMX), and carbamazepine (CBZ). Langmuir isotherm suggested that the maximum adsorption capacities of β-CD polymer were 164.4, 144.1, and 136.4 mg g^{-1} for BPA, PCMX, and CBZ, respectively. The polymer was regenerated five times by methanol soaking, and the removal efficiency of BPA, PCMX, and CBZ was 98.1%, 90.8%, and 65.0%. No significant effect of pH, fulvic acid concentration, or ionic strength on the adsorption of the drugs onto the CD-polymer was found.

Recently, Lahiani-Skiba et al. [234] prepared polymer inclusion membranes incorporating, in varying proportions, an insoluble β-CD polymer as a carrier, poly(vinyl chloride) as a base polymer, and dibuthylphtalate as a plasticizer, for removing ibuprofen and progesterone in water and wastewater. The proportion of β-CD polymer and agitation of wastewater favored drug extraction at 10 ppm. Removal of drugs was also more effective at acidic pH.

14.4 CONCLUDING REMARKS AND FUTURE CHALLENGES

At the heart of the process "to reconcile the economy with our planet, to reconcile the way we produce, the way we consume with our planet and to make it work with our people," following "a new growth strategy that aims to transform the EU into a fair and prosperous society, with a modern, resource-efficient and competitive economy where there are no net emissions of greenhouse gases in 2050" (words of the European Commission President Ursula von der Leyen when launching the Green Deal on December 11, 2019) it is paramount to (1) adapt all types of technology for dealing with the effects of climate change, (2) manage and clean-contaminated environmental resources, and (3) carry out an in-depth analysis and revision of the most effective systems for soil, water, and air remediation.

Intense debate on the best strategies is ongoing with the common notion that best solutions are discovered by combining comprehensive multidisciplinary approaches. It is clear that all Chemistry, Materials Science, and Engineering subfields play a key role in achieving such ambitious goals.

This chapter describes several remediations approaches toward different micropollutants based on the inherent characteristics of CD-polymers and related adsorption/removal processes, addressing structural characteristics, operating variables, pollutant and matrix specificities, and remediation performances.

The most promising combination of technologies for soil, water, and air remediation must be scaled-up and tested in the field. The selection of the best adsorbing materials should be based on the (1) specificity of the target environmental matrix, (2) removal efficiency, (3) nature and concentration of the target pollutants, (4) applicability in situ, (5) technical barriers to up-scale, (6) versatility, and (7) economic feasibility.

In the context of CD contributions, contaminant removal with CD-based materials should be prioritized over ongoing studies at the lab-scale, considering systematic, and comparative procedures for evaluating performances among CDs, and CD-polymers, and other conventional materials. This will allow in achieving effective environmental application regimes.

The design of innovative and effective materials based on supramolecular chemistry and host–guest interactions of CD has provided attractive remediation solutions, especially those directed at cleaning sediments and aqueous environments.

In spite of the increasing number of conclusive studies, interpreting the mechanisms of pollutant adsorption and removal remains an interesting source of debate. Theoretical and computational rationales of the involved interaction patterns/mechanisms are still much needed, as these rationales will be the mainstay of (1) the knowledge on the fundamental mechanisms governing structural modification, performance, and recycling of potential adsorbing materials, as well as on the solubilization and recognition phenomena, (2) the design and optimization of remediation systems, and (3) the next generation of environmental models.

ACKNOWLEDGMENTS

The Coimbra Chemistry Centre (CQC) is supported by the Portuguese Agency for Scientific Research, "Fundação para a Ciência e a Tecnologia" (FCT), through Project UID/QUI/00313/2020.

KEYWORDS

- cyclodextrins
- polymers
- pollutants
- adsorption
- environmental remediation

REFERENCES

1. Abd-Elhamid, A I, Nayl, A A, El. Shanshory, A A, Soliman, H M AAly, H F (2019). Decontamination of organic pollutants from aqueous media using cotton fiber–graphene oxide composite, utilizing batch and filter adsorption techniques: a comparative study. RSC Adv, 9: 5770–5785. DOI: 10.1039/C8RA10449B.

2. Abdallah, A M (2019). The effect of hydrogel particle size on water retention properties and availability under water stress. Int Soil Water Cons Res. DOI: 10.1016/j.iswcr.2019.05.001.

3. Abdelhameed, R M, El-Zawahry, M, Emam, H E (2018). Efficient removal of organophosphorus pesticides from wastewater using polyethylenimine-modified fabrics. Polymer, 155: 225–234. DOI: 10.1016/j.polymer. 2018.09. 030.

4. Abdolmaleki, A, Mallakpour, S, Borandeh, S (2015). Efficient heavy metal ion removal by triazinyl-β-cyclodextrin functionalized iron nanoparticles. RSC Adv, 5: 90602–90608. DOI: 10.1039/C5RA15134A.

5. Adriano, D C (2001). Trace elements in terrestrial environments: biogeochemistry, bioavailability, and risks of metals. DOI: 10.1007/978-0-387-21510-5.

6. Adriano, D C, Wenzel, W W, Vangrons-veld, J, Bolan, N S (2004). Role of assisted natural remediation in environmental cleanup. Geoderma, 122: 121–142. DOI: 10.1016/j.geoderma. 2004.01.003.

7. Agarwal, S, Wendorff, J H, Greiner, A (2008). Use of electrospinning technique for biomedical applications. Polymer, 49: 5603–5621. DOI: 10.1016/j.polymer.2008.09.014.

8. Aioub, A A A, Li, Y, Qie, X, Zhang, X, Hu, Z (2019). Reduction of soil contamination by cypermethrin residues using phytoremediation with plantago major and some surfactants. Environ Sci Eur, 31: 26. DOI: 10.1186/s12302-019-0210-4.

9. Alaba, P A, Oladoja, N A, Sani, Y M, Ayodele, O B, Mohammed, I Y, Olupinla, S F, Daud, W M W (2018). Insight into wastewater decontamination using polymeric adsorbents. J Environ Chem Eng, 6: 1651–1672. DOI: https://doi.org/10.1016/j.jece.2018.02.019.

10. Alaboudi, K A, Ahmed, B, Brodie, G (2020). Soil washing technology for removing heavy metals from a contaminated soil: a case study. Polish J Environ Studies, 29: 1029–1036. DOI: https://doi.org/10.15244/pjoes/104655.

11. Ali, H, Mukhopadhyay, S, Jana, N R (2019a). Selective electrochemical detection of bisphenol a using a molecularly imprinted polymer nanocomposite. New J Chem, 43: 1536–1543. DOI: 10.1039/C8NJ05883K.

12. Ali, I, Kulsum, U, Saleem, K, Hussain, A (2013), Chiral pollutants, Kirk-Othmer Encyclopedia of Chemical Tech. John Wiley & Sons, Inc. : Hoboken, NJ, USA.

13. Ali, I, Mbianda, X, Burakov, A, Galunin, E, Burakova, I, Mkrtchyan, E, Tkachev, A, Grachev, V (2019b). Graphene based adsorbents for remediation of noxious pollutants from wastewater. Environ Int, 127: 160–180.

14. Ali, R, Mina, O, Masoome, S, Farideh, G (2018). Natural polymers as environmental friendly adsorbents for organic pollutants such as dyes removal from colored wastewater. Curr Org Chem, 22: 1297–1306. DOI: 10.2174/138527 2822666180511123839.

15. Allabashi, R, Arkas, M, Hörmann, GT, Siourvas, D (2007). Removal of some organic pollutants in water employing ceramic membranes impregnated with cross-linked silylated dendritic and cyclodextrin polymers. Water Res, 41: 476–486. DOI: 10.1016/j.watres.2006. 10.011.

16. Alsbaiee, A, Smith, B J, Xiao, L, Ling, Y, Helbling, D E, Dichtel, W R (2016a). Rapid removal of organic micropollutants from water by a porous beta-cyclodextrin polymer. Nature, 529: 190–U146. DOI: 10.1038/nature16185.

17. Alsbaiee, A, Smith, B J, Xiao, L, Ling, Y, Helbling, D E, Dichtel, W R (2016b). Rapid removal of organic micropollutants from water by a porous β-cyclodextrin polymer. Nature, 529: 190–194. DOI: 10.1038/nature16185.

18. Moya-Ortega, M D, Alvarez-Lorenzo, C, Concheiro, A, Loftsson, L (2012). Cyclodextrin-based nanogels for pharmaceutical and biomedical applications. Int J Pharm, 428: 152–163. DOI: 10.1016/j.ijpharm.2012.02.038..

19. An, F, Feng, X, Gao, B (2009). Adsorption of aniline from aqueous solution using novel adsorbent pam/sio2. Chem Eng J, 151: 183–187. DOI: 10.1016/j. cej.2009.02.011.

20. Annadurai, G, Ling, L Y, Lee, J-F (2008). Adsorption of reactive dye from an aqueous solution by chitosan: isotherm, kinetic and thermodynamic analysis. J Hazard Mater, 152: 337–346. DOI: 10.1016/j.jhazmat.2007.07.002.

21. Aoki, N, Nishikawa, M, Hattori, K (2003). Synthesis of chitosan derivatives bearing cyclodextrin and

adsorption of p-nonylphenol and bisphenol A. Carbohydr Polym, 52: 219–223.

22. Arjoon, A, Olaniran, A O, Pillay, B (2013). Co-contamination of water with chlorinated hydrocarbons and heavy metals: challenges and current bioremediation strategies. Int J Environ Sci Technol, 10: 395–412. DOI: 10.1007/s13762-012-0122-y.

23. Arslan, M, Sayin, S, Yilmaz, M (2013). Removal of carcinogenic azo dyes from water by new cyclodextrin-immobilized iron oxide magnetic nanoparticles. Water Air Soil Pollut, 224: 1527. DOI: 10.1007/s11270-013-1527-z.

24. Asghar, A, Raman, A A A, Daud, W M A W (2015). Advanced oxidation processes for in-situ production of hydrogen peroxide/hydroxyl radical for textile wastewater treatment: a review. J Cleaner Prod, 87: 826–838. DOI: 10.1016/j.jclepro.2014.09.010.

25. Asman, S, Mohamad, S, Sarih, N (2015). Exploiting β-cyclodextrin in molecular imprinting for achieving recognition of benzylparaben in aqueous media. Int J Mol Sci, 16: 3656–3676. DOI: 10.3390/ijms16023656.

26. Atteia, O, Del Campo Estrada, E, Bertin, H (2013a). Soil flushing: a review of the origin of efficiency variability. Rev Environ Sci Bio/Technol, 12: 379–389. DOI: 10.1007/s11157-013-9316-0.

27. Atteia, O, Estrada, E D C, Bertin, H (2013b). Soil flushing: a review of the origin of efficiency variability. Rev Environ Sci Bio/Technol, 12: 379–389. DOI: 10.1007/s11157-013-9316-0.

28. Babuponnusami, A, Muthukumar, K (2014). A review on fenton and improvements to the fenton process for wastewater treatment. J Environ Chem Eng, 2: 557–572. DOI: 10.1016/j.jece.2013.10.011.

29. Badruddoza, A Z M, Hazel, G S S, Hidajat, K, Uddin, M S (2010).

Synthesis of carboxymethyl-β-cyclodextrin conjugated magnetic nano-adsorbent for removal of methylene blue. Colloids Surf, A Physicochem Eng Asp, 367: 85–95. DOI: 10.1016/j.colsurfa.2010.06.018.

30. Badruddoza, A Z M, Shawon, Z B Z, Tay, W J D, Hidajat, K, Uddin, M S (2013a). Fe3o4/cyclodextrin polymer nanocomposites for selective heavy metals removal from industrial wastewater. Carbohydr Polym, 91: 322–332. DOI: 10.1016/j.carbpol.2012.08.030.

31. Badruddoza, A Z M, Shawon, Z B Z, Tay, W J D, Hidajat, K, Uddin, M S (2013b). Fe3o4/cyclodextrin polymer nanocomposites for selective heavy metals removal from industrial wastewater. Carbohydr Polym, 91: 322–332. DOI: https://doi.org/10.1016/j.carbpol.2012.08.030.

32. Belgsir, E M, Liaigre, D, Breton, T, El Belgsir, M. Regioselective oxidation of cyclodextrin to percarboxy-cyclodextrin for use e.G. As an inclusion agent, involves reaction with an oxidation mediator, e.G. Oxo-ammonium ions derived from tetramethylpiperidine-1-oxyl, Cent Nat Rech Sci; University of Poitiers; Breese Derambure Majerowicz, Poitiers.

33. Belyakova, L A, Lyashenko, D Y (2014). Nanoporous functional organosilicas for sorption of toxic ions. Russ J Phys Chem A, 88: 489–493. DOI: 10.1134/S0036024414030030.

34. Bennett, J W, Klich, M (2003). Mycotoxins. Clin Microbio Rev, 16: 497. DOI: 10.1128/CMR.16.3.497-516.2003.

35. Berto, S, Bruzzoniti, M C, Cavalli, R, Perrachon, D, Prenesti, E, Sarzanini, C, Trotta, F, Tumiatti, W (2007). Synthesis of new ionic β-cyclodextrin polymers and characterization of their heavy metals retention. J Inclusion Phenom Macrocyclic Chem, 57: 631–636. DOI: 10.1007/s10847-006-9273-0.

36. Beslin, L (2017). Textural features-indicators of pollution. J Environ Anal Toxicol, 7: 2161–0525.1000505. DOI: 10.4172/2161-0525.1000505.

37. Bhardwaj, N, Kundu, S C (2010). Electrospinning: a fascinating fiber fabrication technique. Biotechnol Adv, 28: 325–347. DOI: 10.1016/j.biotechadv.2010.01.004.

38. Bhattarai, B, Muruganandham, M, Suri, R P S (2014). Development of high efficiency silica coated beta-cyclodextrin polymeric adsorbent for the removal of emerging contaminants of concern from water. J Hazard Mater, 273: 146–154. DOI: 10.1016/j.jhazmat.2014.03.044.

39. Binello, A, Cravotto, G, Nano, G M, Spagliardi, P (2004). Synthesis of chitosan–cyclodextrin adducts and evaluation of their bitter-masking properties. Flavour Fragance J, 19: 394–400. DOI: 10.1002/ffj.1434.

40. Biswas, A, Appell, M, Liu, Z, Cheng, H N (2015). Microwave-assisted synthesis of cyclodextrin polyurethanes. Carbohydr Polym, 133: 74–79. DOI: https://doi.org/10.1016/j.carbpol.2015.06.044.

41. Blanford, W J, Pecoraro, M P, Heinrichs, R, Boving, T B (2018). Enhanced reductive de-chlorination of a solvent contaminated aquifer through addition and apparent fermentation of cyclodextrin. J Contam Hydrol, 208: 68–78. DOI: https://doi.org/10.1016/j.jconhyd.2017.10.006.

42. Bolan, N, Choppala, G, Kunhikrishnan, A, Park, J Ravi, N 2013, Microbial Transformation of Trace Elements in Soils in Relation to Bioavailability and Remediation, in M D Whitacre (ed.), Reviews of Environmental Contamination and Toxicology, Springer: New York, NY, 1–16.

43. Borhani, S (2015). Removal of reactive dyes from wastewater using cyclodextrin functionalized polyacrylonitrile nanofibrous membranes. J Tex Polym, 4: 45–52. DOI.

44. Bose, P K, Polavarapu, P L (1999). Acetate groups as probes of the stereochemistry of carbohydrates: a vibrational circular dichroism study. Carbohydr Res, 322: 135–141. DOI: 10.1016/S0008-6215(99)00211-6.

45. Boving, T, Blanford, W, McCray, J, Divine, C, Brusseau, M L (2008). Comparison of line-drive and push-pull flushing schemes. Ground Water Monit Rem, 28: 75–86. DOI: 10.1111/j.1745-6592.2007.00182.x.

46. Boving, T B Brusseau, M L (2000). Solubilization and removal of residual trichloroethene from porous media: comparison of several solubilization agents. J Contam Hydrol, 42: 51–67.

47. Boving, T B, McCray, J E (2000). Cyclodextrin-enhanced remediation of organic and metal contaminants in porous media and groundwater. Remed J, 10: 59–83. DOI: 10.1002/rem.3440100206.

48. Boving, T B, Wang, X, Brusseau, M L (1999). Cyclodextrin-enhanced solubilization and removal of residual-phase chlorinated solvents from porous media. Environ Sci Technol, 33: 764–770. DOI: 10.1021/es980505d.

49. Brusseau, M, Boving, T, Blanford, W, Klingel, E, McCray, J, Wang, X (2003). Extraction of pollutants from underground water. US Patent US20030146172A1, published 2003-08-07, assigned to Cerestar Holding BV..

50. Brusseau, M L, Wang, X, Hu, Q (1994). Enhanced transport of low-polarity organic compounds through soil by cyclodextrin. Environ Sci Technol, 28: 952–956. DOI: 10.1021/es00054a030.

51. Brusseau, M L, Wang, X, Wang, W-Z (1997). Simultaneous elution of heavy metals and organic compounds from soil by cyclodextrin. Environ

Sci Technol, 31: 1087–1092. DOI: 10.1021/es960612c.

52. Bueno, P V, Matsushita, A F, Rubira, A F, Muniz, E C, Durães, L, Murtinho, D, Valente, A J (2018). Synthesis, characterization and sorption studies of aromatic compounds by hydrogels of chitosan blended with β-cyclodextrin-and pva-functionalized pectin. RSC Adv, 8: 14609–14622.

53. Burger, C, Hsiao, B S, Chu, B (2006). Nanofibrous materials and theris applications. Annu Rev Mater Res, 36: 333–368. DOI: 10.1146/annurev.matsci.36.011205.123537.

54. Buvári, Á, Barcza, L (1979). B-cyclodextrin complexes of different type with inorganic compounds. Inorg Chim Acta, 33: L179–L180. DOI: 10.1016/S0020-1693(00)89441-4.

55. Buvári, Á, Barcza, L (1989). Complex formation of inorganic salts with β-cyclodextrin. J Inclusion Phenom Mol Recognit Chem, 7: 379–389. DOI: 10.1007/BF01076992.

56. Cai, D, Zhang, T, Zhang, F, Luo, X (2017). Quaternary ammonium β-cyclodextrin-conjugated magnetic nanoparticles as nano-adsorbents for the treatment of dyeing wastewater: synthesis and adsorption studies. J Environ Chem Eng, 5: 2869–2878. DOI: https://doi.org/10.1016/j.jece.2017.06.001.

57. Cao, X T, Showkat, A M, Kang, I, Gal, Y-S, Lim, K T (2016). β-cyclodextrin multi-conjugated magnetic graphene oxide as a nano-adsorbent for methylene blue removal. J Nanosci Nanotechnol, 16: 1521–1525. DOI: 10.1166/jnn.2016.11987.

58. Cao, X T, Showkat, A M, Kim, D W, Jeong, Y T, Kim, J S, Lim, K T (2015). Preparation of β-cyclodextrin multi-decorated halloysite nanotubes as a catalyst and nanoadsorbent for dye removal. J Nanosci Nanotechnol, 15: 8617–8621. DOI: 10.1166/jnn.2015.11482.

59. Carrazana, J, Reija, B, Cabrer, P R, Al-Soufi, W, Novo, M, Tato, J V (2004). Complexation of methyl orange with ß-cyclodextrin: detailed analysis and application to quantification of polymer-bound cyclodextrin. Supramol Chem, 16: 549–559. DOI: 10.1080/10610270412331315235.

60. Domenech, N A, Serraima, C C, Puche, J C, Montiel, A F, Sanz, N G, Van den Nest, W (2014). Peptides used in the treatment and/or care of the skin, mucous membranes and/or scalp and their use in cosmetic or pharmaceutical compositions. Australian Patent AU2010213094B2, published 2014-07-24, assigned to Lipotec SA.

61. Carvalho, L B d, Carvalho, T G, Magriotis, Z M, Ramalho, T d C, Pinto, L d M A (2014). Cyclodextrin/silica hybrid adsorbent for removal of methylene blue in aqueous media. J Inclusion Phenom Macrocyclic Chem, 78: 77–87. DOI: 10.1007/s10847-012-0272-z.

62. Castelo-Grande, T, Augusto, P A, Monteiro, P, Estevez, A M, Barbosa, D (2010). Remediation of soils contaminated with pesticides: a review. Int J Environ Anal Chem, 90: 438–467. DOI: 10.1080/03067310903374152.

63. Celebioglu, A, Yildiz, Z I, Uyar, T (2017). Electrospun crosslinked poly-cyclodextrin nanofibers: highly efficient molecular filtration thru host-guest inclusion complexation. Sci Rep, 7: 7369. DOI: 10.1038/s41598-017-07547-4.

64. Chabalala, M B, Seshabela, B C, Hulle, S W H V, Mamba, B B, Mhlanga, S D, Nxumalo, E N (2018). Cyclodextrin-based nanofibers and membranes: fabrication, properties and applications, cyclodextrin—a versatile ingredient. InTech.

65. Chain, E Panel o C i t F, Knutsen, H-K, Alexander, J, Barregård, L, Bignami,

M, Brüschweiler, B, Ceccatelli, S, Cottrill, B, Dinovi, M, Edler, L, Grasl-Kraupp, B, Hogstrand, C, Hoogen-boom, L, Nebbia, C S, Petersen, A, Rose, M, Roudot, A-C, Schwerdtle, T, Vleminckx, C, Vollmer, G, Wallace, H, Dall'Asta, C, Dänicke, S, Eriksen, G-S, Altieri, A, Roldán-Torres, R, Oswald, I P (2017). Risks for animal health related to the presence of zearalenone and its modified forms in feed. EFSA J, 15: e04851. DOI: 10.2903/j.efsa.2017.4851.

66. Chalasani, R, Vasudevan, S (2012). Cyclodextrin functionalized magnetic iron oxide nanocrystals: a host-carrier for magnetic separation of non-polar molecules and arsenic from aqueous media. J Mater Chem, 22: 14925–14931. DOI: 10.1039/C2JM32360E.

67. Chatain, V, Hanna, K, De Brauer, C, Bayard, R, Germain, P (2004). Enhanced solubilization of arsenic and 2,3,4,6 tetrachlorophenol from soils by a cyclodextrin derivative. Chemosphere, 57: 197–206. DOI: 10.1016/j.chemosphere.2004.07.002.

68. Chauke, V P, Maity, A, Chetty, A (2015). High-performance towards removal of toxic hexavalent chromium from aqueous solution using graphene oxide-alpha cyclodextrin-polypyrrole nanocomposites. J Mol Liq, 211: 71–77. DOI: https://doi.org/10.1016/j.molliq.2015.06.044.

69. Chen, B, Chen, S, Zhao, H, Liu, Y, Long, F, Pan, X (2019a). A versatile β-cyclodextrin and polyethyleneimine bi-functionalized magnetic nanoadsorbent for simultaneous capture of methyl orange and Pb(II) from complex wastewater. Chemosphere, 216: 605–616. DOI: 10.1016/j.chemosphere.2018.10.157.

70. Chen, C-Y, Chen, C-C, Chung, Y-C (2007). Removal of phthalate esters by α-cyclodextrin-linked chitosan bead. Bioresource Tech, 98: 2578–2583. DOI: https://doi.org/10.1016/j.biortech.2006.09.009.

71. Chen, F, Li, X, Ma, J, Qu, J, Yang, Y, Zhang, S (2019b). Remediation of soil co-contaminated with decabromodiphenyl ether (BDE-209) and copper by enhanced electrokinetics-persulfate process. J Hazard Mater, 369: 448–455. DOI: 10.1016/j.jhazmat.2019.02.043.

72. Chen, F, Luo, Z, Liu, G, Yang, Y, Zhang, S, Ma, J (2017). Remediation of electronic waste polluted soil using a combination of persulfate oxidation and chemical washing. J Environ Manage, 204: 170–178. DOI: 10.1016/j.jenvman.2017.08.050.

73. Chen, M, Ding, W, Wang, J, Diao, G (2013a). Removal of azo dyes from water by combined techniques of adsorption, desorption, and electrolysis based on a supramolecular sorbent. Ind Eng Chem Res, 52: 2403–2411. DOI: 10.1021/ie300916d.

74. Chen, Q, Wen, Y, Cang, Y, Li, L, Guo, X, Zhang, R (2013b). Selective removal of phenol by spherical particles of α-, β- and γ-cyclodextrin polymers: kinetics and isothermal equilibrium. Front Chem Sci Eng, 7: 162–169. DOI: 10.1007/s11705-013-1318-5.

75. Chen, Y, Ma, Y, Lu, W, Guo, Y, Zhu, Y, Lu, H, Song, Y (2018). Environmentally friendly gelatin/β-cyclodextrin composite fiber adsorbents for the efficient removal of dyes from wastewater. Molecules, 23: 2473. DOI: 10.3390/molecules23102473.

76. Cheng, J, Chang, P R, Zheng, P, Ma, X (2014). Characterization of magnetic carbon nanotube–cyclodextrin composite and its adsorption of dye. Ind Eng Chem Res, 53: 1415–1421. DOI: 10.1021/ie402658x.

77. Cheng, Y, Nie, J, Li, Z, Yan, Z, Xu, G, Li, H, Guan, D (2017). A molecularly imprinted polymer synthesized using

β-cyclodextrin as the monomer for the efficient recognition of forchlorfenuron in fruits. Anal Bioanal Chem, 409: 5065–5072. DOI: 10.1007/s00216-017-0452-1.

78. Choi, J, Jeong, D, Cho, E, Yu, J-H, Tahir, M, Jung, S (2017). Pentynyl ether of β-cyclodextrin polymer and silica micro-particles: a new hybrid material for adsorption of phenanthrene from water. Polymers, 9: 10. DOI: 10.3390/polym9010010.

79. Choi, J M, Jeong, D, Piao, J, Kim, K, Nguyen, A B L, Kwon, N-J, Lee, M-K, Yu, J-H, Jung, S (2015). Hydroxypropyl cyclic β-(1→ 2)-D-glucans and epichlorohydrin β-cyclodextrin dimers as effective carbohydrate-solubilizers for polycyclic aromatic hydrocarbons. Carbohydr Res, 401: 82–88. DOI: 10.1016/j.carres.2014.10.025.

80. Churchill, D, Cheung, J C F, Park, Y S, Smith, V H, VanLoon, G, Buncel, E (2006). Complexation of diazinon, an organophosphorus pesticide, with α-, β-, and γ-cyclodextrin nmr and computational studies. Can J Chem, 84: 702–708. DOI: 10.1139/v06-053.

81. Cova, T F, Cruz, S M, Valente, A J, Abreu, P E, Marques, J M Pais, A A 2018a, Aggregation of cyclodextrins: fundamental issues and applications. In : Cyclodextrin Fundamentals, Reactivity and Analysis. Springer Nature: Basel, 45–65. DO I: 10.5772/intechopen. 73532.

82. Cova, T F, Milne, B F Pais, A A C C (2019). Host flexibility and space filling in supramolecular complexation of cyclodextrins: a free-energy-oriented approach. Carbohydr Polym, 205: 42–54. DOI: 10.1016/j.carbpol.2018.10.009.

83. Cova, T F, Murtinho, D, Pais, A A C C Valente, A J M (2018b). Combining cellulose and cyclodextrins: fascinating designs for materials and

pharmaceutics. Front Chem, 6. DOI: 10.3389/fchem.2018.00271.

84. Cova, T F G G, Milne, B F, Nunes, S C C Pais, A A C C (2018c). Drastic stabilization of junction nodes in supramolecular structures based on host–guest complexes. Macromolecules, 51: 2732–2741. DOI: 10.1021/acs.macromol.8b00154.

85. Cova, T F G G, Murtinho, D, Pais, A A C C, Valente, A J M (2018d). Cyclodextrin-based materials for removing micropollutants from wastewater. Curr Org Chem, 22: 2150–2181. DOI: 10.21 74/1385272822666181019125315.

86. Hu, X, Xu, G, Zhang, H, Li, M, Tu, Y, Xie, X, Zhu, Y, Jiang, L, Zhu, X, Ji, X, Li, Y, Li, A (2020). Multifunctional β-cyclodextrin polymer for simultaneous removal of natural organic matter and organic micropollutants and detrimental microorganisms from water. ACS Appl Mater Interfaces, 12: 12165–12175. DOI: https://doi.org/10.1021/acsami.0c00597.

87. Crini, G (2005). Recent developments in polysaccharide-based materials used as adsorbents in wastewater treatment. Prog Pol Sci, 30: 38–70. DOI: 10.1016/j.progpolymsci.2004.11.002.

88. Crini, G (2008). Kinetic and equilibrium studies on the removal of cationic dyes from aqueous solution by adsorption onto a cyclodextrin polymer. Dyes Pigm, 77: 415–426. DOI: 10.1016/j.dyepig.2007.07.001.

89. Crini, G, Badot, P M (2008). Application of chitosan, a natural aminopolysaccharide, for dye removal from aqueous solutions by adsorption processes using batch studies: a review of recent literature. Prog Pol Sci, 33: 399–447. DOI: 10.1016/j.progpolymsci. 2007.11.001.

90. Crini, G, Bertini, S, Torri, G, Naggi, A, Sforzini, D, Vecchi, C, Janus, L, Lekchiri, Y Morcellet, M (1998a).

Sorption of aromatic compounds in water using insoluble cyclodextrin polymers. J App Pol Sci, 68: 1973–1978. DOI: 10.1002/(sici)1097-4628(19980620)68:12<1973::aid-app11>3.0.co;2-t.

91. Crini, G, Cosentino, C, Bertini, S, Naggi, A, Torri, G, Vecchi, C, Janus, L, Morcellet, M (1998b). Solid state NMR spectroscopy study of molecular motion in cyclomaltoheptaose (β-cyclodextrin) crosslinked with epichlorohydrin. Carbohydr Res, 308: 37–45. DOI: https://doi.org/10.1016/S0008-6215(98)00077-9.

92. Crini, G, Exposito Saintemarie, A, Rocchi, S, Fourmentin, M, Jeanvoine, A, Millon, L Morin-Crini, N (2017). Simultaneous removal of five triazole fungicides from synthetic solutions on activated carbons and cyclodextrin-based adsorbents. Heliyon, 3: e00380. DOI:10.1016/j.heliyon.2017.e00380.

93. Crini, G, Fourmentin, S, Fenyvesi, É, Torri, G, Fourmentin, M Morin-Crini, N 2018, Fundamentals and applications of cyclodextrins. In : Cyclodextrin Fundamentals, Reactivity and Analysis. Springer: Berlin, 1–55.

94. Crini, G, Janus, L, Morcellet, M, Torri, G, Naggi, A, Bertini, S, Vecchi, C (1998c). Macroporous polyamines containing cyclodextrin: synthesis, characterization, and sorption properties. J App Polym Sci, 69: 1419–1427. DOI: 10.1002/(sici)1097-4628(19980815)69:7<1419::aid-app17>3.0.co;2-o.

95. Crini, G, Morcellet, M (2002). Synthesis and applications of adsorbents containing cyclodextrins. J Sep Sci, 25: 789–813. DOI: 10.1002/1615-9314(20020901)25:13<789::aid-jssc789>3.0.co;2-j.

96. Crini, G, Morin, N, Rouland, J-C, Janus, L, Morcellet, M, Bertini, S (2002a). Adsorption de béta-naphtol sur des gels de cyclodextrine–carboxyméthylcellulose réticulés. Eur Pol J, 38: 1095–1103. DOI: 10.1016/S0014-3057(01)00298-1.

97. Li, X, Zhao, B, Zhu, K, Xao, X (2011). Removal of nitrophenols by adsorption using β-cyclodextrin modified zeolites. Sep Sci Tech, 19: 938–943. DOI: https://doi.org/10.1016/S1004-9541(11)60075-X.

98. Crini, G, Peindy, H (2006). Adsorption of C.I. Basic blue 9 on cyclodextrin-based material containing carboxylic groups. Dyes Pigm, 70: 204–211. DOI: 10.1016/j.dyepig.2005.05.004.

99. Crini, G, Peindy, H, Gimber, F, Robert, C (2007). Removal of C.I. Basic green 4 (malachite green) from aqueous solutions by adsorption using cyclodextrin-based adsorbent: kinetic and equilibrium studies. Sep Purif Technol, 53: 97–110. DOI: 10.1016/j.seppur.2006.06.018.

100. Crini, G, Torri, G, Guerrini, M, Martel, B, Lekchiri, Y, Morcellet, M (1997). Linear cyclodextrin-polyvinylamine): synthesis and characterization. Eur Polym J, 33: 1143–1151. DOI: 10.1016/S0014-3057(96)00169-3.

101. Cruz, S, Marques, P, Valente, A (2019). Supramolecular graphene-based systems for drug delivery, in C Ozkan (Ed.), Handbook of Graphene, Scrivener Publishing LLC: New York, NY, 443–480.

102. Cukierman, A L, Platero, E, Fernandez, M E, Bonelli, P R (2016), Potentialities of graphene-based nanomaterials for wastewater treatment, Smart Materials for Waste Water Applications, John Wiley & Sons, Inc.: Hoboken, NJ, 47–86.

103. Dai, Z, Sun, Y, Zhang, H, Ding, D, Li, L (2019). Highly efficient removal of uranium(vi) from wastewater by polyamidoxime/polyethyleneimine magnetic graphene oxide. J Chem Eng Data. DOI: 10.1021/acs.jced.9b00759.

104. Danquah, M K, Aruei, R C, Wilson, L D (2018). Phenolic pollutant uptake properties of molecular templated polymers containing beta-cyclodextrin. J Phys Chem B, 122: 4748–4757. DOI: 10.1021/acs.jpcb.8b01819.

105. Ding, L, Li, Y, Jia, D, Deng, J, Yang, W (2011). B-cyclodextrin-based oil-absorbents: preparation, high oil absorbency and reusability. Carbohydr Polym, 83: 1990–1996. DOI: 10.1016/j.carbpol.2010.11.005.

106. Dong, K, Qiu, F, Guo, X, Xu, J, Yang, D, He, K (2013a). Adsorption behavior of azo dye eriochrome black T from aqueous solution by β-cyclodextrins/polyurethane foam material. Polym Plast Technol Eng, 52: 452–460. DOI: 10.1080/03602559.2012.748805.

107. Dong, K, Qiu, F, Guo, X, Xu, J, Yang, D, He, K (2013b). Adsorption behavior of azo dye eriochrome black T from aqueous solution by β-cyclodextrins/polyurethane foam material. Polym Plast Technol Eng, 52: 452–460. DOI: 10.1080/03602559.2012.748805.

108. Dsouza, R N, Pischel, U, Nau, W M (2011). Fluorescent dyes and their supramolecular host/guest complexes with macrocycles in aqueous solution. Chem Rev, 111: 7941–7980. DOI: 10.1021/cr200213s.

109. Duffus, J H (2002). "Heavy metals" a meaningless term? (iupac technical report). Pure Appl Chem, 74: 793–807. DOI: 10.1351/pac200274050793.

110. Duru, İ, Ege, D, Kamali, A R (2016). Graphene oxides for removal of heavy and precious metals from wastewater. J Mater Sci, 51: 6097–6116. DOI: 10.1007/s10853-016-9913-8.

111. Ehsan, S, Prasher, S O, Marshall, W D (2007). Simultaneous mobilization of heavy metals and polychlorinated biphenyl (PCB) compounds from soil with cyclodextrin and EDTA in admixture. Chemosphere, 68: 150–158. DOI: https://doi.org/10.1016/j.chemosphere.2006.12.018.

112. Elanchezhiyan, S S D, Meenakshi, S (2017). Facile fabrication of metal ions-incorporated chitosan/β-cyclodextrin composites for effective removal of oil from oily wastewater. ChemistrySelect, 2: 11393–11401. DOI: 10.1002/slct.201702147.

113. Faisal, Z, Fliszár-Nyúl, E, Dellafiora, L, Galaverna, G, Dall'Asta, C, Lemli, B, Kunsági-Máté, S, Szente, L, Poór, M (2019). Cyclodextrins can entrap zearalenone-14-glucoside: interaction of the masked mycotoxin with cyclodextrins and cyclodextrin bead polymer. Biomolecules, 9: 354. DOI: 10.3390/biom9080354.

114. Fan, L, Li, M, Lv, Z, Sun, M, Luo, C, Lu, F, Qiu, H (2012a). Fabrication of magnetic chitosan nanoparticles grafted with β-cyclodextrin as effective adsorbents toward hydroquinol. Colloids Surf B: Biointerfaces, 95: 42–49. DOI: 10.1016/j.colsurfb.2012.02.007.

115. Fan, L, Luo, C, Sun, M, Qiu, H, Li, X (2013). Synthesis of magnetic β-cyclodextrin−chitosan/graphene oxide as nanoadsorbent and its application in dye adsorption and removal. Coll Surf B: Biointerfaces, 103: 601–607. DOI: https://doi.org/10.1016/j.colsurfb.2012.11.023.

116. Fan, L, Zhang, Y, Luo, C, Lu, F, Qiu, H, Sun, M (2012b). Synthesis and characterization of magnetic β-cyclodextrin−chitosan nanoparticles as nano-adsorbents for removal of methyl blue. Int J Biol Macromol, 50: 444–450. DOI: https://doi.org/10.1016/j.ijbiomac.2011.12.016.

117. Fei, Y, Dexian, C, Jie, M (2017). Synthesis of cyclodextrin-based adsorbents and its application for organic pollutant removal from water. Curr Org Chem, 21: 1976–1990. DOI: 10.2174/1385272821666170503110023.

118. Fekete-Kertész, I, Molnár, M, Atkári, Á, Gruiz, K, Fenyvesi, É (2013). Hydrogen peroxide oxidation for in situ remediation of trichloroethylene—from the laboratory to the field. Period Polytech Chem Eng, 57: 41–51. DOI.

119. Fennell Evans, D, Wennerstrom, H, Rajagopalan, R (1995). The colloidal domain: where physics, chemistry, biology, and technology meet. J Colloid Interface Sci, 172: 541–541. DOI.

120. Fenyvesi, É, Barkács, K, Gruiz, K, Varga, E, Kenyeres, I, Záray, G, Szente, L (2020). Removal of hazardous micropollutants from treated wastewater using cyclodextrin bead polymer—a pilot demonstration case. J Hazard Mater, 383: 121181.

121. Flores, J, Jiménez, V, Belmar, J, Mansilla, H D, Alderete, J B (2005). Inclusion complexation of phenol derivatives with a β-cyclodextrin based polymer. J Inclusion Phenom Macrocyclic Chem, 53: 63–68. DOI: 10.1007/s10847-005-0994-2.

122. Folch-Cano, C, Yazdani-Pedram, M, Olea-Azar, C (2014). Inclusion and functionalization of polymers with cyclodextrins: current applications and future prospects. Molecules, 19: 14066–14079.

123. Forgacs, E, Cserhati, T, Oros, G (2004). Removal of synthetic dyes from wastewaters: a review. Environ Int, 30: 953–971.

124. Gao, H, Miles, M S, Meyer, B M, Wong, R L, Overton, E B (2012). Assessment of cyclodextrin-enhanced extraction of crude oil from contaminated porous media. J Environ Monit, 14: 2164–2169. DOI: 10.1039/C2EM30223C.

125. Gao, S, Jiang, J-Y, Liu, Y-Y, Fu, Y, Zhao, L-X, Li, C-Y, Ye, F (2019). Enhanced solubility, stability, and herbicidal activity of the herbicide diuron by complex formation with β-cyclodextrin. Polymers (Basel), 11: 1396.

126. García-Zubiri, Í X, González-Gaitano, G, Isasi, J R (2007). Isosteric heats of sorption of 1-naphthol and phenol from aqueous solutions by β-cyclodextrin polymers. J Colloid Interface Sci, 307: 64–70. DOI: https://doi.org/10.1016/j.jcis.2006.10.076.

127. Ghemati, D, Aliouche, D (2014). Study of the sorption of synthetic dyes from aqueous solution onto cellulosic modified polymer. J Water Chem Technol, 36: 265–272. DOI.

128. Gibson, L T (2014). Mesosilica materials and organic pollutant adsorption: part B removal from aqueous solution. Chem Soc Rev, 43: 5173–5182. DOI: 10.1039/C3CS60095E.

129. Gidwani, B, Vyas, A (2014). Synthesis, characterization and application of epichlorohydrin-β-cyclodextrin polymer. Colloid Surf B: Biointerfaces, 114: 130–137. DOI: https://doi.org/10.1016/j.colsurfb.2013.09.035.

130. Giraldo, L F, López, B L, Pérez, L, Urrego, S, Sierra, L, Mesa, M (2007). Mesoporous silica applications. Macromol Symp, 258: 129–141. DOI: 10.1002/masy.200751215.

131. Girek, T, Kozlowski, C A, Koziol, J J, Walkowiak, W, Korus, I (2005). Polymerisation of β-cyclodextrin with succinic anhydride. Synthesis, characterisation, and ion flotation of transition metals. Carbohydr Polym, 59: 211–215. DOI: https://doi.org/10.1016/j.carbpol.2004.09.011.

132. Girek, T, Shin, D-H, Lim, S-T (2000). Polymerization of β-cyclodextrin with maleic anhydride and structural characterization of the polymers. Carbohydr Polym, 42: 59–63. DOI: https://doi.org/10.1016/S0144-8617(99)00138-1.

133. Gou, P-F, Zhu, W-P, Shen, Z-Q (2010). Synthesis, self-assembly, and drug-loading capacity of well-defined cyclodextrin-centered drug-conjugated amphiphilic A(14)B(7) miktoarm

star copolymers based on poly(ε-caprolactone) and poly(ethylene glycol). Biomacromolecules, 11: 934–943. DOI: 10.1021/bm901371p.

134. Gross, K C, Seybold, P G (2000). Substituent effects on the physical properties and pk_a of aniline. Int J Quantum Chem, 80: 1107–1115. DOI: 10.1002/1097-461X (2000) 80:4/5< 1107::AID-QUA60>3.0.CO;2-T.

135. Gruiz, K, Meggyes, T, Fenyvesi, É (2015). Engineering Tools for Environmental Risk Management: 2. Environmental Toxicology. CRC Press: Boca Raton, FL.

136. Guerra, F D, Attia, M F, Whitehead, D C, Alexis, F (2018a). Nanotechnology for environmental remediation: materials and applications. Molecules, 23: 1760.

137. Guerra, F D, Attia, M F, Whitehead, D C, Alexis, F (2018b). Nanotechnology for environmental remediation: materials and applications. Molecules, 23: 1760. DOI: 10.3390/molecules23071760.

138. Guo, L J, Qu, J R, Miao, S S, Geng, H R, Yang, H (2013). Development of a molecularly imprinted polymer for prometryne clean-up in the environment. J Sep Sci, 36: 3911–3917. DOI: 10.1002/jssc.201300914.

139. Hamai, S, Satou, H (2000). Inclusion complexes of cyclodextrins with methylene blue and acid orange 7 in aqueous solutions. Bull Chem Soc Jpn, 73: 2207–2214. DOI: 10.1246/Bcsj.73.2207.

140. Han, J, Qiu, W, Tiwari, S, Bhargava, R, Gao, W, Xing, B (2015). Consumer-grade polyurethane foam functions as a large and selective absorption sink for bisphenol a in aqueous media. J Mater Chem A, 3: 8870–8881.

141. Stojić, N, Štrbac, S, Prokić, D (2019). Soil pollution and remediation. In: Hussain C (eds) Handbook of Environmental Materials Management.

Springer, Cham. DOI: https://doi. org/10.1007/978-3-319-73645-7_81.

142. Hao, Z, Yi, Z, Bowen, C, Yaxing, L, Sheng, Z (2019). Preparing γ-cyclodextrin-immobilized starch and the study of its removal properties to dyestuff from wastewater. Pol J Environ Stud, 28: 1701–1711. DOI: 10.15244/pjoes/90028.

143. Hiew, B Y Z, Lee, L Y, Lee, X J, Thangalazhy-Gopakumar, S, Gan, S, Lim, S S, Pan, G-T, Yang, T C-K, Chiu, W S, Khiew, P S (2018). Review on synthesis of 3d graphene-based configurations and their adsorption performance for hazardous water pollutants. Process Saf Environ Prot, 116: 262–286. DOI: 10.1016/j.psep.2018.02.010.

144. Hirai, H, Toshima, N, Uenoyama, S (1981). Inclusion complex formation of cyclodextrin with large dye molecule. Polym J, 13: 607.

145. Hodge, M, Gyanwali, G, Villines, C, White, J L (2015). Synthesis of amphiphilic polymer networks with guest-host properties. J Polym Sci A: Polym Chem, 53: 1824–1831. DOI: 10.1002/pola.27637.

146. Hollas, M, Chung, M-A, Adams, J (1998). Complexation of pyrene by poly(allylamine) with pendant β-cyclodextrin side groups. J Phys Chem B, 102: 2947–2953. DOI: 10.1021/jp9800719.

147. Hu, X-j, Liu, Y-g, Wang, H, Zeng, G-m, Hu, X, Guo, Y-m, Li, T-t, Chen, A-w, Jiang, L-hGuo, F-y (2015). Adsorption of copper by magnetic graphene oxide-supported β-cyclodextrin: effects of pH, ionic strength, background electrolytes, and citric acid. Chem Eng Res Des, 93: 675–683. DOI: 10.1016/j. cherd.2014.06.002.

148. Huang, W, Hu, Y, Li, Y, Zhou, Y, Niu, D, Lei, Z, Zhang, Z (2018). Citric acid-crosslinked β-cyclodextrin for simultaneous removal of bisphenol A,

methylene blue and copper: the roles of cavity and surface functional groups. J Taiwan Inst Chem Eng, 82: 189–197. DOI: https://doi.org/10.1016/j.jtice.2017.11.021.

149. Huang, Z, Liu, S, Zhang, B, Wu, Q (2014). Preparation and swelling behavior of a novel self-assembled β-cyclodextrin/acrylic acid/sodium alginate hydrogel. Carbohydr Polym, 113: 430–437. DOI: https://doi.org/10.1016/j.carbpol.2014.07.009.

150. Huq, R, Mercier, L, Kooyman, P J (2001). Incorporation of cyclodextrin into mesostructured silica. Chem Mater, 13: 4512–4519. DOI: 10.1021/cm010171i.

151. Hüttermann, A, Zommorodi, M, Reise, K (1999). Addition of hydrogels to soil for prolonging the survival of pinus halepensis seedlings subjected to drought. Soil Tillage Res, 50: 295–304.

152. Iijima, S (1991). Helical microtubules of graphitic carbon. Nature, 354: 56–58. DOI: 10.1038/354056a0.

153. Janus, L, Crini, G, El-Rezzi, V, Morcellet, M, Cambiaghi, A, Torri, G, Naggi, A, Vecchi, C (1999). New sorbents containing beta-cyclodextrin. Synthesis, characterization, and sorption properties. React Funct Polym, 42: 173–180. DOI: https://doi.org/10.1016/S1381-5148(98)00066-2.

154. Jia, S, Tang, D, Peng, J, Sun, Z, Yang, X (2019). b-cyclodextrin modified electrospinning fibers with good regeneration for efficient temperature-enhanced adsorption of crystal violet. Carbohydr Polym, 208: 486–494. DOI: 10.1016/j.carbpol.2018.12.075.

155. Jiang, Y, Liu, B, Xu, J, Pan, K, Hou, H, Hu, J, Yang, J (2018). Cross-linked chitosan/beta-cyclodextrin composite for selective removal of methyl orange: adsorption performance and mechanism. Carbohydr Polym, 182: 106–114. DOI: 10.1016/j.carbpol.2017.10.097.

156. Jozefaciuk, G, Muranyi, A, Fenyvesi, E (2003). Effect of randomly methylated β-cyclodextrin on physical properties of soils. Environ Sci Technol, 37: 3012–3017.

157. Junthip, J, Jumrernsuk, N, Klongklaw, P, Promma, W, Sonsupap, S (2018). Removal of paraquat herbicide from water by textile coated with anionic cyclodextrin polymer. SN Appl Sci, 1: 106. DOI: 10.1007/s42452-018-0102-z.

158. Jurecska, L, Dobosy, P, Barkács, K, Fenyvesi, É, Záray, G (2015). Reprint of "characterization of cyclodextrin containing nanofilters for removal of pharmaceutical residues". J Pharm Biomed Anal, 106: 124–128. DOI: https://doi.org/10.1016/j.jpba.2015.01.024.

159. Kakhki, R M (2015). Application of magnetic nanoparticles modified with cyclodextrins as efficient adsorbents in separation systems. J Inclusion Phenom Macrocyclic Chem, 82: 301–310. DOI: 10.1007/s10847-015-0512-0.

160. Kang, Y-L, Zhang, J, Wu, G, Zhang, M-X, Chen, S-C, Wang, Y-Z (2018). Full-biobased nanofiber membranes toward decontamination of wastewater containing multiple pollutants. ACS Sustainable Chem Eng, 6: 11783–11792. DOI: 10.1021/acssuschemeng.8b01996.

161. Kayaci, F, Aytac, Z, Uyar, T (2013). Surface modification of electrospun polyester nanofibers with cyclodextrin polymer for the removal of phenanthrene from aqueous solution. J Hazard Mater, 261: 286–294. DOI: https://doi.org/10.1016/j.jhazmat.2013.07.041.

162. Kayaci, F, Uyar, T (2014). Electrospun polyester/cyclodextrin nanofibers for entrapment of volatile organic compounds. Polym Eng Sci, 54: 2970–2978. DOI: 10.1002/pen.23858.

163. Kiji, J, Konishi, H, Okano, T, Terashima, T, Motomura, K (1992). Adsorption

of organic species by a cyclodextrin epichlorohydrin network polymer. Die Angew Makromol Chem, 199: 207–210. DOI: 10.1002/apmc.1992.051990116.

164. Kono, H, Onishi, K, Nakamura, T (2013). Characterization and bisphenol a adsorption capacity of beta-cyclodextrin-carboxymethylcellulose-based hydrogels. Carbohydr Polym, 98: 784–792. DOI: 10.1016/j.carbpol.2013.06.065.

165. Koopmans, C, Ritter, H (2008). Formation of physical hydrogels via host–guest interactions of β-cyclodextrin polymers and copolymers bearing adamantyl groups. Macromolecules, 41: 7418–7422. DOI: 10.1021/ma801202f.

166. Kopperi, M, Riekkola, M-L (2016). Non-targeted evaluation of selectivity of water-compatible class selective adsorbents for the analysis of steroids in wastewater. Anal Chim Acta, 920: 47–53. DOI: 10.1016/j.aca.2016.03.036.

167. Krause, R W, Mamba, B B, Bambo, F M, Malefetse, T J (2010a). Cyclodextrin polymers: synthesis and application in water treatment. In: Cyclodextrins: Chemistry and Physics, Hu, J, Ed; Transworld Res Net: Kerala, India, 1–25.

168. Krause, R W M, Mamba, B B, Dlamini, L N, Durbach, S H (2010b). Fe–Ni nanoparticles supported on carbon nanotube-co-cyclodextrin polyurethanes for the removal of trichloroethylene in water. J Nanopart Res, 12: 449–456. DOI: 10.1007/s11051-009-9659-1.

169. Kuo, C-Y, Wu, C-H, Wu, J-Y (2008). Adsorption of direct dyes from aqueous solutions by carbon nanotubes: determination of equilibrium, kinetics and thermodynamics parameters. J Colloid Interface Sci, 327: 308–315.

170. Kyzas, G, Bikiaris, D (2015). Recent modifications of chitosan for adsorption applications: a critical and systematic review. Mar Drugs, 13: 312–337. DOI: 10.3390/md13010312.

171. Kyzas, G Z, Deliyanni, E A, Matis, K A (2014a). Graphene oxide and its application as an adsorbent for wastewater treatment. J Chem Technol Biotechnol, 89: 196–205. DOI: 10.1002/jctb.4220.

172. Kyzas, G Z, Deliyanni, E A, Matis, K A (2014b). Graphene oxide and its application as an adsorbent for wastewater treatment. J Chem Technol Biotechnol, 89: 196–205. DOI: 10.1002/jctb.4220.

173. Lambert, J B, Liu, C, Boyne, M T, Zhang, A P, Yin, Y (2003). Solid phase host–guest properties of cyclodextrins and calixarenes covalently attached to a polysilsesquioxane matrix. Chem Mater, 15: 131–145. DOI: 10.1021/cm020751v.

174. Landy, D, Mallard, I, Ponchel, A, Monflier, E, Fourmentin, S (2012). Remediation technologies using cyclodextrins: an overview. Environ Chem Lett, 10: 225–237. DOI: 10.1007/s10311-011-0351-1.

175. Lau, E V, Gan, S, Ng, H K, Poh, P E (2014). Extraction agents for the removal of polycyclic aromatic hydrocarbons (pahs) from soil in soil washing technologies. Environ Pollution, 184: 640–649. DOI: 10.1016/j.envpol.2013.09.010.

176. Lee, J H, Kwak, S Y (2019). Branched polyethylenimine-polyethylene glycol-β-cyclodextrin polymers for efficient removal of bisphenol A and copper from wastewater. J Appl Polym Sci, 48475. DOI: 10.1002/app.48475.

177. Leudjo Taka, A, Pillay, K, Yangkou Mbianda, X (2017). Nanosponge cyclodextrin polyurethanes and their modification with nanomaterials for the removal of pollutants from waste water: a review. Carbohydr Polym, 159: 94–107. DOI: https://doi.org/10.1016/j.carbpol.2016.12.027.

178. Li, C, Klemes, M J, Dichtel, W R, Helbling, D E (2018a). Tetrafluoroterephlithalomtrile-crosslinked beta-cyclodexfrin polymers for efficient extraction and recovery of organic

micropollutants from water. J Chromatogr A, 1541: 52–56. DOI: 10.1016/j.chroma.2018.02.012.

179. Li, D, Ma, M (1999). Nanoporous polymers: new nanosponge absorbent media. Filtr Sep, 36: 26–28. DOI: 10.1016/S0015-1882(00)80050-6.

180. Li, D, Ma, M (2000). Nanosponges for water purification. Cleaner Prod Processes, 2: 112–116. DOI: 10.1007/s100980000061.

181. Li, L, Fan, L, Duan, H, Wang, X, Luo, C (2014). Magnetically separable functionalized graphene oxide decorated with magnetic cyclodextrin as an excellent adsorbent for dye removal. RSC Adv, 4: 37114–37121. DOI: 10.1039/C4RA06292B.

182. Li, L, Fan, L, Sun, M, Qiu, H, Li, X, Duan, H, Luo, C (2013a). Adsorbent for chromium removal based on graphene oxide functionalized with magnetic cyclodextrin–chitosan. Colloids Surf B, Biointerfaces, 107: 76–83. DOI: 10.1016/j.colsurfb.2013.01.074.

183. Li, L, Fan, L, Sun, M, Qiu, H, Li, X, Duan, H, Luo, C (2013b). Adsorbent for hydroquinone removal based on graphene oxide functionalized with magnetic cyclodextrin–chitosan. Int J Biol Macromol, 58: 169–175. DOI: https://doi.org/10.1016/j.ijbiomac.2013.03.058.

184. Li, S, Zhang, Y, You, Q, Wang, Q, Liao, G, Wang, D (2018b). Highly efficient removal of antibiotics and dyes from water by the modified carbon nanofibers composites with abundant mesoporous structure. Colloids Surf, A Physicochem Eng Asp, 558: 392–401. DOI: 10.1016/j.colsurfa.2018.09.002.

185. Li, X, Zhou, M, Jia, J, Jia, Q (2018c). A water-insoluble viologen-based β-cyclodextrin polymer for selective adsorption toward anionic dyes. React Funct Polym, 126: 20–26. DOI: https://doi.org/10.1016/j.reactfunctpolym.2018.03.004.

186. Li, Y, Zhou, Y, Zhou, Y, Lei, J, Pu, S (2018d). Cyclodextrin modified filter paper for removal of cationic dyes/Cu ions from aqueous solutions. Water Sci Technol, 78: 2553–2563. DOI: 10.2166/wst.2019.009.

187. Liang, J, Song, C, Deng, J (2014). Optically active microspheres constructed by helical substituted polyacetylene and used for adsorption of organic compounds in aqueous systems. ACS Appl Mater Interfaces, 6: 19041–19049. DOI: 10.1021/am504943x.

188. Lim, C T (2017). Nanofiber technology: current status and emerging developments. Prog Polym Sci, 70: 1–17.

189. Lindman, B, Karlstrom, G, Stigsson, L (2010). On the mechanism of dissolution of cellulose. J Mol Liq, 156: 76–81. DOI: 10.1016/j.molliq.2010.04.016.

190. Lingamdinne, L P, Koduru, J R, Karri, R R (2019). A comprehensive review of applications of magnetic graphene oxide based nanocomposites for sustainable water purification. J Environ Manage, 231: 622–634. DOI: https://doi.org/10.1016/j.jenvman.2018.10.063.

191. Liu, B, Huang, Y (2011). Polyethyleneimine modified eggshell membrane as a novel biosorbent for adsorption and detoxification of Cr(VI) from water. J Mater Chem, 21: 17413. DOI: 10.1039/c1jm12329g.

192. Liu, C, Lambert, J B, Fu, L (2003). A novel family of ordered, mesoporous inorganic/organic hybrid polymers containing covalently and multiply bound microporous organic hosts. J Am Chem Soc, 125: 6452–6461. DOI: 10.1021/ja0213930.

193. Liu, C, Naismith, N, Economy, J (2004). Advanced mesoporous organosilica material containing microporous β-cyclodextrins for the removal of humic acid from water. J Chromatogr A, 1036: 113–118. DOI: https://doi.org/10.1016/j.chroma.2004.02.076.

194. Liu, H, Cai, X, Wang, Y, Chen, J (2011). Adsorption mechanism-based screening of cyclodextrin polymers for adsorption and separation of pesticides from water. Water Res, 45: 3499–3511. DOI: https://doi.org/10.1016/j.watres.2011.04.004.

195. Liu, W, Jiang, X, Chen, X (2014a). A novel method of synthesizing cyclodextrin grafted multiwall carbon nanotubes/iron oxides and its adsorption of organic pollutant. Appl Surf Sci, 320: 764–771. DOI: https://doi.org/10.1016/j.apsusc.2014.09.165.

196. Liu, X, Yan, L, Yin, W, Zhou, L, Tian, G, Shi, J, Yang, Z, Xiao, D, Gu, Z, Zhao, Y (2014b). A magnetic graphene hybrid functionalized with beta-cyclodextrins for fast and efficient removal of organic dyes. J Mater Chem A, 2: 12296–12303. DOI: 10.1039/C4TA00753K.

197. Liu, Y, Chen, Y, Li, B, Wada, T, Inoue, Y (2001). Cooperative multipoint recognition of organic dyes by bis (β-cyclodextrin)s with 2,2'-bipyridine-4,4'-dicarboxy tethers. Chemistry, 7: 2528–2535.

198. Liu, Y, Huang, S, Zhao, X, Zhang, Y (2018). Fabrication of three-dimensional porous β-cyclodextrin/chitosan functionalized graphene oxide hydrogel for methylene blue removal from aqueous solution. Colloids and Surf A: Physchem Eng Asp, 539: 1–10. DOI: https://doi.org/10.1016/j.colsurfa.2017.11.066.

199. Liu, Y, Liu, M, Jia, J, Wu, D, Gao, T, Wang, X, Yu, J, Li, F (2019). b-cyclodextrin-based hollow nanoparticles with excellent adsorption performance towards organic and inorganic pollutants. Nanoscale, 11: 18653–18661. DOI: 10.1039/C9NR07342F.

200. Liu, Z-G, Xu, M, Yang, Z, Wang, Y-X, Wang, S-Q, Wang, H-X (2017). Efficient removal of organic dyes from water by β-cyclodextrin functionalized graphite carbon nitride composite. ChemistrySelect, 2: 1753–1758. DOI: 10.1002/slct.201602032.

201. Ma, J, He, Y, Zeng, G, Yang, X, Chen, X, Zhou, L, Peng, L, Sengupta, A (2018). High-flux PVDF membrane incorporated with β-cyclodextrin modified halloysite nanotubes for dye rejection and Cu(II) removal from water. Pol Adv Technol, 29: 2704–2714. DOI: 10.1002/pat.4356.

202. Mahmoud, H R, Ibrahim, S MEl-Molla, S A (2016). Textile dye removal from aqueous solutions using cheap mgo nanomaterials: adsorption kinetics, isotherm studies and thermodynamics. Adv Powder Technol, 27: 223–231.

203. Mak, Y W, Leung, W W-F (2019). Crosslinking of genipin and autoclaving in chitosan-based nanofibrous scaffolds: structural and physioChem properties. J Mater Sci, 54: 10941–10962. DOI: 10.1007/s10853-019-03649-8.

204. Malayoglu, U (2018). Removal of heavy metals by biopolymer (chitosan)/nanoclay composites. Sep Sci Technol, 53: 2741–2749. DOI: 10.1080/01496395.2018.1471506.

205. Mallard Favier, I, Baudelet, D, Fourmentin, S (2011). VOC trapping by new crosslinked cyclodextrin polymers. J Inclusion Phenom Macrocyclic Chem, 69: 433–437. DOI: 10.1007/s10847-010-9776-6.

206. Mamba, B B, Krause, R W, Malefetse, T J, Nxumalo, E N (2007). Monofunctionalized cyclodextrin polymers for the removal of organic pollutants from water. Environ Chem Lett, 5: 79–84. DOI: 10.1007/s10311-006-0082-x.

207. Mamba, G, Mbianda, X Y, Govender, P P (2013). Phosphorylated multiwalled carbon nanotube-cyclodextrin polymer: synthesis, characterisation and potential application in water purification. Carbohydr Polym, 98: 470–476. DOI:

https://doi.org/10.1016/j.carbpol.2013.06.034.

208. Mani, S, Chowdhary, P, Bharagava, R N 2019, Textile wastewater dyes: toxicity profile and treatment approaches. In: Emerging and Eco-friendly Approaches for Waste Management. Springer: Berlin, 219–244.

209. Maretschek, S, Greiner, A, Kissel, T (2008). Electrospun biodegradable nanofiber nonwovens for controlled release of proteins. J Control Rel, 127: 180–187. DOI: 10.1016/j.jconrel.2008.01.011.

210. Marican, A, Durán-Lara, E F (2018). A review on pesticide removal through different processes. Environ Sci Pollut Res, 25: 2051–2064. DOI: 10.1007/s11356-017-0796-2.

211. Martel, B, Devassine, M, Crini, G, Weltrowski, M, Bourdonneau, M, Morcellet, M (2001). Preparation and sorption properties of a β-cyclodextrin-linked chitosan derivative. J Polym Sci Part A: Pol Chem, 39: 169–176. DOI: 10.1002/ 1099-0518 (20010101)39:1< 169::aid-pola190> 3.0.co;2-g.

212. Martel, B, Ruffin, D, Weltrowski, M, Lekchiri, Y, Morcellet, M (2005). Water-soluble polymers and gels from the polycondensation between cyclodextrins and poly(carboxylic acid)s: a study of the preparation parameters. J Appl Polym Sci, 97: 433–442. DOI: 10.1002/app.21391.

213. Martinez-Huitle, C A, Rodrigo, M A, Sires, I, Scialdone, O (2015). Single and coupled electrochemical processes and reactors for the abatement of organic water pollutants: a critical review. Chem Rev, 115: 13362–13407.

214. Martins, A F, Bueno, P V A, Almeida, E A M S, Rodrigues, F H A, Rubira, A F, Muniz, E C (2013). Characterization of N-trimethyl chitosan/alginate complexes and curcumin release. Int J Biol Macromol, 57: 174–184. DOI: 10.1016/j.ijbiomac.2013.03.029.

215. Matias, T, Marques, J, Quina, M J, Gando-Ferreira, L, Valente, A J M, Portugal, A, Durães, L (2015). Silica-based aerogels as adsorbents for phenol-derivative compounds. Colloids Surf, A Physicochem Eng Asp, 480: 260–269. DOI: https://doi.org/10.1016/j.colsurfa.2015.01.074.

216. Matsui, Y, Kurita, T, Date, Y (1972). Complexes of copper(ii) with cyclodextrins. Bull Chem Soc Jpn, 45: 3229–3229. DOI: 10.1246/bcsj.45.3229.

217. Matsui, Y, Kurita, T, Yagi, M, Okayama, T, Mochida, K, Date, Y (1975). The formation and structure of copper(ii) complexes with cyclodextrins in an alkaline solution. Bull Chem Soc Jpn, 48: 2187–2191. DOI: 10.1246/bcsj.48.2187.

218. McCray, J E, Brusseau, M L (1998). Cyclodextrin-enhanced in situ flushing of multiple-component immiscible organic liquid contamination at the field scale: Mass removal effectiveness. Environ Sci Tech, 32: 1285–1293. DOI: 10.1021/es970579+.

219. McNamara, M, Russell, N R (1991). FT-IR and raman spectra of a series of metallo-β-cyclodextrin complexes. J Inclusion Phenom Macrocyclic Recognit Chem, 10: 485–495. DOI: 10.1007/BF01061078.

220. Mezohegyi, G, van der Zee, F P, Font, J, Fortuny, A, Fabregat, A (2012). Towards advanced aqueous dye removal processes: a short review on the versatile role of activated carbon. J Environ Manage, 102: 148–164.

221. Mirsal, I A 2004, Sources of soil pollution. In: Soil Pollution, Springer: Berlin, 72–110.

222. Mizobuchi, Y, Tanaka, M, Shono, T (1980). Preparation and sorption behaviour of cyclodextrin polyurethane resins. J Chromatogr A, 194: 153–161. DOI: https://doi.org/10.1016/S0021-9673(00)87291-X.

223. Mocanu, G, Vizitiu, D, Carpov, A (2001). Cyclodextrin polymers. J Bio Comp Polym, 16: 315–342.

224. Mochida, K, Matsui, Y (1976). Kinetic study on the formation of a binuclear complex between copper(II) and cyclo-dextrins. Chem Lett, 5: 963–966. DOI: 10.1246/cl.1976.963.

225. Mohamed, M H, Wilson, L D, Headley, J V (2011a). Design and characteriza-tion of novel β-cyclodextrin based copolymer materials. Carbohydr Res, 346: 219–229. DOI:10.1016/j.carres. 2010.11.022.

226. Mohamed, M H, Wilson, L D, Headley, J V, Peru, K M (2011b). Sequestration of naphthenic acids from aqueous solution using β-cyclodextrin-based polyure-thanes. Phys Chem Chem Phy, 13: 1112–1122. DOI: 10.1039/C0CP00421A.

227. Mohammadi, A, Veisi, P (2018). High adsorption performance of β-cyclodex-trin-functionalized multi-walled carbon nanotubes for the removal of organic dyes from water and industrial waste-water. J Environ Chem Eng, 6: 4634–4643. DOI:10.1016/j.jece.2018.07.002.

228. Morales-Sanfrutos, J, Javier Lopez-Jara-millo, F, Elremaily, M A A, Hernandez-Mateo, F, Santoyo-Gonzalez, F (2015). Divinyl sulfone cross-linked cyclodex-trin-based polymeric materials: synthesis and applications as sorbents and encapsu-lating agents. Molecules, 20: 3565–3581. DOI: 10.3390/molecules 20033565.

229. Morin-Crini, N, Crini, G (2013a). Environmental applications of water-insoluble β-cyclodextrin–epichloro-hydrin polymers. Prog Polym Sci, 38: 344–368. DOI: 10.1016/j.prog-polymsci. 2012.06.005.

230. Morin-Crini, N, Crini, G (2013b). Environmental applications of water-insoluble β-cyclodextrin–epichloro-hydrin polymers. Prog Polym Sci, 38: 344–368. DOI: https://doi.org/10.1016/j.progpolymsci. 2012.06.005.

231. Morin-Crini, N, Fourmentin, M, Four-mentin, S, Torri, G, Crini, G (2019). Synthesis of silica materials containing cyclodextrin and their applications in wastewater treatment. Environ Chem Lett, 17: 683–696. DOI: 10.1007/s10311-018-00818-0.

232. Morin-Crini, N, Winterton, P, Four-mentin, S, Wilson, L D, Fenyvesi, É, Crini, G (2018). Water-insoluble β-cyclodextrin–epichlorohydrin poly-mers for removal of pollutants from aqueous solutions by sorption processes using batch studies: a review of inclu-sion mechanisms. Prog Polym Sci, 78: 1–23. DOI: https://doi.org/10.1016/j.progpolymsci.2017.07.004.

233. Moulahcene, L, Kebiche-Senhadji, O, Skiba, M, Lahiani-Skiba, M, Oughlis-Hammache, F, Benamor, M (2016). Cyclodextrin polymers for ibuprofen extraction in aqueous solution: recovery, separation, and characterization. Desalin Water Treat, 57: 11392–11402. DOI: 10.1080/19443994. 2015.1048734.

234. Moulahcene, L, Skiba, M, Bounoure, F, Benanamor, M, Milon, N, Hall-ouard, F, Lahiani-Skiba, M (2019). New polymer inclusion membrane containing β-cyclodextrin polymer: application for pharmaceutical pollutant removal from waste water. Int J Environ Res Public Health, 16: 414.

235. Moulahcene, L, Skiba, M, Senhadji, O, Milon, N, Benamor, M, Lahiani-Skiba, M (2015). Inclusion and removal of pharmaceutical residues from aqueous solution using water-insoluble cyclo-dextrin polymers. Chem Eng Res Des, 97: 145–158. DOI: https://doi.org/10.1016/j.cherd.2014.08.023.

236. Mourtzis, N, Cordoyiannis, G, Nounesis, G, Yannakopoulou, K (2003). Single and double threading of congo red into γ-cyclodextrin. Solution structures and thermodynamic parameters of 1:1 and 2:2 adducts, as obtained from NMR

spectroscopy and microcalorimetry. Supramol Chem, 15: 639–649. DOI: 10.1080/10610270310001605223.

237. Mphahlele, K, Onyango, M S, Mhlanga, S D (2015a). Adsorption of aspirin and paracetamol from aqueous solution using fe/n-cnt/β-cyclodextrin nano-comopsites synthesized via a benign microwave assisted method. J Environ Chem Eng, 3: 2619–2630. DOI: https://doi.org/10.1016/j.jece.2015.02.018.

238. Mphahlele, K, Onyango, M S, Mhlanga, S D (2015b). Kinetics, equilibrium, and thermodynamics of the sorption of bisphenol A onto N- CNTs-β-cyclodextrin and Fe/N-CNTs-β-cyclodextrin nanocomposites. J Nanomater, 2015: 3.

239. Murai, S, Imajo, S, Maki, Y, Takahashi, K, Hattori, K (1996). Adsorption and recovery of nonionic surfactants by β-cyclodextrin polymer. J Coll Inter Sci, 183: 118–123. DOI: https://doi.org/10.1006/jcis.1996.0524.

240. Murcia-Salvador, A, Pellicer, J A, Fortea, M I, Gómez-López, V M, Rodrí-guez-López, M I, Núñez-Delicado, E, Gabaldón, J A (2019). Adsorption of direct blue 78 using chitosan and cyclo-dextrins as adsorbents. Polymers, 11: 1003. DOI: 10.3390/polym 11061003.

241. Naeem, K, Ouyang, F (2013). Influence of supports on photocatalytic degra-dation of phenol and 4-chlorophenol in aqueous suspensions of titanium dioxide. J Environ Sci, 25: 399–404.

242. Nair, B U, Dismukes, G C (1983). Models for the photosynthetic water oxidizing enzyme. 1. A binuclear manganese(III)-β-cyclodextrin complex. J Am Chem Soc, 105: 124–125. DOI: 10.1021/ja00339a027.

243. Narayan, R, Nayak, U Y, Raichur, A M, Garg, S (2018). Mesoporous silica nanoparticles: a comprehensive review on synthesis and recent advances. Phar-maceutics, 10: 118.

244. Ni, M, Tian, S, Huang, Q, Yang, Y (2018). Electrokinetic-fenton remedia-tion of organochlorine pesticides from historically polluted soil. Environ Sci Pollut Res, 25: 12159–12168.

245. Tânia, F G G C, Dina, M, Alberto, A C C P, Artur, J M V (2018). Cyclodextrin-based materials for removing micro-pollutants from wastewater. Curr Org Chem, 22: 2150–2181. DOI: http://dx.doi.org/10.2174/138527282266618 1019125315.

246. Nichol, J W, Koshy, S T, Bae, H, Hwang, C M, Yamanlar, S, Khademhosseini, A (2010). Cell-laden microengineered gelatin methacrylate hydrogels. Bioma-terials, 31: 5536–5544. DOI: 10.1016/j.biomaterials.2010.03.064.

247. Nieboer, E, Richardson, D H S (1980). The replacement of the nondescript term 'heavy metals' by a biologically and chemically significant classifica-tion of metal ions. Environ Pollut Ser B, Chem Phys, 1: 3–26. DOI: 10.1016/0143-148X(80)90017-8.

248. Nkambule, T I, Krause, R W, Mamba, B B, Haarhoff, J (2009). Removal of natural organic matter from water using ion-exchange resins and cyclodextrin polyurethanes. Phys Chem Earth, Parts A/B/C, 34: 812–818. DOI: https://doi.org/10.1016/j.pce.2009.07.013.

249. Nojavan, S, Yazdanpanah, M (2017). Micro-solid phase extraction of benzene, toluene, ethylbenzene and xylenes from aqueous solutions using water-insoluble beta-cyclodextrin polymer as sorbent. J Chromatogr A, 1525: 51–59. DOI: 10.1016/j.chroma.2017.10.027.

250. Norkus, E, Grincien, G, Vaitkus, R (2002). Interaction of lead(ii) with β-cyclodextrin in alkaline solutions. Carbohydr Res, 337: 1657–1661. DOI: 10.1016/S0008-6215(02)00044-7.

251. O'Connell, D W, Birkinshaw, CO' Dwyer, T F (2008). Heavy metal adsor-bents prepared from the modification of

cellulose: a review. Bioresour Technol, 99: 6709–6724.

252. Okoli, C P, Adewuyi, G O, Zhang, Q, Diagboya, P N, Guo, Q (2014). Mechanism of dialkyl phthalates removal from aqueous solution using gamma-cyclodextrin and starch based polyurethane polymer adsorbents. Carbohydr Polym, 114: 440–449. DOI: 10.1016/j.carbpol.2014.08.016.

253. Olteanu, A A, Arama, C-C, Bleotu, C, Lupuleasa, D, Monciu, C M (2015). Investigation of cyclodextrin based nanosponges complexes with angiotensin I converting enzyme inhibitors (enalapril, captopril, cilazapril). Farmacia, 63: 492–503.

254. Omurtag, G Z (2008). Fumonisins, trichothecenes and zearalenone in cereals. Int J Mol Sci, 9: 2062–2090.

255. Orprecio, R, Evans, C H (2003). Polymer-immobilized cyclodextrin trapping of model organic pollutants in flowing water streams. J Appl Polym Sci, 90: 2103–2110. DOI: 10.1002/app.12818.

256. Ozmen, E Y, Sezgin, M, Yilmaz, A, Yilmaz, M (2008). Synthesis of β-cyclodextrin and starch based polymers for sorption of azo dyes from aqueous solutions. Bioresour Technol, 99: 526–531. DOI: 10.1016/j.biortech.2007.01.023.

257. Ozmen, E Y, Sirit, A, Yilmaz, M (2007). A calix[4]arene oligomer and two beta-cyclodextrin polymers: synthesis and sorption studies of azo dyes. J Macromol Sci, Part A, 44: 167–173. DOI: 10.1080/10601320601031333.

258. Ozmen, E Y, Yilmaz, M (2007). Use of β-cyclodextrin and starch based polymers for sorption of congo red from aqueous solutions. J Hazard Mater, 148: 303–310. DOI: https://doi.org/10.1016/j.jhazmat.2007.02.042.

259. Paleos, C, Tsiourvas, D, Sideratou, O (2006). Multifunctional dendrimers and hyperbranched polymers as drug and gene delivery systems. US Patent US20060204472A1, published 2006-09-14, assigned to National Centre for Scientific Research Demokritos.

260. Pan, J, Zou, X, Wang, X, Guan, W, Li, C, Yan, Y, Wu, X (2011). Adsorptive removal of 2,4-didichlorophenol and 2,6-didichlorophenol from aqueous solution by β-cyclodextrin/attapulgite composites: Equilibrium, kinetics and thermodynamics. Chem Eng J, 166: 40–48. DOI: https://doi.org/10.1016/j.cej.2010.09.067.

261. Park, J H, Kim, J M, Jin, M, Jeon, J-K, Kim, S-S, Park, S H, Kim, S C, Park, Y-K (2012). Catalytic ozone oxidation of benzene at low temperature over mnox/al-sba-16 catalyst. Nanoscale Res Lett, 7: 14. DOI: 10.1186/1556-276X-7-14.

262. Park, K, Choi, J, Cho, E, Jung, S (2018). Enhanced solubilization of fluoranthene by hydroxypropyl β-cyclodextrin oligomer for bioremediation. Polymers, 10: 111.

263. Patel, B 2011, Natural dyes. In: Handbook of Textile and Industrial Dyeing. Elsevier: Amsterdam, 395–424.

264. Pearson, R G (1968). Hard and soft acids and bases, hsab, part 1: fundamental principles. J Chem Educ, 45: 581. DOI: 10.1021/ed045p581.

265. Pellicer, J, Rodríguez-López, M, Fortea, M, Lucas-Abellán, C, Mercader-Ros, M, López-Miranda, S, Gómez-López, V, Semeraro, P, Cosma, P, Fini, P, Franco, E, Ferrándiz, M, Pérez, E, Ferrándiz, M, Núñez-Delicado, E, Gabaldón, J (2019a). Adsorption properties of β- and hydroxypropyl-β-cyclodextrins cross-linked with epichlorohydrin in aqueous solution. a sustainable recycling strategy in textile dyeing process. Polymers, 11: 252. DOI: 10.3390/polym11020252.

266. Pellicer, J A, Rodríguez-López, M I, Fortea, M I, Gabaldón Hernández, J A, Lucas-Abellán, C, Mercader-Ros,

M T, Serrano-Martínez, A, Núñez-Delicado, E, Cosma, P, Fini, P, Franco, E, García, R, Ferrándiz, M, Pérez, E, Ferrándiz, M (2018). Removing of direct red 83:1 using α- and hp-α-cds polymerized with epichlorohydrin: kinetic and equilibrium studies. Dyes Pigm, 149: 736-746. DOI: 10.1016/j.dyepig.2017.11.032.

267. Pellicer, J A, Rodríguez-López, M I, Fortea, M I, Lucas-Abellán, C, Mercader-Ros, M T, López-Miranda, S, Gómez-López, V M, Semeraro, P, Cosma, P, Fini, P (2019b). Adsorption properties of β-and hydroxypropyl-β-cyclodextrins cross-linked with epichlorohydrin in aqueous solution. a sustainable recycling strategy in textile dyeing process. Polymers, 11: 252.

268. Pereva, S, Nikolova, V, Angelova, S, Spassov, T, Dudev, T (2019). Water inside beta-cyclodextrin cavity: amount, stability and mechanism of binding. Beilstein J Org Chem, 15: 1592–1600. DOI: 10.3762/bjoc.15.163.

269. Pratt, D Y, Wilson, L D, Kozinski, J A, Mohart, A M (2010). Preparation and sorption studies of β-cyclodextrin/epichlorohydrin copolymers. J Appl Polym Sci, 116: 2982–2989. DOI: 10.1002/app.31824.

270. Qin, X, Bai, L, Tan, Y, Li, L, Song, F, Wang, Y (2019). B-cyclodextrin-crosslinked polymeric adsorbent for simultaneous removal and stepwise recovery of organic dyes and heavy metal ions: fabrication, performance and mechanisms. Chem Eng J, 372: 1007–1018. DOI: 10.1016/j.cej.2019.05.006.

271. Qin, X, Zhang, H, Wang, Z, Jin, Y (2018). Magnetic chitosan/graphene oxide composite loaded with novel photosensitizer for enhanced photodynamic therapy. RSC Adv, 8: 10376–10388. DOI: 10.1039/C8RA00747K.

272. Radchenko, O, Sinelnikov, S, Moskalenko, O, Riabov, S (2018a). Nanocomposites based on titanium dioxide, modified by β-cyclodextrin containing copolymers. J Appl Polym Sci, 135: 46373. DOI: 10.1002/app.46373.

273. Radchenko, O, Sinelnikov, S, Moskalenko, O, Riabov, S (2018b). Nanocomposites based on titanium dioxide, modified by β-cyclodextrin containing copolymers. J Appl Polym Sci, 135: 46373.

274. Renard, E, Deratani, A, Volet, G, Sebille, B (1997a). Preparation and characterization of water soluble high molecular weight β-cyclodextrin-epichlorohydrin polymers. Eur Polym J, 33: 49–57. DOI: https://doi.org/10.1016/S0014-3057(96)00123-1.

275. Renard, E, Sebille, B, Barnathan, G, Deratani, A (1997b). Polycondensation of cyclodextrins with epichlorohydrin. Influence of reaction conditions on the polymer structure. Macromol Symp, 122: 229–234. DOI: 10.1002/masy.19971220136.

276. Ribeiro, A, Lobo, V, Valente, A, Simões, S, Sobral, A, Ramos, M, Burrows, H D (2006). Association between ammonium monovanadate and β-cyclodextrin as seen by NMR and transport techniques. Polyhedron, 25: 3581–3587.

277. Rudzinski, W E, Dave, A M, Vaishnav, U H, Kumbar, S G, Kulkarni, A R, Aminabhavi, T (2002). Hydrogels as controlled release devices in agriculture. Des Monomers Polym, 5: 39–65.

278. Rußler, A, Sakakibara, K, Rosenau, T (2011). Cellulose as matrix component of conducting films. Cellulose, 18: 937–944. DOI: 10.1007/s10570-011-9555-6.

279. Saenger, W (1980). Cyclodextrin inclusion compounds in research and industry. Angew Chem Int Ed

English, 19: 344–362. DOI: 10.1002/anie.198003441.

280. Saito, R, Yamaguchi, K (2003). Synthesis of bimodal methacrylic acid oligomers by template polymerization. Macromolecules, 36: 9005–9013. DOI: 10.1021/ma021767+.

281. Salazar, S, Guerra, D, Yutronic, N, Jara, P (2018). Removal of aromatic chlorinated pesticides from aqueous solution using β-cyclodextrin polymers decorated with Fe₃O₄ nanoparticles. Polymers, 10: 1038.

282. Salipira, K L, Mamba, B B, Krause, R W, Malefetse, T J, Durbach, S H (2007). Carbon nanotubes and cyclodextrin polymers for removing organic pollutants from water. Environ Chem Lett, 5: 13–17. DOI: 10.1007/s10311-006-0057-y.

283. Sancey, B, Trunfio, G, Charles, J, Badot, P M, Crini, G (2011). Sorption onto crosslinked cyclodextrin polymers for industrial pollutants removal: an interesting environmental approach. J Inclusion Phenom Macrocyclic Chem, 70: 315–320. DOI: 10.1007/s10847-010-9841-1.

284. Sawicki, R, Mercier, L (2006). Evaluation of mesoporous cyclodextrin-silica nanocomposites for the removal of pesticides from aqueous media. Environ Sci Technol, 40: 1978–1983. DOI: 10.1021/es051441r.

285. Schäfer, A I, Stelzl, K, Faghih, M, Sen Gupta, S, Krishnadas, K R, Heißler, S, Pradeep, T (2018a). Poly(ether sulfone) nanofibers impregnated with β-cyclodextrin for increased micropollutant removal from water. ACS Sustain Chem Eng, 6: 2942–2953. DOI: 10.1021/acssuschemeng.7b02214.

286. Schäfer, A I, Stelzl, K, Faghih, M, Sen Gupta, S, Krishnadas, K R, Heißler, S, Pradeep, T (2018b). Poly(ether sulfone) nanofibers impregnated with β-cyclodextrin for increased

micropollutant removal from water. ACS Sustain Chem Eng, 6: 2942–2953. DOI: 10.1021/acssuschemeng.7b02214.

287. Schmidt, A, Buschmann, H J, Knittel, D, Schollmeyer, E (2003). Method for producing reactive cyclodextrins, textile material provided with same, and use of said cyclodextrin derivatives. International Patent WO2003042449A1, published 2003-05-22, assigned to Ciba Specialty Chemicals Holding Inc.

288. Schmidt, B V K J, Barner-Kowollik, C (2017). Dynamic macromolecular material design—the versatility of cyclodextrin-based host–guest Chem. Angewandte Chemie Inter Edition, 56: 8350–8369. DOI: 10.1002/anie.201612150.

289. Seema, K, Mamba, B, Njuguna, J, Bakhtizin, R, Mishra, A K (2018). Removal of lead (II) from aqeouos waste using (cd-pcl-tio2) bio-nanocomposites. Int J Biol Macromol, 109: 136–142.

290. Sevillano, X, Isasi, J R, Penas, F J (2008). Feasibility study of degradation of phenol in a fluidized bed bioreactor with a cyclodextrin polymer as biofilm carrier. Biodegradation, 19: 589–597. DOI: 10.1007/s10532-007-9164-0.

291. Shao, D, Sheng, G, Chen, C, Wang, X, Nagatsu, M (2010). Removal of polychlorinated biphenyls from aqueous solutions using β-cyclodextrin grafted multiwalled carbon nanotubes. Chemosphere, 79: 679–685. DOI: https://doi.org/10.1016/j.chemosphere.2010.03.008.

292. Shao, S, Si, J, Tang, J, Sui, M, Shen, Y (2014). Jellyfish-shaped amphiphilic dendrimers: synthesis and formation of extremely uniform aggregates. Macromolecules, 47: 916–921. DOI: 10.1021/ma4025619.

293. Shao, Y, Martel, B, Morcellet, M, Weltrowski, M, Crini, G (1996a). Sorption of textile dyes on β-cyclodextrin-epichlorhydrin gels. J Inclusion Phenom

Mol Recognit Chem, 25: 209–212. DOI: 10.1007/BF01041570.

294. Shao, Y, Martel, B, Morcellet, M, Weltrowski, M, Crini, G (1996b). Sorption of textile dyes on β-cyclodextrin-epichlorhydrin gels. J Inclusion Phenom Mol Recognit Chem, 25: 209–212.

295. Shen, H-M, Zhu, G-Y, Yu, W-B, Wu, H-K, Ji, H-B, Shi, H-X, Zheng, Y-F, She, Y-B (2015). Surface immobilization of β-cyclodextrin on hybrid silica and its fast adsorption performance of p-nitrophenol from the aqueous phase. RSC Adv, 5: 84410–84422. DOI: 10.1039/C5RA15592D.

296. Shier, W T, Shier, A C, Xie, W, Mirocha, C J (2001). Structure-activity relationships for human estrogenic activity in zearalenone mycotoxins. Toxicon, 39: 1435–1438. DOI: https://doi.org/10.1016/S0041-0101(00)00259-2.

297. Sikder, M T, Islam, M S, Kikuchi, T, Suzuki, J, Saito, T, Kurasaki, M (2014). Removal of copper ions from water using epichlorohydrin cross-linked beta-cyclodextrin polymer: characterization, isotherms and kinetics. Water Environ Res, 86: 296–304. DOI: 10.2175/106143013x13807328848054.

298. Sikder, M T, Rahman, M M, Jakariya, M, Hosokawa, T, Kurasaki, M, Saito, T (2019). Remediation of water pollution with native cyclodextrins and modified cyclodextrins: a comparative overview and perspectives. Chem Eng J, 355: 920–941. DOI: https://doi.org/10.1016/j.cej.2018.08.218.

299. Singh, V P, Vaish, R (2019). Candle soot coated polyurethane foam as an adsorbent for removal of organic pollutants from water. Eur Phys J Plus, 134: 419.

300. Sivaram, N, Barik, D 2019, Toxic waste from leather industries. In: Energy from Toxic Organic Waste for Heat and Power Generation. Elsevier: Amsterdam, 55–67.

301. Skold, M E, Thyne, G D, Drexler, J W, McCray, J E (2009). Solubility enhancement of seven metal contaminants using carboxymethyl-β-cyclodextrin (cmcd). J Contam Hydrol, 107: 108–113. DOI: https://doi.org/10.1016/j.jconhyd.2009.04.006.

302. Soleimani, K, Dadkhah Tehrani, A, Adeli, M (2018). Preparation of new go-based slide ring hydrogel through a convenient one-pot approach as methylene blue absorbent. Carbohydr Polym, 187: 94–101. DOI: https://doi.org/10.1016/j.carbpol.2018.01.084.

303. Solms, J, Egli, R H (1965). Harze mit einschlusshohlräumen von cyclodextrin-struktur. Helv Chimica Acta, 48: 1225–1228. DOI: 10.1002/hlca.19650480603.

304. Song, W, Hu, J, Zhao, Y, Shao, D, Li, J (2013). Efficient removal of cobalt from aqueous solution using β-cyclodextrin modified graphene oxide. RSC Adv, 3: 9514–9521. DOI: 10.1039/C3RA41434E.

305. Sreenivasan, K (1996). Grafting of β-cyclodextrin-modified 2-hydroxyethyl methacrylate onto polyurethane. J Appl Polym Sci, 60: 2245–2249. DOI: 10.1002/(sici)1097-4628(19960620)60:12<2245::aid-app23>3.0.co;2-4.

306. Taccari, M, Milanovic, V, Comitini, F, Casucci, C, Ciani, M (2012). Effects of biostimulation and bioaugmentation on diesel removal and bacterial community. Int Biodeter Biodegrad, 66: 39–46.

307. Taka, A L, Pillay, K, Mbianda, X Y (2017). Nanosponge cyclodextrin polyurethanes and their modification with nanomaterials for the removal of pollutants from waste water: a review. Carbohydr Polym, 159: 94–107. DOI: 10.1016/j.carbpol.2016.12.027.

308. Tan, K B, Vakili, M, Horri, B A, Poh, P E, Abdullah, A Z, Salamatinia, B

(2015). Adsorption of dyes by nano-materials: recent developments and adsorption mechanisms. Sep Purif Technol, 150: 229–242. DOI.

309. Tan, P, Hu, Y (2017). Improved synthesis of graphene/β-cyclodextrin composite for highly efficient dye adsorption and removal. J Mol Liq, 242: 181–189. DOI: https://doi.org/10.1016/j.molliq.2017.07.010.

310. Teng, M, Li, F, Zhang, B, Taha, A A (2011). Electrospun cyclodextrin-functionalized mesoporous polyvinyl alcohol/sio2 nanofiber membranes as a highly efficient adsorbent for indigo carmine dye. Colloids Surf, A Physicochem Eng Asp, 385: 229–234. DOI: https://doi.org/10.1016/j.colsurfa.2011.06.020.

311. Tizro, N, Moniri, E, Saeb, K, Panahi, H A, Ardakani, S S (2019). Preparation and application of grafted β-cyclodextrin/thermo-sensitive polymer onto modified fe3o4@sio2 nano-particles for fenitrothion elimination from aqueous solution. Microchem J, 145: 59–67. DOI: 10.1016/j.microc.2018.09.005.

312. Tojima, T, Katsura, H, Nishiki, M, Nishi, N, Tokura, S, Sakairi, N (1999). Chitosan beads with pendant α-cyclodextrin: preparation and inclusion property to nitrophenolates. Carbohydr Polym, 40: 17–22. DOI: 10.1016/S0144-8617(99)00030-2.

313. Topuz, F, Uyar, T (2017). Poly-cyclodextrin cryogels with aligned porous structure for removal of polycyclic aromatic hydrocarbons (pahs) from water. J Hazard Mater, 335: 108–116. DOI: DOI:10.1016/j.jhazmat.2017.04.022.

314. Trellu, C, Mousset, E, Pechaud, Y, Huguenot, D, van Hullebusch, E D, Esposito, G, Oturan, M A (2016). Removal of hydrophobic organic pollutants from soil washing/flushing solutions: a critical review. J Hazard Mater, 306: 149–174. DOI: 10.1016/j.jhazmat.2015.12.008.

315. Trotta, F, Cavalli, R, Tumiatti, W, Zerbinati, O, Roggero, C, Vallero, R (2008). Ultrasound-assisted synthesis of cyclodextrin-based nanosponges. US Patent US20080213384A1, published 2008-09-04, assigned to Sea Marconi Technologies di W. Tumiatti SAS.

316. Trotta, F, Tumiatti, W (2014). Cross-linked polymers based on cyclodextrins for removing polluting agents. US Patent US20050154198A1, published 2005-07-14, assigned to Sea Marconi Technologies di W. Tumiatti SAS.

317. Valente, A J, Ribeiro, A C, Lobo, V M, Jiménez, A (2004). Diffusion coefficients of lead (II) nitrate in nitric acid aqueous solutions at 298 k. J Mol Liq, 111: 33–38.

318. Vareda, J P, Valente, A J M, Durães, L (2016). Heavy metals in iberian soils: removal by current adsorbents/amendments and prospective for aerogels. Adv Colloid Interface Sci, 237: 28–42. DOI: https://doi.org/10.1016/j.cis.2016.08.009.

319. Vareda, J P, Valente, A J M, Durães, L (2019). Assessment of heavy metal pollution from anthropogenic activities and remediation strategies: a review. J Environ Manage, 246: 101–118. DOI: 10.1016/j.jenvman.2019.05.126.

320. Vázquez-Guilló, R, Calero, A, Valente, A J M, Burrows, H D, Reyes Mateo, C, Mallavia, R (2013). Novel electrospun luminescent nanofibers from cationic polyfluorene/cellulose acetate blend. Cellulose, 20: 169–177. DOI: 10.1007/s10570-012-9809-y.

321. Villaverde, J, Pérez-Martínez, J I, Maqueda, C, Ginés, J M, Morillo, E (2005). Inclusion complexes of α-and γ-cyclodextrins and the herbicide norflurazon: I. Preparation and characterisation. II. Enhanced solubilisation and removal from soils. Chemosphere,

60: 656–664. DOI: 10.1016/j.chemo-sphere. 2005.01.030.

322. Wan, J, Wen, H, Hu, Z, Chen, W (2008). Production of cyclodextrin-phthalocy-anine double analog enzyme functional fibre. PRC Patent CN100398741C, published 2007-04-04, assigned to Zhejiang Sci-tech University.

323. Wang, D, Liu, L, Jiang, X, Yu, J, Chen, X, Chen, X (2015a). Adsorbent for p-phenylenediamine adsorption and removal based on graphene oxide func-tionalized with magnetic cyclodextrin. Appl Surf Sci, 329: 197–205. DOI: 10.1016/j.apsusc.2014.12.161.

324. Wang, G, Xu, W, Wang, X, Huang, L (2012a). Glycine-β-cyclodextrin–enhanced electrokinetic removal of atrazine from contaminated soils. Environ Eng Sci, 29: 406–411. DOI: 10.1089/ees.2010.0272

325. Wang, H, Wang, Y, Zhou, Y, Han, P, Lu, X (2014a). A facile removal of phenol in wastewater using crosslinked beta-cyclodextrin particles with ultrasonic treatment. Clean, Soil Air Water, 42: 51–55. DOI: 10.1002/clen.201200605.

326. Wang, H, Wang, Y, Zhou, Y, Han, P, Lü, X (2014b). A facile removal of phenol in wastewater using crosslinked β-cyclodextrin particles with ultrasonic treatment. Clean, Soil, Air, Water, 42: 51–55. DOI: 10.1002/clen.201200605.

327. Wang, J-x, Shuo, C, Xie, Q, Zhao, H-mZhao, Y-z (2006a). Enhanced photo-degradation of phenolic compounds by adding tio2 to soil in a rotary reactor. J Environ Sci, 18: 1107–1112. DOI: 10.1016/S1001-0742(06)60047-8.

328. Wang, J, Chen, C (2006). Biosorp-tion of heavy metals by saccharo-myces cerevisiae: a review. Biotechnol Adv, 24: 427–451. DOI: 10.1016/j. biotechadv.2006.03.001.

329. Wang, J, Huang, T, Zhang, L, Yu, Q J, Hou, L a (2018). Dopamine cross-linked graphene oxide membrane for simultaneous removal of organic pollutants and trace heavy metals from aqueous solution. Environ Technol, 39: 3055–3065. DOI: 10.1080/09593330. 2017.1371797.

330. Wang, Q Z, Chen, X G, Liu, N, Wang, S X, Liu, C S, Meng, X H, Liu, C G (2006b). Protonation constants of chitosan with different molecular weight and degree of deacetylation. Carbohydr Polym, 65: 194–201. DOI: 10.1016/j. carbpol.2006.01.001.

331. Wang, S, Li, Y, Fan, X, Zhang, F, Zhang, G (2015b). B-cyclodextrin functionalized graphene oxide: an effi-cient and recyclable adsorbent for the removal of dye pollutants. Front Chem Sci Eng, 9: 77–83. DOI: 10.1007/ s11705-014-1450-x.

332. Wang, X, Hsiao, B S (2016). Electro-spun nanofiber membranes. Curr Opin Chem Eng, 12: 62–81. DOI: 10.1016/j. coche.2016.03.001.

333. Wang, Y, Wang, X, Antonietti, M (2012b). Polymeric graphitic carbon nitride as a heterogeneous organocata-lyst: from photoChem to multipurpose catalysis to sustainable Chem. Angew Chem Int Ed, 51: 68–89. DOI: 10.1002/ anie.201101182.

334. Wang, Z, Cui, F, Pan, Y, Hou, L, Zhang, B, Li, Y, Zhu, L (2019). Hierarchi-cally micro-mesoporous β-cyclodextrin polymers used for ultrafast removal of micropollutants from water. Carbohydr Polym, 213: 352–360. DOI: 10.1016/j. carbpol.2019.03.021.

335. Wang, Z, Zhang, P, Hu, F, Zhao, Y, Zhu, L (2017a). A crosslinked beta-cyclodextrin polymer used for rapid removal of a broad-spectrum of organic micropollutants from water. Carbohydr Polym, 177: 224–231. DOI: 10.1016/j. carbpol.2017.08.059.

336. Wang, Z, Zhang, P, Hu, F, Zhao, Y, Zhu, L (2017b). A crosslinked β-cyclodextrin polymer used for rapid

removal of a broad-spectrum of organic micropollutants from water. Carbohydr Polym, 177: 224–231. DOI: 10.1016/j.carbpol.2017.08.059.

337. Weckhuysen, B M, Wachs, I E, Schoonheydt, R A (1996). Surface Chem and spectroscopy of chromium in inorganic oxides. Chem Rev, 96: 3327–3350. DOI: 10.1021/cr940044o.

338. Wei, H, Wang, E (2013). Nanomaterials with enzyme-like characteristics (nanozymes): next-generation artificial enzymes. Chem Soc Rev, 42: 6060–6093. DOI: 10.1039/c3cs35486e.

339. Wei, Z, Liu, Y, Hu, H, Yu, J, Li, F (2016). Biodegradable poly(butylene succinate-co-terephthalate) nanofibrous membranes functionalized with cyclodextrin polymer for effective methylene blue adsorption. RSC Adv, 6: 108240–108246. DOI: 10.1039/C6RA22941G.

340. Weltrowski, M, Morcellet, M, Martel, B (2003). Cyclodextrin polymers and/or cyclodextrin derivatives with complexing properties and ion-exchange properties and method for the production thereof. Google Patents.

341. Wilson, L D, Mohamed, M H, Headley, J V (2011). Surface area and pore structure properties of urethane-based copolymers containing β-cyclodextrin. J Colloid Interface Sci, 357: 215–222. DOI: 10.1016/j.jcis.2011.01.081.

342. Wong, Y, Szeto, Y, Cheung, W, McKay, G (2004). Adsorption of acid dyes on chitosan—equilibrium isotherm analyses. Process Biochem, 39: 695–704. DOI: 10.1016/S0032-9592(03)00152-3.

343. Wu, F-C, Liu, B-L, Wu, K-T, Tseng, R-L (2010). A new linear form analysis of redlich–peterson isotherm equation for the adsorptions of dyes. Chem Eng J, 162: 21–27. DOI: 10.1016/j.cej.2010.03.006.

344. Wu, Y, Zhao, Z, Chen, M, Jing, Z, Qiu, F (2018). b-cyclodextrin–graphene oxide–diatomaceous earth material: preparation and its application for adsorption of organic dye. Monatsh Chem, 149: 1367–1377. DOI: 10.1007/s00706-018-2168-0.

345. Xia, Y, Wan, J (2008). Preparation and adsorption of novel cellulosic fibers modified by β-cyclodextrin. Polym Adv Technol, 19: 270–275. DOI: 10.1002/pat.997.

346. Xiao, L, Ling, Y, Alsbaiee, A, Li, C, Helbling, D E, Dichtel, W R (2017). b-cyclodextrin polymer network sequesters perfluorooctanoic acid at environmentally relevant concentrations. J Am Chem Soc, 139: 7689–7692. DOI: 10.1021/jacs.7b02381.

347. Xiao, N, Wen, Q, Liu, Q, Yang, Q, Li, Y (2014). Electrospinning preparation of β-cyclodextrin/glutaraldehyde cross-linked PVP nanofibrous membranes to adsorb dye in aqueous solution. Chem Res Chin Univ, 30: 1057–1062. DOI: 10.1007/s40242-014-4203-y.

348. Xiao, P, Dudal, Y, Corvini, P F X, Shahgaldian, P (2011). Polymeric cyclodextrin-based nanoparticles: synthesis, characterization and sorption properties of three selected pharmaceutically active ingredients. Polym Chem, 2: 120–125. DOI: 10.1039/C0PY00225A.

349. Xiao, P, Weibel, N, Dudal, Y, Corvini, P F X, Shahgaldian, P (2015). A cyclodextrin-based polymer for sensing diclofenac in water. J Hazard Mater, 299: 412–416. DOI: 10.1016/j.jhazmat.2015.06.047.

350. Xinhong, G, Ying, T, Wenjie, R, Jun, M, Christie, P, Yongming, L (2017). Optimization of ex-situ washing removal of polycyclic aromatic hydrocarbons from a contaminated soil using nano-sulfonated graphene. Pedosphere, 27: 527–536. DOI: 10.1016/S1002-0160(17)60348-5.

351. Yamasaki, H, Makihata, Y, Fukunaga, K (2006). Efficient phenol removal of wastewater from phenolic resin plants

using crosslinked cyclodextrin particles. J Chem Technol Biotechnol, 81:
1271–1276. DOI: 10.1002/jctb.1545.

352. Yamjala, K, Nainar, M S, Ramisetti,
N R (2016). Methods for the analysis of azo dyes employed in food
industry—a review. Food Chem, 192:
813–824. DOI: 10.1016/j.foodchem.
2015.07.085.

353. Yan, J, Zhu, Y, Qiu, F, Zhao, H, Yang,
D, Wang, J, Wen, W (2016). Kinetic,
isotherm and thermodynamic studies
for removal of methyl orange using a
novel β-cyclodextrin functionalized
graphene oxide-isophorone diisocyanate composites. Chem Eng Res Des,
106: 168–177. DOI: 10.1016/j.cherd.
2015.12.023.

354. Yang, C, Huang, H, Ji, T, Zhang, K,
Yuan, L, Zhou, C, Tang, K, Yi, J, Chen,
X (2019a). A cost-effective crosslinked
β-cyclodextrin polymer for the rapid
and efficient removal of micropollutants from wastewater. Polymer Int, 68:
805–811. DOI: 10.1002/pi.5771.

355. Wang, Z, Guo, S, Zhang, B, Fang, J,
Zhu, L (2020). Interfacially crosslinked
β-cyclodextrin polymer composite
porous membranes for fast removal
of organic micropollutants from water
by flow-through adsorption. J Hazard
Mater 384, 121187. DOI: 10.1016/j.
jhazmat.2019.121187.

356. Yang, J S, Yang, L (2013). Preparation and application of cyclodextrin
immobilized polysaccharides. J Mater
Chem B, 1: 909–918. DOI: 10.1039/
C2TB00107A.

357. Yaseen, D, Scholz, M (2019). Textile
dye wastewater characteristics and
constituents of synthetic effluents:
a critical review. Int J Environ Sci
Technol, 16: 1193–1226. DOI: 10.1007/
s13762-018-2130-z.

358. Ye, M, Sun, M, Kengara, F O, Wang,
J, Ni, N, Wang, L, Song, Y, Yang, X,
Li, H, Hu, F (2014a). Evaluation of soil

washing process with carboxymethyl-
β-cyclodextrin and carboxymethyl
chitosan for recovery of pahs/heavy
metals/fluorine from metallurgic plant
site. J Environ Sci, 26: 1661–1672.
DOI: 10.1016/j.jes.2014.06.006.

359. Ye, M, Sun, M, Kengara, F O, Wang, J,
Ni, N, Wang, L, Song, Y, Yang, X, Li, H,
Hu, F, Jiang, X (2014b). Evaluation of soil
washing process with carboxymethyl-β-
cyclodextrin and carboxymethyl chitosan
for recovery of pahs/heavy metals/fluo-
rine from metallurgic plant site. J Environ
Sci, 26: 1661–1672. DOI: 10.1016/j.
jes.2014.06.006.

360. Yoshikazu, M, Minoru, T, Yoshihiro, K,
Toshiyuki, S (1981). Sorption behavior
of low molecular weight organic vapors
on β-cyclodextrin polyurethane resins.
Bull Chem Soc Jpn, 54: 2487–2490.
DOI: 10.1246/bcsj.54.2487.

361. Yu, D, Wu, L L, Wang, J F, Tao, Y W,
Shen, Y T, Liang, B Y, Wang, H (2013).
Preparation of β-cyclodextrin/chitosan
membranes and its application in the
wastewater treatment of acid dyes. Adv
Mater Res, 726–731: 2558–2562. DOI:
10.4028/www.scientific.net/AMR.
726-731.2558.

362. Yu, F, Chen, D, Ma, J (2017). Synthesis
of cyclodextrin-based adsorbents and
its application for organic pollutant
removal from water. Curr Org Chem,
21: 1976–1990. DOI: 10.2174/138527
2821666170503110023.

363. Yu, Z, Li, F, Sun, L (2015). Recent
advances in dye-sensitized photoelec-
trochemical cells for solar hydrogen
production based on molecular components. Energy Environ Sci, 8: 760–775.
DOI: 10.1039/C4EE03565H.

364. Yuan, G, Prabakaran, M, Qilong, S,
Lee, J S, Chung, I-M, Gopiraman, M,
Song, K-H, Kim, I S (2017). Cyclodextrin functionalized cellulose nanofiber
composites for the faster adsorption
of toluene from aqueous solution. J

Taiwan Inst Chem Eng, 70: 352–358. DOI: 10.1016/j.jtice.2016.10.028.

365. Yue, X, Huang, J, Jiang, F, Lin, H, Chen, Y (2019). Synthesis and characterization of cellulose-based adsorbent for removal of anionic and cationic dyes. J Eng Fibers Fabr, 14: 155892501982819. DOI: 10.1177/1558925019828194.

366. Zemel, H, Koch, M B (1990). Preparation of crosslinked cyclodextrin resins with enhanced porosity. Google Patents.

367. Zengin, A, Tamer, U, Caykara, T (2018). Sers detection of polyaromatic hydrocarbons on a β-cyclodextrin containing polymer brush. J Raman Spectrosc, 49: 452–461. DOI: 10.1002/jrs.5300.

368. Zha, F, Li, S, Chang, Y (2008). Preparation and adsorption property of chitosan beads bearing β-cyclodextrin crosslinked by 1,6-hexamethylene diisocyanate. Carbohydr Polym, 72: 456–461. DOI: 10.1016/j.carbpol.2007.09.013.

369. Zhang, F, Chen, Y, Lin, H, Lu, Y (2007). Synthesis of an amino-terminated hyperbranched polymer and its application in reactive dyeing on cotton as a salt-free dyeing auxiliary. Color Technol, 123: 351–357. DOI: 10.1111/j.1478-4408.2007.00108.x.

370. Zhang, J, Chen, Y, Wang, X (2015). Two-dimensional covalent carbon nitride nanosheets: synthesis, functionalization, and applications. Energy Environ Sci, 8: 3092–3108. DOI: 10.1039/C5EE01895A.

371. Zhang, K-D, Tsai, F-C, Ma, N, Xia, Y, Liu, H-L, Guo, X, Yu, X-Y, Jiang, T, Chiang, T-C, Chang, C C-J (2017). Removal of azo dye from aqueous solution by host-guest interaction with β-cyclodextrin. Desalin Water Treat, 86. DOI: 10.5004/dwt.2017.21187.

372. Zhang, L, Zeng, Y, Cheng, Z (2016). Removal of heavy metal ions using chitosan and modified chitosan: a review. J Mol Liq, 214: 175–191. DOI: 10.1016/j.molliq.2015.12.013.

373. Zhang, N, Qiu, H, Si, Y, Wang, W, Gao, J (2011). Fabrication of highly porous biodegradable monoliths strengthened by graphene oxide and their adsorption of metal ions. Carbon, 49: 827–837. DOI: 10.1016/j.carbon.2010.10.024.

374. Zhang, S, He, Y, Wu, L, Wan, J, Ye, M, Long, T, Yan, Z, Jiang, X, Lin, Y, Lu, X (2019). Remediation of organochlorine pesticide-contaminated soils by surfactant-enhanced washing combined with activated carbon selective adsorption. Pedosphere, 29: 400–408. DOI: 10.1016/S1002-0160(17)60328-X.

375. Zhang, X, Wang, Y, Yang, S (2014). Simultaneous removal of co(II) and 1-naphthol by core–shell structured fe3o4@cyclodextrin magnetic nanoparticles. Carbohydr Polym, 114: 521–529. DOI: 10.1016/j.carbpol.2014.08.072.

376. Zhang, X M, Peng, C S, Xu, G C (2012). Synthesis of modified β-cyclodextrin polymers and characterization of their fuchsin adsorption. J Inclusion Phenom Macrocyclic Chem, 72: 165–171. DOI: 10.1007/s10847-011-9956-z.

377. Zhang, Y, Ou, H, Liu, H, Ke, Y, Zhang, W, Liao, G, Wang, D (2018). Polyimide-based carbon nanofibers: a versatile adsorbent for highly efficient removals of chlorophenols, dyes and antibiotics. Colloids Surf A, Physicochem Eng Asp, 537: 92–101. DOI: 10.1016/j.colsurfa.2017.10.014.

378. Zhao, C, Dong, Y, Feng, Y, Li, Y, Dong, Y (2019). Thermal desorption for remediation of contaminated soil: a review. Chemosphere, DOI: 10.1016/j.chemosphere.2019.01.079.

379. Zhao, D, Zhao, L, Zhu, C-S, Huang, W-Q, Hu, J-L (2009a). Water-insoluble β-cyclodextrin polymer crosslinked by citric acid: synthesis and adsorption properties toward phenol and methylene blue. J Inclusion Phenom Macrocyclic Chem, 63: 195–201. DOI: 10.1007/s10847-008-9507-4.

380. Zhao, D, Zhao, L, Zhu, C-S, Shen, X, Zhang, X, Sha, B (2009b). Comparative study of polymer containing β-cyclodextrin and –COOH for adsorption toward aniline, 1-naphthylamine and methylene blue. J Hazard Mater, 171: 241–246. DOI: 10.1016/j.jhazmat.2009.05.134.

381. Zhao, D, Zhao, L, Zhu, C-S, Wang, J, Lv, X-H (2012). A novel β-cyclodextrin polymer modified by sulfonate groups. J Inclusion Phenom Macrocyclic Chem, 73: 93–98. DOI: 10.1007/s10847-011-0024-5.

382. Zhao, F, Repo, E, Meng, Y, Wang, X, Yin, D, Sillanpää, M (2016). An edta-β-cyclodextrin material for the adsorption of rare earth elements and its application in preconcentration of rare earth elements in seawater. J Colloid Interface Sci, 465: 215–224. DOI: 10.1016/j.jcis.2015.11.069.

383. Zhao, F, Repo, E, Yin, D, Chen, L, Kalliola, S, Tang, J, Iakovleva, E, Tam, K C, Sillanpää, M (2017). One-pot synthesis of trifunctional chitosan-edta-β-cyclodextrin polymer for simultaneous removal of metals and organic micropollutants. Sci Rep, 7: 15811. DOI: 10.1038/s41598-017-16222-7.

384. Zhao, F, Repo, E, Yin, D, Meng, Y, Jafari, S, Sillanpää, M (2015a). EDTA-cross-linked β-cyclodextrin: an environmentally friendly bifunctional adsorbent for simultaneous adsorption of metals and cationic dyes. Environ Sci Technol, 49: 10570–10580. DOI: 10.1021/acs.est.5b02227.

385. Zhao, F, Repo, E, Yin, D, Meng, Y, Jafari, S, Sillanpää, M (2015b). EDTA-cross-linked β-cyclodextrin: an environmentally friendly bifunctional adsorbent for simultaneous adsorption of metals and cationic dyes. Environ Sci Technol, 49: 10570–10580. DOI: 10.1021/acs.est.5b02227.

386. Zhao, J, Zou, Z, Ren, R, Sui, X, Mao, Z, Xu, H, Zhong, Y, Zhang, L, Wang, B (2018a). Chitosan adsorbent reinforced with citric acid modified β-cyclodextrin for highly efficient removal of dyes from reactive dyeing effluents. Eur Polym J, 108: 212–218. DOI: 10.1016/j.eurpolymj.2018.08.044.

387. Zhao, Q, Chen, Y, Liu, Y (2018b). Cyclodextrin-based supramolecular hydrogel. In: Handbook of Macrocyclic Supramolecular Assembly. Springer, New York. DOI: 10.1007/978-981-13-1744-6_19-1.

388. Zhao, R, Wang, Y, Li, X, Sun, B, Jiang, Z, Wang, C (2015c). Water-insoluble sericin/β-cyclodextrin/PVA composite electrospun nanofibers as effective adsorbents towards methylene blue. Colloids Surf B: Biointerfaces, 136: 375–382. DOI.

389. Zhao, R, Wang, Y, Li, X, Sun, B, Wang, C (2015d). Synthesis of β-cyclodextrin-based electrospun nanofiber membranes for highly efficient adsorption and separation of methylene blue. ACS Appl Mater Interfaces, 7: 26649–26657. DOI: 10.1021/acsami.5b08403.

390. Celebioglu, A, Uyar, T (2010). Cyclodextrin nanofibers by electrospinning. Chem Comm, 46, 6903–6905. DOI: 10.1039/c0cc01484b.

391. Zhou, Y, Cheng, G, Chen, K, Lu, J, Lei, J, Pu, S (2019). Adsorptive removal of bisphenol a, chloroxylenol, and carbamazepine from water using a novel β-cyclodextrin polymer. Ecotoxicol Environ Safety, 170: 278–285. DOI: https://doi.org/10.1016/j.ecoenv.2018.11.117.

392. Zhou, Y, Gu, X, Zhang, R, Lu, J (2014). Removal of aniline from aqueous solution using pine sawdust modified with citric acid and β-cyclodextrin. Ind Eng Chem Res, 53: 887–894. DOI: 10.1021/ie403829s.

393. Zhou, Y, Hu, Y, Huang, W, Cheng, G, Cui, C, Lu, J (2018). A novel amphoteric β-cyclodextrin-based adsorbent for simultaneous removal of cationic/anionic dyes and bisphenol A. Chem Eng J, 341: 47–57. DOI: 10.1016/j.cej.2018.01.155.

394. Zhu, J, Wei, S, Gu, H, Rapole, S B, Wang, Q, Luo, Z, Haldolaarachchige, N, Young, D P, Guo, Z (2012a). One-pot synthesis of magnetic graphene nanocomposites decorated with core@double-shell nanoparticles for fast chromium removal. Environ Sci Tech, 46: 977–985. DOI: 10.1021/es2014133.

395. Zhu, X, Zhou, D, Cang, L, Wang, Y (2012b). Tio 2 photocatalytic degradation of 4-chlorobiphenyl as affected by solvents and surfactants. J Soils Sediments, 12: 376–385. DOI: 10.1007/s11368-011-0464-y.

396. Zou, H, Du, W, Ji, M H, Zhu, R (2016). Enhanced electrokinetic remediation of pyrene-contaminated soil through pH control and rhamnolipid addition. Environ Eng Sci, 33: 507–513. DOI: 10.1089/ees.2016.0019.

397. Zou, W-S, Wang, Y-Q, Wang, F, Shao, Q, Zhang, J Liu, J (2013). Selective fluorescence response and magnetic separation probe for 2,4,6-trinitrotoluene based on iron oxide magnetic nanoparticles. Anal Bioanal Chem, 405: 4905–4912. DOI: 10.1007/s00216-013-6873-6.

CHAPTER 15

MOLECULAR MODELING AND PROPERTIES OF CHELATE AGENTS AND THEIR COMPOSITES FOR TREATMENT OF HEAVY METAL INTOXICATION

ANDREEA IRINA BARZIC* and RALUCA MARINICA ALBU

"Petru Poni" Institute of Macromolecular Chemistry,
Laboratory of Physical Chemistry of Polymers,
41A Grigore Ghica Voda Alley, 700487 Iasi, Romania

Corresponding author. E-mail: irina_cosutchi@yahoo.com.

ABSTRACT

The increasing progress of many industries makes more easy life for consumers but damages the environment by large amounts of waste production. Researchers working in the field of materials science must reorient toward the synthesis of green products with reduced or no toxicity. In the places where wastes are deposited, the health of humans, animals, and plants are negatively affected. This chapter is focused on the use in biomedicine of composite materials containing chelate polymers for obtaining products useful for metal poisoning treatment. Introductive aspects on molecular modeling of several categories of chelate compounds and composite materials based on green polymers of great importance in biomedicine are described. The impact of chemical structure on some physical properties of the green materials with biological potential is analyzed. Applications of the described materials in decontamination of living organisms of toxic metallic particles are briefly reviewed. The most efficient chelation therapies used in medical treatment for reducing the toxic effects of metals are presented.

Current developments in this domain lie at the basis of the future perspectives for patients with superior techniques of health treatments and implicitly better life quality.

15.1   INTRODUCTION

Some human activities and industrial processes produce wastes and residues, which in many cases are disposed in unsafe conditions in the environment [1]. Unfortunately, such actions make possible infiltration of toxic substances in water and soil and this gravely threatens the health of people and the ecosystem [2]. The real problem is that nowadays the levels of contamination are alarming and cannot be neglected. Therefore, immediate solutions must be formulated, given the fact that the pollution of water and other resources are already determining serious diseases and many deaths.

In this context, a particular issue arises from metal toxicity, which is the result of metal overload or exposure to heavy metals from diverse sources [3–5]. Most metals are able to interact via covalent bonds with carbon, thus forming metal-organic compounds. Once entered into a living organism, metals and metal compounds can interfere with the functions of many organs, namely the central nervous system, liver, kidneys, lungs, the haematopoietic system, and so on [3–6]. For instance, the kidney is a target organ in heavy metal poisoning considering its function of filtering that sometimes fails to prevent reabsorbing of divalent ions and in such circumstances leads to acute toxicity and produces the kidney chronic damage [7]. As a consequence, the removal of metals from the affected environment and human body represents a subject of paramount importance.

Diagnostic and investigation of the effects caused by the contamination with heavy metals to reduce and even eliminate the body's burden of these substances enabled the formulation of various kinds of treatments [8, 9]. These were developed by close monitoring of the physiological and pathological processes in the presence and absence of such toxic elements inside the organism. Part of these treatments was analogously created to water cleaning techniques, such as chemical precipitation and membrane filtration [10–14]. But these are not entirely efficient at low concentrations of toxic metals (incomplete removal) and also generate a big amount of sludge and other undesired products that impose careful disposal to avoid further pollution. Biosorption is a better alternative, which is affected by several process parameters such as temperature, pH, initial quantity of the metal ions, biosorbent dose, and speed of agitation [15]. Biomass can be viewed as a potential biosorbent for heavy metal elimination owing to the presence of metal-binding functional groups [16]. In addition, biomass can be modified by physical and chemical procedures before use, allowing regeneration and reusing of the biosorbent after interaction with the heavy metals.

Besides the biosorption procedures, chelation therapy is an important tool for diminishing metal concentrations in the body [17]. The term chelate is linked to the caliper-like groups which are acting as two associating units and fasten to a central atom enabling the formation of heterocyclic rings [17–19]. In other words, this is a special therapy, which is employed for the removal of bad elements from the living organisms by the formation of a chelate complex with adequate chelating ligands. In this way, the chelating ligands complex with the metal ions and facilitate the elimination of excess or toxic metal from the body restoring almost immediately to its initial nontoxic state, reducing the late effects. Based on the aforementioned aspects, the administration of chelating substances for individuals with a metal poisoning symptomatology (tremors, depression, fatigue, bad memory, hair loss, headache) or a known exposure to these compounds could be the key to restore the healthy physiology of the organism [3, 20, 21]. This is

majorly focused on taking out the toxic particles, or converting them as carriers for targeted drug delivery or for labeling certain molecules for diagnostics. Such elaborated techniques may be sorted under the category of metallo-pharmacology.

Green chelate polymers have attracted great interest in the preparation of composite for membrane science and technology and for biomedical purposes [17]. There is a wide range of natural and synthetic chelating polymers with partial or full biodegradability. Green polymeric composites are obtained by combining biopolymers as matrix and solid particles/fibers as fillers [22, 23]. This approach enables the enhancement of the desired physical and chemical properties of the final product. The chapter describes the preparation methods of the most relevant green composite systems containing chelate polymers reported in the literature. In addition, the modeling data reveal information concerning the properties of such complex materials including optimized conformation in low energy conditions, HOMO–LUMO energy gap, total energy, dipole moment, and the complex stability [24]. Applications in biomedicine of the described materials are also reviewed. Chelation therapy is widely useful for oral insulin delivery, tissue engineering, antitumor, antimicrobial coatings, orthopedics, wound dressing, drug delivery, medical implant, and bioresorbable fixation devices [25–31]. The future directions of this field are briefly presented.

15.2 MOLECULAR MODELING OF CHELATE POLYMERS AND THEIR COMPOSITES

In order to understand the properties developed in a simple or multicomponent polymeric system, it is useful to perform molecular simulations at the molecular level. This can be done using computer software that facilitates the prediction of the most relevant chemical and physical features of the analyzed compound. Moreover, such computer techniques are allowing the estimation of the evolution in time of a system of atoms, molecules, or granules [32]. More precisely, it provides specific data, like atomic positions, velocities, and forces from which the macroscopic characteristics (e.g., energy, pressure, heat capacities) can be extracted through the statistical mechanics. The most commonly used simulation methods include molecular mechanics (MM), molecular dynamics (MD), and Monte Carlo approach [32].

Modeling of chelate polymers and their composites at this level is principally directed toward the kinetics and thermodynamic aspects of the formation, macromolecular structure and interactions. As expected, in case of the polymer composites, the constituents are not dissolving or entirely merging. Thus, the composite components present an interface between them, which can be viewed as a distinct material with separate mechanical properties.

The literature describes some modeling studies concerning polymers with chelating groups, but not all of them are entirely green materials [33–36]. One of the most studied green chelating polymers is chitosan—an amino polysaccharide largely produced in nature. It is well-known that this biopolymer is characterized by a fibrillar architecture with a considerable degree of crystallinity and also polymorphism [33]. Molecular modeling of chitosan combined with X-ray investigations suggests that its chains display an extended twofold helix in a zigzag structure, while the crystal phase the macromolecular coils are disposed in an antiparallel fashion, analogously to the anhydrous form of the α-chitin [33, 37, 38]. The chemical structure of the α and β forms is distinct in terms of the arrangement of the piles of chains, which are antiparallel in an alternate manner for α-chitin and all parallel in β-chitin [33]. The comparison of the crystallographic aspects of chitin and chitosan has indicated that both biopolymers present anhydrous forms [39, 40]. The free amino groups in the chitosan's structure allow distinction of four types of helical conformations in acid [33]: type I (anhydrous), type II (hydrated), type IIa (hydrated), and type III (anhydrous) that can adopt a helical chain in a twofold helix, relaxed twofold helix, a 4/1 helix, and a fivefold helix, respectively. These aspects were confirmed also by molecular modeling data [33]. When dealing with solid-state, the twofold helix feature is maintained stable by O3-HO3••• O5' intrachain hydrogen bonds along the glycosidic part [33]. In order to check these helical features in an aqueous solution, MD modeling experiments were done for chitin and chitosan [41, 42]. This extracted information revealed that chitin chains have a prevalent twofold helix shape, which indeed is strongly settled by the O3-HO3••• O5' hydrogen bond interactions. In any case, chitosan macromolecules are able to embrace a huge number of distinct conformations that comprise the helical shapes noted in the solid phase. Helical forms and conformational transformations were proved to be influenced by the degree of acetylation of the chitosan structure [33].

The MD simulations reported by Lange et al. [34] show the behavior of chelate copolymer structure, namely poly(lactic-co-glycolic) acids (PLGAs). Their investigation was focused on the evaluation of certain interaction parameters of the Consistent Valence Force-Field (CVFF), which are essential for the analysis of thermodynamic and transport properties of oxaliplatin. The latter is known to be a colorectal anti-cancer drug in PLGAs matrices. The proposed methodology to confirm the parameters for PLGAs relied on the calculation of vitrification temperature and connecting these data with other structural properties like fractional free volume, material density, and cohesive energy density using the MD simulations. In case of the oxaliplatin, the metal-dependent and separate forces were introduced into CVFF and corroborated with an ab-initio approach (RHF/LanL2DZ) [34]. The differences noticed in the final equilibrium geometries of oxaliplatin are linked to bond lengths, but not to angles.

Houshmand et al. [35] analyzed via molecular modeling the stability of chitosan–MX (3-chloro-4-(dichloromethyl)-5-hydroxy-2(5H)-furanone) complexes in aqueous media. For this purpose, they determined interaction energy, the thermodynamic parameter and reactivity. The simulations of nanoadsorbent chitosan were done in the presence of MX—a mutagenic halogenated disinfection compound that is found in potable water. The structural characteristics and electronic properties of chitosan-MX system are computed during metal functionalization as shown by the data attained from density functional theory calculations. Chitosan and chitosan-based systems were optimized and their properties were estimated. The data showed that the features of linking sites recognize the important outcome of the functionalization step. The thermodynamic characteristics related to hydrolysis of examined structures in aqueous media reveal that metal absorbed materials tend to be more steady in solution in regard to other ones. The HOMO energy for isolated chitosan and that of the MX system in regard to the values for the MX/chitosan or MX/chitosan (Fe) demonstrates that complex systems are noted to have higher stability than single systems. Degradation of MX and its analogous in the aqueous environment and also the probability of absorption of MX by biopolymer in solution were also discussed through the distinct level of theory.

MD experiments were also used for the analysis of adsorption of cadmium(II) on amidoxime-chelating cellulose (ACCS) [36]. More specifically, the adsorption structure of ACCS was obtained by quantum molecular (QM) and molecular orbital experiments. The impact of the functional groups of ACCS on Cd(II) adsorption were examined by estimating the energy of interaction among of Cd(II) and its radial distribution functions of particular groups that can be understood as the probability Cd(II) presence at a cutoff distance in a totally casual distribution. By investigation of Mulliken charges and MD data,

the amidoxime unit was successfully demonstrated to be the prevalent group, which is in concordance with the purpose of chelating modification. Then it was important to discern which groups display essential roles in adsorption using binding energy data. Moreover, the binding energy data clearly proved that Cd(II) enables binding with $C-NH_2$ and less with $C=N-OH$.

The quantitative structure–activity relationships (QSAR) are paramount for the evaluation of molecular properties of various compounds. In order to get a clear image of how the molecular properties of a chelate polymer are changing when forming a composite, molecular modeling was performed for some single chelate macromolecules as well as for certain reinforcement agents used in biomedicine. Then, the extracted information was corroborated with those resulted from the simulation of the composite system. Table 15.1 displays QSAR calculations for some selected chelate biocompounds. QSAR approach is essential for correlating the molecular structure or properties extracted from the molecular structure of a chelate substance, with a certain type of chemical activity. Among the estimated QSAR features one can mention: surface areas, surface-bounded molecular volume, hydration energy, polarizability, and refractivity. Such theoretical calculations are empirical and more quickly achieved in regard to other methods.

TABLE 15.1 QSAR Calculations of the Chelate Compounds: Polylactic Acid (PLA), Chitosan, Poly(glycolic acid) (PGA), Linezoid, and Trimethylene Carbonate (TMC)

Molecule	Surface area, Å^2	Volume, Å^3	Hydration Energy, kcal/mol	Refractivity, Å^3	Polarizability, Å^3
PLA	468.32	720.79	−14.25	49.53	20.09
Chitosan	460.14	1243.40	−26.13	104.63	43.18
PGA	408.59	556.64	−16.40	36.05	14.59
Linezolid	477.41	942.22	−5.06	84.47	32.64
TMC	396.02	507.37	−1.78	33.32	13.14

Table 15.2 presents other molecular physical properties of the compounds from Table 15.1. The chemical structure peculiarities determine distinct values of the listed parameters. Theoretically, the chemical reactivity descriptors, like HOMO, LUMO, and energy gap, can be evaluated. From the HOMO–LUMO energy gap value, the stability index of compounds can be determined. Table 15.3 describes the values of the gap energy (achieved for the highest molecular orbital (HOMO—highest occupied molecular orbital) and lower vacant molecular orbital (LUMO—lowest unoccupied molecular orbital) derived from the PM3 approach.

TABLE 15.2 Calculated Molecule Properties Extracted from MM Simulations

Molecule	Total Energy, kcal/mol	Binding Energy, kcal/mol	Heat of Formation, kcal/mol	Dipole Moment, Debye
PLA	−76990.98	−2977.02	−292.67	0.859
Chitosan	−154090.33	−6392.91	−439.61	3.646
PGA	−66640.85	−2130.95	−271.91	2.646
Linezolid	−99270.25	−4515.66	−143.26	7.009
TMC	−40977.87	−1952.16	−122.92	0.8285

TABLE 15.3 Border Level Energy Values (eV) Determined After Geometric Optimization for the Examined Molecules, in Free Space and Energy Gap Computed with PM3 Method

Molecule	E_{HOMO}	E_{LUMO}	Energy Gap
PLA	−11.19652	0.24300	10.95352
Chitosan	−9.16916	1.88496	7.2842
PGA	−11.34102	−0.03463	11.30639
Lineozid	−8.70807	−0.51232	8.19575
TMC	−10.77433	0.84063	9.9337

Table 15.4 illustrates the optimized geometries and frontier molecular orbitals. The HOMO–LUMO energy gap is strongly connected to the stability of analyzed compounds. Also, this parameter is important for elucidating the chemical and biological activities of chelate polymer systems. The three-dimensional isosurface image of the electrostatic potential (EP) at the surface of considered chelate materials is presented in Table 15.4. The green color indicates positive (ESP) regions and the pink color shows negative (EP) zones.

TABLE 15.4 Optimized Geometries, Molecular Orbital Surface, HOMO–LUMO Energy Gap for HOMO and LUMO Representation and EP of the Considered Green Materials Extracted from PM3 Semiempirical Method

	PLA	Chitosan
Optimized structure		
HOMO		
LUMO		
EP (Electrostatic potential)	+ 2.173 -0.098	+ 0.673 -0.139

TABLE 15.4 *(Continued)*

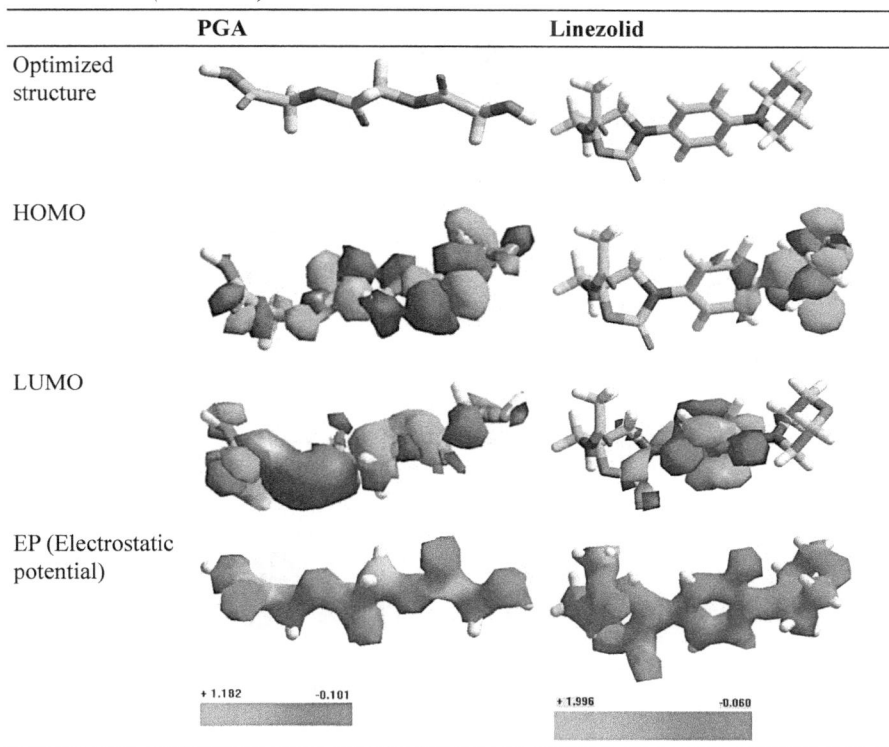

	PGA	Linezolid
Optimized structure		
HOMO		
LUMO		
EP (Electrostatic potential)		
	+ 1.182 -0.101	+ 1.996 -0.060

Molecular modeling computations are also essential for determining the molecular stability of a composite material. This can be performed by comparing the total potential energies of the isolated and complex systems. In the situation, the value of total potential energies of the complex is below that of the potential energies of isolated matrix and filler in the same conformations, the complexed form displays high steadiness and its formation is favorable. Conversely, when the value of total potential energies of complex exceeds that of the individual counterparts, the system is less steady from the thermodynamic point of view. Figure 15.1 is depicted as the conformation of several systems containing PLA chelate polymer. It is observed that PLA adopts a distinct conformation as a function of the structure of the interacting substance, namely chitosan, PGA, Linezoid, and TMC. Table 15.5 shows the data concerning the complex stability: total energy, binding energy, and heat of formation. Also, the dipole moment of the complex systems varies as a function of the structural characteristics of each compound.

PLA/Chitosan complex

PLA/PGA complex

PLA/Lineozid complex

PLA/TCM complex

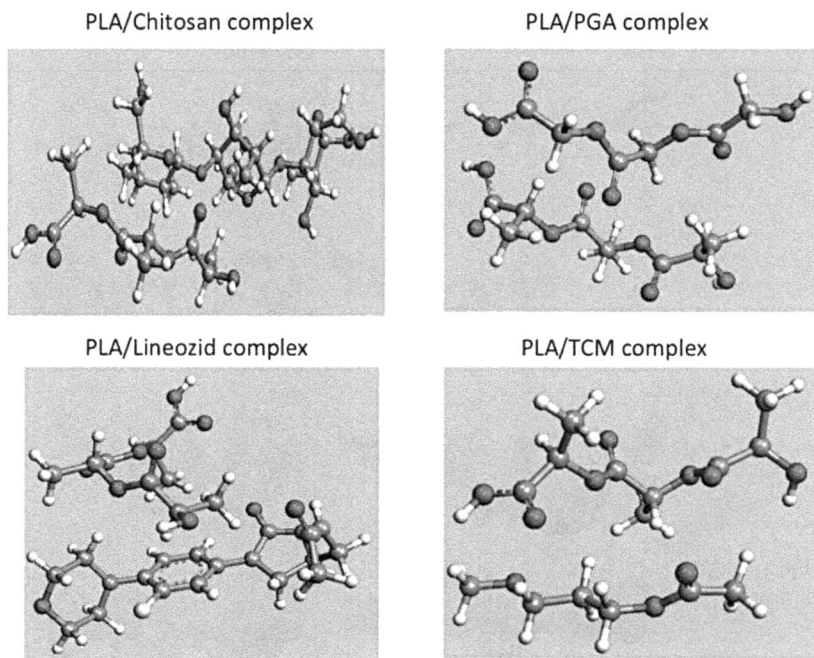

FIGURE 15.1   Visualization of geometrical preferences of green chelate polymer composite systems after molecular simulation in a vacuum.

TABLE 15.5   Molecule Properties of System Composites

System	Total Energy, kcal/mol	Binding Energy, kcal/mol	Heat of Formation, kcal/mol	Dipole Moment
PLA/Chitosan	−231022.89	−9311.50	−673.86	4.503
PLA/PGA	−14360.811	−5106.98	−563.561	3.687
PLA/Linezolid	−176257.67	−7489.13	−432.381	6.810
PLA/TMC	−117970.18	−4930.50	−416.91	3.660

15.3 PREPARATION APPROACHES USED FOR COMPOSITES BASED ON CHELAE POLYMERS

The processes of preparation and processing of composite materials based on chelate polymers are not so different from those applied in the case of other polymer composites. The first step requires selecting from the available green polymers which contain chelating groups. The most important categories of such polymers are as follows:

1. Natural modified polymers
   - chitin
   - chitosan
   - cellulose
   - collagen.
2. Synthetic biodegradable polymers
   - polylactic acid (PLA)
   - poly(ε-caprolactone)
   - poly(lactic-*co*-glycolic acid) (PLGA)
   - poly(glycolic acid) (PGA).

All known traditional methodologies of polymer chemistry that are used for attaining polymers containing chelating fragments are referring to:

1. homopolymerization, copolymerization, and grafted polymerization of monomer reactants that have chelate sites/nodes;
2. polycondensation of respective compounds, which determine the occurrence of chelating polymer ligand; and
3. postpolymerization modification of certain functional groups from macromolecule structure to achieve fabrication of the chelating fragments.

The green composites can be prepared by chemical or physical blending of chelate polymers with reinforcement agents that might have or not chelating properties, such as:

1. Natural fibers and nanofibers extracted from the following:
   - vegetable resources: seed, fruit, leaf, wood, grass,
   - animals: silk, wool/ hair, and

   - mineral resources.
2. Chelating polymers,
3. Low molecular chelating compounds, and
4. Nanoparticles.

15.4 APPLICATIONS OF CHELATE POLYMER COMPOSITES IN TREATMENT OF HEAVY METAL INTOXICATION

15.4.1 CHELATORS USED IN THE PHARMACEUTICAL FIELD FOR HEAVY METAL DISINTOXICATION

Metal-binding proteins are known for their potential chelating ability for heavy toxic metals, exhibiting an interesting impact on the natural reaction of the organism to these toxic substances [43]. Glutathione represents another material with chelating characteristics that determine the specific cellular response, transport, and separation of metal cations [44]. Moreover, this compound is a biomarker for toxic metal excess [44]. Chelating compounds are synthesized by plants and animals. When dealing with metal poisoning, the formulated treatments are based on substances extracted from sources that are known to diminish the absorption or reabsorption of undesired particles and to sustain natural detoxification pathways. Among these it is worth mentioning [45, 46]:

1) natural polymers: algal polysaccharides alginate, chlorella, poly
   (γ-glutamic acid),
2) sulfur-based peptides found in
   foods like alliums (e.g., garlic)
   and brassicas (e.g., broccoli),
3) dietary fibers encountered in
   edible materials, including bran
   from grains,
4) cilantro (leaves of Coriandrum
   sativum),
5) sulfur-derived amino acids:
   taurine and methionine,
6) alpha lipoic acid is useful for
   regeneration of certain antioxidants,
7) N-acetyl-cysteine which is a
   chelator compound responsible
   for glutathione synthesis,
8) Selenium, and
9) other medicinal herbs: *Ginkgo
   biloba, Curcuma longa, Phyto-
   chelatins, Emblica officinalis,
   green algae*

There are some reports that are
focused on the impact of micronutrients (like antioxidants, vitamins,
sulfur-derived amino acids, and
essential minerals) on the kinetics
and negative effects of poisonous
substances [47–51]. Nutritional status
is very important when analyzing
uptake, knowing that toxic cations
are carried by proteins providing
important nutrients, including Zn,
Fe, and Mg. Thus, persons who are
malnourished are subjected to huge
risks of toxicity [52]. It was noticed
that in the case of animals, calcium

deprivation favors the absorption
of cadmium and lead, whereas zinc
and magnesium in excess lower the
absorption of cadmium. Administration of calcium diminishes lead
mobilization from the mother's
bones, sheltering the child [53, 54].
As for kids, the medicamentation-
containing iron leads to a lower
accumulation of lead. In any case,
mineral administration, combined
with home and school meals must
not deflect attention from the goal of
healthy food that excludes undesired
exposure [55].

Pharmaceuticals that can remove
metal ions in a liquid environment are small organic molecules,
which mainly make coordination
complexes implying oxygen, sulfur,
and/or nitrogen atoms. Literature [45,
46] mentions five types of chelating
agents that are widely encountered
in the treatment of humans suffering
from intoxication with heavy metals
and metalloids. Their structures as
resulted from molecular modeling
are presented in Figure 15.2.

The most relevant features of
each substance used in pharmaceuticals products or drugs containing
chelating agents are as follows:

1) (2S)-2-amino-3-methyl-3-sulfa-
   nylbutanoic acid: makes possible
   chelation of the following
   elements: Zn, Hg, Cu (Wilson's
   disease), Pb, As, as a result of
   its binding groups (hydroxyl,
   oxygen, amine, and sulfhydryl).

2,3-bis(sulfanyl)butanedioic acid

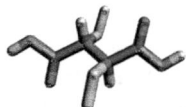

2-[2-[bis(carboxymethyl)amino]ethyl-(carboxymethyl)amino]aceticacid

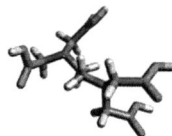

Sodium 2,3-bis(sulfanyl)propane-1-sulfonate

2,3-bis(sulfanyl)propan-1-ol

(2S)-2-amino-3-methyl-3-sulfanylbutanoic acid

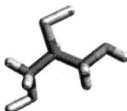

FIGURE 15.2   Optimized molecular geometries of the main important types of chelating agents used in treatments of metal intoxication.

2) 2,3-bis(sulfanyl)butanedioic acid: makes possible chelation of the following elements: Pb, Sn, As, Ag, Hg, Cd, Cu as a result of its binding groups (oxygen and sulfhydryl).

3) sodium 2,3-bis(sulfanyl)propane-1-sulfonate: makes possible chelation of the following elements: Pb, Sn, Hg, Ag, Cd, Mg as a result of its binding groups (oxygen and sulfhydryl).

4) 2,3-bis(sulfanyl)propan-1-ol: makes possible chelation of the following elements: Hg, As, Pb, Au, as a result of its binding groups (sulfhydryl and hydroxyl).

5) 2-[2-[bis(carboxymethyl)amino] ethyl-(carboxymethyl)amino] acetic acid: makes possible chelation of the following elements: Zn, Pb, Cd as a result of its binding group (oxygen).

15.4.2   CHELATE THERAPIES USED IN CASES OF METAL POISONING

The basic principles of dealing with metal poisoning consist of a relatively simple protocol that relies on the following three steps:

1) avoiding further metal incorporation in the organism,

2) excretion of the toxic element from the circulation system, and

3) inactivation of the metal found in the body.

The current therapeutic approaches are majorly relying on reduction of the metal amounts in the affected organs and whole organism, enabling the appearance of some novel branches, like metallotoxicology. The regain the usual healthy physiology of the body can be done via the following procedures [21]:

1) by direct absorption of essential metals,
2) by chelating out surplus of metallic particles, and
3) by employing metals in drug delivery formulations or for labeling certain biomolecules for diagnostics.

In this chapter, the presentation will be mainly directed toward chelation therapy since it represents a significant tool for controlling the metal amounts in the living organism. A good chelating agent must be characterized by a series of features, such as [21]:

1) quick removal of the toxic metal,
2) formation of nontoxic complexes,
3) high affinity, diminished toxicity,
4) capacity to penetrate cell membranes,
5) good water solubility, and
6) similar distribution to that of the metal.

It was proven that the success of clinical administration of a chelating compound in a living organism is influenced by several aspects:

1) competing metals and ligands,
2) the manner of administration,
3) compartmentalization of metal and chelating compound,
4) metabolism and/or elimination of the chelating material,
5) variations in toxicity of the metal either "free" or chelated, and
6) the toxicity degree corresponding to the chelating compound.

Chelating materials can interfere and diminish the metal poisoning via mobilizing the toxic substance, particularly into the urine. Upon interaction with the heavy metal, a stable complex is formed, so the chelating agent is protecting biological molecules from interaction with the metal ion. In this way, th e local toxicity is reduced. For example, desferrioxamine is a chelator for iron which is able to completely cover Fe^{3+} during the complex occurrence and thus avoids iron-catalyzed free radical reactions [21, 56]. There are situations when the chelator might expose the metal to the biological medium and consequently raise the toxicity of the metal. Ethylenediaminetetraacetic acid was found to be less capable of ecranate the surface of the Fe^{3+}, however, leads to an open complex (basket complex) which enhances the catalytic feature of Fe^{3+} for rendering oxidative stress [21].

The chelation therapies have numerous advantages, namely, they are very efficient in cases of acute intoxication, they are able to create nontoxic complexes, they excrete the metal from soft tissues and they are adequate for oral administration. However, there are also some drawbacks, like no elimination of metal atoms/ions from intracellular places, low clinical recovery, redistribution of poisoning elements, occurrence of prooxidant effects, and other symptoms as a result of hepatotoxicity, nephrotoxicity, nausea, headache, and so on.

There is still a great need for newer chelate antidotes if considering the following aspects:

1) there is no ideal treatment concerning the As intoxication,
2) there are few oral drugs,
3) most therapies are given intravenously and the patient is not capable to self-aid,
4) there are some limitations as a result of secondary effects since certain substances are contraindicated in some instances of heavy metal poisoning, and
5) there is no fast drug that could enable quick and immediate elimination of metal particles from tissues and blood.

Having all these in view, it is imperative to improve the properties of conventional chelators for achieving a more secure and adequate treatment for heavy metal exposure. Newer tendencies in chelation therapy are directed toward the utilization of a set of two structurally different chelating compounds. The notion of combination therapy is based on the idea that two prescribed active substances will operate via a distinct mechanism of action. Therefore, the approach will result in supplementary effects, or the chelate drugs could sustain the individual mechanism of action creating a synergism. This concept is constructed on assumption that many chelating substances are more able to mobilize poisoning particles from various tissue compartments, leading to better outcomes concerning patient health. Using this treatment approach, favors increased metal mobilization and diminish in the amount of potential toxic chelators, while avoids redistribution of toxic substance in the organism [21]. The basic mechanism concerning the administration of many structurally distinct chelating materials is focused on using a lipophilic substance, working on intracellular metal, combined with a lipophobic compound acting on extracellular metal—all-determining enhanced metal excretion, lowering in the level of the dose, and low redistribution of pursued trace elements [57].

Therapeutic approaches to solve the limitations in conventional chelation therapy are based on the following strategies [21]:

1) synthesis of new chelating substances having superior therapeutic efficiency, specificity, and better access to intracellularly linked metals [58–60],
2) combination therapy involving more than one chelating substance with reduced adverse impact and enhanced chelation efficiency, while impeding metal redistribution,
3) mixing chelating agent with antioxidants is adequate for metal chelation and shielding from reactive oxygen species [61, 62],
4) blending chelating agent with micronutrients is an approach that changes the toxicokinetics of metals, whereas it balances the essential metal loss, and
5) combining chelating agents with herbal extract is beneficial since plant substances are able to enhance the efficacy of chelating compounds and have the advantages of natural chelator properties and those of antioxidants [63–65].

large number of available chelate substances for metal retention but their performance is still far from that of an ideal chelator. The majority of available chelate compounds present nonspecific binding and some undesired effects. The lack of detailed clinical trials regarding chelation therapy in the treatment of metal exposure still leaves nonelucidated aspects of its clinical therapeutic benefits. Inspite of all these aspects, it is essential to understand the necessity of more advanced chelate agents that address the unanswered poisonings, such as Cd, As, or Hg toxicity but also to attain full clinical recovery for patients suffering from other disorders.

Future investigations should be focused on the development of newer therapeutic strategies that lead to better therapeutic outcomes. Application of the combination therapy and/or prescribing antioxidants or micronutrients should be viewed as vital recommendations of chelation therapy.

15.5 CONCLUSIONS AND FUTURE PERSPECTIVES

The presence in excess of certain metals in a living organism can produce serious toxic manifestations. Chelation therapy has led to successful results in treating patients of metal poisoning since it facilitates the elimination of the toxic elements from the body. There are a

KEYWORDS

- chelation
- green polymers
- composites
- surface processing
- bioapplications

REFERENCES

1. Hermann, H.R.; Are we our own worst enemy? In *Dominance and Aggression in Humans and Other Animals. The Great Game of Life.* Academic Press: Amsterdam, **2017**, 251–288.
2. Buczyńska, A.; Rolecki, R.; Tarkowski S.; Industrial wastes and health hazards (article in Polish). *Med. Pr.* **1999**, *50*, 179–190.
3. Jaishankar, M.; Tseten, T.; Anbalagan, N.; Mathew, B.B.; Beeregowda, K.N.; Toxicity, mechanism and health effects of some heavy metals. *Interdiscip. Toxicol.* **2014**, *7*, 60–72.
4. Singh, J.; Kalamdhad, A.S.; Effects of heavy metals on soil, plants, human health and aquatic life. *Int. J. Res. Chem. Environ.* **2011**, *1*, 15–21.
5. Tchounwou, P. B.; Yedjou, C. G.; Patlolla, A. K.; Sutton, D. J.; Heavy metal toxicity and the environment. *Exp. Suppl.* **2012**, *101*, 133–164.
6. Engwa, G.A.; Ferdinand, P.U.; Nwalo, F.N.; Unachukwu, M.N.; Mechanism and health effects of heavy metal toxicity in humans. In: *Poisoning in the Modern World—New Tricks for an Old Dog?* Karcioglu, O.; Arslan, B.; Eds.; InTech: Rijeka, **2018**, 1–23.
7. Orr, S.E.; Bridges, C.C.; Chronic kidney disease and exposure to nephrotoxic metals. *Int. J. Mol. Sci.* **2017**, *18*, 1039(1–35).
8. Heavy metal poisoning. https://raredis-eases.org/rare-diseases/heavy-metal-poisoning/ (accessed on 19 January 2020).
9. Jan, A.T.; Azam, M.; Siddiqui, K.; Ali, A.; Choi, I.; Haq, Q.M.; Heavy metals and human health: mechanistic insight into toxicity and counter defense system of antioxidants. *Int. J. Mol. Sci.* **2015**, *16*, 29592–29630.
10. Khulbe, K.C., Matsuura, T.; Removal of heavy metals and pollutants by membrane adsorption techniques. *Appl. Water Sci.* **2018**, *8:19*, 1–30.
11. Eccles, H.; Treatment of metal-contaminated wastes: Why select a biological process? *Trends Biotechnol.* **1999**, *17*, 462–465.
12. Baraka, M.A.; New trends in removing heavy metals from industrial wastewater. *Arab. J. Chem.* **2011**, *4*, 361–377.
13. Kurniawan, T.A.; Chan, G.Y.S.; Lo, W.H.; Babel, S.; Physico-chemical treatment techniques for wastewater laden with heavy metals. *Chem. Eng. J.* **2006**, *118*, 83–98.
14. Ates, N.; Uzal, N.; Removal of heavy metals from aluminum anodic oxidation wastewaters by membrane filtration. *Environ. Sci. Pollut. Res. Int.* **2018**, *25*, 22259–22272.
15. Farooq, U.; Kozinski, J.A.; Khan, M.A.; Athar, M.; Biosorption of heavy metal ions using wheat based biosorbents—a review of the recent literature. *Biores. Technol.* **2010**, *101*, 5043–5053.
16. Lindholm-Lehto, P.; Biosorption of heavy metals by lignocellulosic biomass and chemical analysis. *Bio Resources* **2019**, *14*, 4952–4995.
17. Crisponi, G.; Nurchi, V.M.; Chelating agents as therapeutic compounds—basic principles. In: *Chelation Therapy in the Treatment of Metal Intoxication.* Aaseth, J.; Crisponi, G.; Andersen, O.; Eds.; Academic Press: Amsterdam, **2016**, 35–61.
18. Sartore, L.; Dey, K.; Preparation and heavy metal ions chelating properties of multifunctional polymer-grafted silica hybrid materials. *Adv. Mater. Sci. Eng.* **2019**, *2019*, 1–11.
19. Kalicanin, B.; Rašic. M.T.; The significance of chelation therapy in heavy metal intoxication. *J. Heavy Met. Toxicity Dis.* **2019**, *4*, 1–10.
20. WHO (2004); Guidelines for drinking-water quality. 61st meeting, Rome, **2003**. Joint FAO/WHO expert committee on

food additives. http://ftp.fao.org/es/esn/jecfa/jecfa61sc.pdf.

21. Flora, S.J.S.; Pachauri, V.; Chelation in metal intoxication. *Int. J. Environ. Res. Public Health* **2010**, *7*, 2745–2788.

22. Verardo, V.; Ochando-Pulido, J.M.; Moral, S.P.; Segura-Carretero, A.; Garrido-Frenich, A.; Fernández-Gutiérrez, A.; Martínez-Ferez, A.; Properties and applications of polysaccharide green polymer composites for antibacterial and anti-fogging coatings. In: *Food Green Polymer Composites Technology. Properties and Applications*. Inamuddin, Ed.; CRC Press: Boca Raton, FL, **2017**, 31–48.

23. Rose, C.; Designing for composites: traditional and future views. In: *Green Composites: Polymer Composites and the Environment*. Baillie, C.; Ed.; CRC Press: Boca Raton, FL, **2004**.

24. Reichert, D.E.; Norrby, P.-O.; Welch, M.J.; Molecular modeling of bifunctional chelate peptide conjugates. 1. Copper and indium parameters for the AMBER force field. *Inorg. Chem.* **2001**, *40*, 20, 5223–5230.

25. Yan, Y.; Zhang, J.; Rend, L.; Tang, C.; Metal-containing and related polymers for biomedical applications. *Chem. Soc. Rev.* **2016**, *45*, 5232–5263.

26. Callari, M.; Aldrich-Wright, J.R.; de Souza, P.L.; Stenzel, M.H.; Polymers with platinum drugs and other macromolecular metal complexes for cancer treatment. *Prog. Polym. Sci.* **2014**, *39*, 1614–1643.

27. Pomogailo, A.D.; Kestelman, V.N.; Metallopolymer nanocomposites. Springer-Verlag: Berlin, **2006**.

28. Durant, J.R.; *Cisplatin: A Clinical Overview*. Academic Press: New York, NY, **1980**.

29. Hartinger, C.G.; Dyson, P.J.; Bioorganometallic chemistry—from teaching paradigms to medicinal applications. *Chem. Soc. Rev.* **2009**, *38*, 391–401.

30. Martins, P.; Marques, M.; Coito, L.; Pombeiro, A.J.L.; Baptista, P.V.; Fernandes, A.R.; Organometallic compounds in cancer therapy: past lessons and future directions. *Anticancer Agent Med. Chem.* **2014**, *14*, 1199–1212.

31. Rosenberg, B.; Van Camp, L.; Krigas, T.; Inhibition of cell division in *Escherichia coli* by electrolysis products from a platinum electrode. *Nature* **1965**, *205*, 698–699.

32. Gartner III, T.E.; Jayaraman, A.; Modeling and simulations of polymers: a roadmap. *Macromolecules* **2019**, *52*, 3, 755–786.

33. Cunha, R.A.; Soares, T.A.; Rusu, V.H.; Pontes, F.J.S.; Franca E.F.; Lins, R.D.; The molecular structure and conformational dynamics of chitosan polymers: an integrated perspective from experiments and computational simulations. In: *The Complex World of Polysaccharides*. Karunaratne, D.N.; Ed.; InTech: Rijeka, **2012**, 229–256.

34. Lange, J.; de Souza Junior, F.G.; Nele, M.; Tavares, F.W.; Segtovich, I.S.V.; da Silva, G.C.Q.; Pinto, J.C.; Molecular dynamic simulation of oxaliplatin diffusion in poly(lactic acid-co-glycolic acid). Part A: parameterization and validation of the force-field CVFF. *Macromol. Theory Simul.* **2016**, *25*, 45–62.

35. Houshmand, F.; Neckoudari, H.; Baghdadi, M.; Host-guest interaction in chitosan—MX (3-chloro-4-(dichloromethyl)-5-hydroxy-2(5H)-furanone) complexes in water solution: density functional study. *Asian J. Nanosci. Mater.* **2018**, *2*, 49–65.

36. Zheng, L.; Zhang, S.; Cheng, W.; Zhang, L.; Meng, P.; Zhang, T.; Yu, H.; Peng, D.; Theoretical calculations, molecular dynamics simulations and experimental investigation of the adsorption of cadmium(II) on

amidoxime-chelating cellulose. *J. Mater. Chem. A.* **2019**, *7*, 13714–13726.

37. Okuyama, K.; Noguchi, K.; Miyazawa, T.; Yui, T.; Ogawa, K.; Molecular and crystal structure of hydrated chitosan. *Macromolecules* **1997**, *30*, 5849–5855.

38. Ogawa, K.; Effect of heating an aqueous suspension of chitosan on the crystallinity and polymorphs. *Agric. Biol. Chem.* **1991**, *55*, 2375–2379.

39. Gardner, K.H.; Blackwell, J.; Refinement of structure of beta-chitin. *Biopolymers* **1975**, *14*, 1581–1595.

40. Ogawa, K.; Yui, T.; Okuyama, K.; Three D structures of chitosan. *Int. J. Biol. Macromol.* **2004**, *34*, 1–8.

41. Franca, E.F.; Lins, R.D.; Freitas, L.C.G.; Straatsma, T.P. Characterization of chitin and chitosan molecular structure in aqueous solution. *J. Chem. Theory Comp.* **2009**, *4*, 2141–2149.

42. Franca, E.F.; Lins, R.D.; Freitas, L.C.G.; Chitosan molecular structure as a function of n-acetylation. *Biopolymers* **2011**, *95*, 448–460.

43. Klaassen, C.D.; Liu, J.; Diwan, B.A.; Metallothionein protection of cadmium toxicity. *Toxicol. Appl. Pharmacol.* **2009**, *238,* 215–220.

44. Franco, R.; Sánchez-Olea, R.; Reyes-Reyes, E.M.; Panayiotidis, M.I.; Environmental toxicity, oxidative stress and apoptosis: ménage à trois. *Mutat. Res.* **2009**, *674*, 3–22.

45. Margaret, E.S.; Chelation: Harnessing and enhancing heavy metal detoxification—a review. *Sci. World J.* **2013**, 1–13.

46. Mehrandish, R.; Rahimian, A.; Shahriary, A.; Heavy metals detoxification: a review of herbal compounds for chelation therapy in heavy metals toxicity. *J. Herbmed. Pharmacol.* **2019**; *8*, 69–77.

47. Flora, S.J.S.; Bhadauria, S.; Kannan, G.M.; Singh, N.; Arsenic induced oxidative stress and the role of antioxidant supplementation during chelation: A review. *J. Environ. Biol.* **2007**, *28*, 333–347.

48. Joshi, D.; Mittal, D.; Shrivastav, S.; Shukla, S.; Srivastav, A.K.; Combined effect of N-acetyl cysteine, zinc, and selenium against chronic dimethyl-mercury-induced oxidative stress: a biochemical and histopathological approach. *Arch. Environ. Contam. Toxicol.* **2011**, *61*, 558–567.

49. Peraza, M.A.; Ayala-Fierro, F.; Barber, D.S.; Casarez, E.; Rael, L.T.; Effects of micronutrients on metal toxicity. *Environ. Health Perspect.* **1998**, *106*, 203–216.

50. Aremu, D.A.; Madejczyk, M.S.; Ballatori, N.; N-acetylcysteine as a potential antidote and biomonitoring agent of methylmercury exposure. *Environ. Health Perspect.* **2008**, *116*, 26–31.

51. Ito, Y.; Niiya, Y.; Otani, M.; Sarai, S.; Shima, S.; Effect of food intake on blood lead concentration in workers occupationally exposed to lead. *Toxicol. Lett.* **1987**, *37*, 105–114.

52. Matovic, V.; Buha, A.; Bulat, Z.; Dukić-Ćosić, D.; Cadmium toxicity revisited: focus on oxidative stress induction and interactions with zinc and magnesium. *Arh. Hig Rada Toksikol.* **2011**, *62*, 65–76.

53. Ettinger, A.S.; Hu, H.; Hernandez-Avila, M.; Dietary calcium supplementation to lower blood lead levels in pregnancy and lactation. *J. Nutr. Biochem.* **2007**, *18*, 172–178.

54. Ettinger, A.S.; Téllez-Rojo. M.M.; Amarasiriwardena, C.; Peterson, K.E.; Schwartz, J.; Aro, A.; Hu, H.; Hernández-Avila, M.; Influence of maternal bone lead burden and calcium intake on levels of lead in breast milk over the course of lactation. *Am. J. Epidemiol.* **2006**, *163*, 48–56.

55. Rosado, J.L.; Lopez, P.; Kordas K.; García-Vargas, G.; Ronquillo, D.; Alatorre, J.; Stoltzfuset, R.J.; Iron and/or zinc supplementation did not reduce

blood lead concentrations in children in a randomized, placebo-controlled trial. *J. Nutr.* **2006**, *136*, 2378–2383.

56. Tilbrook, G.S.; Hider, R.C.; Iron chelators for clinical use metal ions. *Biol. Syst.* **1998**, *35*, 691–730.

57. Flora, S.J.S.; Bhattacharya, R.; Vijayaraghavan, R.; Combined therapeutic potential of meso 2,3-dimercaptosuccinic acid and calcium disodium edetate in the mobilization and distribution of lead in experimental lead intoxication in rats. *Fundam. Appl. Toxicol.* **1995**, *25*, 233–240.

58. Mishra, D.; Mehta, A.; Flora, S.J.S.; Reversal of hepatic apoptosis with combined administration of DMSA and its analogues in guinea pigs: role of glutathione and linked enzymes. *Chem. Res. Toxicol.* **2008**, *21*, 400–407.

59. Albu, R.M.; Avram, E.; Stoica, I.; Ioan, S.; Polysulfones with chelating groups for heavy metals retention. *Polym. Compos.* **2012**, *33*, 573–581.

60. Albu, R.M.; Avram, E.; Musteata, V.E.; Ioan; S.; Dielectric relaxation and AC-conductivity of modified polysulfones with chelating groups. *J. Solid State Electrochem.* **2014**, *18*, 785–794.

61. Bhadauria, S.; Flora, S.J.S.; Response of arsenic induced oxidative stress,

DNA damage and metal imbalance to combined administration of DMSA and monoisoamyl DMSA during chronic arsenic poisoning in rats. *Cell. Biol. Toxicol.* **2007**, *23*, 91–104.

62. Flora, S.J.S.; Dubey, R.; Kannan, G.M.; Chauhan, R.S.; Pant, B.P.; Jaiswal, D.K.; Meso-2,3-dimercaptosuccinic acid (DMSA) and monoisoamyl DMSA effect on gallium arsenide induced pathological liver injury in rats. *Toxicol. Lett.* **2002**, *132*, 9–17.

63. Mishra, D.; Gupta, R.; Pant, S.C.; Kushwah, P.; Satish, H.T.; Flora, S.J.S.; Therapeutic potential of combined administration of MiADMSA and Moringa oleifera seed powder on arsenic induced oxidative stress and metal distribution in mouse. *Toxicol. Mech. Methods.* **2008**, *19*, 169–182.

64. Saxena, G.; Flora, S.J.S.; Changes in brain biogenic amines and heme-biosynthesis and their response to combined administration of succimer and Centella asiatica in lead poisoned rats. *J. Pharm. Pharmacol.* **2006**, *58*, 547–559.

65. Flora, S.J.S.; Mehta, A.; Gupta, R.; Prevention of arsenic-induced hepatic apoptosis by concomitant administration of garlic extracts in mice. *Chem. Biol. Inter.* **2009**, *177*, 227–233.

CHAPTER 16

HIGH ENERGY RADIATION: A PROMISING MEDIA FOR THE SYNTHESIS OF HYDROGELS FREE FROM TOXIC CHEMICALS

S. ROOPA*, M. P. CHANDRESH, and SIDDARAMAIAH

Department of Polymer Science and Technology,
Sri Jayachamarajendra College of Engineering,
JSS Science and Technology University, Mysuru 570006, India

Corresponding author. E-mail: roopasm2000@sjce.ac.in.

ABSTRACT

The rapid growth of materials science has provided many types of functional materials. One such type of functional material is a hydrogel. Hydrogels are crosslinked networks of hydrophilic polymer molecules that can swell but do not dissolve in water; however, they can retain a significant amount of water in their structures. Due to their unique characteristics such as swelling, deswelling, hydrophilicity, and biocompatibility, the hydrogels are used in several applications related to tissue engineering, pharmaceutics, biomedical, agriculture, ophthalmic, dental, cosmetics, waste treatment, separation, intelligent/smart textiles, and sensors. The performance characteristics of the hydrogels not only depend on the hydrophilic nature or chemical composition but very much on the type and number of crosslinks formed during the synthesis and development of hydrogels. In the last decade, several scientists have extensively reported the synthesis of hydrogels based on natural and synthetic polymers such as polysaccharides (carboxy methyl cellulose (CMC), chitosan, albumin, poly (hydroxyl alkyl methacrylates), poly (acrylamide), poly(ethyleneoxide) (PEO), poly(N-vinyl -2-pyrrolidone)

(PVP), poly(vinylalcohol) (PVA), polyethylene glycol (PEG), etc.,) by various routes. Hydrogels are generally prepared by (1) polymerization of hydrophilic monomers and crosslinking simultaneously or sequentially (i.e., polymerization followed by crosslinking) and (2) chemical modification or functionalization of the polymers which exhibit desired hydrogel properties. The hydrogels may be physically crosslinked or chemically crosslinked with or without crosslinking agents. Radiation crosslinking is one of the important methods used to prepare hydrogels for various applications due to its technical benefits. This chapter focuses on the research findings reported in the last decade in the area of synthesis or crosslinking of hydrogels using high energy radiation.

16.1 INTRODUCTION

The polymer networks that can absorb a large quantity of water (between 20% and 100% of total weight) or biological fluids showing resemblance to natural tissue without dissolution is termed as hydrogel (HG). These hydrogels are capable of maintaining their chemical structure even after swelling. Furthermore, the absorbed fluid is hardly removed under some pressure [1–3]. When water content exceeds 100%, these hydrogels are called super-absorbent hydrogels (SAHs). The polymeric backbone of these hydrogels is attached with hydrophilic functional groups such as –OH, –CONH, –COOH, –SO$_3$H, and –NH$_2$, which makes the hydrogels to absorb water. The resistance of these hydrogels to dissolution arises from the crosslink density between polymeric network chains. The crosslinks in the hydrogel may be permanent (chemical/irreversible) or temporary (physical/reversible) based on the nature of crosslink/interaction between the polymer molecules and methods of production. Both natural and synthetic polymers are used to prepare hydrogels for various industrial applications and are listed in Table 16.1. Nowadays, synthetic polymer hydrogels are slowly replacing natural polymer hydrogels due to their superior properties with respect to durability, water absorption capability and strength [4].

16.2 CLASSIFICATION AND APPLICATIONS OF HYDROGELS

Hydrogels are classified based on several parameters as schematically represented in Figure 16.1, and some of the commercially available hydrogels and their applications are summarized in Table 16.2 and Figure 16.2.

TABLE 16.1 List of Some of the Commercially Available Natural and Synthetic Polymer-based Hydrogels

Natural Polymers			Synthetic Polymers		
Polymer	Brand Name	Applications	Polymer	Brand Name	Applications
Carboxy methyl cellulose (CMC)	Primellose [8]	• As super-disintegrant for variety of tablet and capsule formulations	Poly methacrylic acid and divinyl benzene as crosslinker	Amberlite IRP64 [5]	• Taste making • Drug stabilization • Carrier for cationic drugs • Controlled release formulation
Gellan gum	Timoptic-XE [9]	• Ophthalmic gel forming solution	Polyacrylamaide (PAA)	Aquamid [6]	• To enhance facial contours • To eliminate wrinkles and folds
Hydroxy propyl methyl cellulose (HPMC)	Lescol [10]	• Controlled drug release	Polyvinyl N-pyrrolidone (Crospovidone)	Kollidon CL [7]	• Super-disintegrants and dissolution enhancers • Controlled release preparations and transdermal systems to regulate the release of active substances
Xanthan gum	Tricor [10]	• In lipid regulating tablet	Sodium polystyrene sulfonate	Kionex [10]	• Cation-exchange resin
Starch (Potato starch)	Primojel [8]	• As a dissolution enhancing agent	Polydimethylsiloxane (PDMS)	Night & day [11]	• Soft contact lens

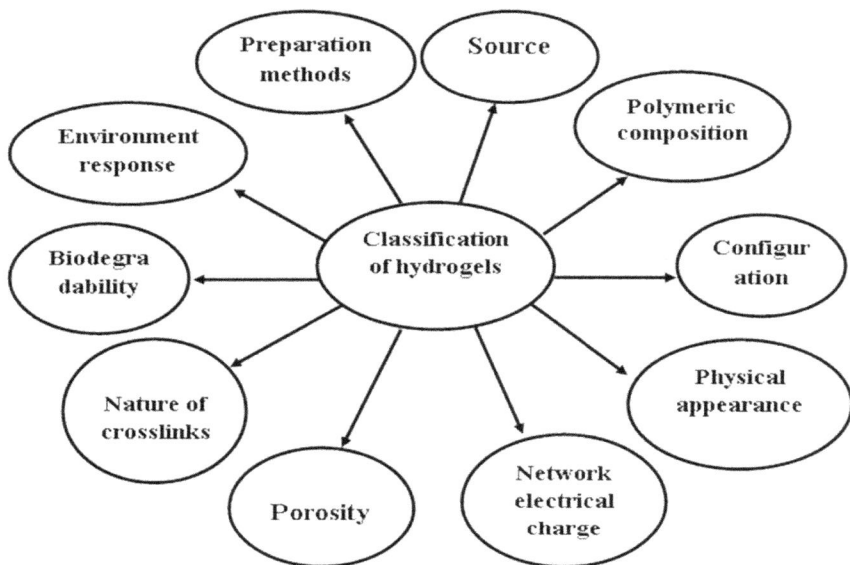

FIGURE 16.1 Classifications of hydrogels.

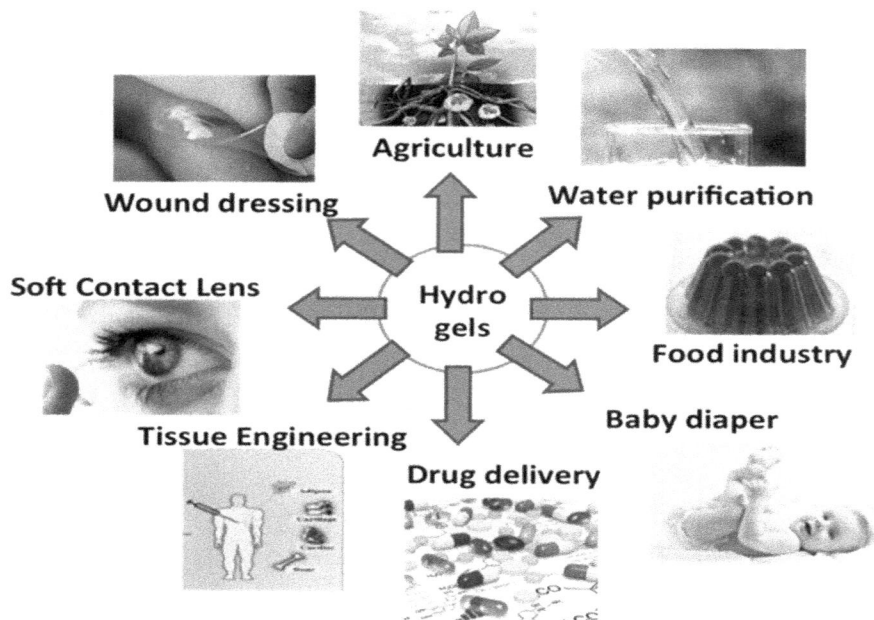

FIGURE 16.2 Applications of hydrogels.

TABLE 16.2 Summary of Classifications of Hydrogels (HG) [Selected from References 12–22]

Parameters	Types	Examples	Remarks
Resource used to prepare hydrogel	Natural HG	Chitosan, dextran, collagen, albumin	Prepared using natural resources
	Synthetic HG	Polyacrylic or methacrylic acid, polystyrene, PVA and PDMS	Prepared using man made monomers/polymers
	Hybrid HG	Chitosan combined with poly(N-isopropylacrylamide) or PVA	Containing both natural and synthetic polymer
Polymeric composition	Homopolymeric HG	Poly(3-hydroxypropyl methacrylate (PHPMA), poly(glyceryl methacrylate), (PGMA) and poly (2-hydroxyethyl methacrylate) (PHEMA)	Derived from one type of monomer
	Copolymeric HG	poly(NVP-co-HEMA), poly(HEMA-co-AA)	Contains more than one type of monomer with at least one being hydrophilic
	Interpenetrating polymer network (IPN) HG	PEG and poly(methacrylic acid)	Network consists of two independent crosslinked polymer.
Physical state or configuration	Amorphous HG	Poly(R-lactide)	—
	Semicrystalline HG	Poly(D-lactide)	
	Crystalline HG	Hydrogel from linear poly(ethyleneimine)	
Physical form or physical appearance	HG film	Transdermal drug delivery patch (polyacrylates)	—
	HG microsphere	Bioadhesive carriers (polycarbophil, sodium carboxymethylcellulose, carrageenan)	
	HG fiber	Optical fiber (polyethylene glycol diacrylate)	
	HG pressed powder matrices	Pills or capsules for oral ingestion (chitosan)	
Network electrical charge	Nonionic HG	Dextran, agarose, pullulan, hydroxyalkylmethacrylates[a] N-vinyl caprolactam[a]	Neutral in nature

TABLE 16.2   *(Continued)*

Parameters	Types	Examples	Remarks
	Ionic HG	Sodium polystyrene sulfonate	Contains anionic or cationic groups
	Amphoteric electrolyte (ampholytic) HG	Hydrogel from maleic anhydride and N-vinylsuccinimide, carboxymethyl chitin	Contains both acidic and basic groups
	Zwitterionic HG	Polybetaines	Each repeating unit contains both anionic and cationic group
Porosity	Nonporous HG	Polyacrylamide	Hydrogel with nonporous, micro-, nano-, and super-porous structure can be produced with same monomer, depending on the presence of foaming agent, foaming aid and foam stabilizer
	Microporous HG		
	Macroporous HG		
	Super-porous HG		
Nature of crosslinks	Physical HG	Gelatin, Chitosan	Strong physical interaction which is reversible, sensitive to temperature, solvent, pH, etc.. (due to hydrogen bonding, double or triple helix formation, ionic interaction)
	Chemical HG	Chemically crosslinked polystyrene, polymethacrylic acid, etc.	Covalent crosslinks between the molecules which are irreversible and permanent
Biodegradability	Biodegradable HG	All natural polymeric hydrogels, PLA, PEO	Degrade under biological environment
	Nonbiodegradable HG	Hydrogels based on synthetic polymers	Does not degrade under biological environment
Environmental response	pH sensitive HG	Hydrogels made of polyelectrolytes	Hydrogels undergo shrinking or swelling in the presence of stimuli like pH, heat, electric field
	Thermal sensitive HG	poly(N-isopropylacrylamide), block copolymer of PEG and PPO	
	Electrical signal sensitive HG	Polyacrylamide hydrogels (partially hydrolyzed)	

TABLE 16.2 (*Continued*)

Parameters	Types	Examples	Remarks
	Photosensitive HG	Poly(*N*-isopropylacrylamide) hydrogels	Chromophores such as trisodium salt of copper chlorophyllin, bis (4-dimethylamino) phenyl methyl leuco cyanide which are sensitive to visible or UV light are used
Preparation methods	Graft copolymerized HG	Triblock-graft copolymers of polyethylene glycol-*b*-[poly(ε-capro lactone)-g-poly(2-(2-methoxy ethoxy) ethyl methacrylate-*co*-oligo (ethylene glycol) methacrylate)]-*b*-polyethylene glycol	Used in tissue engineering applications, graft copolymerization through ring-opening and atom transfer radical polymerization
	Radical polymerized HG	Kolliphor/acrylamide	HG with antiseptic property for wound healing applications
	Condensation polymerized HG	Polyethylene glycol and polybutylene terephthalate multiblock copolymer	
	Enzymatic polymerized HG	Dopamine modified carboxymethyl cellulose	High wet tissue adhesion strength and good biocompatibility for wound healing/masking applications
	Radiation crosslinked HG	Sodium carboxymethyl cellulose/sodium styrene sulfonate	Free from chemical crosslinking agents
	Freeze thawing HG	Polyvinyl alcohol	Freeze thawing technique is used to produce hydrogels without chemical crosslinking agents

Note: [a]Monomers used to prepare neutral hydrogels.

16.3 TECHNICAL REQUIREMENTS OF HYDROGELS

As hydrogels are used in various areas such as biomedical, agriculture, hygiene products, consumer products, water treatment, these hydrogels should satisfy the application requirements. The characteristic properties of an ideal hydrogel are as follows [14, 23]:

1. High absorption capacity in saline.
2. Desired rate of absorption and desorption.
3. High-absorbency under load (e.g., baby diapers, feminine napkins).
4. Minor amount of water-soluble content and residual monomer.
5. Should not be expensive.
6. Longer service life and higher stability in the swelling environment and during the storage.

7. High rate of biodegradation and free from the release of toxic byproducts during degradation (e.g., drug delivery and some biomedical applications).
8. pH neutrality after swelling in water.
9. Colorlessness, odorless, nontoxic.
10. Thermal and light stability, reproducibility.
11. Able to reabsorb water (e.g., in agricultural application).

16.4 SYNTHESIS OF HYDROGELS

Many researchers have reported various preparation techniques of polymeric hydrogels, namely, physical crosslinking, free-radical polymerization, chemical crosslinking of polymers, graft copolymerization, and radiation crosslinking of polymers [24]. Figure 16.3 summarizes

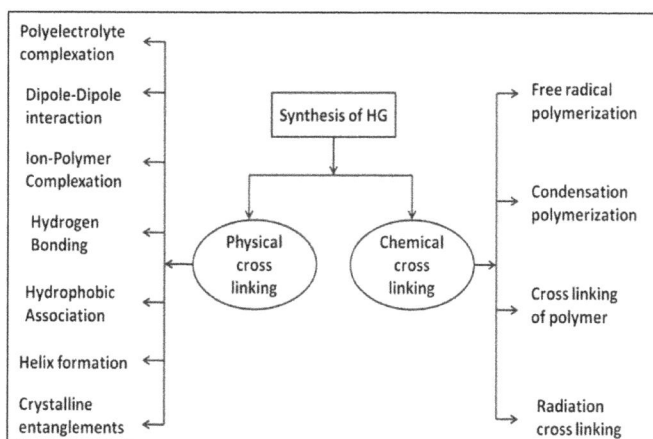

FIGURE 16.3   Different approaches to the synthesis of hydrogels.

the various routes used for the synthesis of hydrogels.

Physical hydrogels (reversible) are formed when the polymer molecules are having strong interaction between them. These interactions are also termed as physical crosslinks, which are sensitive to temperature, pH, and solvent environment. The physical crosslinks can be formed through polyelectrolyte complexation, dipole–dipole interaction, hydrogen bonding, hydrophobic association, crystalline entanglements, helix-formation, and ion-polymer complexation. Physical hydrogels have gained substantial interest in the food, pharmaceutical, cosmeceutical, and biomedical applications due to the ease of production and the advantage of not using crosslinking agents during their synthesis protocol [25].

Chemical hydrogels are synthesized from various routes: free-radical polymerization and crosslinking of hydrophilic monomer(s) using multifunctional crosslinking agents, chemical crosslinking of polymers, graft copolymerization followed by crosslinking, and radiation crosslinking of polymers.

Ionizing radiation has long been recognized as the most suitable tool for the formation of hydrogels. The main advantages of radiation synthesis are ease of process control, able to combine the formation of hydrogel and sterilization in one step, no need for initiators and crosslinkers, having no waste, free from toxic byproducts,

and relatively low running costs. Due to these benefits, the irradiation source is used to synthesize hydrogels, particularly for biomedical and food industries [26–30].

16.4.1 RADIATION SOURCES

The term radiation refers to the energy that travels through space or matter in the form of energetic waves or particles. Radiation can be ionizing or nonionizing in nature. Ionizing radiation has enough energy to ionize simple molecules either in air or water. Ionizing radiations can be high energy electromagnetic radiations such as ultraviolet (UV), X-ray, gamma radiation, α-radiation, and β-radiation or high energy particle radiations, such as electrons or protons. Gamma and electron beam (EB) radiations are the most commonly used ionizing radiations for the synthesis of hydrogels as well as for the modification of polymer properties.

Gamma-rays are generated by a radioactive isotope ^{60}Co and they cover a wide range of energies from 100 keV to 10 MeV. The dose rate of γ-rays is much lower than that of EB and this is a limiting factor for the throughput of the radiation processing. Compared to EB irradiation, γ-irradiation has high penetration power, which gives advantages for irradiating bulky products with large volumes or odd shapes.

EB radiation is a form of ionizing energy that is generally characterized by its low penetration and high dosage rates. The EBs are composed of a highly charged stream of electrons. EB is generated using an electron accelerator by energizing and accelerating the electrons through an electromagnetic or electrostatic field. The industrial electron accelerator system consists of an electron gun, accelerating tube, power source system, control system, vacuum system, beam window, and a scanner. Commercially, various types of EB accelerators are available, which are used as radiation sources for the synthesis of hydrogels and chemical modification of polymers. The acceleration can be controlled by either direct current power or radiofrequency power. The intensity of the EB radiation is dependent on accelerating voltage and beam current. The energy of electrons is equivalent to the accelerating voltage, which controls its penetration power. The commonly used electron energy ranges from 100 keV to 10 MeV and the corresponding power lies in the range 0.5–200 kW. Based on the energy levels, the accelerators are classified into low energy accelerators (80–300 keV), medium energy accelerators (300 keV–5 MeV), and high energy accelerators (>5 MeV) [31, 32].

16.4.2 RADIATION SYNTHESIS

Synthesis of hydrogels using high energy radiation involves the treatment of monomer/prepolymer/polymeric materials with ionizing radiation to modify their physical, chemical, and biological properties. Homo-polymerization, grafting, and crosslinking reaction takes place when these materials are exposed to radiation. These reactions can be carried out either in bulk or solution state (commonly aqueous solution).

When we look at the history, radioisotope sources and electronic accelerators that could provide high-energy ionizing radiation became available to many scientists during the 1940s. The utilization of radiation sources for the modification of polymers started during the 1950s. In the year 1955, the first report on the crosslinking of polyvinyl N-pyrrolidone (PVP) was reported by Charlesby et al. [33]. Later many scientists started working on radiochemical crosslinking of various water-soluble polymers [34]. Over several years of research by many scientists [31, 32, 34–36], it has been proved that the synthesis of hydrogels using high energy radiation is one of the important methods especially for biomedical, food-grade, and other hygienic application areas due to various advantages. However, this method also has some shortcomings. The advantages and disadvantages of radiation processing technique are tabulated in Table 16.3.

In the last decade, several research groups have utilized high-energy radiation to synthesize a variety of hydrogels namely, conventional, responsive, nanocomposite and SAHs for biomedical, wastewater treatment, agriculture, sensor, and catalytic applications which will be addressed in the forthcoming sections.

16.5 RADIATION SYNTHESIS OF HYDROGELS FOR BIOMEDICAL APPLICATIONS

As hydrogels can be synthesized by a variety of natural, synthetic, and combination of these polymers, the chemical, physical, and biological properties of the hydrogels can be tailor-made as per the intended applications. Also, hydrogels can

TABLE 16.3 Advantages and Disadvantages of Radiation Synthesis of Hydrogels [Selected from References 31, 32, 34–36]

Advantages	Disadvantages
• Radiation is a very convenient tool for the synthesis of hydrogels	• Irradiator system is typically expensive
• Requires minimal sample preparation	• It is not economically viable when the volume is not high as compared to conventional methods
• No need for catalysts or additives to initiate the reaction	• Sometimes, the properties of the hydrogels prepared by radiation processing are inferior as compared to chemical modification
• Strong interaction between chains through covalent bonds	• Safety procedures need to be in place.
• Polymerization can be initiated at low temperature and hence porosity of the hydrogels can be controlled easily	• The dearth of awareness and acceptance of radiation technology by the community is also a hindrance
• Hydrogels are free from residual contaminants like toxic initiators, crosslinkers	
• Solves the problem of sterilization	
• Ionizing irradiation treatment is fast	
• Does not induce a significant increase in temperature	
• More homogeneous network structures can be synthesized	
• Physical properties can be tuned by varying irradiation parameters, such as irradiation dose, irradiation dose rate, pulse length, and frequency (for EB only), temperature and pressure	

be fabricated with a wide range of morphologies as well as different physical forms, namely, powder, film, coatings, porous sponge, nonporous, and fiber form. Due to these reasons as well as other unique benefits of hydrogels (Table 16.4), the hydrogels are being used in biomedical applications like tissue engineering, ophthalmic, drug delivery, wound dressing, dentures, biosensors, and coatings for sutures and catheters.

Radiation-induced crosslinking is a well-established technique for developing sterile and pure hydrogels for biomedical applications. In this technique, hydrogel formation does not require a chemical initiator and crosslinker. It is a single-step process wherein gel formation and sterilization take place simultaneously. It is an easy technique to regulate drug delivery by controlling crosslinking density and swelling parameters of the hydrogels. It is a superior, cost-effective, and efficient method of hydrogel synthesis as compared to the chemical method [37–39].

16.5.1   CONVENTIONAL HYDROGELS

It is well known that wet dressing is more effective in wound healing when compared with a conventional dry dressing. It is also established that the hydrogel is an inexpensive, effective, convenient material for

TABLE 16.4   Advantages and Disadvantages of Hydrogels with Respect to Biomedical Applications [Selected from References 37–39]

Advantages	Disadvantages
• Can be biocompatible, biodegradable, and bioabsorbable	• High cost
• Physical properties of hydrogels resemble living tissues compared to any other class of biomaterials	• Low mechanical strength
• Soft and rubbery nature of some hydrogels reduces frictional irritation to surrounding cells and tissues	• Sometimes, it is hard to incorporate some drugs/nutrients in the hydrogel
• Can be injected in-vivo as a liquid, then gels at body temperature	• Nonadherent
• Good transport properties (such as nutrients to cells or cell products from cells, low molecular weight metabolites, enzymes, genes, and ions can easily diffuse through hydrogels)	• Can be tough to handle
• Controlled release of medicines or nutrients	
• Ease of chemical modification	

wet dressing, due to its biocompatibility, water absorption capability, moisture retention, and ventilation properties. Poly(N-vinyl pyrrolidone) is a transparent, biocompatible synthetic polymer but has some limitations to use due to its fragile mechanical properties and swelling capability. To widen the application window of PVP, it is blended with other polymers.

Sen et al. [40] prepared the interpenetrating network (IPN) hydrogel based on PVP and carrageenan using gamma radiation at different doses of 15, 25, and 35 kGy. From the preliminary laboratory tests such as swelling behavior in pseudo extracellular fluid solution, pH of aqueous extracts, mechanical, and bioadhesion, it has been reported that the prepared hydrogels have properties that are required for an ideal wound dressing application. They also suggested that the hydrogels prepared at relatively higher irradiation dose, and potassium chloride content is suitable for wound dressing because of their ability to be peeled off from the wound easily.

In the other study [41], PVP is blended with carboxymethyl cellulose (CMC). Three different compositions of PVP/CMC hydrogels, namely, 5:5, 6:4, and 7:3 (PVP:CMC, w/w) were synthesized by gamma radiation using ^{60}Co source in the dose range of 2.5–30 kGy. The properties of the obtained PVP/CMC hydrogels were compared with that of homopolymeric PVP and CMC hydrogels. The authors reported that the PVP/CMC hydrogels have better mechanical properties than CMC hydrogel and better flexibility and swelling behavior than that of prestine PVP hydrogel. They have also investigated the important property of wound dressing, moisture retention of PVP/CMC hydrogel, and compared with that of commercially available PVA-based hydrogel wound dressing. Both the hydrogels showed similar water loosing rate (-5.0×10^{-4} g/min) and a comparable water-holding ratio (R_h). Based on the results, the authors stated that PVP/CMC blend hydrogel prepared using radiation is a good candidate for wound dressing application.

In order to use biomaterial in tissue engineering applications, the material should fulfill various requirements. One of the important requirements is antibacterial and antifouling property, otherwise, the hydrogel employed in tissue engineering will suffer from microbial infections caused by bacteria and fungi. Several research groups have developed antibacterial and antifungal polymeric hydrogels, more details on this can be found in the recent review reported by González-Henríquez et al. [42]. Recently, Arslan et al. [43] synthesized a biocidal hydrogel by gamma radiation with a dose of 30 kGy using acrylamide and 4-vinyl pyridine monomers. The hydrogels were chemically modified (quaternization) using *N*-aromatic alkyl quaternizing agent of chloromethyl benzene. The antibacterial and antifungal activities of modified and unmodified hydrogels have been established using Gram-positive bacteria *Staphylococcus aureus* (ATCC 25923), Gram-negative bacteria *Pseudomonas aeruginosa* (ATCC 27853), *Escherichia coli* (ATCC 25922), and fungal strain *C. albicans* (ATCC 10231). From these studies, it was noticed that the quaternized hydrogels possess antibacterial and antifungal characteristics but not true with the pristine hydrogel.

El-Arnaouty [44] synthesized copolymeric hydrogel based on butyl methacrylate and poly(acrylamide) using gamma radiation and also studied the release profile of heparin drug (used to prevent surface-induced thrombosis). The properties of copolymeric hydrogels were compared with their homopolymeric hydrogels. The effect of radiation dose and the copolymer composition on the gelation % was also studied. It is reported that with an increase in the radiation dose, the gelation % reached a maximum at 30 kGy, and with a further increase in radiation dose to 40 kGy, the gelation % reduced. The copolymeric hydrogels showed better thermal stability and hydrophilic character than their homopolymers.

In order to understand the effect of electron irradiation on the pore size, rheological properties, and cytocompatibility, Riedel et al. developed the collagen-based hydrogel using EB radiation [45]. Confocal laser scanning microscopy was used to analyze the network structure of the collagen hydrogel. The result reveals that the higher the radiation dose, the pore size has reduced due to the formation of the denser inter and intrachain crosslinks. The rheology study reveals that the storage and loss modulus of the hydrogel has been increased with an increase in the radiation dose due to a decrease in viscosity and an increase in the stiffness of the network. The authors also demonstrated that the collagen hydrogel crosslinked with EB has excellent cellular acceptance and cytocompatibility.

An and coworkers [46] have synthesized the minocycline hydrochloride (MH) loaded β-glucan hydrogel for periodontal disease treatment via radiation crosslinking method. Further, MH-loaded β-glucan hydrogels have been evaluated for their cytotoxicity and antibacterial activity. The gel content and compressive strength of the β-glucan hydrogels improved with an increase in β-glucan content and absorbed dosage (up to 7 kGy). In addition, the MH-loaded β-glucan hydrogels showed no cytotoxicity and exhibited good antibacterial activity against

Porphyromonas gingivalis. Hence, it demonstrated the potential to prevent the invasion of bacteria and to treat wounds.

González-Torres et al. [47] have synthesized gamma-radiation-induced C–PVP–PEG hydrogel and evaluated its structure, morphology, and cell viability in fibroblasts. The morphology of the hydrogel strongly depends on the feed ratios of the polymers. The in-vitro studies of the hydrogels suggested that at 50 kGy, the presence of collagen reduced the cell viability for all formulations, except for C–PVP. Furthermore, when PEG and PVP were added to the collagen, the cell's viability enhanced with reference to the irradiated collagen sample.

Taşdelen et al. [48] developed a new chitosan (CS)/hyaluronic acid (HA)/hydroxyapatite (HAP) hydrogel by using gamma irradiation for anticancer oral drug delivery. In this research investigation, a model anticancer drug, 5-fluorouracil (5-FU) was used to study the drug release efficiency of the prepared hydrogel. The properties such as drug uptake and release behavior were estimated and it was found that the presence of HA and HAP in the hydrogel improved the drug uptake as well as release efficiency of the chitosan-based hydrogel.

Recently, Singh et al. developed few hydrogels of moringa gum (MO) with poly(acrylic acid) (PAAc) [49], N-vinyl imidazole [50], acrylamide [51], and tragacanth gum/acacia gum [52] by radiation induced cross-linking for biomedical applications.

The model drug "ciprofloxacin," which is an antibiotic used for the treatment of various bacterial infections in human beings was used to study the drug release capability of MO-co-poly(AAc) hydrogel. The effect of network formation between PAAc and MO on some of the biomedical properties such as blood compatibility, antioxidant activity, mucoadhesion, and gel strength of the hydrogels were also evaluated. From these studies, it was noticed that they are mucoadhesive, nonthrombogenic, nonhemolytic, pH responsive, and antioxidant in nature. Furthermore, these hydrogels may be used for controlled and sustained release of the drug delivery for colon and GIT system [49].

The effect of radiation grafting of N-vinyl imidazole/MO copolymer on blood compatibility, antioxidant activity, and mucoadhesion, and gel strength have been studied. It was found that these graft copolymeric hydrogels are nonhemolytic, mucoadhesive, and antioxidant in nature and are suitable for gastrointestinal drug delivery system. Furthermore, an antibiotic drug levofloxacin release profile has been reported [50].

The effect of radiation dose on the properties of acrylamide/MO graft hydrogel has been evaluated, and it was found that with an increase in irradiation dose, the pore size decreased and the cross-linking density was increased. The drug release profiles of antibiotic drug levofloxacin were studied and it was noticed that slow release of drug delivery had no burst effect from the drug-loaded hydrogels. The acrylamide/MO graft hydrogels were found to be mucoadhesive, nonhemolytic, and antioxidant in nature. It can be attributed to the presence of moringa gum in the polymer gel matrix. On the basis of results, it is concluded that these pristine and sterile hydrogels can be used for gastrointestinal drug delivery system [51].

Tragacanth gum/acacia gum hydrogel loaded with gentamicin for wound dressings were found to be mucoadhesive, impermeable to the microorganisms, and permeable to H_2O/O_2. These hydrogel wound dressings absorbed simulated wound fluid 4.62 ± 0.31 g/g gel. The sterile hydrogel wound dressing exhibited antioxidant activity in addition to antibacterial activity, which is due to the presence of both the gums and antibiotic drug. The result clearly indicates that these dressings could be used to enhance the wound healing potential [52].

Very recently, to improve the biomedical properties such as gel strength, drug loading capability, and release profile of the polysaccharide-based hydrogels, the sterculia gum (karaya gum), was copolymerized in presence of N-vinyl imidazole, a biocompatible hydrophilic monomer [53], carbopol and graphene oxide (GO) [54]. The anticancer drug, gemcitabine was used to study the drug loading capability and release profile of the GO-sterculia-cl-poly(N-vinylimidazole) and GO-sterculia-cl-carbopol hydrogels. The gel strength of the hydrogel was determined using a texture analyzer. The GO-containing hydrogels exhibited high gel strength (47.96 N mm) as compared to that of pristine hydrogel (25.27 N mm). From the drug release study, it was found that the presence of GO and N-vinyl imidazole improves the drug release profile and loading capacity of the hydrogel. The biomedical properties of the GO-sterculia-cl-poly(N-vinylimidazole) hydrogel was highly influenced by the GO content and the irradiation dose used to prepare the hydrogel.

16.5.2 STIMULUS-RESPONSIVE HYDROGELS

Stimuli-responsive hydrogels change their volume and elasticity abruptly in response to a change in the surrounding, such as temperature, pH, solvent composition, ionic strength, supply of electric field and light [31, 55]. Due to this nature, the responsive hydrogels are used as smart or intelligent material for many applications such as artificial muscles, sensors, intelligent drug delivery and bioseparation. Hence, in recent years, there has been increasing interest in stimuli-responsive hydrogels. This section aims to provide an overview of stimuli-responsive hydrogels particularly the hydrogels prepared using high-energy radiations.

Many investigations have been carried out on pH-sensitive hydrogels, prepared from high energy radiation for controlled drug release application [56–59]. The release of Atorvastatin drug from acrylic acid/polyethyleneimine (PEI) as pH-sensitive IPN hydrogel, obtained by radiation induced polymerization was investigated by El-Din et al. [56]. It was noticed that the drug release from PAA/PEI hydrogel is higher in the acidic medium than in the basic medium. The swelling characteristics and the drug release of Ketoprofen, as a model drug from CMC and PEI smart hydrogels, prepared by gamma radiation was investigated in-vitro [57]. The swelling study reveals that the CMC/PEI hydrogels were temperature-responsive as well as pH-sensitive. The in-vitro release of drug had extended beyond 3 hours and follows non-Fickian diffusion process.

Insulin loaded methacrylic acid (MAA), *N,N*-dimethyl aminoethyl methacrylate (DMAEMA) copolymeric, multifunctional, smart hydrogels for oral delivery of insulin was prepared by Taleb et al. [58]. Gamma radiation was used to fabricate these hydrogels. The influence of copolymer composition and pH value of the surrounding medium on the swelling behavior as well as insulin release profile has been studied. They have observed that as pH increased from 1, the equilibrium swelling % of MMA/DMAEMA copolymeric hydrogel reduced and reached minimum value at around 4 pH, and then it raised again with further increase in pH. The amount of drug released was found to be much higher in pH-7.2 (simulated intestinal fluid) than in pH-1.5 (simulated gastric fluid). It was noticed that the hydrogel with higher MMA content (90%) is the right candidate for the controlled and targeted release of insulin drug.

Fluorouracil, as a model drug, was incorporated into pH-sensitive hydrogel based on cyclodextrin grafted PEG and acrylic acid copolymeric hydrogel [59]. These hydrogels were fabricated using EB at different dosages, namely, 80, 120, 180, and 200 kGy. The in-vitro drug release study was performed using a UV-visible spectrophotometer. The results were compared with PAA hydrogel. The results of controlled-release showed that the release rate of PEG-*co*-AA copolymeric hydrogel is slower than polyacrylic acid hydrogel and it followed non-Fickian diffusion.

The effect of oligo (ethylene glycol) dimethacrylate (OEGDMA) monomer to solvent feed ratio on radiation induced synthesis of thermoresponsive OEGDMA-based hydrogels have been investigated by Suljovrujic et al. [60]. The obtained results clearly indicate that all OEGDMA-based hydrogels exhibited a high gel content, inverse thermo response, and volume phase transition temperature (VPTT). Furthermore, by varying the monomer to solvent ratios in the prepolymerization formulation, a large diversity in the microstructure, swelling capacity, and VPTT have been achieved.

In order to address the issues such as higher viscosity and lower solubility of high molecular weight chitosan-based hydrogels, Hafeez et al. [61] developed a novel pH-sensitive, γ-irradiated low molecular weight chitosan/poly(vinyl alcohol) injectable hydrogel for drug delivery application. The prepared hydrogel was loaded with the model drug (montelukast sodium) during the preparation phase and the release capability of the hydrogel was investigated. These hydrogels exhibited higher swelling at the basic condition and less swelling in acidic and neutral pH environment. The chitosan/PVA hydrogel films also exhibit improved antibacterial activity.

The pH and temperature-responsive hydrogels based on PVA and acrylamide (AM) mixture were studied by El-Din et al. [62]. In order to understand the sensitivity of the prepared hydrogels with temperature and pH, the swelling behavior at different temperatures and pH levels have been studied. The results showed that the PVA/AM hydrogels are temperature-sensitive in the range of 40–50 °C due to the formation of hydrophobic interchain bonding and/or interpenetrating polymer network between PVA and AM, and pH sensitive is lying in the range 5–7, as amides undergo reversible hydrolysis reaction in acidic or alkaline solutions. It was noticed that the PVA/AM hydrogel sensitivity is dependent on AM content and the nature of molecular entanglement.

Thermo-responsive hydrogels were synthesized by radiation induced crosslinking polymerization in bulk based on oligo ether methacrylates of side-chain lengths of 2–19 EG; monomer units by Piechocki et al. [63]. Side-chain lengths have been shown to have a strong effect on synthesis parameters and hydrogel properties. Networks based on oligo ether methacrylates of sufficiently long side chains are capable of partial crystallization, and the crystalline phase is chemically linked to the polymer network

Branca et al. [64] developed the hydrogels containing polyethylene oxide (PEO)/chitosan mixture using EB radiation at a dose rate of 4 Gy/s. The effect of chitosan content on the crosslink density/gel fraction and the pH sensitivity of swelling media with pH values 1.2–7.5 have been reported. The addition of chitosan induces the formation of more stable hydrogel with less equilibrium degree of swelling due to higher crosslink density in the hydrogel network as compared to PEO hydrogel. The hydrogels containing chitosan showed pH sensitivity but that is not true with the PEO hydrogel. It was also noticed that the higher chitosan content exhibit larger pH dependence on the swelling capacity.

Hydrogels from polyacrylic acid (PAA) has the ability to absorb water many times to its dry weight. These hydrogels find applications in diapers and personal hygiene products, ion exchange resins, membranes for hemodialysis, and ultrafiltration and controlled release devices [65]. The hydrogel properties such as mechanical strength and swelling ratio depend on the degree of crosslink, which in turn depends on the type and amount of crosslinking agents used, the method and conditions employed to synthesize the hydrogel. In order to study the effect of crosslinking agent and the absorbed dose, Yang et al. [66] synthesized the pH-sensitive PAA hydrogels using gamma irradiation at different doses between 10 and 22 kGy. The swelling study revealed that the swelling ability of these hydrogels was strongly influenced by the absorbed dose and the amount of crosslinking agent. It was noticed that the swelling ratio was decreased with increasing absorbed dose and the crosslinking agent. It was also noticed that the pH sensitivity and the swelling ratio of the hydrogels in alkaline solution are much higher than that in acid solution.

16.5.3 NANOPARTICLE FILLED HYDROGELS

Nanoparticles (NPs) dimension usually lies in the range of 1–100 nm and these NPs have unique properties as compared to their bulk state. The properties of these materials changes as the dimensions of the particles are reduced to the atomic level. The NPs possess unique physicochemical, optical, electrical, thermal, and biological properties that can be tuned as per the requirement. Presently, the metallic NPs such as Cu, Ag, Au, and Pt are thoroughly being explored and comprehensively investigated as potential antimicrobials [67–71]. The high surface-to-volume ratio of the NPs enhances their interaction with the microbes to bring out a broad range of possible antimicrobial activities. Nanocomposite hydrogels are simply the crosslinked three-dimensional networks that are formed in the presence of nanostructured particles [72]. The combination of NPs with polymer hydrogels provides superior functional properties to the composite hydrogels because of their smart characteristics with diverse applications, such as nanomedicine, drug delivery, and biosensing. This creates ample scope for researchers to study the performance of nanocomposite hydrogels in biomedical applications.

Several studies have been proved that silver (Ag) exhibits antimicrobial activity [73–75] and these are considered nontoxic and environmentally friendly materials. This has led to the development of Ag NP hydrogel composites for both in-vivo and in-vitro biomedical applications [72, 76–78].

Silver NPs embedded PVA/PVP-based hydrogels have been devised for wound dressing application via gamma radiation [76]. The effect of PVP content and radiation dosage on gelation was studied. The study reveals that there was a steep increase in the gelation (%) up to 30 kGy and a slight increase at 40 kGy. Consequently, the percentage of swelling decreases with an increase in the radiation dose. The authors also noticed that the addition of Ag NPs in the hydrogel network enhances the swelling capacity due to an increase in the overall porosity in the hydrogel.

In the other study, the Ag NPs were incorporated in the polyvinyl alcohol/cellulose acetate/gelatin (PVA/CA/Gel) hydrogel matrix [77]. These

hydrogel nanocomposites have been synthesized using gamma radiation induced crosslinking. The presence of Ag NPs has been confirmed by UV–Vis spectroscopy, X-ray diffraction, transmission electron microscope, energy dispersive X-ray analysis, and Fourier transform infrared spectroscopy. The antifungal activity of nano-Ag/PVA/CA/Gel hydrogels against *Aspergillus fumigatus, Geotrichum candidum, Candida albicans, Syncephalastrum racemosum*, and antibacterial activity against *S. aureus, Bacillus subtilis, Pseudomonas aeruginoca* and *E. coli* has been studied. The antibacterial and antifungal activity was increased with an increase in Ag content from 0 to 5 mM, and it was concluded that this system can be used for wound dressing application due to its antibacterial and antifungal activity.

Micic et al. [78] demonstrated the antimicrobial activity of novel copolymeric silver/poly(2-hydroxyethyl methacrylate/itaconic acid) nanocomposite hydrogels for wound dressing applications. Silver NPs were synthesized by in-situ reduction of silver nitrate using gamma radiation. In this study, the authors have investigated the effect of silver salt concentrations on the size and distribution of NPs. To prove the antimicrobial properties of Ag/P(HEMA/IA) nanocomposites, the Gram-negative bacterium (*E. coli*), Gram-positive bacterium (*S. aureus*), and fungus (*C. albicans*) were used. It was demonstrated that the hydrogel nanocomposites are highly effective with respect to antibacterial and antifungal activity even at small silver concentrations.

HAP is used in biomedical applications because of its bioactivity, biocompatibility, and moderate resorbability with time. It has been frequently used as a genetic carrier and for drug delivery. HAP NPs have a stronger antitumor effect than macromolecules with minimal side effects and can produce a strong cooperative effect with chemotherapy medicine [79–81]. The sodium alginate/chitosan/nano hydroxyl apatite hydrogel nanocomposite system for liver cancer drug delivery was prepared using radiation induced free-radical copolymerization and crosslinking [82]. Doxorubicin, a chemotherapy drug is loaded to nanocomposite hydrogel using the swelling equilibrium method. The swelling characteristics and drug release of the nanocomposite were highly sensitive to the pH, and it was concluded that this system can be a potential candidate for anticancer drug delivery system.

Raafat et al. [83] fabricate a nanocomposite hydrogel based on xanthan (Xan) as a biocompatible natural polymer and PVA as a synthetic polymer with good mechanical properties reinforced with ZnO NPs using γ-ray irradiation. The author prepared a series of xanthan-polyvinyl alcohol (Xan-PVA)/ZnO nanocomposite hydrogels for wound dressing application using ^{60}Co γ-ray irradiation facility as a clean source of initiation. The crosslinking of the hydrogels was accomplished by the sterilization method. The presence of ZnO NPs reconstructed the internal structure of the hydrogel network which aids in a homogenous porous structure as shown by SEM images. The porosity along with the presence of ZnO NPs in Xan-PVA hydrogels controls the fluid uptake ability, water retention, and water vapor transmission rate. In-vitro cytotoxicity and hemolytic potency studies reveal that Xan-PVA hydrogels are biocompatible.

16.5.4 SUPER-ABSORBENT HYDROGELS

SAHs belongs to a class of hydrogels that can absorb and retain extraordinarily large amounts of water or aqueous solution as high as 1000%–100,000%, whereas the conventional hydrogels' absorption capacity is not more than 100% [84]. This unique property of SAHs has engrossed the attention of researchers/scientists and technologists and found extensive applications in biomedical, pharmaceutical, agriculture, and separation processes [85]. As it is well-known that the radiation induced polymerization method is simple and additive or chemical-free and the degree of crosslinking can be easily controlled by irradiation conditions, scientists have attempt to investigate the performance of SAHs synthesized using high energy radiation. The basic materials used in these studies are cellulose derivatives, acrylic acid, and acrylamides [86–93]. The forthcoming sections summarize the findings of these SAHs.

SAHs based on cellulose derivatives such as CMC sodium salt, methyl cellulose (MC), hydroxyl ethyl cellulose (HEC), and hydroxyl propyl cellulose (HPC) were prepared by radiation induced crosslinking in an aqueous solution [86]. The effect of absorption dose and solute concentration on the swelling behavior of the hydrogels were studied. It was found that the gel fraction of hydrogels based on CMC and HEC was increased with an increase in absorption dose up to 80 kGy, and a further increase in absorption dose decreased the gel fraction. The swelling capability of these hydrogels showed a reverse trend with the absorption dose and HEC showed the best absorbing capacity.

The effect of synthesis parameters on the performance of hydrogels based on cellulose derivatives in presence of the crosslinking agent, N,N'-methylene-bis-acrylamide (MBA) was investigated by Fekete et al. [87]. In this study, it has been proved that the MBA concentration has a significant effect on the properties (gel fraction and swelling capability) of the CMC and HEC hydrogels, whereas, with the HPC and MC hydrogels, the presence of MBA showed an insignificant effect. The effect of absorbed dose at different MBA content for CMC gels were also investigated and it has been reported that the gelation occurred at a very lower dose of 0.25 kGy in presence of MBA, whereas in the absence of MBA, the minimum dose required for gelation to occur was 1 kGy. Also, the radiation dose required for maximum gelation shifted to lower values with an increase in MBA concentration in the CMC gel.

Erizal et al. have prepared poly (acrylamide-co-acrylic acid)/sodium alginate (NaAlg) SAHs via gamma radiation crosslinking method [88]. The effect of sodium alginate content and irradiation dosage on the properties of gels have been evaluated. The experimental results showed that the swelling ratio was maximum (800 g/g) when NaAlg content was 0.1%. On further increase in NaAlg content from 0.1% to 0.7%, the swelling ratio decreased to 400 g/g. Also, the swelling ratio of hydrogels decreased from 800 to 200 g/g with increasing irradiation dose from 20 to 40 kGy.

Awadallah et al. [89] investigated the effect of comonomer composition and absorption dose on the gelation and swelling capability of poly (acrylic acid/2-acrylamido-2-methyl propane sulfonic acid) copolymeric hydrogel. In order to increase the swelling capacity of the hydrogels, the prepared SAPs were chemically treated with various chemical reagents such as n-glycylglycine, sodium bisulfite, hydroxylamine hydrochloride, thiosemicarbazide, and hydrazine hydrate. In this study, the authors have reported that the swelling ratio of the hydrogel is strongly dependent on the pH, concentration and composition of the comonomer, irradiation dose, and ionic strength of salts. These hydrogels widen the application window, such as use in hygienic products, agriculture, drug delivery systems, sealing, coal dewatering, and artificial snow.

Tomar et al. [90] investigated the copolymeric SAHs based on acrylamide and acrylic acid using gamma radiation at the dose rate of 15 kGy. The effect of acrylamide content on the crosslink density and equilibrium swelling has been studied by varying the acrylamide content from 30% to 70%. It was noticed that the water absorption capacity increased with an increase in acrylamide content up to 38.87 mol% and a further increase in acrylamide content slightly reduces the water absorption capacity of the resultant hydrogel. Since these SAHs are widely used in agricultural applications, the water retention property of the soil and the growth of ladyfinger seeds in the soil mixed with the acrylamide-co-acrylic acid hydrogel have been studied. This study reveals that the addition of the SAHs has a significant effect on the water retention property of the soil and is evident from the good growth of ladyfinger plant in the soil containing SAHs as compared to the soil without SAHs.

Polysaccharides in their dry or solution form tend to degrade when exposed to ionizing radiation [91] and the chain scission strongly depends on the concentration of the polymer. When the concentration of the polymer goes beyond a certain limit, the crosslinking reaction dominates over chain scission. If polysaccharides in the solid state are exposed to radiation in presence of alkyne gas, the chemical structure can be modified in a controlled manner [92]. These two approaches were utilized by Hayrabolulu et al. to prepare carboxylated locust bean gum hydrogels using ^{60}Co γ radiation [93]. The properties of hydrogels prepared by these two methods were compared by the investigators and it was observed that the hydrogel prepared in the presence of acetylene gas exhibit higher swelling behavior as compared to that of hydrogel obtained in the paste-like state.

Recently, Villegas et al. [94] developed SAHs by chemical modification of PVA with glyoxylic acid to obtain poly(vinyl glyoxylic acid) as biodegradable SAHs. The swelling behavior of the radiation crosslinked PVAG superabsorbent was compared with that of a chemically crosslinked hydrogel. From the study, it was noticed that the radiation-crosslink route exhibits the swelling behavior of hydrogel in ~215% as compared to chemically crosslinked when swelled in pure water.

Honget et al. [95] adopted a new method to synthesize the polysaccharide-based grafted sodium styrene sulfonate SAHs in an aqueous solution by γ-radiation under ambient conditions. The process parameters, namely gamma irradiation dose and the ratio of feed composition, have been optimized to synthesize hydrogels with best-swelling capability. The obtained hybrid hydrogel exhibits the most optimum swelling capacity at neutral pH whereas equilibrium swelling of SAHs was noticed within 5 hours.

16.6 RADIATION SYNTHESIS OF HYDROGELS FOR WASTEWATER TREATMENT

The industrial effluents from the industries such as the paint industry, textiles, paper, food industry, battery manufacturing industry, when discharged improperly to the environment, tends to pollute the natural resources. The dyes and heavy metal ions present in these industrial effluents are dangerous to biological organisms and they are harmful to human health. Various techniques such as membrane separation, reverse osmosis, ion exchange, coagulation–flocculation, biological treatment, and adsorption have been developed and used for the treatment of contaminated/wastewater [96–98].

Adsorption technique is one of the industrially preferred methods over other techniques used for wastewater treatment due to its design flexibility and economical advantages [99]. Activated carbon, zeolites, chitosan, clay minerals, and hydrogels are being used as effective adsorbents for water treatment and recycling. More and more researchers have paid attention to develop hydrogels as an effective adsorbent for water treatment because of their high water retention property and low cost.

Abdel-Aal et al. [100] prepared electrolyte-type hydrogels based on 2-hydroxy ethyl-methacrylate (HEMA) containing mono-, di-, and tri-protic acid moieties by using gamma radiation at various doses. The atomic absorption spectroscopy was used to evaluate the metals (Fe, Ni, Co, and Cu) and a UV–visible spectrophotometer was used to investigate the adsorption of dyes (Maxilon C.I. Basic dye and acid-fast yellow G). These hydrogels exhibited the good adsorption of metal ions and dyes from aqueous solutions. The authors also reported that the adsorption capacity of these hydrogels is dependent on the pH of the aqueous metal or dye solution.

Carboxy methylated chitosan based hydrogels were synthesized using gamma radiation for the absorption of Fe(III) ions [101]. The adsorption behaviors of the prepared hydrogels have been evaluated using the adsorption kinetics experiment. The results confirmed that the adsorption rate of Fe(III) ions was very fast, and the adsorption amount of Fe(III) ions per unit weight of the gel reached the maximum after 20 min. The adsorption of Fe(III) ions was favorable due to the presence of amino, hydroxyl, and carboxyl coordination groups in carboxymethylated chitosan structures.

Molecular imprinting technology is used to prepare the molecular imprinted polymers (MIPs), where the formation of a complex between the template and the functional monomer takes place after polymerization and crosslinking, the template is removed from the system. Molecularly imprinted polymers have a high affinity toward a templated target molecule and are used in the area of sensors and separation process [102, 103]. El-Arnaouty [104] developed a nickel ion-imprinted acrylamide/citric acid polymer (AAm/CA) hydrogels by using gamma rays. The conditions such as irradiation dose, monomer concentration, and metal ion concentration were optimized. The effect of irradiation dose and acrylamide concentration on the gel fraction was studied. The nickel imprinted hydrogels were characterized for their thermal stability, porosity/morphology, and adsorption capacity toward other metal ions (Co^{2+}, Cu^{2+}) using TGA, SEM, and UV spectrophotometer, respectively. It was noticed that the gel content increases by increasing the monomer concentration and the irradiation dose up to 25 kGy. From these studies, it was concluded that the adsorption capacity of the Ni^{2+} imprinted hydrogel was considerably enhanced due to the structural similarities of Co^{2+} and Cu^{2+} ions.

The MIP technology was used by Mahmoud et al. [105] to develop the imprinted hydrogels in order to remove the cobalt and lead metal ions from wastewater with high selectivity. N-vinyl-2-pyrrolidone and acrylamide monomers were copolymerized using gamma irradiation in the presence of cobalt or lead metal ions. The effect of pH, time, and initial feed concentration on the extraction efficiency was investigated. It was observed that the high affinity of the hydrogels exists toward the Pb^{2+} metal ions.

Neri et al. [106] prepared a pH-sensitive comb-type hydrogel based on PAAc and 4-vinylpyridine (4VP) by gamma radiation. The effect of radiation dose and VP monomer concentration on the percentage of grafting were evaluated and it was noticed that there was an increasing order of grafting percentage with an increase in irradiation dose and the reverse trend with an increase in 4VP concentration. These hydrogels also exhibited enhanced swelling response, reversibility, and better apparent mechanical properties. Due to these characteristics, the authors claimed that this comb-type system can be used for the immobilization of metal ions in wastewater.

Mahmoud et al. [107] synthesized a series of hydrogels based on starch, acrylic acid (AAc), and HEMA hydrogels using gamma radiation at a different radiation dose in order to remove acid dye from wastewater. The adsorption capacity of these hydrogels has been increased with increasing dye concentration and decreasing pH of the medium. The adsorption capacity for AAc/starch and NaAc/starch hydrogels increased with an increase in temperature but AAc/HEMA/starch hydrogels showed lower adsorption capacity with an increase in temperature.

The hydrogels based on a blend of pectin, acrylamide (AAm), and 2-acrylamido-2-methyl-1-propane sulfonic acid (AMPS) were synthesized by gamma radiation. Hydrogels were investigated for selective adsorption of five elements namely, Al, Cr, Fe, Ga, and In from the multielement solution, as well as adsorption from

the manually prepared solution of five elements. The effect of pH on the adsorption capacity of hydrogel has been investigated. The desorption efficiency of the hydrogel was found to be around 80%. The obtained results reveal that the hydrogel can be used for the recovery of all trivalent metal ions investigated effectively [108].

A novel environmental remediation process to absorb/recover uranium from seawater was proposed by Hara et al. [109]. Acrylonitrile and MAA monomers were polymerized using gamma irradiation of 10–40 kGy and N,N'-MBA was used as the crosslink agent. From the kinetic study of seawater absorption, it was found that the absorbent rapidly absorbs the seawater and the equilibrium swelling occurred in about 30 min. The prepared gel exhibits the absorption capacity of 409 mg/g, when it was exposed to seawater containing 2140 ppm uranium for one week. This clearly indicates that these hydrogels can be effectively used to recover uranium from seawater.

16.7   SUMMARY

From the forgone discussions, it can be concluded that the hydrogels synthesized by radiation route are very much required and suitable for certain applications that demand chemical-free, nontoxic, and clean or sterile conditions, especially like biomedical and food industries. The high energy radiations, either the electromagnetic (as UV-, X-, α-, β-, or γ- radiations) or the particle (as electrons or protons), do ionize the molecular mixture to generate polymeric hydrogels from monomer or prepolymer or the macromolecular stages. Further, by altering the parameters such as irradiation dose and rate, pulse length and frequency, system temperature and pressure, the tailor-made molecular structures could be developed. The most efficient radiations are found to be the Gamma and EB with an energy range from 100 keV to 10 MeV when the synthesis is carried out in an aqueous medium. Other advantages offered by radiation methods are minimal preparation, faster rates of synthesis, no requirement or attainment of higher temperatures, ability to develop the products in any of the required physical forms (powder, film, etc.), and homogeneous network structures. However, the limitations of the approach are its high initial investments, lack of awareness among consumers, and the safety precautions that are required to be adapted during the process.

Though the applications of hydrogels can be in any field wherever the fluid retention plays an important role, the widely investigated ones in this chapter are the medical and wastewater treating hydrogels, which seems to be most important from personal health treatment and potable water point of view. For which, any particular hydrogel alone is not pervasive and hence blending with suitable polymers or copolymerizing with suitable repeat units make them effective in various characteristics, namely, wound healing, antibacterial, antifouling, drug intake and controlled delivery, biocompatible, selectively permeable, etc., along with the required strengths (physical, mechanical, thermal, chemical, optical, biological, etc.) that are offered by specific moieties due to variation in polarity/hygroscopic nature/secondary forces of attraction and degree of entanglement or crosslinking or crystallinity.

Yet, there is always a complex requirement for hydrogels to be functional and efficient in many practical conditions of applications and those characteristics are achieved by incorporation of specific additives or molecules that has made hydrogels more interesting like stimuli-responsive hydrogels and superabsorbent hydrogels. That is, they exhibiting smartness by being condition sensitive (temperature, pH, contact medium, etc.), or they exhibiting increased efficiency (extent of mass transport, rate, etc.) that has proved to serve in diverse techniques such as "membrane separation, reverse osmosis, ion exchange, coagulation-flocculation, biological treatment and adsorption" used for the treatment of water. Among all the above-mentioned research approaches, few have turned out to have design flexibility and economical advantage, hence, they are commercially viable.

KEYWORDS

- hydrogels
- radiation synthesis
- super-absorbent polymers
- biomedical applications
- responsive hydrogels

REFERENCES

1. Cushing, M. C., & Anseth, K. S. (2007). Hydrogel cell cultures. Science, 316, 1133–1134.
2. Kumar, A., Srivastava, A., Galaev, I. Y., & Mattiasson, B. (2007). Smart polymers: physical forms and bioengineering applications. Progress in Polymer Science, 32, 1205–1237.
3. Caló, E., & Khutoryanskiy, V. V. (2015). Biomedical applications of hydrogels: a review of patents and commercial products. European Polymer Journal, 65, 252–267.
4. AMBERLITE™ IRP64. (2006). Pharmaceutical grade ion exchange resin. Retrieved from http://www.dow.com/assets/attachments/business/process_chemicals/amberlite_and_duolite_pharmaceutical_grade_resins/amberlite_irp64/

tds/amberlite_irp64.pdf (accessed on 26 May 2019).
5. Qinyuan, C., Yang, J., & Xinjun, Y. (2017). Hydrogels for biomedical applications: their characteristics and the mechanisms behind them. Gels, 3(6), 2–15.
6. Aquamid, Your beuty secret to keep (n.d) retrieved from https://www.aquamid.com/about-aquamid-patients/ (accessed on 22 February 2020)
7. Kollidon-CL (n.d) retrieved from https://pharmaceutical.basf.com/en/Drug-Formulation/Kollidon-CL.html (accessed on 26 May 2019)
8. Primellose (n.d) retrieved from https://www.dfepharma.com/Excipients/Expertise/Oral-Solid-Dose/Superdisintegrants/Primellose (accessed on 26 May 2019)
9. Sterile ophthalmic gel forming solution Timoptic-XE (2008, October) retrieved from https://www.merck.com/product/usa/pi_circulars/t/timoptic/timoptic_xe_pi.pdf (accessed on 26 May 2019)
10. Tricor (n.d) retrieved from https://www.rxlist.com/tricor-drug.htm#description (accessed on 26 May 2019)
11. Airoptix, Night & Day Aqua contact lenses (n.d) retrieved from https://airoptix.myalcon.com/contact-lenses/air-optix/products/air-optix-night-and-day/ (accessed on 19 March 2020).
12. Laftah, W. A., Hashim, S., & Ibrahim, A. N. (2011). Polymer hydrogels: a review. Polymer-Plastics Technology and Engineering, 50(14), 1475–1486.
13. Ahmed, E. M. (2015). Hydrogel: preparation, characterization, and applications: a review. Journal of Advanced Research, 6(2), 105–121.
14. Singh, S. K., Dhyani, A., & Juyal, D. (2017). Hydrogel: preparation, characterization and applications. The Pharma Innovation, 6 (6, Part A), 25.
15. Laftah, W. A., Hashim, S., & Ibrahim, A. N. (2011). Polymer hydrogels: a

review. Polymer-Plastics Technology and Engineering, 50(14), 1475–1486.

16. Omidian, H., Roccaa, J. G., & Park, K., (2005). Advances in superporous hydrogels. Journal of Controlled Release, 102, 3–12.

17. Qiu, Y., & Park, K. (2001). Environment-sensitive hydrogels for drug delivery. Advanced Drug Delivery Reviews, 53(3), 321–339.

18. An, Y. M., Liu, T., Tian, R., Liu, S. X., Han, Y. N., Wang, Q. Q., & Sheng, W. J. (2015). Synthesis of novel temperature responsive PEG-b-[PCL-gP (MEO2MA-co-OEGMA)]-b-PEG (tBG) triblock-graft copolymers and preparation of tBG/graphene oxide composite hydrogels via click chemistry. Reactive and Functional Polymers, 94, 1–8.

19. Varaprasad, K., & Sadiku, R. (2015). Development of microbial protective Kolliphor-based nanocomposite hydrogels. Journal of Applied Polymer Science, 132(46), 42781–42784.

20. Zhong, Y., Wang, J., Yuan, Z., Wang, Y., Xi, Z., Li, L., Liu, Z., & Guo, X. (2019). A mussel-inspired carboxymethyl cellulose hydrogel with enhanced adhesiveness through enzymatic cross-linking. Colloids and Surfaces B: Biointerfaces, 179, 462–469.

21. Tran, T. H., Okabe, H., Hidaka, Y., & Hara, K. (2017). Removal of metal ions from aqueous solutions using carboxymethyl cellulose/sodium styrene sulfonate gels prepared by radiation grafting. Carbohydrate Polymers, 157, 335–343.

22. Schulze, J., Hendrikx, S., Schulz-Siegmund, M., & Aigner, A. (2016). Microparticulate poly(vinyl alcohol) hydrogel formulations for embedding and controlled release of polyethylenimine (PEI)-based nanoparticles. Acta Biomaterialia, 45, 210–222.

23. Nagam, S. P., Jyothi, A. N., Poojitha, J., Aruna, S. A. N. T. H. O. S. H., &

Nadendla, R. R. (2016). A comprehensive review on hydrogels. International Journal of Current Pharmaceutical Research, 8(1), 19–23.

24. Jacqueline I Kroschwitz, H. F. Mark (Eds.). (2003). Encyclopedia of Polymer Science and Technology, Wiley-Interscience: Hoboken, NJ, Vol 2, 691–722.

25. Varaprasad, K., Raghavendra, G. M., Jayaramudu, T., Yallapu, M. M., & Sadiku, R. (2017). A mini review on hydrogels classification and recent developments in miscellaneous applications. Materials Science and Engineering: C, 79, 958–971.

26. Rosiak, J. M., & Ulański, P. (1999). Synthesis of hydrogels by irradiation of polymers in aqueous solution. Radiation Physics and Chemistry, 55(2), 139–151.

27. Hoffman, A. S. (1987). Applications of thermally reversible polymers and hydrogels in therapeutics and diagnostics. Journal of Controlled Release, 6(1), 297–305.

28. Hoffman, A. S. (2012). Hydrogels for biomedical applications. Advanced Drug Delivery Reviews, 64, 18–23.

29. Abd El-Mohdy, H. L., & Hegazy, E. S. A. (2008). Preparation of polyvinyl pyrrolidone-based hydrogels by radiation-induced crosslinking with potential application as wound dressing. Journal of Macromolecular Science, Part A, 45(12), 995-1002.

30. Chourasia, M. K., & Jain, S. K. (2004). Polysaccharides for colon targeted drug delivery. Drug Delivery, 11(2), 129–148.

31. Makuuchi, K., & Cheng, S. (2012). Radiation processing of polymer materials and its industrial applications. John Wiley & Sons: Hoboken, NJ.

32. Kharisov, B. I., Kharissova, O. V., & Méndez, U. O. (Eds.). (2016). Radiation Synthesis of Materials and

Compounds. CRC Press: New York, NY.

33. Charlesby, A., & Alexander, P. (1955). Reticulation of polymers in aqueous solution by gamma rays. Chimie Physique Rev. Gen. Colloids, 52.

34. Gerlach, G., & Arndt, K. F. (Eds.). (2009). Hydrogel Sensors and Actuators: Engineering and Technology (Vol. 6). Springer Science & Business Media: Berlin.

35. Carenza, M. (1992). Recent achievements in the use of radiation polymerization and grafting for biomedical applications. International Journal of Radiation Applications and Instrumentation. Part C. Radiation Physics and Chemistry, 39(6), 485–493.

36. Andrade, J. D. (Ed.). (1976). Hydrogels for medical and related applications. American Chemical Society, Washington, DC.

37. Peppas, N. A. (2010). Biomedical Applications of Hydrogels Handbook. Springer Science & Business Media: Berlin.

38. Rimmer, S. (Ed.). (2011). Biomedical Hydrogels: Biochemistry, Manufacture and Medical Applications. Elsevier: London.

39. Şen, M., & Avcı, E. N. (2005). Radiation synthesis of poly(N-vinyl-2-pyrrolidone)-κ-carrageenan hydrogels and their use in wound dressing applications. I. Preliminary laboratory tests. Journal of Biomedical Materials Research Part A, 74(2), 187–196.

40. Wang, M., Xu, L., Hu, H., Zhai, M., Peng, J., Nho, Y., Li, J., & Wei, G. (2007). Radiation synthesis of PVP/CMC hydrogels as wound dressing. Nuclear Instruments and Methods in Physics Research Section B: Beam Interactions with Materials and Atoms, 265(1), 385–389.

41. González-Henríquez, C. M., Sarabia-Vallejos, M. A., & Rodriguez-Hernandez, J. (2017). Advances in the fabrication of antimicrobial hydrogels for biomedical applications. Materials, 10(3), 232.

42. Arslan, M., Saraydin, D., Öztop, A. Y., & Şahiner, N. (2017). Radiation-induced acrylamide/4-Vinyl pyridine biocidal hydrogels: synthesis, characterization, and antimicrobial activities. Polymer-Plastics Technology and Engineering, 56(12), 1295–1306.

43. El-Arnaouty, M. B., Ghaffar, A. M., Aboulfotouh, M. E., Taher, N. H., & Taha, A. A. (2015). Radiation synthesis and characterization of poly(butyl methacrylate/acrylamide) copolymeric hydrogels and heparin controlled drug release. Polymer Bulletin, 72(11), 2739–2756.

44. Riedel, S., Hietschold, P., Krömmelbein, C., Kunschmann, T., Konieczny, R., Knolle, W., & Mayr, S. G. (2019). Design of biomimetic collagen matrices by reagent-free electron beam induced crosslinking: structure–property relationships and cellular response. Materials and Design, 168, 107606.

45. An, S., Jeong, S., & Khil, M. (2018). Preparation and evaluation of β-glucan hydrogel prepared by the radiation technique for drug carrier applications. International Journal of Biological Macromolecules, 118, 333–339.

46. González-Torres, M., Leyva-Gómez, G., Rivera, M., Krötzsch, E., Rodríguez-Talavera, R., Rivera, A. L., & Cabrera-Wrooman, A. (2018). Biological activity of radiation-induced collagen–polyvinylpyrrolidone–PEG hydrogels. Materials Letters, 214, 224–227.

47. Taşdelen, B., Erdoğan, S., & Bekar, B. (2018). Radiation synthesis and characterization of chitosan/hyaluronic acid/hydroxyapatite hydrogels: drug uptake and drug delivery systems. Materials Today: Proceedings, 5(8), 15990–15997.

48. Singh, B., & Kumar, A. (2018). Network formation of Moringaoleifera gum by radiation induced crosslinking:

evaluation of drug delivery, network parameters and biomedical properties. International Journal of Biological Macromolecules, 108, 477–488.

49. Singh, B., & Kumar, A. (2018). Radiation–induced graft copolymerization of N-vinyl imidazole onto moringa gum polysaccharide for making hydrogels for biomedical applications. International Journal of Biological Macromolecules, 120, 1369–1378.

50. Singh, B., & Kumar, A. (2018). Hydrogel formation by radiation induced crosslinked copolymerization of acrylamide onto moringa gum for use in drug delivery applications. Carbohydrate Polymers, 200, 262–270.

51. Singh, B., & Rajneesh. (2018). Gamma radiation synthesis and characterization of gentamicin loaded polysaccharide gum based hydrogel wound dressings. Journal of Drug Delivery Science and Technology, 47, 200–208.

52. Singh, B., & Singh, B. (2019). Developing a drug delivery carrier from natural polysaccharide exudate gum by graft-copolymerization reaction using high energy radiations. International Journal of Biological Macromolecules, 127, 450–459.

53. Singh, B., & Singh, B. (2018). Modification of sterculia gum polysaccharide via network formation by radiation induced crosslinking polymerization for biomedical applications. International Journal of Biological Macromolecules, 116, 91–99.

54. R.M. Ottenbrite et al. (eds.). (2010). Biomedical Applications of Hydrogels Handbook. Springer Science Business Media: Berlin.

55. El-Din, H. M. N., & El-Naggar, A. W. M. (2012). Radiation synthesis of acrylic acid/polyethyleneimine interpenetrating polymer networks (IPNs) hydrogels and its application as a carrier of atorvastatin drug for controlling

cholesterol. European Polymer Journal, 48(9), 1632–1640.

56. El-Din, H. M., El-Naggar, A. W., & Fadle, F. I. (2013). Radiation synthesis of pH-sensitive hydrogels from carboxymethyl cellulose/poly(ethylene oxide) blends as drug delivery systems. International Journal of Polymeric Materials, 62(13), 711–718.

57. Taleb, M. F. (2013). Radiation synthesis of multifunctional polymeric hydrogels for oral delivery of insulin. International Journal of Biological Macromolecules, 62, 341–347.

58. Chen, J., Rong, L., Lin, H., Xiao, R., & Wu, H. (2009). Radiation synthesis of pH-sensitive hydrogels from β-cyclodextrin-grafted PEG and acrylic acid for drug delivery. Materials Chemistry and Physics, 116(1), 148–152.

59. Suljovrujic, E., Miladinovic, Z. R., Micic, M., Suljovrujic, D., & Milicevic, D. (2019). The influence of monomer/solvent feed ratio on POEGDMA thermoresponsive hydrogels: radiation-induced synthesis, swelling properties and VPTT. Radiation Physics and Chemistry, 158, 37–45.

60. Hafeez, S., Islam, A., Gull, N., Ali, A., Khan, S. M., Zia, S., & Jamil, T. (2018). γ-Irradiated chitosan based injectable hydrogels for controlled release of drug (Montelukast sodium). International Journal of Biological Macromolecules, 114, 890–897.

61. El-Din, H. M., Alla, S. G., & El-Naggar, A. W. (2007). Radiation synthesis and characterization of hydrogels composed of poly(vinyl alcohol) and acrylamide mixtures. Journal of Macromolecular Science, Part A, 44(1), 47–54.

62. Piechocki, K., Kozanecki, M., Kadlubowski, S., Pacholczyk-Sienicka, B., Ulanski, P., & Biela, T. (2018). Controlling the properties of radiation-synthesized thermoresponsive oligo ether methacrylate hydrogels by varying

the monomer side-chain length; self-composite network containing crystalline phase. Polymer, 150, 275–288.

63. Branca, C., Auditore, L., Loria, D., Trimarchi, M., & Wanderlingh, U. (2012). Radiation synthesis and characterization of poly(ethylene oxide)/chitosan hydrogels. Journal of Applied Polymer Science, 127(1), 217–223.

64. Elliott, J. (2004). Structure and swelling of poly(acrylic acid) hydrogels: effect of pH, ionic strength, and dilution on the crosslinked polymer structure. Polymer, 45(5), 1503–1510.

65. Yang, M., Song, H., Zhu, C., & He, S. (2007). Radiation synthesis and characterization of polyacrylic acid hydrogels. Nuclear Science and Techniques, 18(2), 82–85.

66. Cioffi, N., Torsi, L., Ditaranto, N., Tantillo, G., Ghibelli, L., Sabbatini, L., Bleve-Zacheo, T., D'Alessio, M., Zambonin, P.G., & Traversa, E. (2005). Copper nanoparticle/polymer composites with antifungal and bacteriostatic properties. Chemistry of Materials, 17, 5255–5262.

67. Zaporojtchenko, V., Podschun, R., Schurmann, U., Kulkarni, A., & Faupel, F. (2006). Physico-chemical and antimicrobial properties of co-sputtered Ag-Au/PTFE nanocomposite coatings. Nanotechnology, 17, 4904–4908.

68. Dowling, D. P., Betts, A. J., Pope, C., McConnell, M. L., Eloy, R., & Arnaud, M. N. (2003). Anti-bacterial silver coatings exhibiting enhanced activity through the addition of platinum. Surface and Coatings Technology, 163, 637–640.

69. Petica, A., Gavriliu, S., Lungu, M., Buruntea, N., & Panzaru, C. (2008). Colloidal silver solutions with antimicrobial properties. Materials Science and Engineering: B, 152(1-3), 22–27.

70. Vittal Ravishankar Bai Aswathanarayan Jamuna. (2011). Nanoparticles

and their potential application as antimicrobials. Méndez-Vilas, A. Ed. In: Science Against Microbial Pathogens: Communicating Current Research and Technological Advances. Formatex Research Center: Badajoz.

71. Ebrahim, S., Usha, K., Singh, B., & Mendez-Vilas, A. (2011). Science Against Microbial Pathogens: Communicating Current Research and Technological Advances. Formatex Research Center: Badajoz, 1043–1054.

72. Eid, M., El-Arnaouty, M. B., Salah, M., Soliman, E. S., & Hegazy, E. S. A. (2012). Radiation synthesis and characterization of poly(vinyl alcohol)/poly(N-vinyl-2-pyrrolidone) based hydrogels containing silver nanoparticles. Journal of Polymer Research, 19(3), 9835.

73. Rai, M., Yadav, A., & Gade, A. (2009). Silver nanoparticles as a new generation of antimicrobials. Biotechnology Advances, 27(1), 76–83.

74. Kim, J. S., Kuk, E., Yu, K. N., Kim, J. H., Park, S. J., Lee, H. J., Kim, S. H., Park, Y. K., Park, Y. H., Hwang, C.-Y., & Kim, Y. K. (2007). Antimicrobial effects of silver nanoparticles. Nanomedicine: Nanotechnology, Biology and Medicine, 3(1), 95–101.

75. Vashist, A., Kaushik, A., Ghosal, A., Bala, J., Nikkhah-Moshaie, R., A Wani, W., Manickam, P., & Nair, M. (2018). Nanocomposite hydrogels: advances in nanofillers used for nanomedicine. Gels, 4(3), 75.

76. El-Mohdy, H. A. (2013). Radiation synthesis of nanosilver/polyvinyl alcohol/cellulose acetate/gelatin hydrogels for wound dressing. Journal of Polymer Research, 20(6), 177.

77. Hu, J., Liu, Z. S., Tang, S. L., & He, Y. M. (2007). Effect of hydroxyapatite nanoparticles on the growth and p53/c-Myc protein expression of implanted hepatic VX2 tumor in rabbits by

intravenous injection. World Journal of Gastroenterology, 13(20), 2798.

78. Micic, M., Milic, T. V., Mitric, M., Jokic, B., & Suljovrujic, E. (2013). Radiation synthesis, characterization and antimicrobial application of novel copolymeric silver/poly(2-hydroxy-ethyl methacrylate/itaconic acid) nanocomposite hydrogels. Polymer Bulletin, 70(12), 3347–3357.

79. Bauer, I. W., Li, S. P., Han, Y. C., Yuan, L., & Yin, M. Z. (2008). Internalization of hydroxyapatite nanoparticles in liver cancer cells. Journal of Materials Science: Materials in Medicine, 19(3), 1091–1095.

80. Pezzatini, S., Solito, R., Morbidelli, L., Lamponi, S., Boanini, E., Bigi, A., & Ziche, M. (2006). The effect of hydroxyapatite nanocrystals on microvascular endothelial cell viability and functions. Journal of Biomedical Materials Research Part A: An Official Journal of The Society for Biomaterials, The Japanese Society for Biomaterials, and The Australian Society for Biomaterials and the Korean Society for Biomaterials, 76(3), 656–663.

81. Li, J., Yin, Y., Yao, F., Zhang, L., & Yao, K. (2008). Effect of nano-and micro-hydroxyapatite/chitosan-gelatin network film on human gastric cancer cells. Materials Letters, 62(17-18), 3220–3223.

82. Taleb, M. F. A., Alkahtani, A., & Mohamed, S. K. (2015). Radiation synthesis and characterization of sodium alginate/chitosan/hydroxyapatite nanocomposite hydrogels: a drug delivery system for liver cancer. Polymer Bulletin, 72(4), 725–742.

83. Raafat, A. I., El-Sawy, N. M., Badawy, N. A., Mousa, E. A., & Mohamed, A. M. (2018). Radiation fabrication of Xanthan-based wound dressing hydrogels embedded ZnO nanoparticles: in vitro evaluation. International Journal of Biological Macromolecules, 118, 1892–1902.

84. Zohourian, M. M., & Kabiri, K. (2008). Superabsorbent polymer materials: a review, Iranian Polymer Journal, 17(6), 451–477.

85. Gonçalves, A. A. L., Fonseca, A. C., Fabela, I. G. P., Coelho, J. F. J., & Serra, A. C. (2016). Synthesis and characterization of high performance superabsorbent hydrogels using bis [2-(methacryloyloxy) ethyl] phosphate as crosslinker. Express Polymer Letters, 10(3), 248.

86. Fekete, T., Borsa, J., Takács, E., & Wojnárovits, L. (2014). Synthesis of cellulose derivative based superabsorbent hydrogels by radiation induced crosslinking. Cellulose, 21(6), 4157–4165.

87. Fekete, T., Borsa, J., Takács, E., & Wojnárovits, L. (2016). Synthesis of cellulose-based superabsorbent hydrogels by high-energy irradiation in the presence of crosslinking agent. Radiation Physics and Chemistry, 118, 114–119.

88. Budianto, E., Mahendra, A., & Yudianti, R. (2013). Radiation synthesis of superabsorbent poly(acrylamide-co-acrylic acid)-sodium alginate hydrogels. In Advanced Materials Research (Vol. 746, pp. 88-96). Trans Tech Publications Ltd.

89. Awadallah-F, A., & Mostafa, T. B. (2014). Synthesis and characterization studies of γ-radiation crosslinked poly(acrylic acid/2-acrylamido-2-methyl propane sulfonic acid) hydrogels. Journal of Polymer Engineering, 34(5), 459–469.

90. Tomar, R. S., Gupta, I., Singhal, R., & Nagpal, A. K. (2007). Synthesis of poly(acrylamide-co-acrylic acid)-based super-absorbent hydrogels by gamma radiation: study of swelling behaviour and network parameters. Designed Monomers and Polymers, 10(1), 49–66.

91. Wasikiewicz, J. M., Yoshii, F., Nagasawa, N., Wach, R. A., & Mitomo, H.
(2005). Degradation of chitosan and
sodium alginate by gamma radiation,
sonochemical and ultraviolet methods.
Radiation Physics and Chemistry,
73(5), 287–295.

92. Al-Assaf, S., Phillips, G. O., &
Williams, P. A. (2006). Controlling the
molecular structure of food hydrocolloids. Food Hydrocolloids, 20(2-3),
369–377.

93. Hayrabolulu, H., Şen, M., Çelik, G.,
& Kavaklı, P. A. (2014). Synthesis
of carboxylated locust bean gum
hydrogels by ionizing radiation.
Radiation Physics and Chemistry, 94,
240–244. doi:10.1016/j.radphyschem.
2013.05.048.

94. Villegas, G. M. E., Morselli, G. R.,
González-Pérez, G., & Lugão, A. B.
(2018). Enhancement swelling properties of PVGA hydrogel by alternative
radiation crosslinking route. Radiation
Physics and Chemistry, 153, 44–50.

95. Hong, T. T., Okabe, H., Hidaka, Y., &
Hara, K. (2018). Radiation synthesis
and characterization of super-absorbing
hydrogel from natural polymers and
vinyl monomer. Environmental Pollution, 242, 1458–1466.

96. Vashi, H., Iorhemen, O. T., & Tay, J. H.
(2018). Aerobic granulation: a recent
development on the biological treatment of pulp and paper wastewater.
Environmental Technology & Innovation, 9, 265–274.

97. Wang, X., Wang, Z., Chen, H., & Wu,
Z. (2017). Removal of Cu(II) ions
from contaminated waters using a
conducting microfiltration membrane.
Journal of Hazardous Materials, 339
(Supplement C), 182–190.

98. Wang, M., Payne, K. A., Tong, S., &
Ergas S. J. (2018). Hybrid algal photosynthesis and ion exchange (HAPIX)
process for high ammonium strength

wastewater treatment. Water Research,
142, 65–74.

99. Venkatesham, V., Mohsin Khan, Shri-
kanth, Y. N., Shashi Kumar, M. B., &
Manjunath, B. V. (2018). Removal of
heavy metals and dyes from wastewater using hydrogels. Asian Journal of
Applied Science and Technology, 2(2),
1046–1065.

100. Abdel-Aal, S. E., Hegazy, E. A., Taleb,
M. F., & Dessouki, A. M. (2007). Radiation synthesis and characterization of
2-hydroxyethyl-methacrylate-based
hydrogels containing di- and tri-protic
acid and its application on wastewater
treatment. Journal of Applied Polymer
Science, 107(3), 1759–1776.

101. Wang, M., Xu, L., Zhai, M., Peng, J.,
Li, J., & Wei, G. (2008). γ-ray radiation-induced synthesis and Fe(III)
ion adsorption of carboxymethylated
chitosan hydrogels. Carbohydrate
Polymers, 74(3), 498–503.

102. Bolisay, L. D., & Kofinas, P. (2010).
Imprinted polymer hydrogels for the
separation of viruses. Macromolecular
Symposia, 291-292(1), 302–306.

103. Vasapollo, G., Sole, R. D., Mergola, L.,
Lazzoi, M. R., Scardino, A., Scorrano,
S., & Mele, G. (2011). Molecularly
imprinted polymers: present and future
prospective. International Journal of
Molecular Sciences, 12(9), 5908–5945.

104. El-Arnaouty, M. B. (2010). Radiation
synthesis and characterization study of
imprinted hydrogels for metal ion adsorption. Polymer-Plastics Technology and
Engineering, 49(10), 963–971.

105. Mahmoud, G. A., Hegazy, E. A.,
Badway, N. A., Salam, K. M. M., &
Elbakery, S. M. (2016). Radiation
synthesis of imprinted hydrogels for
selective metal ions adsorption. Desali-
nation and Water Treatment, 57(35),
16540–16551.

106. Neri, R., Burillo, G., & Castillo-Rojas,
S. (2010). Gamma radiation synthesis

of comb-type graft hydrogels based on poly(acrylic acid) and 4-vinylpyridine. Journal of Radioanalytical and Nuclear Chemistry, 287(3), 787–793.

107. Mahmoud, G. A., Abdel-Aal, S. E., Badway, N. A., Farha, S. A., & Alshafei, E. A. (2013). Radiation synthesis and characterization of starch-based hydrogels for removal of acid dye. Starch-Stärke, 66(3–4), 400–408.

108. Bhuyan, M. M., Adala, O. B., Okabe, H., Hidaka, Y., & Hara, K. (2019). Selective adsorption of trivalent metal ions from multielement solution by using gamma radiation-induced pectin-acrylamide-(2-acrylamido-2-methyl-1-propanesulfonic acid) hydrogel. Journal of Environmental Chemical Engineering, 7(1), 102844.

109. Wongjaikham, W., Wongsawaeng, D., Hosemann, P., Kanokworakan, C., & Ratnitsai, V. (2018). Enhancement of uranium recovery from seawater using amidoximated polymer gel synthesized from radiation-polymerization and crosslinking of acrylonitrile and methacrylic acid monomers. Journal of Environmental Chemical Engineering, 6(2), 2768–2777.

CHAPTER 17

CHEMISTRY OF SCHIFF BASE SYNTHESIS AND THEIR APPLICATIONS: A GREENER APPROACH

PRITI YADAV[1], DEEPAK PODDAR[1], PURNIMA JAIN[1], AMIT KUMAR SINGH[2], and ANJANA SARKAR[1*]

[1]Department of Chemistry, Netaji Subhas Institute of Technology, Dwarka Sector 3, University of Delhi, New Delhi 110021, India

[2]Department of Chemistry, Dyal Singh College, University of Delhi, New Delhi 110021, India

*Corresponding author. E-mail: anjisarkar@gmail.com.

ABSTRACT

Schiff bases and their metal complexes are widely used in both industrial as well as in the biological significance. Schiff bases are formed by the condensation reaction of an amino group with the carbonyl compounds, which leads to the formation of imide groups ($-C=N-$). In the recent era, the macrocyclic ligands of Schiff bases in the inorganic chemistry are the motivating topic and continuously gaining popularity in the research for younger scientists due to their versatile applications in industries such as polymer, dyes as well as pharmaceutical. In this chapter, we focused on the greener approach to synthesize the Schiff base legends. Tremendously, well organized and simple green synthesis method of Schiff bases and their metal complexes are discussed in this chapter. The significance of green chemistry and efficient, practical techniques like microwave irradiation, water-based reaction, UV radiation, and simple mortal—pastel method (mechanochemistry), Ultrasound sonication method, sonication (sonochemistry), along with few conventional techniques were explored and discussed. Few of the characterization methods have also been discussed, such as

NMR and physicochemical studies. This chapter compiles the different aspects and principles of green chemistry in the Schiff base ligands and their metal complexes.

17.1   INTRODUCTION

Greener approaches for the synthesis of organic compounds are the need of an hour to save the earth. These types of approaches are beneficial to remove the dangerous and expensive solvents that are used to synthesize different types of organic compounds. In recent years, a wide range of green methods is observed for the synthesis of different types of organic compounds. Presently, fruit juice is obtained to use as a solvent that has an excellent ability in the synthesis of organic compounds [1–3]. There are different types of difficulties for the chemists to synthesize eco-friendly compounds using the green synthesis routes. They face the challenges such as finding a way for the reaction conditions of organic compounds in nontoxic ways, the selection of solvents which improve quality and quantity of the synthesized organic compounds, no toxic solvent are used by which their waste is not hazardous for environment and the chemicals used are also safe for the synthesis as well as the environment. Hence for the safety of mankind and the environment, it is better to use natural resources,

but at this time, there is a lack of natural resources, therefore "greener methods and technologies" are used to form the chemical which is safe for both mankind and the environment [4]. For the conversion in the organic synthesis reactions, the easiest way is the oxidation reactions. Therefore, this type of reaction is needful because, in other organic reactions, hazardous chemicals are used, which can also produce a huge amount of quantity of by-products that are not as kind to nature. At the opposite in the oxidation reactions, oxygen is used which present in the air, thus this type of synthesis reaction is called aerobic oxygenation which is not harmful to both economy as well as the environment. Therefore, natural oxygen used in these processes, and the water was released as by-products, which are useful in many ways and not hazardous. Hence, the aerobic oxygenations reactions are the best greener routes for the synthesis of organic compounds. Benzothiazoles are the important organic compounds that are most useful in the pharmaceutical due to their significant curing and biological applications. As we know that Alzheimer's disease was a type of dementia, which is a worldwide issue due to which around 15 million peoples were affected by the disease. The ratio was doubled every 20 years. Benzothiazoles derivatives are used in the therapy of Alzheimer's disease due to their medicinal applications. The benzothiazoles derivative can also be used in the treatment of cancer disease because due to their curative applications these derivatives are used in the anti-tumor drugs [5–7]. Schiff bases are the aromatic imines having nitrogen atom that shows remarkable applications in many areas. These are the organic compounds that are formed by various type reactions in which hazardous chemical is used, which is a serious issue for our nature. But now in days, these types of imines can also be synthesized by eco-friendly methods. As we see that these types of imines can also be synthesized by solvent-free reactions by using the catalysts like p-toluidine, o-vanillin, CaO, CeCl$_3$ under microwave conditions, and a huge amount of products is formed. Due to such type of synthesis processes, a wide range of Schiff bases are formed. These are the greener routes in which natural catalysts are used and through which no dangerous chemicals are used or produced by the reactions [8–13]. In the present situation, there is a need for greener methods of synthesis to reduce the environmental influence and minimize hazardous raw materials. This can be possible by using natural and less toxic solvents. There are some proposals for the greener routes or methods using for the synthesis of organic compounds, as shown in Figure 17.1.

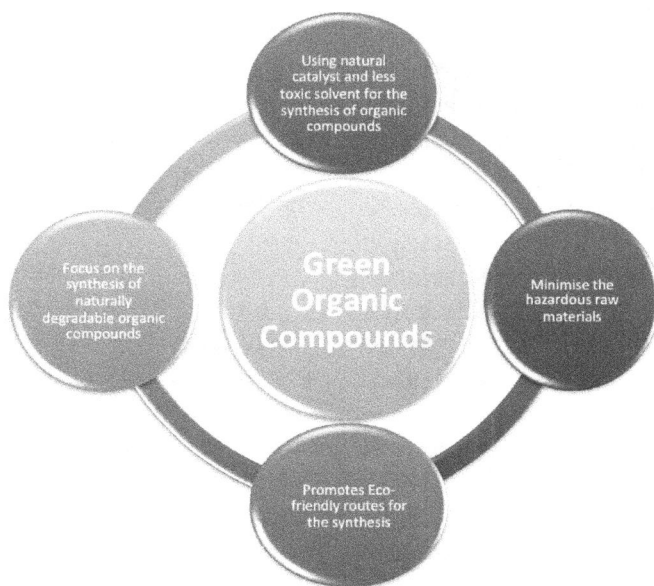

FIGURE 17.1 Few approaches developed for the synthesis of organic compounds using the greener method.

This chapter discussed the use of green methods for the synthesis of imines organic compounds such as by solvent-free processes, using natural solvents like fruit juices, aerobic oxidations, etc. This chapter also discussed the different types of natural synthesized Schiff bases and their remarkable application in pharmaceutical and industrials and the purpose of making these eco-friendly organic compounds.

17.2   NATURAL CATALYSTS USED IN THE SYNTHESIS OF ORGANIC COMPOUNDS

In the present scenario of the environment, we all worried about it due to this a need for the development of greener or natural processes that are eco-friendly in nature, and even hazardous raw materials are not desirable. Hence for the development of the new synthesis processes, an ecological point of view must be taken for them. In green chemistry, the fruits are the natural catalysts that are very useful in the synthesis of organic compounds and can also get attraction and interest of young scientist toward them and provide a solvent-free synthesis reaction. This chapter discusses various types of fruit juices that are used as a catalyst in the conversion and formation of imines compounds and

can also promote the use of juices as a catalyst. In the organic synthesis coconut, pineapple, lemon, star fruit, and tamarind are thoroughly used in the organic synthesis. Using the fruit juices in the organic synthesis is also cost-effective because fruits are easily available in the market, and their extraction of juice is also very simple. The various types of fruits are used as catalysts which are given below.

17.2.1   STAR FRUIT JUICE

It can also be called carambola which is sweet with slightly sour in taste. It has a star-like shape and yellow or green in color. Star fruit juice is useful in the synthesis of organic compounds and easily available. It is catalyzed the Knoevenagel condensation and is the major beneficial method in the formation of a carbon–carbon double bond and the example has been shown in Figure 17.2.

Due to this method, a mechanism of the reaction of Knoevenagel condensation can also be predicted as shown in Figure 17.3. The environment-friendly way for the condensation reaction of an aldehyde with malononitrile is used in visible light. It is a short-time taken reaction, which produces a huge amount of yield and can easily be separated by the simple filtration process [14].

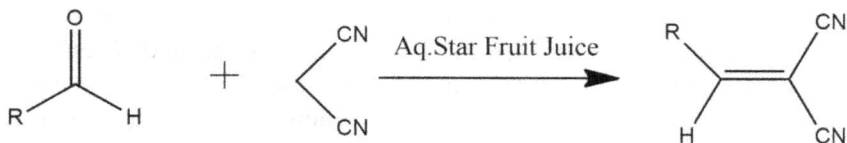

FIGURE 17.2   Knoevenagel condensation under visible light catalyzed by star fruit [14].

Mechanism:

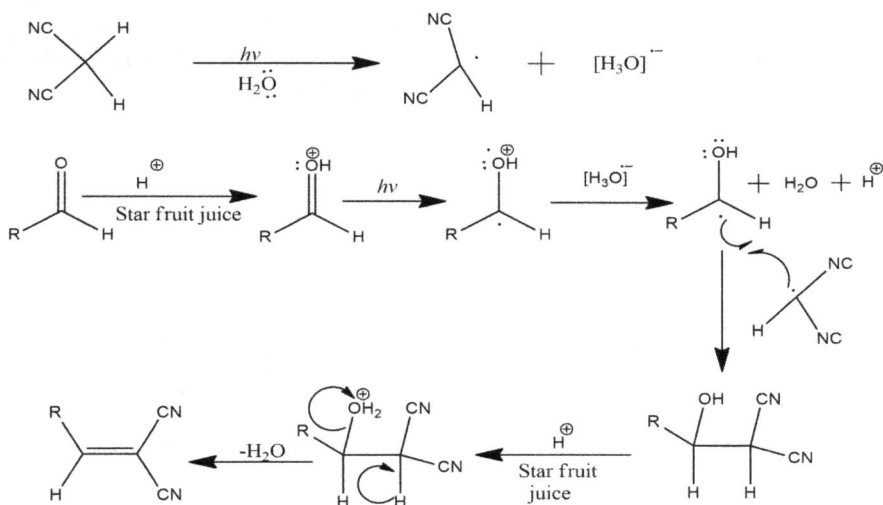

FIGURE 17.3   Plausible mechanistic pathway for the photochemical Knoevenagel conden-sation of aldehydes and malononitrile catalyzed by starfruit juice [14].

17.2.2   LEMON JUICE

Lemon is a small green and yellow in color fruit, which is sour in taste. It is a familiar fruit that is used in many ways, like for medicines, culinary at home, and also for industrial purposes. In the present day, lemon juice plays a significant role in organic synthesis reactions such as shown in Figure 17.4. Lemon juice is used in the condensation reaction

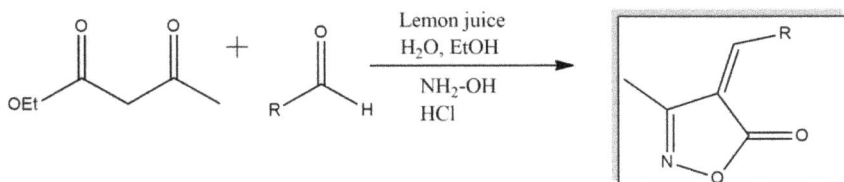

FIGURE 17.4   Reaction involves lemon juice as a catalyst in Schiff base synthesis [15].

of an aromatic aldehyde with a methylene group that is a type of Knoevenagel condensation reaction [15, 16] as shown in Figure 17.5.

17.2.3   TAMARIND FRUIT JUICE

Tamarind is a pod-like edible fruit that is sour-sweet in taste. The fruit juice is useful for metal polishes and in traditional medicines. It can also be used in organic synthesis to promote a greener approach like bis-, tris-(indolyl) methane, and di-bis-(indolyl) methane. These are formed by the pot condensation of indoles and substituted aldehydes and are naturally catalyzed by tamarind fruit juice [17] as shown in Figure 17.6.

Mechanism:

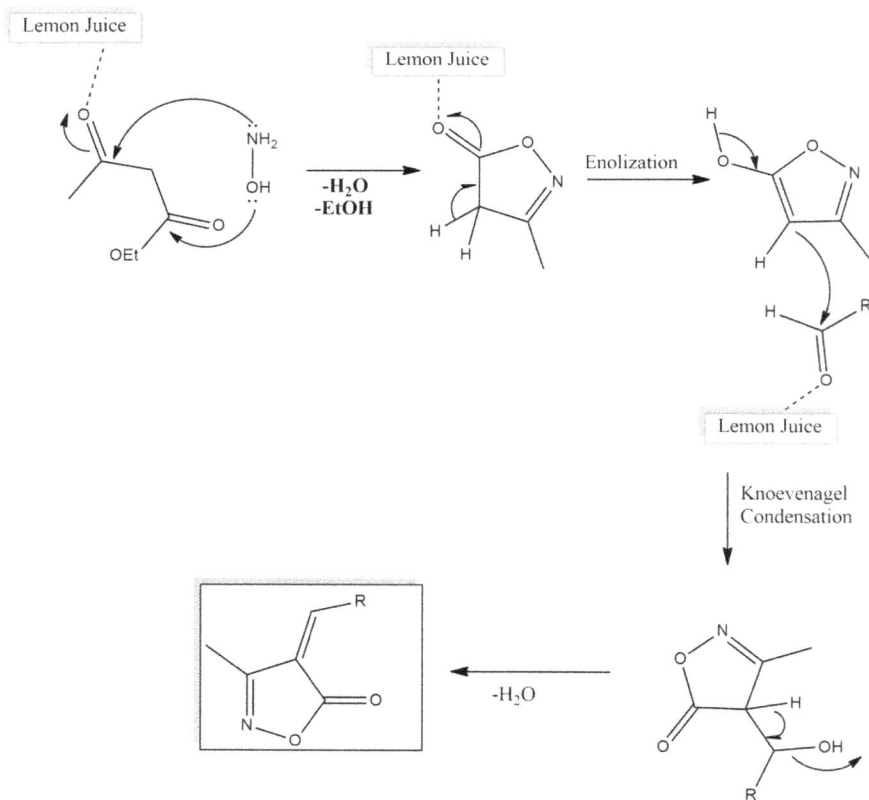

FIGURE 17.5   Reaction mechanism showing the role of lemon juice in reaction [16].

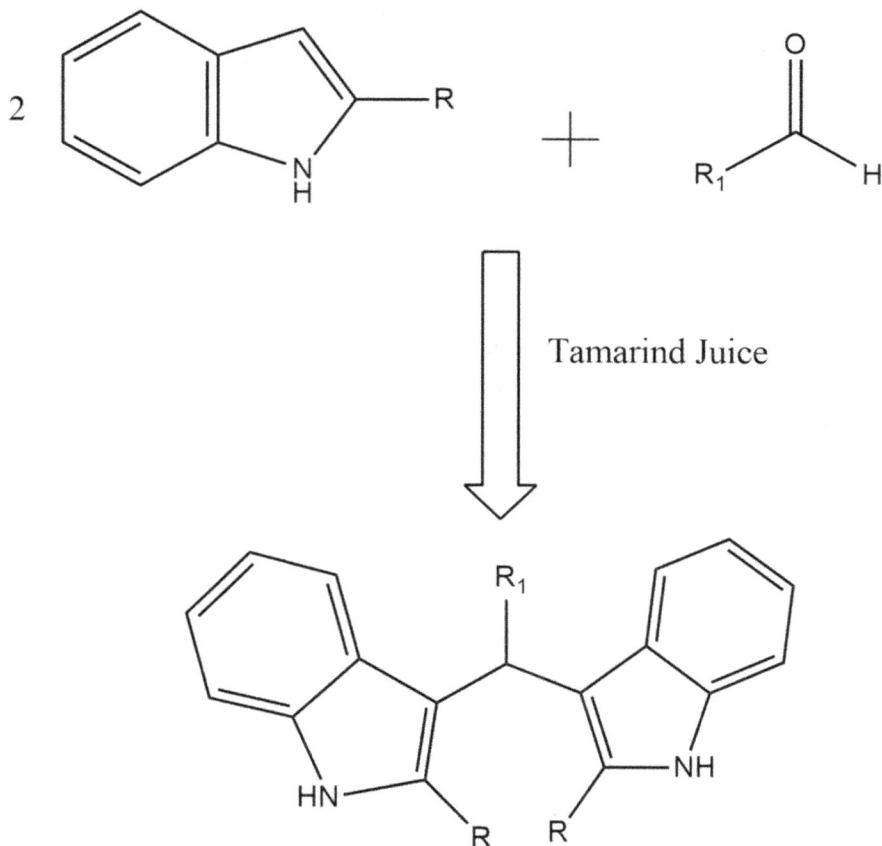

FIGURE 17.6   Reaction involves tamarind juice as a catalyst in Schiff base synthesis [17].

17.2.4   *PINEAPPLE FRUIT JUICE*

Sometimes pineapple is also called a king of fruit and is a delicious and healthy fruit. It is a pack of nutrients, antioxidants, and includes enzymes that are fighting inflammation and many types of disease. It can also be beneficial for health like boosting immunity, helpful in the recovery of surgery, digestion due to having digestive enzymes, etc. Pineapple juice is a natural catalyst that shows remarkable application in the synthesis of organic compounds and produces a good amount yield. It can be used to catalyze a number of dihydroxy-(rimidinone) (DHPMs) derivative. This is a solvent-free method that is nonpolluting, which has a number of benefits like these are the less time-consuming reactions, simple synthesis method, and help to reduce the environmental impact [18] as shown in Figure 17.7.

17.2.5   *SAPINDUM TRIFOLISTUS FRUIT JUICE*

The fruit was leathery, yellow ripening blackish having three seeds. The shell of the fruit is red color, and the color becomes dark when they are dried. Its species are used in food plants by the larvae of moths and butterflies. It can also be used as a dyeing agent for the coloring of cotton and silk. They can also show remarkable pharmaceutical applications. *S. trifolistus* reaction involves tamarind juice as a catalyst in Schiff base synthesis as shown in Figure 17.8.

Fruit juice was catalyzed by the chemo-selective organic synthesis. They are used as a catalyst in the synthesis of aldimines by aromatic aldehydes and amines. This is a safer and greener route of organic synthesis that has not environmental impact [19].

17.2.6   *COCONUT FRUIT JUICE*

Coconut juice is a fruit of *Cocos nucifera* and used as refreshing

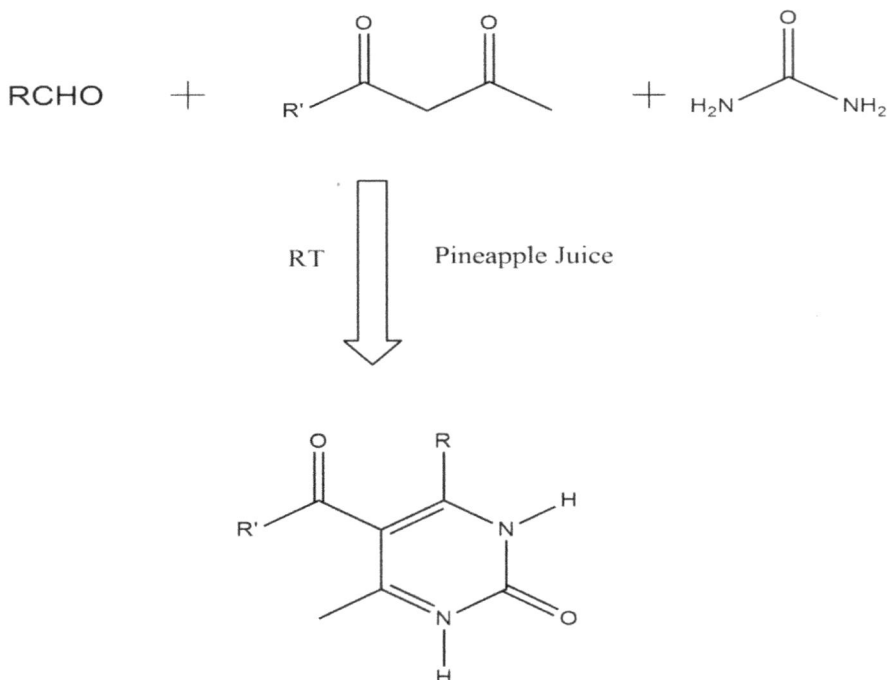

RCHO   +   [chemical structure]   +   H$_2$N [chemical structure] NH$_2$

RT   Pineapple Juice

[chemical structure]

FIGURE 17.7   Reaction involves pineapple juice as a catalyst in Schiff base synthesis [18].

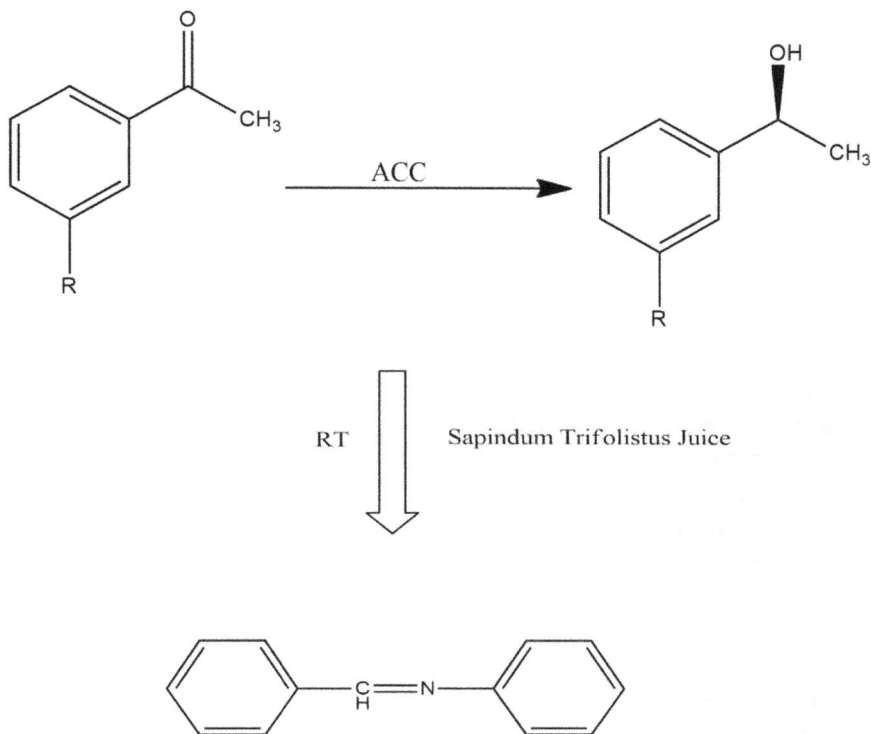

FIGURE 17.8   Reaction involves *S. trifolistus* juice as a catalyst in Schiff base synthesis [19].

beverages. Its juice is generally called "coconut water" and is a soft drink of flavorful or sweet. Its juice contains antioxidants and is used for the treatment of hypertension and high blood pressure and diarrhea-related dehydration. The coconut juice is also performing as biocatalysts in organic synthesis reactions. It can also hydrolyze the esters, amides, and anilides under mild conditions. Coconut juice was used as a catalyst in the reduction of aromatic and aliphatic carbonyl compounds. Its juice was also used in the reduction of a series of ketones and aldehyde groups. In this method, all ketones, aldehydes, aromatic, and aliphatic compounds are treated with fresh coconut juice called ACC (água-de-coco do Ceará) [20] as shown in Figure 17.9.

17.3   GREENER METHODS USED FOR THE SYNTHESIS

In the current scenario where environmental pollution increases in drastic ways, there is a demand for

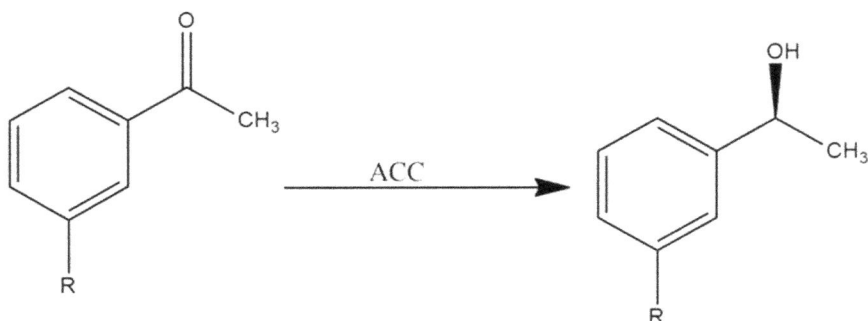

FIGURE 17.9   Reaction involves tamarind juice as a catalyst in Schiff base synthesis [20].

the development of greener and safe resources to reduce the harmful effects of chemicals on the environment as well as the living ones. Nowadays, pollution is a global issue, and it is necessary to include it in the point of view of all over processes. Hence, it is necessary to develop nonhazardous, innovative environmentally safe, and green synthesis processes, a selective improved solvent that is eco-friendly; natural catalysts and safer chemicals are used and reduced the formation of by-products. The greener methods are used for the synthesis of the imine group are as follows:

1) microwave irradiation
2) ultrasonication or sonochemistry
3) phase transfer catalysis
4) ionic liquids.

17.3.1   MICROWAVE IRRADIATION

At present, need has reduced environmental impact. Several types of greener routes are used for the synthesis of imines compounds like microwave-assisted methods used for these processes. A number of imine group or Schiff bases are formed by the condensation of aromatic/heterocyclic of aldehyde and amines as shown in Figure 17.10. The microwave method is very important in the greener synthesis of organic

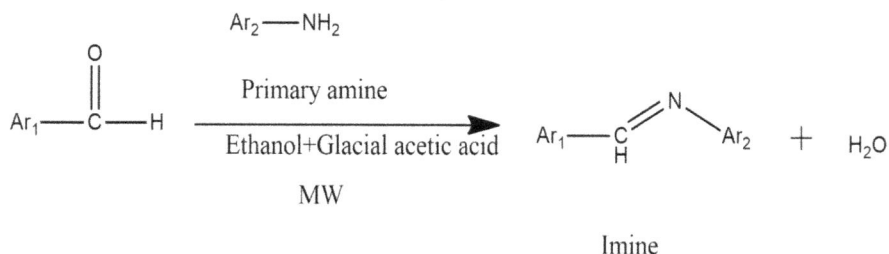

FIGURE 17.10   Microwave-assisted synthesis of imine compound or Schiff base [21].

compounds; it is occurring in mild reaction condition that is the very short time-taken method, and due to this, a huge amount of yield of product was formed. It is a source of nonconventional energy which has got the interest of young scientist for the activation of reactions and also for their various types of applications such as short time-consuming method, a huge amount of yield, excellent selectivity, simple workup, the cleaner reaction products, etc. [21–23].

17.3.2   ULTRASONICATION OR SONOCHEMISTRY

The ultrasonication method is a green and safer method, which was very convenient for the synthesis of organic synthesis of Schiff bases. This technique is more convenient as compare to the traditional methods such as hydrothermal autoclave of chemical synthesis as it shorten the time taken by the method and produces more yields. This was an easy and effective method for the synthesis of various types of thiazolidinones compounds as shown in Figure 17.11. In ultrasonication synthesis, the one-step and two-step methodology were used for the synthesis of Schiff bases and their derivatives. In one-step method, the whole reaction, from start to completion was carried using sonication method only whereas in two-step

method the use of sonication was done for synthesis and in second-step method such as evaporation was employed for the collection of the residue.

Heteroaryl was used in the formation of heteroaryl azetidinones under the ultrasonication irradiation method. It is a convenient and efficient method that speeds up the organic synthesis reaction in significant ways [24, 25].

17.3.3   PHASE TRANSFER CATALYSIS

It is a process in which the reactant migrates from one phase to another phase. Catalysts are useful tools for the removal of waste and also in the prevention of pollution. As we know the condition of the environment due to pollution and other reasons, so there is a need to make the green approaches for it. For the safety of the environment, two approaches are developed for the safety of the environment, which is "clean up" and another one "end of the pipe."

The "end of pipe" is used for the safety of the environment for many years due to its removal of hazardous chemicals and gases like ammonia, nitrogen oxide, and volatile organic compounds. The other approach "clean up" in which catalyst showed a remarkable effect in the production proceedings like clean up proceeding

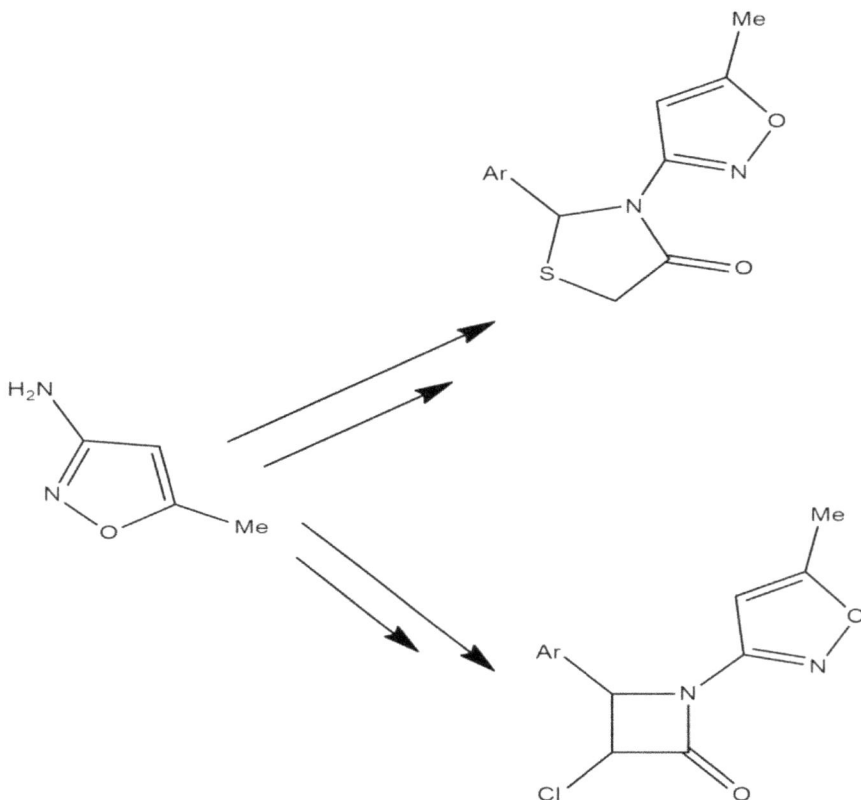

FIGURE 17.11   Synthesis of thiazolidinones compounds under ultrasonication [24].

for the environment. Hence, both the processes of these strategies are called "green chemistry," which was the base of the structure of chemical substances and proceedings that are useful to reduce or remove the uses of hazardous substances. As we took the examples, the different types of routes are used for the liquid-phase oxidation, such as the homolytic route as a radical route and the heterolytic are peroxometal route or oxometal route [26] as shown in Figure 17.12.

17.3.4   IONIC LIQUIDS METHODS

The ionic liquid method is the most advanced method that is used in the synthesis of organic compounds. It was a green and safe method for the synthesis in the current scenario. Ionic liquid has negligible vapor pressure due to this. They have minor toxicity than the other organic solvents. Therefore, they are called green solvent and are eco-friendly in

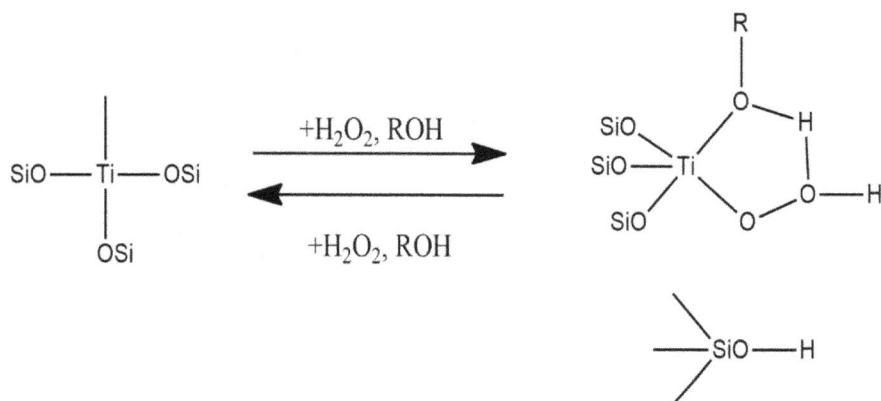

FIGURE 17.12   Formation of solvated peroxo species of titanium [26].

nature. They show remarkable applications like are electrolytes (electrically conducting fluids), sealants due to low vapor pressure. Quinoline synthesis is an example of ionic liquid methods because in quinoline synthesis, its condensation reaction (between isatin and ketones), imidazolium cation was used as a catalyst, which was a basic ionic liquid as shown in Figure 17.13. This method has many remarkable significances like aldol condensation, omission, and in this method, there is no need for transition metals as a catalyst in selectivity, it is a green method done in mild conditions. [27].

17.4   GREENER MOIETIES AND THEIR REMARKABLE APPLICATIONS

The increasing concern for the environment and the continued depletion of fossil fuels, renewable energy, and natural resources are pushing the research toward greener alternatives of conventional materials. The greener alternative of the Schiff base compounds is one of the intensively surveyed topics in the field of Schiff base synthesis and application. The greener alternatives for both the counterparts (NH and CO) in Schiff base molecules are vital from the

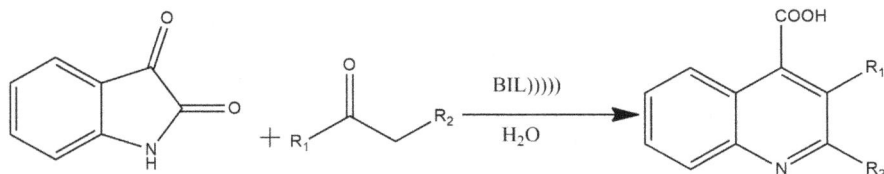

FIGURE 17.13   Example of ionic liquid use [27].

environmental point of view, and it provides the versatility in application. A lot of research has been conduction on the greener moieties, and few a huge portion of available compounds has been explored like amino acids, cellulose, chitosan, vanillin, and many more. We will be discussing the role of chitosan in the field of Schiff base and its potential applications.

17.4.1   CHITOSAN

The N-deacetylated, which was a product of chitin named chitosan, is formed by the units of n-acetyl-β-D glucosamine and β-D glucosamine linked with a 1,4-linkage, it is a natural biopolymer which gets the attention due to their significant properties like low toxicity, biocompatibility, biodegradability, immunological, etc. Hence, these types of natural products consider the various type of significance in all over branches such as in cosmetics, medicines, textiles, and also in the industrial areas. Chitosan have $-NH_2$ and $-OH$ groups due to which metal surface coordinate with it, and therefore it can show the anticorrosive properties [28].

17.4.1.1   BIOLUBRICANTS

There are several examples of the use of chitosan as an alternative of NH counterpart in the Schiff base

synthesis. Singh et al. have used the chitosan-based Schiff base by reacting to the chitosan with 3,5-di-*tert*-butyl-4-hydroxybenzaldehyde as shown in Figure 17.14. A modified (acylated) Schiff base also been synthesized by the addition of lauroyl chloride into the above reaction[29]. The use of the chitosan-based compound has been demonstrated in the field of green multifunctional lubricating oil additive or biolubricant against the conventional N-butyl palmitate/stearate, where the anticorrosive, antioxidant, antifriction, antiwar properties have been tested along with many more. Both the modification of the chitosan was characterized with FTIR, TGA, SEM XRD to ensure the synthesis and the thermal stability and solubility of acrylate chitosan Schiff base was found to be better along with the antifriction coefficient of the oil was also decreased from 0.104 to 0.084 and results suggested the possible use in biolubricants.

17.4.1.2   SELECTIVE SEPARATION

The use of the magnetic chitosan Schiff base has been demonstrated in the field of selective separation of mercury(II) [30]. For the formation of a magnetic Schiff base, chitosan was reacted with thiourea in the presence of Fe_3O_4 magnetite particle as shown in Figure 17.15 and the mixture was cross-linked with glutaraldehyde.

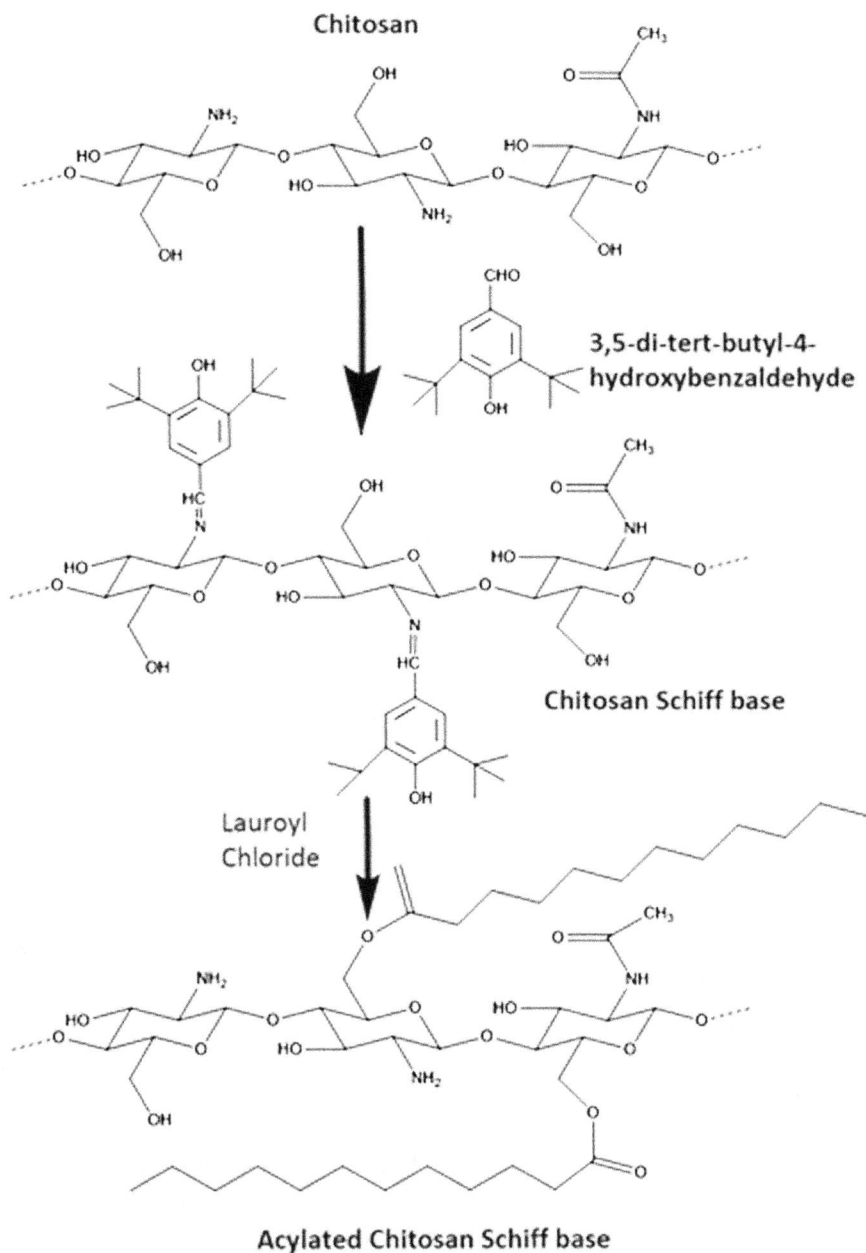

FIGURE 17.14 Scheme of synthesis of chitosan Schiff base using chitosan and 3,5-di-*tert*-butyl-4-hydroxybenzaldehyde [29].

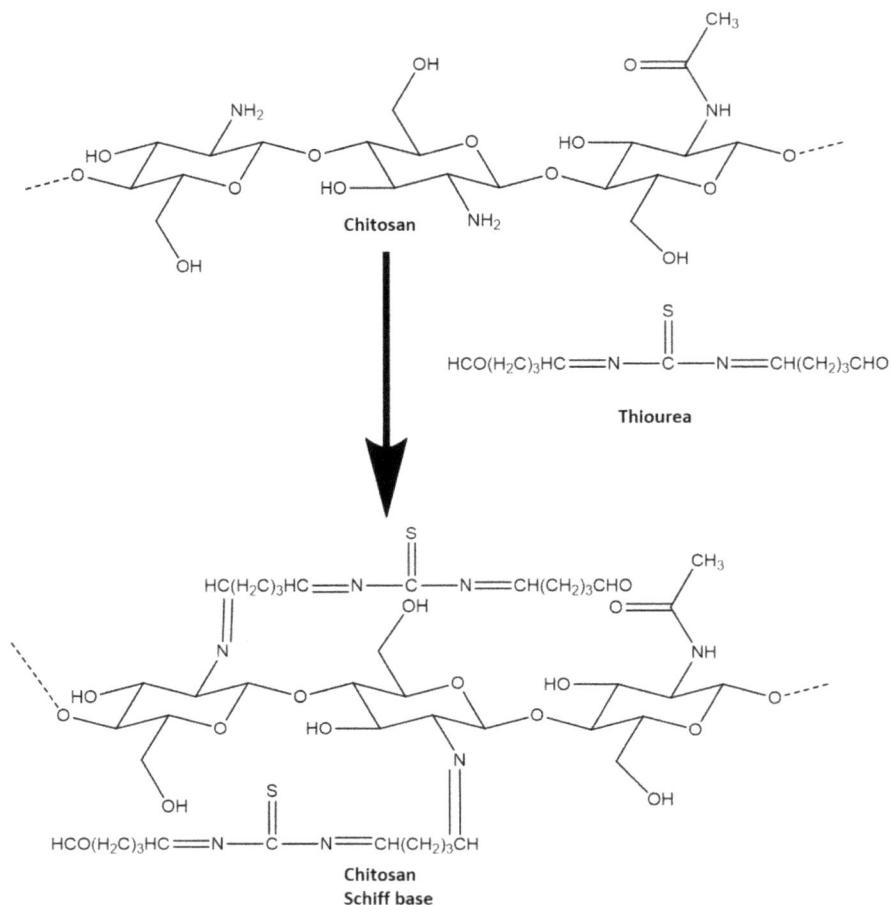

FIGURE 17.15   Scheme of synthesis of chitosan Schiff base using chitosan and thiourea [30].

The formed Schiff base was characterized using FTIR, and further Schiff was tested for the separation of Hg(II) in both batch and column experiments. The nature of the interaction of magnetic Schiff base was found to be dependent on the acidity of the medium in which the experiment has been performed and it has been found at pH = 1, magnetic Schiff base can selectively separate Hg(II) from the mixture of copper, lead, cadmium, zinc, and magnesium. The absorption process, according to Langmuir isotherm, has been found to be a pseudo-second-order exothermic, spontaneous reaction. The results also suggested the uptake efficiency of the compound is comparable to commercial Dowex-D3303.

17.4.1.3 ANTICORROSIVE APPLICATION

Another interesting application has been demonstrated when the chitosan was reacted with another natural compound (cinnamaldehyde) to form the Schiff base, which has a potential application in the acid corrosion industries. The storage and use of acids are very common in different applications like acid picking, cleaning, and descaling. The stored acid use of corroding the pipes and the passages. Chugh et al. had established the use of chitosan cinnamaldehyde Schiff base as a preventive agent for the corrosion of mild steel when the Schiff base was added into the storage container in the minimum quantity (200 ppm) [31] the scheme is shown in Figure 17.16. They have also shown the effect of the degree of substitution of cinnamaldehyde on the corrosion inhibition efficiency. The synthesized compound went through the series of characterization, FTIR, NMR, DSC, along with the efficiency of the compound in corrosion inhibition was estimated with the help of AFM, SEM-EDX, SCM, and EIS studies. Results suggested increasing the order of substitution at 200 ppm shows the improved results.

17.4.1.4 ANTIMICROBIAL ACTIVITY

Antimicrobial and antifungal activities are cover one of the major parts of the applications in the field of Schiff base. Chitosan is itself consider to be consisting of antimicrobial ability, but when it combines with another counterpart (carbonyl) for the synthesis of Schiff base, the process enhances the antimicrobial ability of the compound to a great extent. The author has synthesis the Schiff base by reacting to the chitosan with the citral under high-intensity ultrasound, the scheme is shown in Figure 17.17. The author also investigated the effect of the mole ratio of chitosan to citral along with the common factors like reaction time and temperature, and the yield was monitored against the all [32]. The result suggested the mole ratio is found to be 1:6 in addition with reaction temperature 50 °C and reaction time 10 h. The maximum yield obtained was 86.4%. The antimicrobial activities have been investigated against *Escherichia coli, Staphylococcus*, and *Aspergillus niger*. The series of the experiment suggested that the Schiff base showed the better antimicrobial activity compared to chitosan and the MICs of the Schiff base has found to 0.1% (w/v), 0.1% (w/v), and 0.5% (w/v) against *E. coli, S. aureus*, and *A. niger*, respectively.

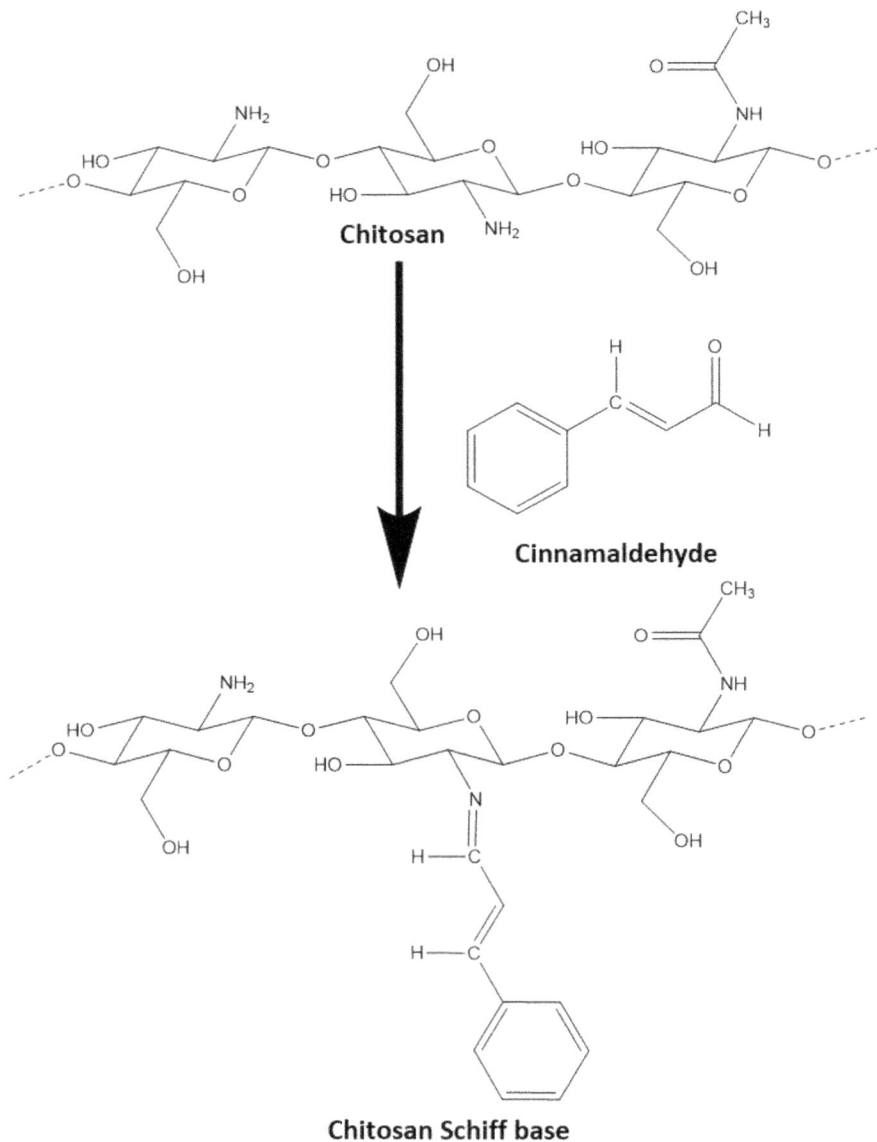

FIGURE 17.16   Scheme of synthesis of chitosan Schiff base using chitosan and cinnamal-dehyde [31].

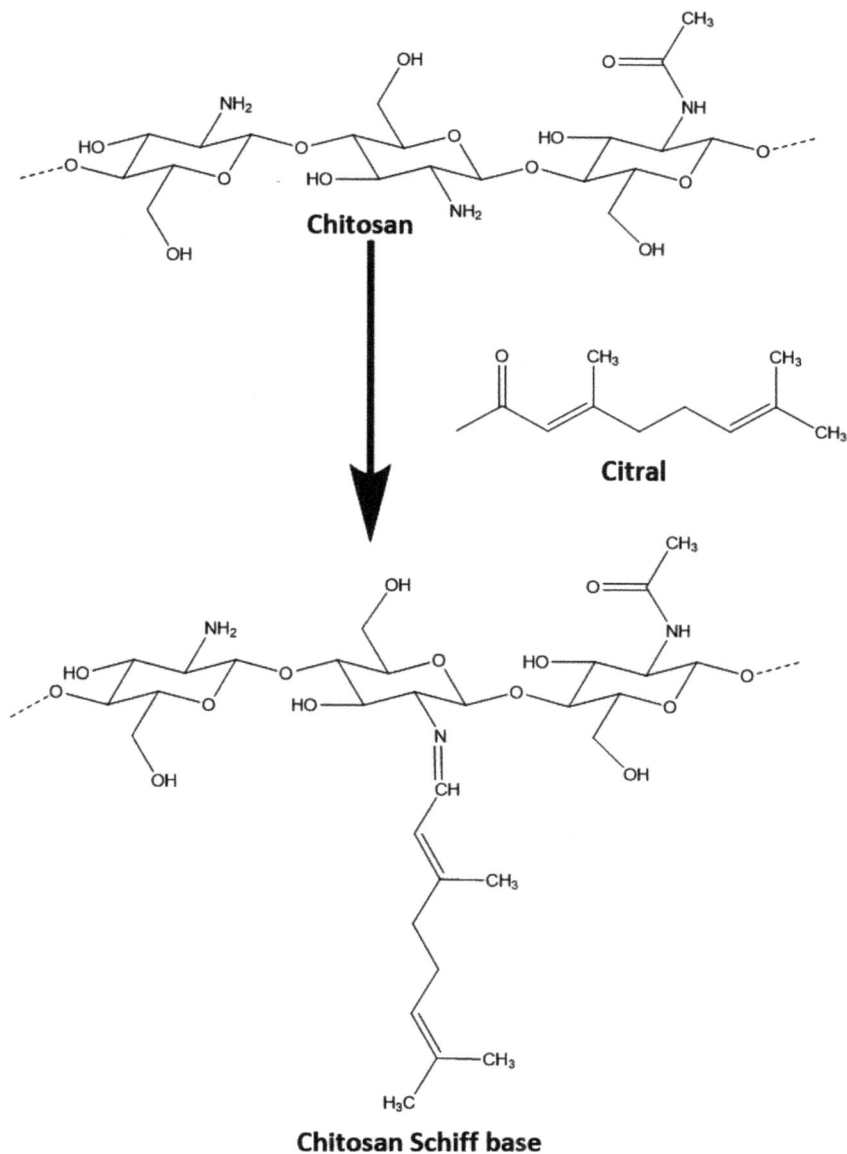

FIGURE 17.17 Scheme of synthesis of chitosan Schiff base using chitosan and citral [32].

17.5   CONCLUSION

Schiff bases form an important class of organic compounds that have widespread applications. Schiff base can adopt a greener approach in multiple ways for example one can start with greener moieties or adopt the synthesis methods which are green and do not leave side products that can be hazardous to the environment.

In this chapter, the author has summarized the synthesis process that can be involved in the preparation of Schiff bases citing examples from various proven literature by numerous scholars. In addition to this, there are various naturally occurring substances that can be used up as catalysts in the reaction for synthesis of Schiff bases. The chapter discussed a few of these naturally occurring catalysts stating their role in the reaction and mechanism involved.

There are many biodegradables naturally occurring precursors that help in the preparation of Schiff bases and one of them is chitosan which is widely used natural and biodegradable polymer and forms Schiff bases with distinguished properties. The application of chitosan has also been discussed in short emphasizing its importance in the preparation of the organic compounds.

It is hoped that this chapter though presented in concise form will help readers to know about the Schiff bases and its applications viz greener perspective.

KEYWORDS

- **Schiff base**
- **metal complexes**
- **green chemistry**
- **macrocyclic ligand**
- **pharmaceuticals**
- **microwave**
- **sonochemistry**
- **mechanochemistry**

REFERENCES

1. Bakht, M. A. (2015). Lemon juice catalyzed ultrasound assisted synthesis of Schiff's base: a total green approach. *Bulletin of Environment, Pharmacology and Life Sciences, 4*, 94–100.
2. Bendale, A. R., Bhatt, R., Nagar, A., Jadhav, A. G., & Vidyasagar, G. (2011). Schiff base synthesis by unconventional route: an innovative green approach. *Der Pharma Chemica, 3*(2), 34–38.
3. Mermer, A., Demirbas, N., Uslu, H., Demirbas, A., Ceylan, S., & Sirin, Y. (2019). Synthesis of novel Schiff bases using green chemistry techniques; antimicrobial, antioxidant, antiurease activity screening and molecular docking studies. *Journal of Molecular Structure, 1181*, 412–422.
4. Naqvi, A., Shahnawaaz, M., Rao, A. V., Seth, D. S., & Sharma, N. K. (2009). Synthesis of Schiff bases via environmentally benign and energy-efficient greener methodologies. *Journal of Chemistry, 6*(S1), S75–S78.
5. Fan, L. Y., Shang, Y. H., Li, X. X., & Hua, W. J. (2015). Yttrium-catalyzed heterocyclic formation via aerobic

oxygenation: a green approach to benzo-thiazoles. *Chinese Chemical Letters, 26* (1), 77–80.

6. Geng, J., Li, M., Wu, L., Ren, J., & Qu, X. (2012). Liberation of copper from amyloid plaques: making a risk factor useful for Alzheimer's disease treatment. *Journal of Medicinal Chemistry, 55*(21), 9146–9155.

7. Mortimer, C. G., Wells, G., Crochard, J. P., Stone, E. L., Bradshaw, T. D., Stevens, M. F., & Westwell, A. D. (2006). Antitumor benzothiazoles. 26. 2-(3,4-dimethoxyphenyl)-5-fluroben-zothiazole (GW 610, NSC 721648), a simple fluorinated 2-arylbenzothiazole, shows potent and selective inhibitory activity against lung, colon, and breast cancer cell lines. *Journal of Medicinal Chemistry, 49*(1), 179–185.

8. Ravishankar, L., Patwe, S. A., Gosarani, N., & Roy, A. (2010). Cerium(III)-cata-lyzed synthesis of Schiff bases: a green approach. *Synthetic Communications, 40*(21), 3177–3180.

9. Patil, S., Jadhav, S. D., & Shinde, S. K. (2012). CES as an efficient natural catalyst for synthesis of Schiff bases under solvent-free conditions: an inno-vative green approach. *Organic Chem-istry International*, Article ID 153159, 5 pages.

10. Thomas, A. B., Tupe, P. N., Badhe, R. V., Nanda, R. K., Kothapalli, L. P., Paradkar, O. D., Sharma, P. A., & Deshpande, A. D. (2009). Green route synthesis of Schiff's bases of isonico-tinic acid hydrazide. *Green Chemistry Letters and Reviews, 2*(1), 23–27.

11. Rao, V. K., Reddy, S. S., Krishna, B. S., Naidu, K. R. M., Raju, C. N., & Ghosh, S. K. (2010). Synthesis of Schiff's bases in aqueous medium: a green alternative approach with effec-tive mass yield and high reaction rates. *Green Chemistry Letters and Reviews, 3*(3), 217–223.

12. Gupta, N. K., Quraishi, M. A., Verma, C., & Mukherjee, A. K. (2016). Green Schiff's bases as corrosion inhibitors for mild steel in 1 M HCl solution: experi-mental and theoretical approach. *RSC Advances, 6*(104), 102076–102087.

13. Ganie, P. A., Wagay, N. A., & Bhat, A. R. Green synthetic route for novel derivatives of Schiff bases and their characterization. *International Journal of Advance Research in Science and Engineering, 7*, 2319–8354.

14. Pal, R., & Sarkar, T. (2014). Visible light induced Knoevenagel condensa-tion catalyzed by starfruit juice of Aver-rhoa carambola. *International Journal of Organic Chemistry, 4*, 105–115.

15. Deshmukh, M. B., Patil, S. S., Jadhav, S. D., & Pawar, P. B. (2012). Green approach for Knoevenagel condensa-tion of aromatic aldehydes with active methylene group. *Synthetic Communi-cations, 42*(8), 1177–1183.

16. Vekariya, R. H., Patel, K. D., & Patel, H. D. (2016). Fruit juice of citrus limon as a biodegradable and reusable cata-lyst for facile, eco-friendly and green synthesis of 3, 4-disubstituted isox-azol-5 (4H)-ones and dihydropyrano [2,3-c]-pyrazole derivatives. *Research on Chemical Intermediates, 42* (10), 7559–7579.

17. Pal, R. (2014). Tamarind fruit juice as a natural catalyst: an excellent catalyst for efficient and green synthesis of bis-, tris-, and tetraindolyl compounds in water. *International Journal of Chemistry, 53*, 763–768.

18. Patil, S., Jadhav, S. D., & Mane, S. Y. (2011). Pineapple juice as a natural catalyst: an excellent catalyst for Bigi-nelli reaction. *International Journal of Organic Chemistry, 1*(3), 125.

19. Pore, S., Rashinkar, G., Mote, K., & Salunkhe, R. (2010). Aqueous extract of the pericarp of Sapindus trifoliatus fruits: a novel 'Green' catalyst for the

aldimine synthesis. *Chemistry & Biodiversity*, *7*(7), 1796–1800.

20. Fonseca, A. M., Monte, F. J. Q., Maria da Conceição, F., de Mattos, M. C., Cordell, G. A., Braz-Filho, R., & Lemos, T. L. (2009). Coconut water (Cocos nucifera L.)—a new biocatalyst system for organic synthesis. *Journal of Molecular Catalysis B: Enzymatic*, *57*(1–4), 78–82.

21. Pandey, V., Chawla, V., & Saraf, S. K. (2012). Comparative study of conventional and microwave-assisted synthesis of some Schiff bases and their potential as antimicrobial agents. *Medicinal Chemistry Research*, *21*(6), 844–852.

22. Ramesh, E., & Raghunathan, R. (2009). Microwave-assisted K-10 montmorillonite clay–mediated Knoevenagel hetero-diels–alder reactions: a novel protocol for the synthesis of polycyclic pyrano [2, 3, 4-kl] xanthene derivatives. *Synthetic Communications*, *39* (4), 613–625.

23. Bhuiyan, M. M. H., Hossain, M. I., Alam, M. A., & Mahmud, M. M. (2012). Microwave assisted Knoevenagel condensation: synthesis and antimicrobial activities of some arylidene-malononitriles. *Chemistry Journal*, *2*(1), 30–36.

24. Ramachandra Reddy, P., Padmaja, A., & Padmavathi, V. (2015). Synthesis of heteroaryl thiazolidinones and azetidinones under conventional and ultrasonication methods. *Journal of Heterocyclic Chemistry*, *52*(5), 1474–1482.

25. Abdel-Rahman, L. H., Abu-Dief, A. M., El-Khatib, R. M., & Abdel-Fatah, S. M. (2016). Sonochemical synthesis, DNA binding, antimicrobial evaluation and in vitro anticancer activity of three new nano-sized Cu(II), Co(II) and Ni(II) chelates based on tri-dentate NOO imine ligands as precursors for metal oxides. *Journal of Photochemistry and Photobiology B: Biology*, *162*, 298–308.

26. Ziolek, M. (2004). Catalytic liquid-phase oxidation in heterogeneous system as green chemistry goal—advantages and disadvantages of MCM-41 used as catalyst. *Catalysis Today*, *90*(1–2), 145–150.

27. Kowsari, E., & Mallakmohammadi, M. (2011). Ultrasound promoted synthesis of quinolines using basic ionic liquids in aqueous media as a green procedure. *Ultrasonics Sonochemistry*, *18*(1), 447–454.

28. Haque, J., Srivastava, V., Chauhan, D. S., Lgaz, H., & Quraishi, M. A. (2018). Microwave-induced synthesis of chitosan Schiff bases and their application as novel and green corrosion inhibitors: experimental and theoretical approach. *ACS Omega*, *3*(5), 5654–5668.

29. Singh, R. K., Kukrety, A., Chatterjee, A. K., Thakre, G. D., Bahuguna, G. M., Saran, S., & Atray, N. (2014). Use of an acylated chitosan Schiff base as an ecofriendly multifunctional biolubricant additive. *Industrial & Engineering Chemistry Research*, *53*(48), 18370–18379.

30. Donia, A. M., Atia, A. A., & Elwakeel, K. Z. (2008). Selective separation of mercury (II) using magnetic chitosan resin modified with Schiff's base derived from thiourea and glutaraldehyde. *Journal of Hazardous Materials*, *151*(2–3), 372–379.

31. Chugh, B., Singh, A. K., Poddar, D., Thakur, S., Pani, B., & Jain, P. (2020). Relation of degree of substitution and metal protecting ability of cinnamaldehyde modified chitosan. *Carbohydrate Polymers*, *234*, 115945.

32. Jin, X., Wang, J., & Bai, J. (2009). Synthesis and antimicrobial activity of the Schiff base from chitosan and citral. *Carbohydrate Research*, *344*(6), 825–829.

CHAPTER 18

ECO-COMPATIBLE SYNTHETIC STRATEGIES: A PARADIGM SHIFT IN ORGANIC SYNTHESIS

MOHD DANISH ANSARI[1], HOZEYFA SAGIR[2*], I. R. SIDDIQUI[1*], and ABU DARDA[3]

[1]Laboratory of Green Synthesis, Department of Chemistry, University of Allahabad, Allahabad 211002, India

[2]Department of Chemistry, Paliwal P.G. College, Shikohabad 205135, India

[3]Department of Applied Science and Technology, Faculty of Engineering and Technology, Jamia Millia Islamia, New Delhi 110025, India

*Corresponding author. E-mail: hozeyfa003@gmail.com, dr.irsiddiqui@gmail.com

ABSTRACT

During decades, progress in the field of chemistry has changed the way people lived, by supplying enormous products aimed at improving the quality of human life. Also the greatest perceived benefits came from pharmaceutical industries with development drugs such as antibiotics that affected mankind for centuries. However, because of the development of pesticides and fertilizers the world's food supply has seen an explosion. Therefore there is practically no facet in material life in which chemistry does not play an important role, either to supply consumer products or to improve services addressed to society in general. Herein, many chemicals and procedures are very hazardous, toxic, and not at all suitable for the environment and nature. Thus today the challenge in organic synthesis lies less in the synthesis of organic compounds than in the development of efficient and environmentally benign transformations. Concern about environmental and human health issues necessitated a paradigm shift in organic synthesis from traditional synthetic protocols to greener and sustainable synthetic processes. The development of benign and nonwasteful alternatives not only reduces the cost of waste treatment but also strengthens economic competitiveness via more efficient use of raw materials. Furthermore, the decrease in process hazards materials greatly reduces risk to both workers as well as the consumer. Thus the scope of this research is very wide and incorporated in numerous areas such as the design of safer chemicals and environmentally benign solvents or catalysts, development of renewable feedstocks, alternative energy sources, and others, which reflects the enormity and complexity of this field.

18.1   INTRODUCTION

In the second half of the 20th century, it has been noticed that there was an accelerated progress that directed to substantial economic development and an escalation in living standards in developed parts of the world. In science, the chemistry of organic synthesis attracts particular attention due to the huge supply of products aimed at improving our quality of life. The study of the production of new molecules from simple, the easier, and commercially available precursor to new materials describes as organic synthesis. The first sensible total synthesis by Friedrich Wohler in 1828 of an organic compound is commonly considered to be that of urea. Synthetic chemistry can strictly be considered a requirement of our modern society [1]. With the help of this discipline, many valuable resources in this world have empowered us to manufacture plenty of fertilizer needed to feed a growing population of the world and the numerous modified materials without which society could not develop. Notably, synthetic chemistry has had an astronomical impact on public health where cures for almost any ailment can be developed resulting in a constant increase in life expectancy [2]. By the curiosity of peers of scientists, all such developments have been empowered and lead to a constant search for new solutions to the assembly of functional molecules. Therefore there is practically no facet in material life in which chemistry does not play an important role, either to supply consumer products or to improve services addressed to society in general.

18.1.1 NEEDS FOR ECO-COMPATIBLE SYNTHETIC STRATEGIES

In spite of all these benefits, the chemical industry is often considered by the people as causing more harm than good. Many chemicals and procedures are very hazardous, toxic, and may have a negative impact on human health and the surrounding environment. However, such growth in living standard has also caused extensive environmental degradation that is exhibited by nondestructive organic pollutants and more distinct climate change in all segments of the biosphere. The environment appeared in the late 1950s and early 1960s has taken under concern and public distress over how these chemicals have a negative impact on human health. Harmful effects of DDT and some other pesticides on the eggs of different birds and the issue of their biomagnification through the food chain was published in 1962. In Japan, another very popular instance of the environmental influence of industries arises where methyl mercury released in the industrial wastewater and biomagnified in fish, which when consumed by the local habitants resulted in the death of more than 100 and the paralysis of around thousand people. Also in Europe there was a scare in 1961, the thalidomide a drug used to alleviate the effects of morning sickness at an early stage of pregnancy has taken by ladies as a result where many children were born with badly deformed or missing limbs. All these examples indicate that not only the environment is in danger but pollution also has an adverse effect on human beings as well. Such environmental concerns necessitate a paradigm shift from the traditional concept of efficiency in organic synthesis, based on chemical output to one that gives importance to maximizing resource efficiency, avoiding the use of harmful and toxic chemicals and reducing waste. In the mid-1990s at the United States, Environmental Protection Agency (EPA) addressed that the environmental concerns of both the products and processes involving chemicals by which they are manufactured and they were evolving the concept of benign by design that is designing environmentally benign products and processes. This soon revealed that waste prevention, designed by EPA for environmental protection not only eliminates the cost of waste treatment but actually strengthens economic effectiveness through more efficient use of raw materials and in the climax the term "green chemistry" has introduced. Some hazardous contaminants and how people are affected with these pollutants are summarized in Table 18.1.

TABLE 18.1   List of Some Hazardous Contaminants

	Contaminants	Examples of Sources	How People Are Exposed
Volatile Organic Compounds	Naphthalene	Vehicle exhaust, deodorizers, paints, glues	Outdoor and indoor air, drinking water, workplaces
	Perchloroethylene	Dry cleaning solvent, degreasing products	Treated clothing, proximity to dry cleaners, workplaces
	Benzene	Gasoline, glues, detergents, vehicle exhaust	Outdoor air, workplaces
Agricultural Products	Organophosphates	Pesticides, flea, and tick pet products	Food, proximity to agriculture, field work, indoor air
	Atrazine	Herbicide	Food, water, proximity to agriculture, field work
Persistent Organic Pollutants	Polybrominated diphenyl ethers (PBDEs)	Flame retardants in furniture and electronics	Food, indoor air, and dust
	PFOA/PFOS	Nonstick and stain resistant coatings	Consumer products, food, water, workplaces
	Dioxins and Furans	By-product of waste incineration, paper mills, manufacturing	Food, outdoor air, drinking water
Plastics Components	Bisphenol A	Hard plastic containers, canned food linings	Food, water
	Phthalates	Cosmetics, detergents, household cleaners, vinyl materials, lacquers	Cosmetics, detergents, household cleaners, vinyl materials, lacquers
Heavy Metals	Lead	Paint, electronics, batteries, fossil fuels	Toys, food, soil, drinking water, workplaces
	Cadmium	Batteries, fertilizer production, waste incineration, plastics, metal coatings	Food, air, water, workplaces

"We're all in the same boat, and we only have one boat."—*Paul Anastas*

18.1.2   GREEN CHEMISTRY

Green chemistry can be stated as "design of chemical products and processes to reduce or eliminate the use and production of hazardous substances" [3, 4]. In other words, it is an ideology that focuses on the increase in avoiding pollution by-products and decreasing dependence on finite nonrenewable resources available. Around 20 years ago that is at the beginning of the 1990s, the concept and definition of green chemistry were first developed by Poul T. Anastas in a special program launched by the EPA to implement sustainable development in chemistry and chemical technology by academia, government and industry. Green chemistry involves two main components. First, it deals with the safety, health, and ecological issues concomitant with disposal or reuse of chemical products use, and manufacture. Second, it addresses the minimization of waste and the efficient utilization of resources. The three main green components required for each reaction to be termed as "green" are reagent/catalyst, solvent, and energy consumption. No one can do design by accident, which is a statement of human objective. It includes systematic conception, planning, and novelty. Nowadays, green chemistry goes beyond the research works in the laboratory and has touched education, environment, industry, and the general public. Green chemistry facilitates creativity and the development of innovative research and gives responsible multidisciplinary access to science, based on ecological, chemical, and social responsibility [5, 6]. There are a number of examples of successful utilization of award-winning and economically competitive technologies that have been applied on all industry sectors like aerospace, cosmetic, automobile, electronics, household products, pharmaceutical, and agriculture [7]. The "design rules" that help the chemists to achieve the voluntary aim of sustainability is the 12 principles of green chemistry. It can be used to create or recreate materials, molecules, processes, and reactions that are safer for the surrounding environment and human health.

18.1.3 TWELVE PRINCIPLES OF GREEN CHEMISTRY

The 12 principles of green chemistry can guide chemists toward fulfilling their role in accomplishing eco-compatible etiquettes as articulated by Anastas and Warner in 1998 [8] Principles of green chemistry are a guiding architecture for the composition of new chemical products and processes [9] They are applying to all features of the process life cycle from the safety of the transformation and raw materials used for the efficiency, biodegradability of products reagents used, and the toxicity. It is impossible at the same time to satisfy all the requirements of twelve principles of the process but during certain stages of synthesis it tries to utilize as many principles as possible. The 12 principles of green chemistry that can be paraphrased as:

1. *Prevention:* Once the waste is created, it can only be treated or cleaned up, so it is better that waste generation is prevented.

2. *Atom economy:* Synthetic methods should be developed to maximize the incorporation of all materials used in the process into the final product.

3. *Safer synthesis:* Wherever possible, synthetic techniques should be developed to use and manufacture substances that possess less or no toxicity to human health and the surrounding atmosphere.

4. *Designing safer chemicals:* The chemical products should be designed to affect their desired function while minimizing their toxicity.

5. *Safer solvents and auxiliaries:* The use of auxiliary substances like solvents separation agents should be made unnecessary wherever possible and innocuous when used.

6. *Design for energy efficiency:* Energy requirements of chemical processes should be recognized for their environmental and economic impacts and should be minimized. If possible, synthetic methods should be conducted at ambient temperature and pressure.

7. *Use of renewable feedstock:* Whenever technically and economically practicable the raw material used should be renewable rather than depleting.

8. *Reduce derivatives:* Unnecessary derivatization (use of blocking groups, protection/deprotection and other temporary modification) should be minimized.

9. *Catalysis:* Catalytic reagents (as selective as possible) are superior to stoichiometric reagents.

10. *Design for degradation:* Chemical products should be designed so that at the end of their function they break down into innocuous degradation products and do not persist in the environment.

11. *Real-time analysis for pollution prevention:* Analytical methodologies need to be further developed to allow for real-time, in-process monitoring and control prior to the formation of hazardous substances.

12. *Inherently safer chemistry for accident prevention:* The substance chosen for the chemical process should be able to minimize the potential for chemical accidents including gas releases, explosions, and fires.

Tang et al. proposed a mnemonic, *PRODUCTIVELY* (Figure 18.1) that captures the spirit of the twelve principles of green chemistry [10].

Diagrams of the tree have been used in chemistry to celebrate the multiplicity of applications that can be supported by a specific raw material. From the beginning of modern chemistry many variations on the theme have appeared, for example, "Coal Products Tree"

Condensed Principles of Green Chemistry

P – Prevent wastes
R – Renewable materials
O – Omit derivatisation steps
D – Degradable chemical products
U – Use of safe synthetic methods
C – Catalytic reagents
T – Temperature, Pressure ambient
I – In-Process monitoring
V – Very few auxiliary substrates
E – E-factor, maximise feed in product
L – Low toxicity of chemical products
Y – Yes, it is safe

FIGURE 18.1   PRODUCTIVELY: mnemonic form of 12 principles of green chemistry, for the reproduction of material from green chemistry: [10] Reproduced with permission of The Royal Society of Chemistry (RSC).

[11] and the "Petroleum Tree [12]. Erythropel et al. [13] introduce the metaphor "Green ChemisTREE" (Figure 18.2). Each of the principles of green chemistry shown with the help of branches in Figure 18.2 and the leaves representing techniques available to the Green chemist like procedures, mechanisms, design, and guidelines.

18.2   DIFFERENT ECO-COMPATIBLE STRATEGIES FOR ORGANIC SYNTHESIS

To improve the chemical synthesis and to reduce or even eliminate the impact of hazardous starting materials, by-products and wastes, new alternative green chemistry tools are needed and they are described below.

18.2.1   SOLVENT FREE SYNTHETIC STRATEGIES

In many synthetic processes solvents have countless applications and play multiple roles in various chemical processes. For example, they bring reactants together by dissolving them, they affect chemical reactivity, use in extracting and washing products, promote a facile separation of mixtures, cleaning reaction apparatus. Moreover, in the majority of the reactions it is essential to achieve homogeneity of substrates, it prevents the formation of undesired by-products through dilution, promote fast and safe conversions, and control the heat flow for both endothermic and exothermic transformations [14]. In spite of a numerous incontrovertible merits, these solvents are the

FIGURE 18.2   The Green "ChemisTREE" emphasizing the areas of inquiry and progress relevant to each of the 12 principles of green chemistry. Reproduced with permission of The Royal Society of Chemistry (RSC) [13].

major contributors to environmental pollution and are characterized by several deleterious effects on the human and environment [15]. Many organic solvents are volatile liquid in nature that themselves can be a problem. Because if they released into the rivers, earth, or the ocean, they can cause direct environmental damage, whereas slowly releasing their vapors increases the risks of fire and explosion. Moreover, after release in the atmosphere they cause the ozone depletion, photochemical smog, and global warming. Furthermore, various conventional solvents are possessed highly chronic toxicological properties. For example, benzene was extensively employed as a solvent, hand cleaner, and as an aftershave for decades before its carcinogenicity became appreciated.

Organic solvents pose a particular concern to the chemical industry because of their use in vast quantities. The relevance of this issue can be better understood by the fact that all over the world, the consumption of organic solvents is approximately 20 million tons per year [16]. Therefore significant efforts have been made toward identifying organic solvents with a reduced ecological footprint as compared to traditional reactions. In this context, solvent-free processes are the best solution, particularly when one of the component either substrates or the products is in liquid form and can be served as a solvent [17, 18].It has been said that "the best solvent is no solvent." In comparison to reactions in molecular solvents, solvent-free reactions include the following advantages: (1) they reduced the environmental pollution as well as costs; (2) there is no reaction medium to collect, purify, and recycle that provides simplicity in process and handling. These factors are important from the

industrial point of view; (3) the products formed are often sufficiently pure to avoid tedious purification process like chromatography. In few cases there is not even the need for recrystallization; (4) sequential solvent-free reactions are possible in high yielding systems; (5) mostly there is no need for specialized equipment; (6) these reactions are often rapid and sometimes reaching substantial completion in minutes as compared to several hours with organic solvents; (7) energy usage is generally lower; (8) preformed salts and metal-metalloid complexes; (9) there often is no need for functional group protection and deprotection; (10) there may be lower capital outlay for equipment when setting up industrial processes. Thus considering all these points, the chemo-, regio-, or stereoselective synthesis of high-value chemical entities and parallel synthesis to construct an athenaeum of small molecules will add to the increase of solvent-free organic reactions in the near future. Shaabani et al. [19] and idyacharan et al. [20] (Scheme 18.1) proposed a protocol without solvent and catalyst for the synthesis of pyrazoles and 3-aminoimidazo-fused heterocycles.

18.2.2   USE OF GREEN SOLVENTS

The complete avoidance of chemical solvents is one of the simplest

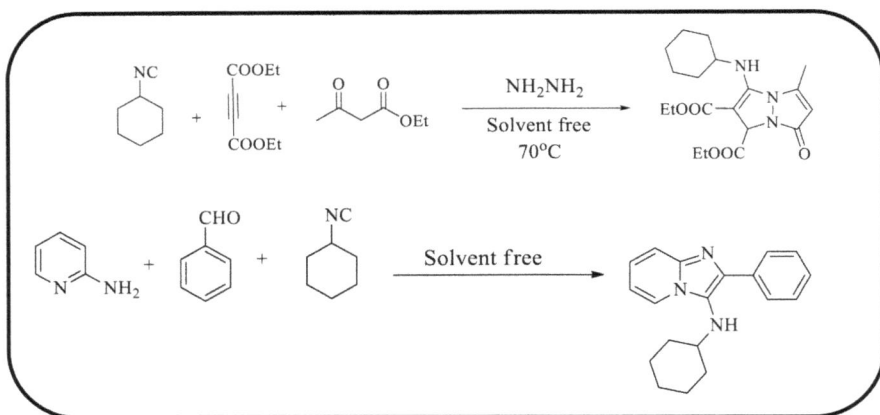

SCHEME 18.1   Solvent and catalyst-free protocol for organic transformation.

solutions to the problem. However, many reactions are not amenable to solvent-free approach, mostly on large industrial scale exothermic reactions can be dangerous. Another drawback of the solvent-free condition is the inefficient mixing of catalyst and reagents, particularly when reagents or products both are solid. Finally, solvents are still often required for extraction and purification of products. So the development and selection of appropriate alternative solvents are becoming one of the crucial aspects to decrease environmental, health, and safety impacts of industrial processes. Researchers in industry and academia are not only facing the issue of substitution of unwanted solvents but also the development of new solvent systems or even new compounds to be used as the solvent. Some of the leading alternative green solvents are discussed below.

18.2.2.1   WATER AS A GREEN SOLVENT

The concept of "green" solvents expresses the goal to reduce the effect on environmental that was resulting from the use of solvents in a process, thus identifying green solvents is of topmost priority for the organic chemist. As we know nature selected the water as a solvent, to carry out all kinds of biochemical transformations. In the organic synthesis, water was rediscovered as a solvent in the 1980s [21] and largely popularized in the 1990s. Water as a solvent is highly economical, nontoxic, nonflammable, it does not contribute to greenhouse emissions, does not demand to synthesize, and the energy required for its isolation in the pure form is negligible. Additional properties of water are tunable acidity and high polarity. Heat capacity and heat of evaporation of water are also very high that allow easy control of exothermic reactions and coexistence of hydrogen bond donor and acceptor functionalities that often make catalysis easier. Moreover, the unique physicochemical properties and structure of water lead to particular interactions like hydrophobic effect, hydrogen bonding, and trans-phase interactions that might significantly influence the reaction mechanism. There are some more advantages that not only established the broad scope of reactions in water but also lead to additional sustainability benefits of water that enhance the overall environmental impact of a given process. In one word it is the green solvent par excellence [22]

18.2.2.1.1 Improve Reactivity and Selectivity

Water enhances the rates and affects the selectivity of a wide variety. The Possibility of using water as the solvent for organic synthesis with surprising and anticipated results has appeared in the literature. Stavber et al. described the efficient conversion of tertiary benzyl alcohols into their respective vicinal halo-substituted derivatives using *N*-halosuccinimides [23] Interestingly, when water was used as a solvent the performance of the reaction was better thus revealing a new sample of an accelerated organic

reaction in aqueous media. One more remarkable result on cycloaddition rate acceleration was reported by the group of Engberts in their research of the Diels–Alder reaction of cyclopentadiene and 3-aryl-1-(2-pyridyl)-2-propen-1-ones [24]. They showed that the reaction carried out in the water as solvent was 287-fold faster than the same reaction in acetonitrile (Scheme 18.2).

18.2.2.1.2 Catalyst Recycling and Efficient Product Isolation

Evolution of high atom efficient catalytic processes is one more possible opportunity for organic synthesis in water that leads to simplified catalyst recycling and product isolation. Ideally, the water-solubility of reactant and the product should be zero very little. The result of this is that the catalyst can be recycled easily and the product can be isolated by simple phase separations. For example, the group of *Mizuno* described the homocoupling of 2-naphthol derivatives in water catalyzed by a supported ruthenium catalyst (Scheme 18.3) [25]. In this aqua mediated reaction catalyst could be separated easily by filtration and recycled up to seven runs without considerable failure of activity, affording a total turnover number of 160 which was superior to those achieved with other available processes.

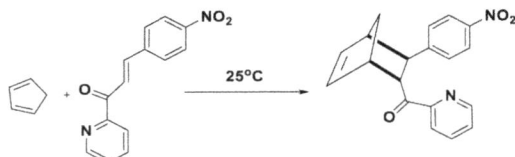

Condition	K_{rel}
Acetonitrile	1
Water	287
$Cu(OSO_3C_{12}H_{25})2.2.4Mm$ Water	$1.8 \cdot 10^6$

SCHEME 18.2   Influence of water in the rate increase of a Diels–Alder reaction.

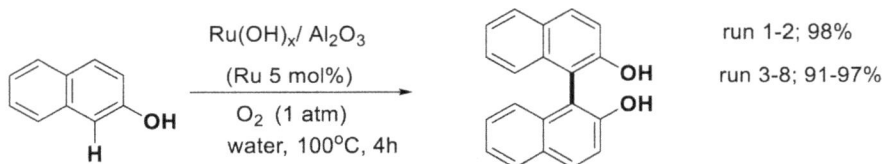

SCHEME 18.3   Oxidative homocoupling of 2-naphthol using a supported ruthenium catalyst in water.

18.2.2.1.3   Mild Reaction Conditions

From a green chemistry point of view, the development of mild reaction conditions is a crucial issue, not only because it can lead to safer processes but also because reagents that are less reactive, generally more easily available and requiring less upstream synthetic procedures. For instance, Li et al. described [26] very mild and convenient reaction conditions for the formation of α-hydroxy ketones at room temperature under air. The oxidation product is obtained with good yields, without the use of any stoichiometric oxidants (Scheme 18.4). The water turned out to be a unique solvent for this reaction since no product is obtained in benzene, THF, acetone, or ethanol.

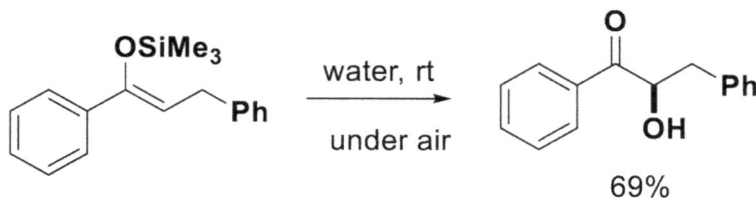

SCHEME 18.4   Catalyst-free formation of α-hydroxy ketones under the aqueous condition.

18.2.2.1.4 Workup Improvement

Even though the use of water as a medium for the reaction is advantageous because of its low toxicity and hazard, it does not necessarily allow us to eliminate organic solvents from the whole process. Indeed, the workup procedure, through chromatography or extractions purifications for example, may be responsible for the utilization of a large quantity of solvent in comparison to the recovered mass of the product. As the organic product is hardly water soluble in most of the cases, efficient procedures for the organic solvent-free purification of reactions conducted with water as solvent are the filtration or phase separation. In

this context, the Butler and coworkers showed that the 1,3-dipolar cycloaddition of phthalazinium-2-dicyanomethanide with various alkenes led to sparingly water-soluble adducts that can be separated from the reaction mixture by simple filtration (Scheme 18.5) [27].

New theoretical frameworks continue to emerge as the structure and properties of water are being studied continuously by scientists in almost all fields of knowledge. In addition, a series of most of the fundamentally useful reactions like aldol condensation, cycloaddition, allylation, epoxidation, hydrogenation, oxidation, and organometallic reactions have also been performed in aqueous media with identical or enhanced reaction rate, yields,

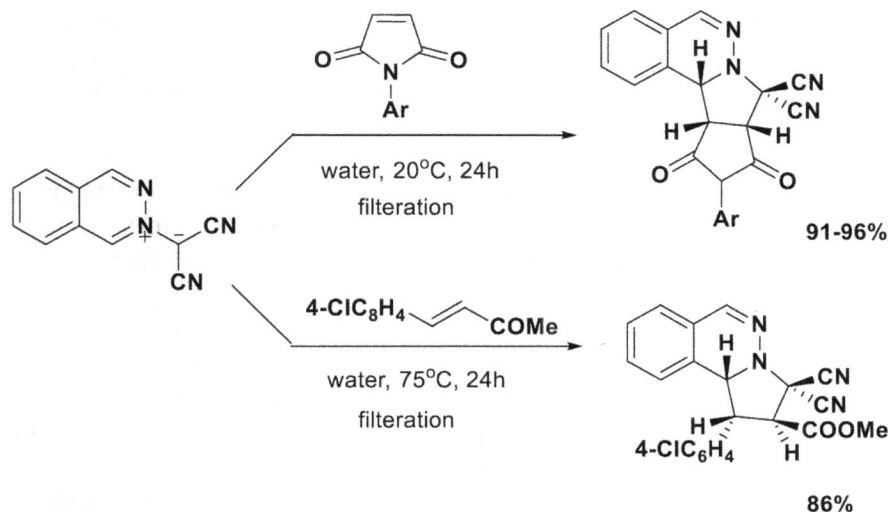

SCHEME 18.5 "On water" 1,3-dipolar cycloaddition of phthalazinium-2 dicyano methanide.

and selectivity as compared to the corresponding reaction in organic solvents [28]

18.2.2.2 IONIC LIQUIDS: AN ENVIRONMENTALLY BENIGN SOLVENT

Another class of chemicals that has been extensively studied as a greener and safer alternative for a wide range of solvent is ionic liquids (ILs) [29–31]. ILs have been previously defined as molten salts comprised of a cation and an anion with melting points around 100 °C. ILs exhibit several crucial properties making them very interesting in academia and industry [32, 33]. First of all, they are nonvolatile and have negligible vapor pressure at near ambient conditions. Thereby minimize the risks of atmospheric contamination and significantly reduce health and environmental concerns. Second, ILs exhibit properties such as the high dissolution power, polarity, and hydrophilicity or lipophilicity, which are rather specific and different to those of conventional organic solvents. These properties depend on the nature of the cationic or anionic group that composes the IL. In general, cations control physical properties, whereas anions are responsible for the chemical properties and reactivity [34]. Thus because of this reason IL has ability to fine-tune the structure, to tailor the properties for

a particular purpose, and has been named as "designer solvents." [35]. It is possible, to design ILs that are able to optimize the relative solubilities of reactants and products, the reaction kinetics, the intrinsic catalytic behavior of the media, and even the liquid range of the solvent and the air stability of the system [36].

In addition the ionic nature of IL confirms that catalysts that are ionic or possess polar fragments can be easily immobilized, separated, and recycled via a biphasic operation without any tedious catalyst modification or work-up process, thus providing a convenient solution to both problems, solvent emission as well as catalytic recycling. Furthermore, they have the ability to generate internal pressure and accelerate the association of reactants in the solvent cavity during the activation process [37]. Third, the IL are nonflammable, noncorrosive, and have a high thermal stability (up to 200 °C). They have densities higher than the density of water thus occur as the lower phase in most biphasic systems. Moreover, they can be recycled and reused many times without loss of reactivity.

The another most attractive feature of ILs is their ability to solubilize both metal salts and organic compounds; in particular, they are able to dissolve biomacromolecules such as carbohydrates [38], cellulose, [39] silk fibroin, [40] wool keratin [41], chitin, and chitosan [42] that are linked together by intermolecular hydrogen bonds, thereby offering the new possibility of exploitation of this type of renewable biomaterials. For example, very harsh extraction and derivatization conditions are required for the industrial exploitation of cellulose as only a few solvents have the ability to dissolve the fibers. On the other hand, ILs are able to dissolve cellulose and the many other biopolymers in high amount under mild conditions, thus allow an easy functionalization of the material. Moreover, they are simply recoverable by precipitation with water or other organic solvents. Therefore we can say it represents a very good "green" alternative to the traditional solvents.

There are a large number of research papers for the use of ILs in synthetic routes and various applications present in the scientific literature. For example, they have been investigated as a substitute for water in Diels–Alder reactions [43], substitutes for conventional organic solvents in alkylation reactions of β-naphtol and indole [44], for the immobilization of homogeneous catalysts for hydrogenation reactions, for Heck reactions [45], hydro-formylation reactions, olefin dimeri-zation, and polymerization processes [46].

1,3-dialkylimidazolium   1-alkylpyridinium   1,1-dialkylpyrolidinium

FIGURE 18.3   Some common cations used for ionic liquids.

18.2.2.3 GLYCEROL AS A GREEN SOLVENT

Glycerol, also known as glycerin, is a simple polyol (1,2,3-propane-triol) compound. In the structure of triglycerides, it is present naturally, which are fatty acid esters of this alcohol. There are more than 2000 applications found for glycerol in various fields like the pharmaceutical food industry or cosmetic, where it is mainly engaged as a lubricant, sweetener, thickener, and humectant [47]. Glycerol appears as a concomitant, In the production of biodiesel, representing 10% by weight of the total output. The research on new uses of glycerol has been stimulated in the last few years, reforming for hydrogen production and its transformation into fuel additives [48, 49].

One of the most important constituents in any chemical process is reaction solvents and they are used in huge amounts. Synthetic organic chemistry is generally done in solution to ease the intimate contact of reactants and catalysts. Nowadays, the chemical community across the world has increasingly identified safety and environmental concerns. In today's time, the twelve principles of "green chemistry" are considered as an essential driving force in the expedition for sustainable chemical processes. As per these principles,

a green solvent should satisfy many criteria such as nonflammability, low toxicity, widespread availability among others, and nonvolatility. Additionally, these green solvents have to be easy to handle, recycle, and cheap [50]. In this framework, glycerol has become of particular interest as a solvent for catalysis or organic chemistry. It should be prominent that some glycerol derivatives have also been proposed as potential green solvents like glycerol formal [51], glycerol carbonate [52], glycerol ethers, [53] and others [54].

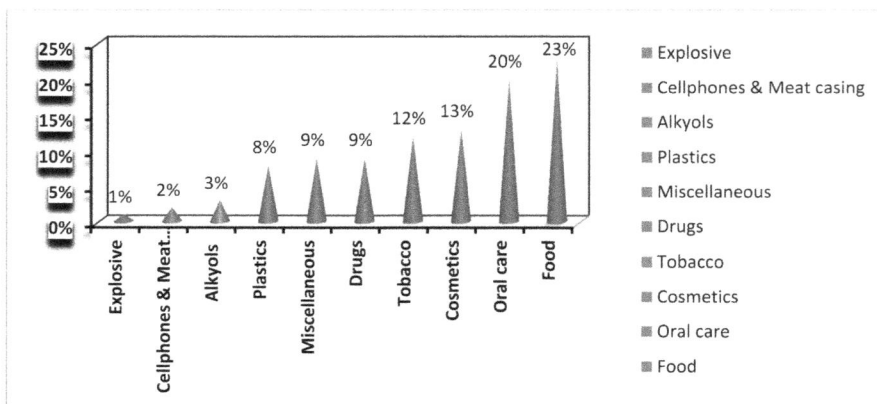

FIGURE 18.4   The market of glycerol.

Glycerol is a sweet-tasting, colorless, odorless, clear, and viscous liquid. It is a polar protic solvent with a dielectric constant of 42.5 (at 25 °C) because glycerol is a trihydric alcohol. It is insoluble in hydrocarbons, sparingly soluble in many common organic solvents such as dichloromethane, ethyl acetate, and

diethyl ether and completely soluble in water and short-chain alcohols. It forms crystals at <17.8 °C (low temperatures). It is used as a green solvent due to specific points which maximize as much as possible its solvent properties:

1. *Solubility:* Glycerol is able to ease dissolution of acids, inorganic

salts, enzymes, bases, and many transition metal complexes. Additionally, it also facilitates the miscibility of organic compounds in water. Many hydrophobic solvents (ethers and hydrocarbons) are immiscible in glycerol that facilitates simple liquid–liquid phase extraction to remove the reaction product.

2. *Volatility and boiling point:* Glycerol has a high boiling point (290 °C) and nonvolatile under normal atmospheric pressure, which provides a feasible separation technique of the reaction products, distillation. Its high boiling point also enables acceleration of the reactions that do not take place in low boiling point solvents.

3. *Safety:* Glycerol is a biodegradable, nonflammable, and non-toxic solvent and because of these properties no special handling precaution or storage is required.

4. *Availability:* Glycerol meets all the benchmarks of the green solvent that is cheap and available on a large scale.

Gu et al., in 2008, reported [55] aza-Michael reaction (Scheme 18.6) between *p*-anisidine and *n*-butyl acrylate that can successfully proceed under catalyst-free conditions using glycerol as an exclusive solvent, a remarkable benefit over the conventional solvent system. This reaction validated the necessity of using glycerol as a promoting medium for organic reactions and avoided the use of catalyst.

Thioacetalization with benzenethiol or 1,2-ethanedithiol just by heating at 90 °C with glycerol and give the desired product good to excellent yield without the abetment of any Brønsted or Lewis acid catalyst (Scheme 18.7) [56]. Additionally, glycerol is almost equally effective up to the 4th cycle in the synthesis of the target compound.

Glycerol has been used for various applications, as a co-solvent, as a single solvent, playing the double role of a solvent, and a reagent or being part of a deep eutectic mixture. It has been found a remarkable reaction media for the organic transformation both with and without

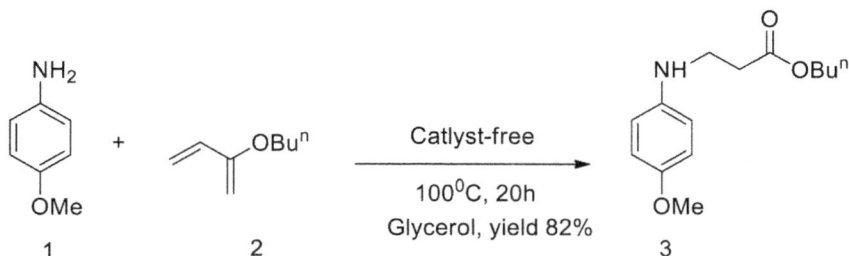

SCHEME 18.6 Aza-Michael reaction.

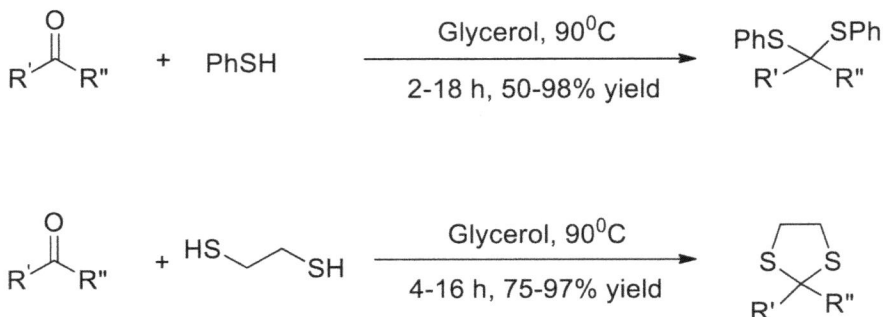

SCHEME 18.7   Catalyst-free thioacetalization of carbonyl compounds in glycerol.

the help of a catalyst. In spite of all these benefits problems concomitant with the presence of three reactive and metal-coordinating hydroxyl groups should be taken into account prior to its utilization as a solvent for a particular reaction and its high viscosity that could stimulate poor substrate diffusion in the medium.

18.2.3   SYNTHETIC STRATEGIES INVOLVING NANOCATALYSIS

The area of catalysis is referred as a "foundational pillar" of green chemistry [57]. By using the suitable catalysts, one can reduce the temperature, reduce reagent-based waste, and decrease separations because of increased selectivity that potentially avoids the unwanted by-products and pollutants, leading to green technology. Homogeneous and heterogeneous are two well-known categories of catalysis. Both the catalysis has its advantages and disadvantages (Figure 18.5) such as

homogenous catalyst allows easier interactions between the components that in turn results in better activity, [58, 59] but it is difficult to separate the catalyst from the final product. Whereas in the case of heterogeneous catalysts, the most attractive feature is their ease of separation from the reaction mixture and recyclability. However, as compared to their homogeneous counterparts traditional heterogeneous catalyst systems have two main drawbacks. First, the reduced surface area that is accessible to the reactant, thereby limiting their catalytic activities [60] and second, it leads to an unnecessarily high consumption of expensive catalyst materials [61]

Thus it is clear that there is an urgent requirement to develop a new catalytic system, which incorporates the merits of both homogeneous as well as heterogeneous catalyst [62]. Todays "nanocatalysis" is an emerging tool for the green technologies in catalysis science. However, the application of nanoparticles as

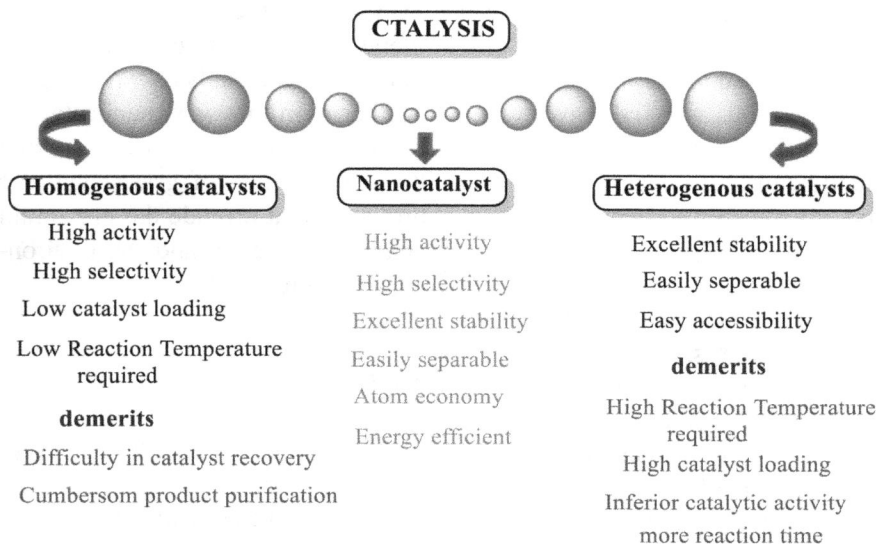

FIGURE 18.5 Comparative study of heterogeneous, homogeneous and nanocatalysis.

a catalyst is not new, but our ability to image and characterize it in the context of its catalytic activity has been increased nowadays. Concept of nanocatalysis is known since the 1950s when the term nanotechnology was not even known [63]. First time in 1986, Haruta et al. [64] reported the catalytic activity of gold (Au) nanoparticles in the oxidation of hydrogen and carbon monoxide at low temperatures. After this reporting, a very old traditional idea about gold for its inertness toward chemical reaction has been completely changed and a new door to interesting applications of nanoscale materials in catalysis is opened.

Nanocatalysis combines the best attributes of both homogeneous as well as heterogeneous catalyst. Because of the small size (1–100 nm), the active metal atoms in nanocatalyst are exposed to the surface thereby increasing the contact between reactants and catalyst like homogeneous catalysis. Additionally, they are insoluble in the solvents therefore can be recovered easily from the reaction mixture like a heterogeneous catalyst [65]. Moreover, these nanoparticles have great potential for improving the selectivity, efficiency, and yield of catalytic processes. In particular, they have the ability to disperse into the solution and provide a higher surface-to-volume ratio. Higher selectivity of the nanocatalysts toward reaction proceeds through less waste generation and fewer impurities that could lead to safer reaction condition and reduced environmental impact. Therefore they are recognized as the most important industrial catalyst and have wider application ranging from chemical manufacturing to energy conversion and storage. An additional virtue of nanocatalysts is the easy control over size, shape, and morphology that makes possible to design the materials that are specifically required for a particular catalytic application. Deoxygenation of epoxides catalyzed by gold nanocatalysts,[66a]photodegradation of 2,4,6-trinitrophenol by calcium oxide nanocatalysts,[66b]palladium nanoparticle catalyzed C–C coupling reactions,[66c]zinc oxide nanoparticle catalyzed synthesis of caumarins[66d], and synthesis of 3-aryl-4H-benzo[1,4]thiazin-2-amines[66e]

are few examples documented in the literature that highlights the application of nanocatalysts in organic synthesis. In this way, nanocatalysis can play a prominent role in guiding the development of green synthetic protocols that provide the maximum benefit for society and the environment (Figure 18.6).

18.2.4   SYNTHETIC STRATEGIES USING VISIBLE LIGHT AS AN ENERGY SOURCE

Field of green chemistry expresses an area of research that inventing new cleaner and more benign chemical methods for the generation of the desired target molecules. Nearly a century ago, Ciamician realized that

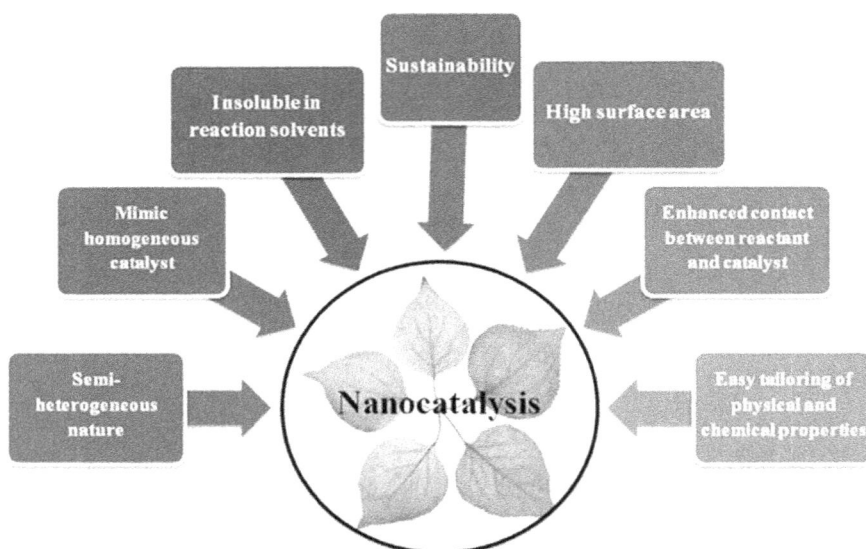

FIGURE 18.6   Benefits of nanocatalysis.

"Visible-light" is a clean, inexpensive, and "almost infinitely" available energy source for performing green chemical reactions [67]. Since then, the visible light promoted photochemistry and photocatalysis have found broad utility in organic synthesis [68, 69]. Such a light-driven process (so-called photocatalysis) directly converts solar energy into chemical energy and provides a greenway for reducing problems like energy consumption as well as risk to human health and the environment. Because of its low cost, cleanness, easily handled, it is significantly advantageous over the processes required for specialized equipment like high-energy ultraviolet light (UV) [70]. Moreover, in such photochemical process mild reaction conditions are required for substrate activation, mostly without the introduction of extra-functional groups and highly reactive radical initiators that makes them attractive for the chemical community. However, one fundamental impediment that limits the application of photochemical processes is the inability of many organic molecules to absorb light in the visible region of the electromagnetic spectrum. To overcome this barrier, intensive research has been devoted toward the development of photocatalysts that can absorb visible light. These photocatalysts, utilizing their electron/energy transfer processes to sensitize organic molecules and facilitate the required photochemical transformations. This photoredox catalyst absorbs light in the visible region and gives stable, long-lived excited states [71] that may carry out a bimolecular electron-transfer reactions [72] Hence, the conversion of a benign species that are poor single-electron oxidant and reluctant in its ground state to redox-active species, simply on irradiation with simple household light bulbs represents an important chemoselective trigger to initiate a valuable catalytic process.

Nature's ability to use various photocatalysts or chromophores for transforming solar energy to chemical energy has inspired various research community to develop a number of hosts involving photoredox systems in an effort to mimic photosynthesis [73]–[75]. These systems have offered a platform to understand the electron transfer or energy transfer pathways involved in natural photosynthesis. As a result, many organic molecules have been developed that may function as visible light photocatalysts such as organic dyes eosin Y [76], 9-10 dicyanoanthracene, [77] and triphenylpyrylium salts [78]. Pyridyl-based complexes such as $Ru(bpy)_3^{2+}$,[79] $[Ir(ppy)_3]$ and Ir $(ppy)_2$ $(dtb-bpy)^+$ (where, bpy = 2,2'-bipyridine, ppy = 2-phenylpyridine, dtb-bpy = 4,4'-di-tert-butyl-2,2'-bipyridine) for the mediation of redox processes are exclusively exploited and utilized in numerous photochemical reactions like dehalogenation, reduction, oxidation, and asymmetric alkylation (Figure 18.7) [80]

These catalysts show absorption in the visible range and have been employed to achieve a vast range of bond-forming reactions. Additionally, their ability to produce a diverse array of reactive intermediates has also proven to be a valuable tool for the construction of complex molecules. Total syntheses of gliocladin C, heitziamide A, and aplyviolene clearly demonstrated their application. The ability of these visible-light photocatalysts has been recognized and extensively investigated for applications in synthetic organic chemistry. Some photocatalyzed conversions are summarized in Scheme 18.8 [81]–[83].

Along with these applications, photoredox catalysts have been also utilized to accomplish the splitting of water into hydrogen and oxygen [84], the reduction of carbon dioxide to methane [85], degradation of chemical pollutants, [86] and in several other designed systems for the conversion of solar energy into electrical current and fuel [87].

18.2.5 MULTICOMPONENT SYNTHETIC STRATEGIES

Preparation of complex organic molecules generally involves a lot of synthetic operations with tedious extraction and purification processes in each step. This conventional multistep procedure not only decreases synthetic efficiency but also generates huge amounts of waste. Recently, academic chemists have revived their interest in multicomponent reaction (MCR), prove to be one of the key tools for efficient and speedy assembly of structurally complex and highly functionalized "drug-like" molecules. They provide a robust and straightforward approach toward the assembly using easily available starting materials [88]. It is defined as the one-pot process in which at least three different reactants join together through covalent bonds and

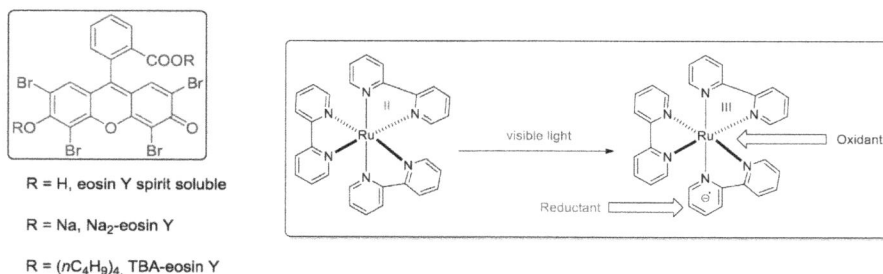

R = H, eosin Y spirit soluble

R = Na, Na$_2$-eosin Y

R = (nC$_4$H$_9$)$_4$, TBA-eosin Y

FIGURE 18.7   EosinY and Ru(bpy)$_3{}^{2+}$: photoredox catalysts.

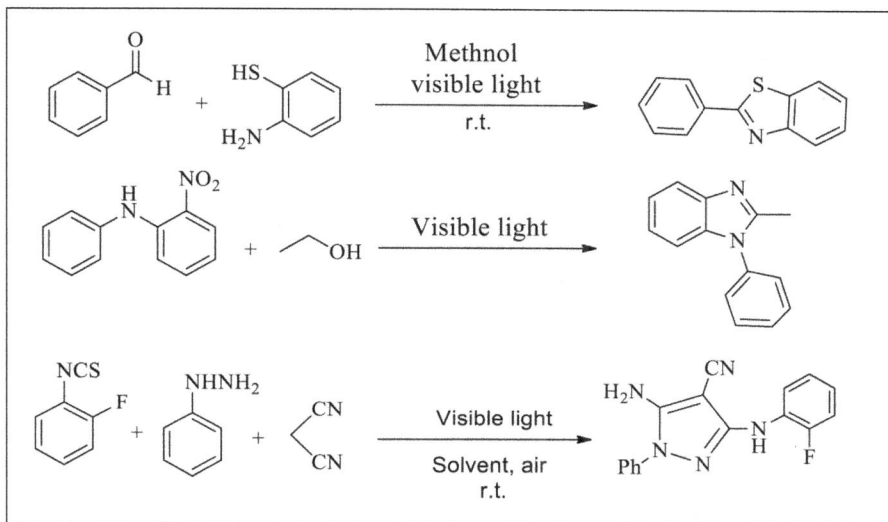

SCHEME 18.8   Some example of photocatalyzed conversions.

allow the construction of several new bonds in a single operation [89]. Thereby offering several remarkable advantages such as convergence, operational simplicity, shorter reaction time, high atom economy, high selectivity, and higher overall yields than multiple-step syntheses [90]. MCRs also avoid the unnecessary steps of workup and extraction processes and hence minimize the waste generation, energy consumption, as well as manpower, rendering the transformations green [91].

It constitutes the pillar of both combinatorial chemistry and diversity-oriented synthesis and thus has played a vital role in the development of the modern synthetic process for pharmaceutical and drug discovery research [92]. Therefore the design

of both new MCRs, as well as improvement of already known MCRs with green methodology, is of substantial interest in current synthetic chemistry. Therefore literature enumerates a number of such multicomponent reaction the preparation of biologically active complex molecules, Some of them are shown in Scheme 18.9. [93, 94].

18.3   CONCLUSION AND OUTLOOK

The present chapter clearly establishes the importance of synthetic strategies through eco-compatible methods that are of academic and pharmaceutical industry interest. As expected, green chemistry has

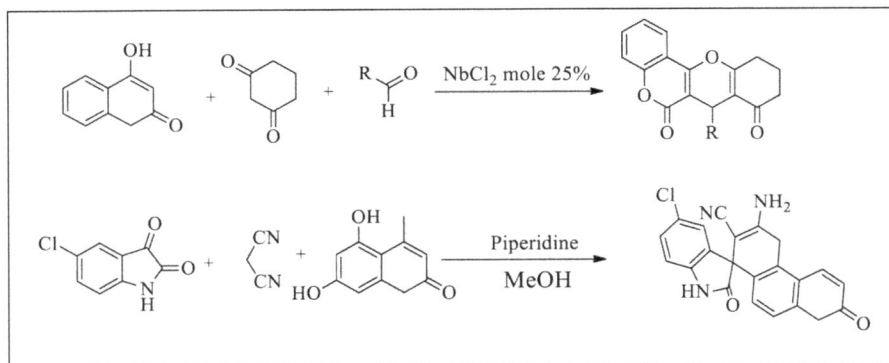

SCHEME 18.9   Some example of multicomponent reaction.

emerged as a significant internationally engaged focus area within chemistry. Each organic synthesis that carries environmental negatives has an alternative and can be replaced with less polluting or nonpolluting alternatives. The main aim of this chapter is to highlight the potential of chemistry to solve many of the global environmental challenges we now face. All referred strategies are aimed at replacing toxic and hazardous substances in many chemical processes. Also, there have been some successful uses of alternatives that represent a new way to meet the challenge of energy and sustainability. It is always interesting to search for an application of our knowledge about easy and green methods for synthetic strategies. However, a continuous significant focused research is always required in this field.

ACKNOWLEDGMENT

Mohd Danish Ansari is thankful to the Council of Science and Technology, Uttar Pradesh (Project No. CST/D-2276) for financial support in the form of research assistant. The author is also thankful to the Department of Chemistry, University of Allahabad, Allahabad for implementation of this fellowship.

KEYWORDS

- green chemistry
- eco-compatible
- renewable
- organic synthesis
- green solvent

REFERENCES

1. Raw, S. A; Wilfred, C. D. and Taylor. R. J. K. Chem. Commun. 2286: 2003.
2. Sehlstedt, U.; Aich, P.; Bergman, J.; Vallberg, H.; Norden, B.; and Graslund. A. J. Mol. Biol. 278; 31:1998.
3. (a) Anastas, P. T. and Warner, J. C. Green Chemistry: Theory and Practice, Oxford University Press, New York, 1998; (b) Horvath, I. and Anastas, P. T. Chem. Rev., 2007, 107, 2167.
4. Anastas, P. T. and Williamson, T. C. Green Chemistry: Designing Chemistry for the Environment, American Chemical Series Books, Washington, DC, 1996, pp. 1–20.
5. Collins, T. J. in Green Chemistry, Macmillan Encyclopedia of Chemistry, Simon and Schuster Macmillan, New York, 1997, vol. 2, pp. 691–697.
6. Riđanović, L.; Ćatović, F. and Riđanović, S. (2013): The Green Chemistry-Ecological Revolution in the Classroom. 8th Research/Expert Conference with International Participations "QUALITY 2013", Neum, B&H, June 06 08, 447-452, In Bosnian.
7. Office of Pollution Prevention and Toxics, The Presidential Green Chemistry Challenge, Award Recipients, 1996–2009, US Environmental Protection Agency, Washington, DC, EPA 744K09002, 2009.
8. Anastas, P. and Warner, J. C. Green Chemistry: Theory and Practice, Oxford University Press, Oxford, 1998.
9. Benign by Design: Alternative Synthetic Design for Pollution Prevention, ACS Symp. Ser. nr. 577, ed. Anastas, P. T. and Farris, C. A. American Chemical Society, Washington DC, 1994.
10. Tang, S. L. Y.; Smith, R. L. and Poliakoff, M. Green Chem., 2005, 7, 761.
11. Lima, M. The Book of Trees. Visualizing Branches of Knowledge, Princeton Architectural Press, New York, 2014.
12. The Raleigh Register, Coal Products Tree, http://williamsonlibrary.lib.wv.us/WV20Facts/Coal20mining/coaltree.htm, (accessed October 2017).
13. Erythropel et al., Green Chem., 2018, 20, 1929–1961.
14. Reichardt, C. Solvents and Solvent Effects in Organic Chemistry; Wiley-VCH Verlag GmbH & Co. KGaA: Weinheim, Germany, 2004, i–xxvi.
15. Constable, D. J. C.; Jiménez-González, C. and Henderson, R. K. Perspective on solvent use in the pharmaceutical industry. Org. Process Res. Dev. 2007, 11, 133.
16. Adams, D. J.; Dyson, P. J. and Taverner, S. J. Chemistry in Alternative Reaction Media, Wiley 2004.
17. Walsh, P. J.; Li, H. and de Parrodi, C. A. Chem. Rev., 2007, 107, 2503–2545.
18. Varma, R. S. Pure Appl. Chem., 2001, 73, 193–198.
19. Shaabani, A.; Sepahvand, H.; Nejad, M. K. Tetrahedron Lett., 2016, 57, 1435.
20. Vidyacharan, S.; Shinde, A. H.; Satpathia, B. and Sharada, D. S. Green Chem., 2014, 16, 1168.
21. Rideout, D. C. and Breslow, R. J. Am. Chem. Soc., 1980, 102, 7816.
22. Clark, J. and Green, H. Chem. 2006, 8, 17.
23. Ajvazi, N. and Stavber, S. Molecules 2016, 21, 1325.
24. Otto, S.; Engberts, J. B. F. N. and Kwak, J. C. T. J. Am. Chem. Soc., 1998, 120, 9517.
25. Matsushita, M.; Kamata, K.; Yamaguchi, K. and Mizuno, N. J. Am. Chem. Soc., 2005, 127, 6632.
26. Li, H. J.; Zhao, J. L.; Chen, Y. J.; Liu, L.; Wang, D. and Li, C. J. Green Chem., 2005, 7, 61.
27. Butler, R. N.; Coyne, A. G. and Moloney, E. M. Tetrahedron Lett., 2007, 48, 3501.
28. Li, C. J. Chem. Rev. 2005, 105, 3095.
29. Hallett, J. P. and Welton, T.; Chem. Rev., 2011, 111, 3508–3576.

30. Pârvulescu, V. I. and Hardacre, C. Chem. Rev., 2007, 107, 2615–2665.
31. Plechkova, N. V. and Seddon, K. R. Chem. Soc. Rev., 2008, 37, 123–150.
32. (a) Buu, O. N. V.; Aupoix, A. V.; Hong, N. D. T.; Vo-Thanh, G. New J. Chem., 2009, 33, 2060–2072 (b) Mi, X.; Luo, S.; Cheng, J. P. J. Org. Chem. 2005, 70, 2338–2341; (c) Guo, H. M.; Cun, L. F.; Gong, L. Z.; Mi, A. Q. and Jiang, Y. Z. Chem. Commun. 2005, 1450–1451; (d) Wang, R.; Twamley, B. and Shreeve, J. M. J. Org. Chem. 2006, 71, 426–429.
33. Willis, M. C.; Taylor, D. and Gillmore, A. T. Org. Lett. 2004, 6, 4755–4757.
34. Gordon, C. M. and Muldoon, M. J.; in Ionic liquids in Synthesis, Wasserscheid, P. and Welton, T. Eds., Wiley-VCH Verlags GmbH & Co., Weinheim, 2008.
35. Seddon, K. R. J. Chem. Technol. Biotechnol., 1997, 68, 351–356.
36. Earle, M. J. and Seddon, K. R. 2000. Ionic liquids. Green solvents for the future. Pure Appl. Chem., 72, 1391–1398.
37. Ranke, J.; Stolte, S.; Stormann, R.; Arning, J. and Jastorff, B. Chem. Rev., 2007, 107, 2183–2206.
38. Liu, Q.; Janssen, M. H. A.; van Rantwijk, F. and Sheldon, R. A. 2005. Room-temperature ionic liquids that dissolve carbohydrates in high concentrations. Green Chem., 7, 39–42.
39. Zhu, S.; Wu, Y.; Chen, Q.; Yu, Z.; Wang, C.; Jin, S.; Dinga, Y. and Wu, G. 2006. Dissolution of cellulose with ionic liquids and its application: a mini-review. Green Chem., 8, 325–327.
40. Phillips, D.; Drummy, L.; Conrady, D.; Fox, D.; Naik, R.; Stone, M.; Trulove, P.; De Long, H. and Mantz, R. 2004. Dissolution and regeneration of Bombyx mori silk fibroin using ionic liquids. JACS, 126, 44, 14350–14351.
41. Xie, H.; Li, S. and Zhang, S. 2005. Ionic liquids as novel solvents for the dissolution and blending of wool keratin fibres. Green Chem., 7, 606–608.
42. Xie, H.; Zhang, S. and Li, S. 2006. Chitin and chitosan dissolved in ionic liquids as reversible sorbents of CO2. Green Chem., 8, 630–633.
43. Jaeger, D. A. and Tucker, C. E. Tetrahedron Lett., 1989, 30, 1785–1788.
44. Badri, M.; Brunet, J. J. and Perron, R. Tetrahedron Lett., 1992, 33, 4435–4438.
45. Kaufmann, D. E.; Nouroozian, M. and Henze, H. Synlett, 1996, 1091.
46. Carlin, R. T. and Wilkes, J. S. J. Mol. Catal., 1990, 63, 125.
47. Claude, S. Lipid/Fett, 1999, 101, 101–104.
48. For recent reviews on this topic, see: (a) Vaidya, P. D. and Rodrigues, A. E. Chem. Eng. Technol., 2009, 32, 1463; (b) Adhikari, S.; Fernando, S. D. and Haryanto, A. Energy Convers. Manage., 2009, 50, 2600.
49. For a recent review on this topic, see: Rahmat, N.; Abdullah, A. Z. and Mohamed, A. R. Renew. Sustain. Energy Rev., 2010, 14, 987.
50. Capello, C.; Fisher, U. and Kunger-buhler, K. Green Chem., 2007, 9, 927–934.
51. Estevez, C. C.; Ferrer, N. B.; Castells, B. J.; and Echeverria, B. B. PCT Int. Appl., 2008, WO 2008080601.
52. (a) Herault, D.; Eggers, A.; Strube, A. and Reinhardt, J. Ger. Offen., 2002, DE 10110855; (b) Notari, M. and Rivetti, F.; PCT Int. Appl., 2004, WO 2004052874.
53. Garcia-Marin, H.; van der Toorn, J. C.; Mayoral, J. A.; Garcia, J. I. and Arends, I. W. C. E. Green Chem., 2009, 11, 1605–1609.
54. (a) Bellina, F.; Bertoli, A.; Melai, B.; Scalesse, S.; Signori, F. and Chiappe, C. Green Chem., 2009, 11, 622–629; (b) Wolfson, A.; Saidkarimov, D.; Dlugy, C. and Tavor, D. Green Chem. Lett. Rev., 2009, 2, 107–110; (c) Aschenbrenner, O.; Supasitmongkol, S.; Taylor, M. and Styring, P. Green Chem., 2009, 11, 1217–1221.

55. Gu, Y.; Barrault, J. and Jeróme, F. ^ Adv. Synth. Catal., 2008, 350, 2007–2012.

56. Perin, G.; Mello, L. G.; Radatz, C. S.; Savegnago, L.; Alves, D.; Jacob, R. G. and Lenardao, E. J. Tetrahedron Lett., 2010, 51, 4354.

57. (a) Hemalatha, K.; Madhumitha, G.; Kajbafvala, A.; Anupama, N.; Sompalle, R. and Roopan, S. M. J. Nanomater, 2013, Article ID 341015, 23 pages, http://dx.doi.org/10.1155/2013/341015. (b) Ansari, M. D.; Sagir, H.; Yadav, V. B.; Yadav, N.; Verma, A. and Siddiqui, I. R. J. Mol. Struct., 2019, 1196, 54–57.

58. Bhaduri, S. and Doble, M. Homogeneous catalysis: Mechanism and industrial Applications, Wiley Interscience, ISBNs: 0-471-37221-8 (Hardsssback); 0-471-22038-8 (Electronic) 2000.

59. van Leeuwen, P. W. N. M. Homogeneous Catalysis: Understanding The Art, Kluwer Academic Publisher, 2004, ISBN 1-4020-2000-7.

60. Zach, M.; Hagglund, C.; Chakarov, D. and Kasemo, B. Curr. Opin. Solid State Mater. Sci., 2006, 10, 132–143.

61. Zahmakiran, M. and Ozkar, S. Nanoscale, 2011, 3, 3462–3481.

62. Gawande, M. B.; Brancoa, P. S. and Varma, R. S. Chem. Soc. Rev., 2013, 42, 3371.

63. Kalidindi, S. B. and Jagirdar, B. R. ChemSusChem, 2012, 5, 65–75.

64. Haruta, M.; Yamada, N.; Kobayashi, T. and Lijima, S. J. Catal., 1989, 115, 301–309.

65. Fukui, T.; Murata, K.; Ohara, S.; Abe, H.; Naito, M. and Nogi, K. J. Power Sources, 2004, 125, 17.

66. (a) Noujima, A.; Mitsudome, T.; Mizugaki, T.; Jitsukawa, K. and Kaneda, K. Molecules, 2011, 16, 8209–8227; (b) Imtiaz, A.; Farrukh, M. A.; K.-ur-rahman, M. and Adnan, R. "Micelle-assisted synthesis of Al_2O_3•CaO nanocatalyst: optical properties and their applications in photodegradation

of 2,4,6-trinitrophenol," Sci. World J., 2013, Volume 2013, http://dx.doi.org/10.1155/2013/641420; (c) Balanta, A.; Godard, C. and Claver, C. Chem. Soc. Rev., 2011, 40, 4973–4985; (d) Kumar, B. V.; Naik, H. S. B.; Girija, D. and Kumar, B. V. J. Chem. Sci., 2011,123, 615–621; (e) Sagir, H.; Rahila, Rai, P.; Singh, P. K. and Siddiqui, I. R. New J.Chem., 2016, 40, 6819.

67. Ciamician, G. Science, 1912, 36, 385–394.

68. Hoffmann, N. Chem. Rev., 2008, 108, 1052–1103.

69. Fagnoni, M.; Dondi, D.; Ravelli, D. and Albini, A. Chem. Rev., 2007, 107, 2725–2756.

70. Albini, A. and Fagnoni, M. Chem Sus Chem, 2008, 1, 63.

71. Kalyanasundaram, K. Coord. Chem. Rev. 1982, 46, 159.

72. Juris, A.; Balzani, V.; Belser, P. and von Zelewsky, A. Helv. Chim. Acta 1981, 64, 2175.

73. Gust, D. and Moore, T. A. Science, 1989, 244, 35–41.

74. Meyer, T. J. Acc. Chem. Res., 1989, 22, 163–170.

75. Gust, D.; Moore, T. A. and Moore, A. L. Acc. Chem. Res., 1993, 26, 198–205.

76. (a) Hari, D. P.; Schroll, P. and Konig, B. J. Am. Chem. Soc. 2012, 134, 2958–2961; (b) Hari, D. P. and Konig, B. Chem. Commun. 2014, 6688–6699.

77. Zaklika, K. A.; Kaskar, Bashir and Schaap, A. P., J. Am. Chem. Soc. 1980, 102, 1, 389–391.

78. Amat, A. M.; Arques, A. and Miranda, M. A. Appl. Catal. B Environ., 1999, 23, 205–214.

79. Gao, X. W.; Meng, Q. Y.; Li, J. X.; Zhong, J. J.; Li, T. L. X.; Tung, C. H. and Wu, L. Z. ACS Catal., 2015, 5, 2391−2396.

80. Narayanam, J. M. R.; Tucker, J. W. and Stephenson, C. R. J. J. Am. Chem. Soc., 2009, 131, 8756.

81. Das, S.; Samanta, S.; Maji, S. K.; Samanta, P. K.; Dutta, A. K.; Srivastava, D. N.; Adhikary, B. and Biswasa, P. Tetrahedron Lett., 2013, 54, 1090.

82. Selvam, K.; et al, Tetrahedron Lett., 2011, 52, 3386.

83. Sagir, H.; Rai, P.; Ibad, A.; Ibad, F. and Siddiqui, I. R. Catal. Commun., 100, 2017, 153-156.

84. Graetzel, M. Acc. Chem. Res. 1981, 14, 376.

85. Takeda, H.; and Ishitani, O. Coord. Chem. Rev. 2010, 254, 346.

86. Ciamician, G. Science, 1912, 36, 385–394.

87. Sala, X.; Romero, I.; Rodríguez, M.; Escriche, L. and Llobet, A.; Angew. Chem. Int. Ed., 2009, 48, 2842.

88. For reviews on multicomponent reactions (MCRs), see: (a) Toure, B. B. and Hall, D. G. Chem. Rev. 2009, 109, 4439. (b) Hudlicky, T. and Reed, J. W. Chem. Soc. Rev. 2009, 38, 3117. (c) Dömling, A.; Wang, W. and Wang, K. Chem. Rev. 2012, 112, 3083.

89. Ugi, I.; Domling, A. and Horl, W. Endeavour, 1994, 18, 115-122.

90. Graaff, C.; Ruijter, E. and Orru, R. V. A. Chem. Soc. Rev., 2012, 41, 3969–4009.

91. Toure, B. B. and Hall, D. G. Chem. Rev., 2009, 109, 4439.

92. Eckert, H. Molecules, 2012, 17, 1074.

93. Santos, W. H. and Filho, L. C. S. Tetrahedron Lett., 2017, 58, 894.

94. Pal, S. Md. Khan, N.; Karamthulla, S.; and Choudhury, L. H. Tetrahedron Lett., 2015, 56, 359.

INDEX

For Product Safety Concerns and Information please contact our EU
representative GPSR@taylorandfrancis.com
Taylor & Francis Verlag GmbH, Kaufingerstraße 24, 80331 München, Germany